Mathematical
Surveys
and
Monographs

Volume 123

Fundamental Algebraic Geometry

Grothendieck's FGA Explained

Barbara Fantechi
Lothar Göttsche
Luc Illusie
Steven L. Kleiman
Nitin Nitsure
Angelo Vistoli

AMS AMERICAN MATHEMATICAL SOCIETY
Providence, Rhode Island

2000 *Mathematics Subject Classification.* Primary 14–01, 14C20, 13D10, 14D15, 14K30, 18F10, 18D30.

For additional information and updates on this book, visit
www.ams.org/bookpages/surv-123

Library of Congress Cataloging-in-Publication Data

Fundamental algebraic geometry : Grothendieck's FGA explained / Barbara Fantechi... [et al.].
p. cm. — (Mathematical surveys and monographs, ISSN 0076-5376 ; v. 123)
Includes bibliographical references and index.
ISBN 0-8218-3541-6 (pbk. : acid-free paper)
ISBN 0-8218-4245-5 (soft cover : acid-free paper)
1. Geometry, Algebraic. 2. Grothendieck groups. 3. Grothendieck categories. I. Fantechi, Barbara, 1966– II. Mathematical surveys and monographs ; no. 123.

QA564.F86 2005
516.3′5—dc22 2005053614

∞ The paper used in this book is acid-free and falls within the guidelines
established to ensure permanence and durability.
Visit the AMS home page at http://www.ams.org/

12 11 10 9 8 7 6 5 4 24 23 22 21

Contents

Preface

Without question, Alexander Grothendieck's work revolutionized Algebraic Geometry. He introduced many concepts — arbitrary schemes, representable functors, relative geometry, and so on — which have turned out to be astoundingly powerful and productive.

Grothendieck sketched his new theories in a series of talks at the Séminaire Bourbaki between 1957 and 1962, and collected his write-ups in a volume entitled "Fondements de la géométrie algébrique," commonly abbreviated FGA. In [**FGA**], he developed the following themes, which have become absolutely central:

- Descent theory,
- Hilbert schemes and Quot schemes,
- The formal existence theorem,
- The Picard scheme.

(FGA also includes a sketch of Grothendieck's extension of Serre duality for coherent sheaves; this theme is already elaborated in a fair number of works, and is not elaborated in the present book.)

Much of FGA is now common knowledge. Some of FGA is less well known, and few geometers are familiar with its full scope. Yet, its theories are fundamental ingredients in most of Algebraic Geometry.

Mudumbai S. Narasimhan conceived the idea of a summer school at the International Centre for Theoretical Physics (ICTP) in Trieste, Italy, to teach these theories. But this school was to be different from most ICTP summer schools. Most focus on current research: important new results are explained, but their proofs are sketched or skipped. This school was to teach the techniques: the proofs too had to be developed in sufficient detail.

Narasimhan's vision was realized July 7–18, 2003, as the "Advanced School in Basic Algebraic Geometry." Its scientific directors were Lother Göttsche of the ICTP, Conjeeveram S. Seshadri of the Chennai Mathematical Institute, India, and Angelo Vistoli of the Università di Bologna, Italy. The school offered the following courses:

(1) Angelo Vistoli: Grothendieck topologies and descent, 10 hours.
(2) Nitin Nitsure: Construction of Hilbert and Quot schemes, 6 hours.
(3) Lother Göttsche: Local properties of Hilbert schemes, and Hilbert schemes of points, 4 hours.
(4) Luc Illusie: Grothendieck's existence theorem in formal geometry, 5 hours.
(5) Steven L. Kleiman: The Picard scheme, 6 hours.

The school addressed advanced graduate students primarily and beginning researchers secondarily; both groups participated enthusiastically. The ICTP's administration was professional. Everyone had a memorable experience.

This book has five parts, which are expanded and corrected versions of notes handed out at the school. The book is not intended to replace [**FGA**]; indeed, nothing can ever replace a master's own words, and reading Grothendieck is always enlightening. Rather, this book fills in Grothendieck's outline. Furthermore, it introduces newer ideas whenever they promote understanding, and it draws connections to subsequent developments. For example, in the book, descent theory is written in the language of Grothendieck topologies, which Grothendieck introduced later. And the finiteness of the Hilbert scheme and of the Picard scheme, which are difficult basic results, are not proved using Chow coordinates, but using Castelnuovo–Mumford regularity, which is now a major tool in Algebraic Geometry and in Commutative Algebra.

This book is not meant to provide a quick and easy introduction. Rather, it contains demanding detailed treatments. Their reward is a far greater understanding of the material. The book's main prerequisite is a thorough acquaintance with basic scheme theory as developed in the textbook [**Har77**].

This book's contents are, in brief, as follows. Lengthier summaries are given in the introductions of the five parts.

Part 1 was written by Vistoli, and gives a fairly complete treatment of descent theory. Part 1 explains both the abstract aspects — fibered categories and stacks — and the most important concrete cases — descent of quasi-coherent sheaves and of schemes. Part 1 comprises Chapters 1–4.

Chapter 1 reviews some basic notions of category theory and of algebraic geometry. Chapter 2 introduces representable functors, Grothendieck topologies, and sheaves; these concepts are well known, and there are already several good treatments available, but the present treatment may be of greater appeal to a beginner, and can also serve as a warm-up to the more advanced theory that follows.

Chapter 3 is devoted to one basic notion, *fibered category*, which Grothendieck introduced in [**SGA1**]. The main example is the category of quasi-coherent sheaves over the category of schemes. Fibered categories provide the right abstract set-up for a discussion of descent theory. Although the general theory may be unnecessary for elementary applications, it is necessary for deeper comprehension and advanced applications.

Chapter 4 discusses *stacks*, fibered categories in which descent theory works. Chapter 4 treats, in full, the various ways of defining descent data, and it proves the main result of Part 1, which asserts that quasi-coherent sheaves form a stack.

Part 2 was written by Nitsure, and covers Grothendieck's construction of Hilbert schemes and Quot schemes, following his Bourbaki talk [**FGA**, 221], together with further developments by David Mumford and by Allen Altman and Kleiman. Part 2 comprises Chapter 5.

Specifically, given a scheme X, Grothendieck solved the basic problem of constructing another scheme Hilb_X, called the *Hilbert scheme of* X, which parameterizes, in a suitable universal manner, all possible closed subschemes of X. More generally, given a coherent sheaf E on X, he constructed a scheme $\mathrm{Quot}_{E/X}$, called the *Quot scheme of* E, which parameterizes, again in a suitable universal manner, all possible coherent quotients of E. These constructions are possible, in a relative set-up, where X is projective over a suitable base. The constructions make crucial use of several basic tools, including faithfully flat descent, flattening stratification, the semi-continuity complex, and Castelnuovo–Mumford regularity.

Part 3 was written jointly by Barbara Fantechi and Göttsche. It comprises Chapters 6 and 7.

Chapter 6 introduces the notion of an (infinitesimal) deformation functor, and gives several examples. Chapter 6 also defines a tangent-obstruction theory for such a functor, and explains how the theory yields an estimate on the dimension of the moduli space. The theory is worked out in some cases, and sketched in a few more, which are not needed in Chapter 7.

Chapter 7 studies the Hilbert scheme of points on a smooth quasi-projective variety, which parameterizes the finite subschemes of fixed length. The chapter constructs the Hilbert–Chow morphism, which maps this Hilbert scheme to the symmetric power by sending a subscheme to its support with multiplicities. For a surface, this morphism is a resolution of singularities. Finally, the chapter computes the Betti numbers of the Hilbert scheme, and sketches the action of the Heisenberg algebra on the cohomology.

Part 4 was written by Illusie, and revisits Grothendieck's Bourbaki talk [**FGA**, 182], where he presented a fundamental comparison theorem of "GAGA" type between algebraic geometry and formal geometry, and outlined some applications to the theory of the fundamental group and to that of infinitesimal deformations. A detailed account appeared shortly afterward in [**EGAIII1**], [**EGAIII2**] and [**SGA1**]. Part 4 comprises Chapter 8.

After recalling basic facts on locally Noetherian formal schemes, Chapter 8 explains the key points in the proof of the main comparison theorem, and sketches some corollaries, including Zariski's connectedness theorem and main theorem, and Grothendieck's criterion for algebraization of a formal scheme. Then Chapter 8 gives Grothendieck's applications to the fundamental group and to lifting vector bundles and smooth schemes, notably, curves and Abelian varieties. Chapter 8 ends with a discussion of Serre's celebrated examples of varieties in positive characteristic that do not lift to characteristic zero.

Part 5 was written by Kleiman, and develops in detail most of the theory of the Picard scheme that Grothendieck sketched in the two Bourbaki talks [**FGA**, 232, 236] and in his commentaries on them [**FGA**, pp. C-07–C-011]. In addition, Part 5 reviews in brief, in a series of scattered remarks, much of the rest of the theory developed by Grothendieck and by others. Part 5 comprises Chapter 9.

Chapter 9 begins with an extensive historical introduction, which serves to motivate Grothendieck's work on the Picard scheme by tracing the development of the ideas that led to it. The story is fascinating, and may be of independent interest.

Chapter 9 then discusses the four common relative Picard functors, which are successively more likely to be the functor of points of the Picard scheme. Next, Chapter 9 treats relative effective (Cartier) divisors and linear equivalence, two important preliminary notions. Then, Chapter 9 proves Grothendieck's main theorem about the Picard scheme: it exists for any projective and flat scheme whose geometric fibers are integral.

Chapter 9 next studies the union of the connected components of the identity elements of the fibers of the Picard scheme. Then, Chapter 9 proves two deeper finiteness theorems: the first concerns the set of points with a multiple in this union; the second concerns the set of points representing invertible sheaves with a given Hilbert polynomial. Chapter 9 closes with two appendices: one contains detailed answers to all the exercises; the other contains an elementary treatment of basic divisorial intersection theory — this theory is used freely in the proofs of the two finiteness theorems.

Part 1

Grothendieck topologies, fibered categories and descent theory

Angelo Vistoli

Introduction

Descent theory has a somewhat formidable reputation among algebraic geometers. In fact, it simply says that under certain conditions homomorphisms between quasi-coherent sheaves can be constructed locally and then glued together if they satisfy a compatibility condition, while quasi-coherent sheaves themselves can be constructed locally and then glued together via isomorphisms that satisfy a cocycle condition.

Of course, if "locally" were to mean "locally in the Zariski topology" this would be a formal statement, certainly useful, but hardly deserving the name of a theory. The point is that "locally" here means locally in the flat topology; and the flat topology is something that is not a topology, but what is called a *Grothendieck topology.* Here the coverings are, essentially, flat surjective maps satisfying a finiteness condition. So there are many more coverings in this topology than in the Zariski topology, and the proof becomes highly nontrivial.

Still, the statement is very simple and natural, provided that one resorts to the usual abuse of identifying the pullback $(gf)^*F$ of a sheaf F along the composite of two maps f and g with f^*g^*F. If one wants to be fully rigorous, then one has to take into account the fact that $(gf)^*F$ and f^*g^*F are not identical, but there is a canonical isomorphism between them, satisfying some compatibility conditions, and has to develop a theory of such compatibilities. The resulting complications are, in my opinion, the origin of the distaste with which many algebraic geometers look at descent theory (when they look at all).

There is also an abstract notion of "category in which descent theory works"; the category of pairs consisting of a scheme and a quasi-coherent sheaf on it is an example. These categories are known as *stacks.* The general formalism is quite useful, even outside of moduli theory, where the theory of algebraic stacks has become absolutely central (see for example [**DM69**], [**Art74b**] and [**LMB00**]).

These notes were born to accompany my ten lectures on *Grothendieck topologies and descent theory* in the *Advanced School in Basic Algebraic Geometry* that took place at I.C.T.P., 7–18 July 2003. Their purpose is to provide an exposition of descent theory, more complete than the original (still very readable, and highly recommended) article of Grothendieck ([**Gro95b**]), or than [**SGA1**]. I also use the language of Grothendieck topologies, which is the natural one in this context, but had not been introduced at the time when the two standard sources were written.

The treatment here is slanted toward the general theory of fibered categories and stacks: so the algebraic geometer searching for immediate gratification will probably be frustrated. On the other hand, I find the general theory both interesting and applicable, and hope that at at least some of my readers will agree.

Also, in the discussion of descent theory for quasi-coherent sheaves and for schemes, which forms the real reason of being of these notes, I never use the convention of identifying objects when there is a canonical isomorphism between them, but I always specify the isomorphism, and write down explicitly the necessary compatibility conditions. This makes the treatment rigorous, but also rather heavy (for a particularly unpleasant example, see §4.3.3). One may question the wisdom of this choice; but I wanted to convince myself that a fully rigorous treatment was indeed possible. And the unhappy reader may be assured that this has cost more suffering to me than to her.

All of the ideas and the results contained in these notes are due to Grothendieck. There is nothing in here that is not, in some form, either in [**SGA1**] or in [**SGA4**], so I do not claim any originality at all.

There are modern developments of descent theory, particularly in category theory (see for example [**JT84**]) and in non-commutative algebra and non-commutative geometry ([**KR04a**] and [**KR04b**]). One of the most exciting ones, for topologists as well as algebraic geometers, is the idea of "higher descent", strictly linked with the important topic of higher category theory (see for example [**HS**] and [**Str03**]). We will not discuss any of these very interesting subjects.

Contents. In Chapter 1 I recall some basic notions in algebraic geometry and category theory.

The real action starts in Chapter 2. Here first I discuss Grothendieck's philosophy of representable functors, and give one of the main illustrative examples, by showing how this makes the notion of group scheme, and action of a group scheme on a scheme, very natural and easy. All of algebraic geometry can be systematically developed from this point of view, making it very clean and beautiful, and incomprehensible for the beginner (see [**DG70**]).

In Section 2.3 I define and discuss Grothendieck topologies and sheaves on them. I use the naive point of view of pretopologies, which I find much more intuitive. However, the more sophisticated point of view using sieves has advantages, so I try to have my cake and eat it too (the Italian expression, more vivid, is "have my barrel full and my wife drunk") by defining sieves and characterizing sheaves in terms of them, thus showing, implicitly, that the sheaves only depend on the topology and not on the pretopology. In this section I also introduce the four main topologies on the category of schemes, Zariski, étale, fppf and fpqc, and prove Grothendieck's theorem that a representable functor is a sheaf in all of them.

There are two possible formal setups for descent theory, fibered categories and pseudo-functors. The first one seems less cumbersome, so Chapter 3 is dedicated to the theory of fibered categories. However, here I also define pseudo-functors, and relate the two points of view, because several examples, for example quasi-coherent sheaves, are more naturally expressed in this language. I prove some important results (foremost is Yoneda's lemma for fibered categories), and conclude with a discussion of equivariant objects in a fibered category (I hope that some of the readers will find that this throws light on the rather complicated notion of equivariant sheaf).

The heart of these notes is Chapter 4. After a thorough discussion of descent data (I give several definitions of them, and prove their equivalence) I define the central concept, that of *stack*: a stack is a fibered category over a category with a Grothendieck topology, in which descent theory works (thus we see all the three notions appearing in the title in action). Then I proceed to proving the main theorem, stating that the fibered category of quasi-coherent sheaves is a stack in the fpqc topology. This is then applied to two of the main examples where descent theory for schemes works, that of affine morphisms, and morphisms endowed with a canonical ample line bundle. I also discuss a particularly interesting example, that of descent along principal bundles (torsors, in Grothendieck's terminology).

In the last section I give an example to show that étale descent does not always work for schemes, and end by mentioning that there is an extension of the concept of scheme, that of *algebraic space*, due to Michael Artin. Its usefulness is that on one hand algebraic spaces are, in a sense, very close to schemes, and one can define for them most of the concepts of scheme theory, and on the other hand fppf descent always works for them. It would have been a natural topic to include in the notes, but this would have further delayed their completion.

Prerequisites. I assume that the reader is acquainted with the language of schemes, at least at the level of Hartshorne's book ([**Har77**]). I use some concepts that are not contained in [**Har77**], such as that of a morphism locally of finite presentation; but I recall their main properties, with references to the appropriate parts of *Éléments de géométrie algébrique*, in Chapter 1.

I make heavy use of the categorical language: I assume that the reader is acquainted with the notions of category, functor and natural transformation, equivalence of categories. On the other hand, I do not use any advanced concepts, nor do I use any real results in category theory, with one exception: the reader should know that a fully faithful essentially surjective functor is an equivalence.

Acknowledgments. Teaching my course at the *Advanced School in Basic Algebraic Geometry* has been a very pleasant experience, thanks to the camaraderie of my fellow lecturers (Lothar Göttsche, Luc Illusie, Steve Kleiman and Nitin Nitsure) and the positive and enthusiastic attitude of the participants. I am also in debt with Lothar, Luc, Steve and Nitin because they never once complained about the delay with which these notes were being produced.

I am grateful to Steve Kleiman for useful discussions and suggestions, particularly involving the fpqc topology, and to Pino Rosolini, who, during several hikes on the Alps, tried to enlighten me on some categorical constructions.

I have had some interesting conversations with Behrang Noohi concerning the definition of a stack: I thank him warmly.

I learned about the counterexample in [**Ray70**, XII 3.2] from Andrew Kresch.

I also thank the many participants to the school who showed interest in my lecture series, and particularly those who pointed out mistakes in the first version of the notes. I am especially in debt with Zoran Skoda, who sent me several helpful comments, and also for his help with the bibliography.

Joachim Kock read carefully most of this, and send me a long list of comments and corrections, which were very useful. More corrections where provided by the referees. I am grateful to them.

Finally, I would like to dedicate these notes to the memory of my father-in-law, Amleto Rosolini, who passed away at the age of 86 as they were being completed. He would not have been interested in learning descent theory, but he was a kind and remarkable man, and his enthusiasm about mathematics, which lasted until his very last day, will always be an inspiration to me.

CHAPTER 1

Preliminary notions

1.1. Algebraic geometry

In this chapter we recall, without proof, some basic notions of scheme theory that are used in the notes. All rings and algebras will be commutative.

We will follow the terminology of *Éléments de géométrie algébrique*, with the customary exception of calling a "scheme" what is called there a "prescheme" (in *Éléments de géométrie algébrique*, a scheme is assumed to be separated).

We start with some finiteness conditions. Recall if B is an algebra over the ring A, we say that B is *finitely presented* if it is the quotient of a polynomial ring $A[x_1, \ldots, x_n]$ over A by a finitely generated ideal. If A is noetherian, every finitely generated algebra is finitely presented.

If B is finitely presented over A, whenever we write $B = A[x_1, \ldots, x_n]/I$, I is always finitely generated in $A[x_1, \ldots, x_n]$ ([**EGAIV1**, Proposition 1.4.4]).

DEFINITION 1.1 (See [**EGAIV1**, 1.4.2]). A morphism of schemes $f: X \to Y$ is *locally of finite presentation* if for any $x \in X$ there are affine neighborhoods U of x in X and V of $f(x)$ in Y such that $f(U) \subseteq V$ and $\mathcal{O}(U)$ is finitely presented over $\mathcal{O}(V)$.

Clearly, if Y is locally noetherian, then f is locally of finite presentation if and only if it is locally of finite type.

PROPOSITION 1.2 ([**EGAIV1**, 1.4]).

(i) If $f: X \to Y$ is locally of finite presentation, U and V are open affine subsets of X and Y respectively, and $f(U) \subseteq V$, then $\mathcal{O}(U)$ is finitely presented over $\mathcal{O}(V)$.

(ii) The composite of morphisms locally of finite presentation is locally of finite presentation.

(iii) Given a cartesian diagram

$$\begin{array}{ccc} X' & \longrightarrow & X \\ \downarrow & & \downarrow \\ Y' & \longrightarrow & Y \end{array}$$

if $X \to Y$ is locally of finite presentation, so is $X' \to Y'$.

DEFINITION 1.3 (See [**EGAI**, 6.6.1]). A morphism of schemes $X \to Y$ is *quasi-compact* if the inverse image in X of a quasi-compact open subset of Y is quasi-compact.

An affine scheme is quasi-compact, hence a scheme is quasi-compact if and only if it is the finite union of open affine subschemes; using this, it is easy to prove the following.

PROPOSITION 1.4 ([**EGAI**, Proposition 6.6.4]).
Let $f: X \to Y$ be a morphism of schemes. The following are equivalent.

(i) f is quasi-compact.
(ii) The inverse image of an open affine subscheme of Y is quasi-compact.
(iii) There exists a covering $Y = \cup_i V_i$ by open affine subschemes, such that the inverse image in X of each V_i is quasi-compact.

In particular, a morphism from a quasi-compact scheme to an affine scheme is quasi-compact.

REMARK 1.5. It is not enough to suppose that there is a covering of Y by open quasi-compact subschemes V_i, such that the inverse image of each V_i is quasi-compact in X, without additional hypotheses. For example, consider a ring A that does not satisfy the ascending chain condition on radical ideals (for example, a polynomial ring in infinitely many variables), and set $X = \operatorname{Spec} A$. In X there will be an open subset U that is not quasi-compact; denote by Y the scheme obtained by gluing two copies of X together along U, and by $f: X \to Y$ the inclusion of one of the copies. Then Y and X are both quasi-compact; on the other hand there is an affine open subset of Y (the other copy of X) whose inverse image in X is U, so f is not quasi-compact.

PROPOSITION 1.6 ([**EGAI**, 6.6]).

(i) The composite of quasi-compact morphisms is quasi-compact.
(ii) Given a cartesian diagram

$$\begin{array}{ccc} X' & \longrightarrow & X \\ \downarrow & & \downarrow \\ Y' & \longrightarrow & Y \end{array}$$

if $X \to Y$ is quasi-compact, so is $X' \to Y'$.

Let us turn to flat morphisms.

DEFINITION 1.7. A morphism of schemes $f: X \to Y$ is *flat* if for any $x \in X$, the local ring $\mathcal{O}_{X,x}$ is flat as a module over $\mathcal{O}_{Y,f(x)}$.

PROPOSITION 1.8 ([**EGAIV2**, Proposition 2.1.2]). Let $f: X \to Y$ be a morphism of schemes. Then the following are equivalent.

(i) f is flat.
(ii) For any $x \in X$, there are affine neighborhoods U of x in X and V of $f(x)$ in Y such that $f(U) \subseteq V$, and $\mathcal{O}(U)$ is flat over $\mathcal{O}(V)$.
(iii) For any open affine subsets U in X and V in Y such that $f(U) \subseteq V$, $\mathcal{O}(U)$ is flat over $\mathcal{O}(V)$.

PROPOSITION 1.9 ([**EGAIV2**, 2.1]).

(i) The composite of flat morphisms is flat.
(ii) Given a cartesian diagram

$$\begin{array}{ccc} X' & \longrightarrow & X \\ \downarrow & & \downarrow \\ Y' & \longrightarrow & Y \end{array}$$

if $X \to Y$ is flat, so is $X' \to Y'$.

DEFINITION 1.10. A morphism of schemes $f\colon X \to Y$ is *faithfully flat* if it is flat and surjective.

Let B be an algebra over A. We say that B is faithfully flat if the associated morphism of schemes $\operatorname{Spec} B \to \operatorname{Spec} A$ is faithfully flat.

PROPOSITION 1.11 ([**Mat89**, Theorems 7.2 and 7.3]). Let B be an algebra over A. The following are equivalent.

(i) B is faithfully flat over A.
(ii) A sequence of A-modules $M' \to M \to M''$ is exact if and only if the induced sequence of B-modules $M' \otimes_A B \to M \otimes_A B \to M'' \otimes_A B$ is exact.
(iii) A homomorphism of A-modules $M' \to M$ is injective if and only if the associated homomorphism of B-modules $M' \otimes_A B \to M \otimes_A B$ is injective.
(iv) B is flat over A, and if M is a module over A with $M \otimes_A B = 0$, we have $M = 0$.
(v) B is flat over A, and $\mathfrak{m}B \neq B$ for all maximal ideals $\mathfrak{m}$ of A.

The following fact is very important.

PROPOSITION 1.12 ([**EGAIV2**, Proposition 2.4.6]). A flat morphism that is locally of finite presentation is open.

This is not true in general for flat morphisms that are not locally of finite presentation; however, a weaker version of this fact holds.

PROPOSITION 1.13 ([**EGAIV2**, Corollaire 2.3.12]). If $f\colon X \to Y$ is a faithfully flat surjective quasi-compact morphism, a subset of Y is open if and only if its inverse image in X is open in X.

In other words, Y has the topology induced by that of X.

REMARK 1.14. For this we need to assume that f is quasi-compact, it is not enough to assume that it is faithfully flat. For example, let Y be an integral smooth curve over an algebraically closed field, X the disjoint union of the $\operatorname{Spec} \mathcal{O}_{Y,y}$ over all closed points $y \in Y$. The natural projection $f\colon X \to Y$ is clearly flat. However, if S is a subset of Y containing the generic point, then $f^{-1}S$ is always open in X, while S is open in Y if and only if its complement is finite.

PROPOSITION 1.15 ([**EGAIV2**, Proposition 2.7.1]). Let

$$\begin{array}{ccc} X' & \longrightarrow & X \\ \downarrow & & \downarrow \\ Y' & \longrightarrow & Y \end{array}$$

be a cartesian diagram of schemes in which $Y' \to Y$ is faithfully flat and either quasi-compact or locally of finite presentation. Suppose that $X' \to Y'$ has one of the following properties:

(i) is separated,
(ii) is quasi-compact,
(iii) is locally of finite presentation,
(iv) is proper,
(v) is affine,
(vi) is finite,

(vii) is flat,
(viii) is smooth,
(ix) is unramified,
(x) is étale,
(xi) is an embedding,
(xii) is a closed embedding.

Then $X \to Y$ has the same property.

In [**EGAIV2**] all these statements are proved when $Y' \to Y$ is quasi-compact. Using Proposition 1.12, and the fact that all those properties are local in the Zariski topology of Y, it is not hard to prove the statement also when $Y' \to Y$ is locally of finite presentation.

1.2. Category theory

We will assume that the reader is familiar with the concepts of category, functor and natural transformation. The standard reference in category theory is [**ML98**]; also very useful are [**Bor94a**], [**Bor94b**] and [**Bor94c**].

We will not distinguish between small and large categories. More generally, we will ignore any set-theoretic difficulties. These can be overcome with standard arguments using universes.

If $F\colon \mathcal{A} \to \mathcal{B}$ is a functor, recall that F is called *fully faithful* when for any two objects A and A' of $\mathcal{A}$, the function

$$\mathrm{Hom}_{\mathcal{A}}(A, A') \longrightarrow \mathrm{Hom}_{\mathcal{B}}(FA, FA')$$

defined by F is a bijection. F is called *essentially surjective* if every object of $\mathcal{B}$ is isomorphic to the image of an object of $\mathcal{A}$.

Recall also that F is called an *equivalence* when there exists a functor $G\colon \mathcal{B} \to \mathcal{A}$, such that the composite $GF\colon \mathcal{A} \to \mathcal{A}$ is isomorphic to $\mathrm{id}_{\mathcal{A}}$, and $FG\colon \mathcal{B} \to \mathcal{B}$ is isomorphic to $\mathrm{id}_{\mathcal{B}}$.

The composite of two equivalences is again an equivalence. In particular, "being equivalent" is an equivalence relation among categories.

The following well-known fact will be used very frequently.

PROPOSITION 1.16. A functor is an equivalence if and only if it is both fully faithful and essentially surjective.

If $\mathcal{A}$ and $\mathcal{B}$ are categories, there is a category $\mathrm{Hom}(\mathcal{A}, \mathcal{B})$, whose objects are functors $\Phi\colon \mathcal{A} \to \mathcal{B}$, and whose arrows $\alpha\colon \Phi \to \Psi$ are natural transformations. If $F\colon \mathcal{A}' \to \mathcal{A}$ is a functor, there is an induced functor

$$F^*\colon \mathrm{Hom}(\mathcal{A}, \mathcal{B}) \longrightarrow \mathrm{Hom}(\mathcal{A}', \mathcal{B})$$

defined at the level of objects by the obvious rule

$$F^*\Phi \stackrel{\text{def}}{=} \Phi \circ F\colon \mathcal{A}' \longrightarrow \mathcal{B}$$

for any functor $\Phi\colon \mathcal{A} \to \mathcal{B}$. At the level of arrow F^* is defined by the formula

$$(F^*\alpha)_{A'} \stackrel{\text{def}}{=} \alpha_{FA'}\colon \Phi FA' \longrightarrow \Psi FA'$$

for any natural transformation $\alpha\colon \Phi \to \Psi$.

Also for any functor $F\colon \mathcal{B} \to \mathcal{B}'$ we get an induced functor

$$_*\colon \mathrm{Hom}(\mathcal{A}, \mathcal{B}) \longrightarrow \mathrm{Hom}(\mathcal{A}, \mathcal{B}')$$

obtained by the obvious variants of the definitions above.

The reader should also recall that a *groupoid* is a category in which every arrow is invertible.

We will also make considerable use of the notions of fibered product and cartesian diagram in an arbitrary category. We will manipulate some cartesian diagrams. In particular the reader will encounter diagrams of the type

$$\begin{array}{ccccc} A' & \longrightarrow & B' & \longrightarrow & C' \\ \downarrow & & \downarrow & & \downarrow \\ A & \longrightarrow & B & \longrightarrow & C; \end{array}$$

we will say that this is cartesian when both squares are cartesian. This is equivalent to saying that the right hand square and the square

$$\begin{array}{ccc} A' & \longrightarrow & C' \\ \downarrow & & \downarrow \\ A & \longrightarrow & C, \end{array}$$

obtained by composing the rows, are cartesian. There will be other statements of the type "there is a cartesian diagram ...". These should all be straightforward to check.

For any category $\mathcal{C}$ and any object X of $\mathcal{C}$ we denote by $(\mathcal{C}/X)$ the *comma category*, whose objects are arrows $U \to X$ in $\mathcal{C}$, and whose arrows are commutative diagrams

$$\begin{array}{ccc} U & \longrightarrow & V \\ & \searrow \quad \swarrow & \\ & X & \end{array}$$

We also denote by $\mathcal{C}^{\mathrm{op}}$ the *opposite category* of $\mathcal{C}$, in which the objects are the same, and the arrows are also the same, but sources and targets are switched. A *contravariant functor* from $\mathcal{C}$ to another category $\mathcal{D}$ is a functor $\mathcal{C}^{\mathrm{op}} \to \mathcal{D}$.

Whenever we have a fibered product $X_1 \times_Y X_2$ in a category, we denote by $\mathrm{pr}_1 \colon X_1 \times_Y X_2 \to X_1$ and $\mathrm{pr}_2 \colon X_1 \times_Y X_2 \to X_2$ the two projections. We will also use a similar notation for the product of three or more objects: for example, we will denote by

$$\mathrm{pr}_i \colon X_1 \times_Y X_2 \times_Y X_3 \longrightarrow X_i$$

the ith projection, and by

$$\mathrm{pr}_{ij} \colon X_1 \times_Y X_2 \times_Y X_3 \longrightarrow X_i \times_Y X_j$$

the projection into the product of the ith and j^{th} factor.

Recall that a category has finite products if and only if it has a terminal object (the product of 0 objects) and products of two objects.

Suppose that $\mathcal{C}$ and $\mathcal{D}$ are categories with finite products; denote their terminal objects by $\mathrm{pt}_{\mathcal{C}}$ and $\mathrm{pt}_{\mathcal{D}}$. A functor $F \colon \mathcal{C} \to \mathcal{D}$ is said to *preserve finite products* if the following holds. Suppose that we have objects $U_1, \dots, U_r$ of $\mathcal{C}$: the projections $U_1 \times \cdots \times U_r \to U_i$ induce arrows $F(U_1 \times \cdots \times U_r) \to FU_i$. Then the corresponding arrow

$$F(U_1 \times \cdots \times U_r) \longrightarrow FU_1 \times \cdots \times FU_r$$

is an isomorphism in $\mathcal{C}$.

If $f_1 \colon U_1 \to V_1, \ldots, f_r \colon U_r \to V_r$ are arrows in $\mathcal{C}$, the diagram

$$\begin{array}{ccc} F(U_1 \times \cdots \times U_r) & \longrightarrow & FU_1 \times \cdots \times FU_r \\ \Big\downarrow{\scriptstyle F(f_1 \times \cdots \times f_r)} & & \Big\downarrow{\scriptstyle Ff_1 \times \cdots \times Ff_r} \\ F(V_1 \times \cdots \times V_r) & \longrightarrow & FV_1 \times \cdots \times FV_r \end{array}$$

in which the arrows are the isomorphism defined above, obviously commutes.

By a simple induction argument, F preserves finite products if and only if $F\mathrm{pt}_{\mathcal{C}}$ is a terminal object of $\mathcal{D}$, and for any two objects U and V of $\mathcal{C}$ the arrow $F(U \times V) \to FU \times FV$ is an isomorphism (in other words, for F to preserve finite products it is enough that it preserves products of 0 and 2 objects).

Finally, we denote by (Set) the category of sets, by (Top) the category of topological spaces, (Grp) the category of groups, and by (Sch/S) the category of schemes over a fixed base scheme S.

CHAPTER 2

Contravariant functors

2.1. Representable functors and the Yoneda Lemma

2.1.1. Representable functors. Let us start by recalling a few basic notions of category theory.

Let $\mathcal{C}$ be a category. Consider functors from $\mathcal{C}^{\mathrm{op}}$ to (Set). These are the objects of a category, denoted by

$$\operatorname{Hom}(\mathcal{C}^{\mathrm{op}}, (\mathrm{Set})),$$

in which the arrows are the natural transformations. From now on we will refer to natural transformations of contravariant functors on $\mathcal{C}$ as *morphisms*.

Let X be an object of $\mathcal{C}$. There is a functor

$$\mathrm{h}_X : \mathcal{C}^{\mathrm{op}} \longrightarrow (\mathrm{Set})$$

to the category of sets, which sends an object U of $\mathcal{C}$ to the set

$$\mathrm{h}_X U = \operatorname{Hom}_{\mathcal{C}}(U, X).$$

If $\alpha \colon U' \to U$ is an arrow in $\mathcal{C}$, then $\mathrm{h}_X\alpha \colon \mathrm{h}_X U \to \mathrm{h}_X U'$ is defined to be composition with α. (When $\mathcal{C}$ is the category of schemes over a fixed base scheme, h_X is often called the *functor of points of* X)

Now, an arrow $f \colon X \to Y$ yields a function $\mathrm{h}_f U \colon \mathrm{h}_X U \to \mathrm{h}_Y U$ for each object U of $\mathcal{C}$, obtained by composition with f. This defines a morphism $\mathrm{h}_X \to \mathrm{h}_Y$, that is, for all arrows $\alpha \colon U' \to U$ the diagram

$$\begin{array}{ccc} \mathrm{h}_X U & \xrightarrow{\mathrm{h}_f U} & \mathrm{h}_Y U \\ \Big\downarrow{\scriptstyle \mathrm{h}_X \alpha} & & \Big\downarrow{\scriptstyle \mathrm{h}_Y \alpha} \\ \mathrm{h}_X U' & \xrightarrow{\mathrm{h}_f U'} & \mathrm{h}_Y U' \end{array}$$

commutes.

Sending each object X of $\mathcal{C}$ to h_X, and each arrow $f \colon X \to Y$ of $\mathcal{C}$ to $\mathrm{h}_f \colon \mathrm{h}_X \to \mathrm{h}_Y$ defines a functor $\mathcal{C} \to \operatorname{Hom}(\mathcal{C}^{\mathrm{op}}, (\mathrm{Set}))$.

YONEDA LEMMA (WEAK VERSION). *Let X and Y be objects of $\mathcal{C}$. The function*

$$\operatorname{Hom}_{\mathcal{C}}(X, Y) \longrightarrow \operatorname{Hom}(\mathrm{h}_X, \mathrm{h}_Y)$$

that sends $f \colon X \to Y$ to $\mathrm{h}_f \colon \mathrm{h}_X \to \mathrm{h}_Y$ is bijective.

In other words, the functor $\mathcal{C} \to \operatorname{Hom}(\mathcal{C}^{\mathrm{op}}, (\mathrm{Set}))$ is fully faithful. It fails to be an equivalence of categories, because in general it will not be essentially surjective. This means that not every functor $\mathcal{C}^{\mathrm{op}} \to (\mathrm{Set})$ is isomorphic to a functor of the form h_X. However, if we restrict to the full subcategory of $\operatorname{Hom}(\mathcal{C}^{\mathrm{op}}, (\mathrm{Set}))$ consisting of functors $\mathcal{C}^{\mathrm{op}} \to (\mathrm{Set})$ which are isomorphic to a functor of the form h_X, we do get a category which is equivalent to $\mathcal{C}$.

DEFINITION 2.1. A *representable functor* on the category $\mathcal{C}$ is a functor

$$F\colon \mathcal{C}^{\mathrm{op}} \longrightarrow (\mathrm{Set})$$

which is isomorphic to a functor of the form h_X for some object X of $\mathcal{C}$.

If this happens, we say that *F is represented by X.*

Given two isomorphisms $F \simeq \mathrm{h}_X$ and $F \simeq \mathrm{h}_Y$, we have that the resulting isomorphism $\mathrm{h}_X \simeq \mathrm{h}_Y$ comes from a unique isomorphism $X \simeq Y$ in $\mathcal{C}$, because of the weak form of Yoneda's lemma. Hence *two objects representing the same functor are canonically isomorphic.*

2.1.2. Yoneda's lemma. The condition that a functor be representable can be given a new expression with the more general version of Yoneda's lemma. Let X be an object of $\mathcal{C}$ and $F\colon \mathcal{C}^{\mathrm{op}} \to (\mathrm{Set})$ a functor. Given a natural transformation $\tau\colon \mathrm{h}_X \to F$, one gets an element $\xi \in FX$, defined as the image of the identity map $\mathrm{id}_X \in \mathrm{h}_X X$ via the function $\tau_X : \mathrm{h}_X X \to FX$. This construction defines a function $\mathrm{Hom}(\mathrm{h}_X, F) \to FX$.

Conversely, given an element $\xi \in FX$, one can define a morphism $\tau\colon \mathrm{h}_X \to F$ as follows. Given an object U of $\mathcal{C}$, an element of $\mathrm{h}_X U$ is an arrow $f : U \to X$; this arrow induces a function $Ff\colon FX \to FU$. We define a function $\tau_U : \mathrm{h}_X U \to FU$ by sending $f \in \mathrm{h}_X U$ to $Ff(\xi) \in FU$. It is straightforward to check that the τ that we have defined is in fact a morphism. In this way we have defined functions

$$\mathrm{Hom}(\mathrm{h}_X, F) \longrightarrow F(X)$$

and

$$F(X) \longrightarrow \mathrm{Hom}(\mathrm{h}_X, F).$$

YONEDA LEMMA. *These two functions are inverse to each other, and therefore establish a bijective correspondence*

$$\mathrm{Hom}(\mathrm{h}_X, F) \simeq FX.$$

The proof is easy and left to the reader. Yoneda's lemma is not a deep fact, but its importance cannot be overestimated.

Let us see how this form of Yoneda's lemma implies the weak form above. Suppose that $F = \mathrm{h}_Y$: the function $\mathrm{Hom}(X,Y) = \mathrm{h}_Y X \to \mathrm{Hom}(\mathrm{h}_X, \mathrm{h}_Y)$ constructed here sends each arrow $f\colon X \to Y$ to

$$\mathrm{h}_Y f(\mathrm{id}_Y) = \mathrm{id}_Y \circ f\colon X \longrightarrow Y,$$

so it is exactly the function $\mathrm{Hom}(X,Y) \to \mathrm{Hom}(\mathrm{h}_X, \mathrm{h}_Y)$ appearing in the weak form of the result.

One way to think about Yoneda's lemma is as follows. The weak form says that the category $\mathcal{C}$ is embedded in the category $\mathrm{Hom}\bigl(\mathcal{C}^{\mathrm{op}}, (\mathrm{Set})\bigr)$. The strong version says that, given a functor $F\colon \mathcal{C}^{\mathrm{op}} \to (\mathrm{Set})$, this can be extended to the representable functor $\mathrm{h}_F\colon \mathrm{Hom}\bigl(\mathcal{C}^{\mathrm{op}}, (\mathrm{Set})\bigr)^{\mathrm{op}} \to (\mathrm{Set})$: thus, every functor become representable, when extended appropriately. (In practice, the category $\mathrm{Hom}\bigl(\mathcal{C}^{\mathrm{op}}, (\mathrm{Set})\bigr)$ is usually much too big, and one has to restrict it appropriately.)

We can use Yoneda's lemma to give a very important characterization of representable functors.

DEFINITION 2.2. Let $F\colon \mathcal{C}^{\mathrm{op}} \to (\mathrm{Set})$ be a functor. A *universal object* for F is a pair (X, ξ) consisting of an object X of $\mathcal{C}$, and an element $\xi \in FX$, with the property that for each object U of $\mathcal{C}$ and each $\sigma \in FU$, there is a unique arrow $f\colon U \to X$ such that $Ff(\xi) = \sigma \in FU$.

In other words: the pair (X, ξ) is a universal object if the morphism $\mathrm{h}_X \to F$ defined by ξ is an isomorphism. Since every natural transformation $\mathrm{h}_X \to F$ is defined by some object $\xi \in FX$, we get the following.

PROPOSITION 2.3. A functor $F\colon \mathcal{C}^{\mathrm{op}} \to (\mathrm{Set})$ is representable if and only if it has a universal object.

Also, if F has a universal object (X, ξ), then F is represented by X.

Yoneda's lemma ensures that the natural functor $\mathcal{C} \to \mathrm{Hom}\big(\mathcal{C}^{\mathrm{op}}, (\mathrm{Set})\big)$ which sends an object X to the functor h_X is an equivalence of $\mathcal{C}$ with the category of representable functors. From now on we will not distinguish between an object X and the functor h_X it represents. So, if X and U are objects of $\mathcal{C}$, we will write $X(U)$ for the set $\mathrm{h}_X U = \mathrm{Hom}_{\mathcal{C}}(U, X)$ of arrows $U \to X$. Furthermore, if X is an object and $F\colon \mathcal{C}^{\mathrm{op}} \to (\mathrm{Set})$ is a functor, we will also identify the set $\mathrm{Hom}(X, F) = \mathrm{Hom}(\mathrm{h}_X, F)$ of morphisms from h_X to F with FX.

2.1.3. Examples. Here are some examples of representable and nonrepresentable functors.

(i) Consider the functor $\mathrm{P}\colon (\mathrm{Set})^{\mathrm{op}} \to (\mathrm{Set})$ that sends each set S to the set $\mathrm{P}(S)$ of subsets of S. If $f\colon S \to T$ is a function, then $\mathrm{P}(f)\colon \mathrm{P}(T) \to \mathrm{P}(S)$ is defined by $\mathrm{P}(f)\tau = f^{-1}\tau$ for all $\tau \subseteq T$.

Given a subset $\sigma \subseteq S$, there is a unique function $\chi_\sigma\colon S \to \{0,1\}$ such that $\chi_\sigma^{-1}(\{1\}) = \sigma$, namely the *characteristic function*, defined by

$$\chi_\sigma(s) = \begin{cases} 1 & \text{if } s \in \sigma \\ 0 & \text{if } s \notin \sigma. \end{cases}$$

Hence the pair $(\{0,1\}, \{1\})$ is a universal object, and the functor P is represented by $\{0,1\}$.

(ii) This example is similar to the previous one. Consider the category (Top) of all topological spaces, with the arrows being given by continuous functions. Define a functor $\mathrm{F}\colon (\mathrm{Top})^{\mathrm{op}} \to (\mathrm{Set})$ sending each topological space S to the collection $\mathrm{F}(S)$ of all its open subspaces. Endow $\{0,1\}$ with the coarsest topology in which the subset $\{1\} \subseteq \{0,1\}$ is closed; the open subsets in this topology are $\emptyset$, $\{1\}$ and $\{0,1\}$. A function $S \to \{0,1\}$ is continuous if and only if $f^{-1}(\{1\})$ is open in S, and so one sees that the pair $(\{0,1\}, \{1\})$ is a universal object for this functor.

The space $\{0,1\}$ is called *the Sierpinski space*.

(iii) The next example may look similar, but the conclusion is very different. Let (HausTop) be the category of all Hausdorff topological spaces, and consider the restriction $\mathrm{F}\colon (\mathrm{HausTop}) \to (\mathrm{Set})$ of the functor above. I claim that this functor is not representable.

In fact, assume that (X, ξ) is a universal object. Let S be any set, considered with the discrete topology; by definition, there is a unique function $f\colon S \to X$ with $f^{-1}\xi = S$, that is, a unique function $S \to \xi$. This means that ξ can only have one element. Analogously, there is a unique function

$S \to X \setminus \xi$, so $X \setminus \xi$ also has a unique element. But this means that X is a Hausdorff space with two elements, so it must have the discrete topology; hence ξ is also closed in X. Hence, if S is any topological space with a closed subset σ that is not open, there is no continuous function $f\colon S \to X$ with $f^{-1}\xi = \sigma$.

(iv) Take (Grp) to be the category of groups, and consider the functor

$$\mathrm{Sgr}\colon (\mathrm{Grp})^{\mathrm{op}} \longrightarrow (\mathrm{Set})$$

that associates with each group G the set of all its subgroups. If $f\colon G \to H$ is a group homomorphism, we take $\mathrm{Sgr}\, f\colon \mathrm{Sgr}\, H \to \mathrm{Sgr}\, G$ to be the function associating with each subgroup of H its inverse image in G.

This is not representable: there does not exist a group Γ, together with a subgroup $\Gamma_1 \subseteq \Gamma$, with the property that for all groups G with a subgroup $G_1 \subseteq G$, there is a unique homomorphism $f\colon G \to \Gamma$ such that $f^{-1}\Gamma_1 = G_1$. This can be checked in several ways; for example, if we take the subgroup $\{0\} \subseteq \mathbb{Z}$, there should be a unique homomorphism $f\colon \mathbb{Z} \to \Gamma$ such that $f^{-1}\Gamma_1 = \{0\}$. But given one such f, then the homomorphism $\mathbb{Z} \to \Gamma$ defined by $n \mapsto f(2n)$ also has this property, and is different, so this contradicts uniqueness.

(v) Here is a much more sophisticated example. Let (Hot) be the category of CW complexes, with the arrows being given by homotopy classes of continuous functions. If n is a fixed natural number, there is a functor $\mathrm{H}^n\colon (\mathrm{Hot})^{\mathrm{op}} \to (\mathrm{Set})$ that sends a CW complex S to its nth cohomology group $\mathrm{H}^n(S, \mathbb{Z})$. Then it is a highly nontrivial fact that this functor is represented by a CW complex, known as a Eilenberg–Mac Lane space, usually denoted by $\mathrm{K}(\mathbb{Z}, n)$.

But we are really interested in algebraic geometry, so let's give some examples in this context. Let $S = \operatorname{Spec} R$ (this is only for simplicity of notation, if S is not affine, nothing substantial changes).

EXAMPLE 2.4. Consider the affine line $\mathbb{A}^1_S = \operatorname{Spec} R[x]$. We have a functor

$$\mathcal{O}\colon (\mathrm{Sch}/S)^{\mathrm{op}} \longrightarrow (\mathrm{Set})$$

that sends each S-scheme U to the ring of global sections $\mathcal{O}(U)$. If $f\colon U \to V$ is a morphism of schemes, the corresponding function $\mathcal{O}(V) \to \mathcal{O}(V)$ is that induced by $f^\sharp\colon \mathcal{O}_V \to f_*\mathcal{O}_U$.

Then $x \in \mathcal{O}(\mathbb{A}^1_S)$, and given a scheme U over S, and an element $f \in \mathcal{O}(U)$, there is a unique morphism $U \to \mathbb{A}^1_S$ such that the pullback of x to U is precisely f. This means that the functor $\mathcal{O}$ is represented by $\mathbb{A}^1_S$, and the pair $(\mathbb{A}_S, x)$ is a universal object.

More generally, the affine space $\mathbb{A}^n_S$ represents the functor $\mathcal{O}^n$ that sends each scheme S to the ring $\mathcal{O}(S)^n$.

EXAMPLE 2.5. Now we look at $\mathbb{G}_{\mathrm{m},S} = \mathbb{A}^1_S \setminus 0_S = \operatorname{Spec} R[x, x^{-1}]$. Here by 0_S we mean the image of the zero-section $S \to \mathbb{A}^1_S$. Now, a morphism of S-schemes $U \to \mathbb{G}_{\mathrm{m},S}$ is determined by the image of $x \in \mathcal{O}(\mathbb{G}_{\mathrm{m},S})$ in $\mathcal{O}(S)$; therefore $\mathbb{G}_{\mathrm{m},S}$ represents the functor $\mathcal{O}^*\colon (\mathrm{Sch}/S)^{\mathrm{op}} \to (\mathrm{Set})$ that sends each scheme U to the group $\mathcal{O}^*(U)$ of invertible sections of the structure sheaf.

A much more subtle example is given by projective spaces.

EXAMPLE 2.6. On the projective space $\mathbb{P}^n_S = \operatorname{Proj} R[x_0, \dots, x_n]$ there is a line bundle $\mathcal{O}(1)$, with $n+1$ sections $x_0, \dots, x_n$, which generate it.

Suppose that U is a scheme, and consider the set of sequences

$$(\mathcal{L}, s_0, \dots, s_n),$$

where $\mathcal{L}$ is an invertible sheaf on U, $s_0, \dots, s_n$ sections of $\mathcal{L}$ that generate it. We say that $(\mathcal{L}, s_0, \dots, s_n)$ is equivalent to $(\mathcal{L}', s'_0, \dots, s'_n)$ if there exists an isomorphism of invertible sheaves $\phi \colon \mathcal{L} \simeq \mathcal{L}'$ carrying each s_i into s'_i. Notice that, since the s_i generate $\mathcal{L}$, if ϕ exists then it is unique.

One can consider a function $Q_n \colon (\mathrm{Sch}/S) \to (\mathrm{Set})$ that associates with each scheme U the set of sequences $(\mathcal{L}, s_0, \dots, s_n)$ as above, modulo equivalence. If $f \colon U \to V$ is a morphism of S-schemes, and $(\mathcal{L}, s_0, \dots, s_n) \in Q_n(V)$, then there are sections $f^*s_0, \dots, f^*s_n$ of $f^*\mathcal{L}$ that generate it; this makes Q_n into a functor $(\mathrm{Sch}/S)^{\mathrm{op}} \to (\mathrm{Set})$.

Another description of the functor Q_n is as follows. Given a scheme U and a sequence $(\mathcal{L}, s_0, \dots, s_n)$ as above, the s_i define a homomorphism $\mathcal{O}_U^{n+1} \to \mathcal{L}$, and the fact that the s_i generate is equivalent to the fact that this homomorphism is surjective. Then two sequences are equivalent if and only if the represent the same quotient of $\mathcal{O}_S^n$.

It is a very well-known fact, and, indeed, one of the cornerstones of algebraic geometry, that for any sequence $(\mathcal{L}, s_0, \dots, s_n)$ over an S-scheme U, there it exists a unique morphism $f \colon U \to \mathbb{P}^n_S$ such that $(\mathcal{L}, s_0, \dots, s_n)$ is equivalent to $(f^*\mathcal{O}(1), f^*x_0, \dots, f^*x_n)$. This means precisely that $\mathbb{P}^n_S$ represents the functor Q_n.

EXAMPLE 2.7. This example is an important generalization of the previous one.

Here we will let S be an arbitrary scheme, not necessarily affine, $\mathcal{M}$ a quasi-coherent sheaf on S. In Grothendieck's notation, $\pi \colon \mathbb{P}(\mathcal{M}) \to S$ is the relative homogeneous spectrum $\operatorname{Proj}_S \operatorname{Sym}_{\mathcal{O}_S} \mathcal{M}$ of the symmetric sheaf of algebras of $\mathcal{M}$ over $\mathcal{O}_S$. Then on $\mathbb{P}(\mathcal{M})$ there is an invertible sheaf, denoted by $\mathcal{O}_{\mathbb{P}(\mathcal{M})}(1)$, which is a quotient of $\pi^*\mathcal{M}$. This is a universal object, in the sense that, given any S-scheme $\phi \colon U \to S$, with an invertible sheaf $\mathcal{L}$ and a surjection $\alpha \colon \phi^*\mathcal{M} \twoheadrightarrow \mathcal{L}$, there is unique morphism of S-schemes $f \colon U \to \mathbb{P}(\mathcal{M})$, and an isomorphism of $\mathcal{O}_U$-modules $\alpha \colon \mathcal{L} \simeq f^*\mathcal{O}_{\mathbb{P}(\mathcal{M})}(1)$, such that the composite

$$f^*\pi^*\mathcal{M} \simeq \phi^*\mathcal{M} \twoheadrightarrow \mathcal{L} \xrightarrow{\alpha} f^*\mathcal{O}_{\mathbb{P}(\mathcal{M})}(1)$$

is the pullback of the projection $\pi^*\mathcal{M} \twoheadrightarrow \mathcal{O}_{\mathbb{P}(\mathcal{M})}(1)$ ([**EGAI**, Proposition 4.2.3]).

This means the following. Consider the functor $Q_{\mathcal{M}} \colon (\mathrm{Sch}/S)^{\mathrm{op}} \to (\mathrm{Set})$ that sends each scheme $\phi \colon U \to S$ over S to the set of all invertible quotients of the pullback $\phi^*\mathcal{M}$. If $f \colon V \to U$ is a morphism of S-schemes from $\phi \colon U \to S$ to $\psi \colon V \to S$, and $\alpha \colon \phi^*\mathcal{M} \twoheadrightarrow \mathcal{L}$ is an object of $Q_{\mathcal{M}}(U)$, then

$$f^*\alpha \colon \psi^*\mathcal{E} \simeq f^*\phi^*\mathcal{M} \longrightarrow\!\!\!\!\!\rightarrow f^*\mathcal{L}$$

is an object of $Q_{\mathcal{M}}(V)$: this defines the pullback $Q_{\mathcal{M}}(U) \to Q_{\mathcal{M}}(V)$. Then this functor is represented by $\mathbb{P}(\mathcal{M})$.

When $\mathcal{M} = \mathcal{O}_S^{n+1}$, we recover the functor Q_n of the previous example.

EXAMPLE 2.8. With the same setup as in the previous example, fix a positive integer r. We consider the functor $(\mathrm{Sch}/S)^{\mathrm{op}} \to (\mathrm{Set})$ that sends each $\phi \colon U \to S$ to the set of quotients of $\phi^*\mathcal{M}$ that are locally free of rank r. This is also representable by a scheme $\mathbb{G}(r, \mathcal{M}) \to S$.

Finally, let us a give an example of a functor that is not representable.

EXAMPLE 2.9. This is very similar to Example (iii) of §2.1.3. Let κ be a field, (Sch/κ) the category of schemes over κ. Consider the functor $F\colon (\mathrm{Sch}/\kappa)^{\mathrm{op}} \to (\mathrm{Set})$ that associates with each scheme U over κ the set of all of its open subsets; the action of F on arrows is obtained by taking inverse images.

I claim that this functor is not representable. In fact, suppose that it is represented by a pair (X, ξ), where X is a scheme over κ and ξ is an open subset. We can consider ξ as an open subscheme of X. If U is any scheme over κ, a morphism of κ-schemes $U \to \xi$ is a morphism of κ-schemes $U \to X$ whose image is contained in ξ; by definition of X there is a unique such morphism, the one corresponding to the open subset U, considered as an element of FU. Hence the functor represented by the κ-scheme ξ is the one point functor, sending any κ-scheme U to a set with one element, and this is represented by $\operatorname{Spec}\xi$. Hence ξ is isomorphic to $\operatorname{Spec}\kappa$ as a κ-scheme; this means that ξ, viewed as an open subscheme of X, consists of a unique κ-rational point of X. But a κ-rational point of a κ-scheme is necessarily a closed point (this is immediate for affine schemes, and follows in the general case, because being a closed subset of a topological space is a local property). So ξ is also closed; but this would imply that every open subset of a κ-scheme is also closed, and this fails, for example, for $\mathbb{A}^1_\kappa \setminus \{0\} \subseteq \mathbb{A}^1_\kappa$.

REMARK 2.10. There is a dual version of Yoneda's lemma, which will be used in §3.2.1. Each object X of $\mathcal{C}$ defines a functor

$$\mathrm{Hom}_{\mathcal{C}}(X, -)\colon \mathcal{C} \longrightarrow (\mathrm{Set}).$$

This can be viewed as the functor $\mathrm{h}_X\colon (\mathcal{C}^{\mathrm{op}})^{\mathrm{op}} \to (\mathrm{Set})$; hence, from the usual form of Yoneda's lemma applied to $\mathcal{C}^{\mathrm{op}}$ for any two objects X and Y we get a canonical bijective correspondence between $\mathrm{Hom}_{\mathcal{C}}(X, Y)$ and the set of natural transformations $\mathrm{Hom}_{\mathcal{C}}(Y, -) \to \mathrm{Hom}_{\mathcal{C}}(X, -)$.

2.2. Group objects

In this section the category $\mathcal{C}$ will have finite products; we will denote a terminal object by pt.

DEFINITION 2.11. A *group object* of $\mathcal{C}$ is an object G of $\mathcal{C}$, together with a functor $\mathcal{C}^{\mathrm{op}} \to (\mathrm{Grp})$ into the category of groups, whose composite with the forgetful functor $(\mathrm{Grp}) \to (\mathrm{Set})$ equals h_G.

A group object in the category of topological spaces is called a *topological!group*. A group object in the category of schemes over a scheme S is called a *group scheme over S*.

Equivalently: a group object is an object G, together with a group structure on $G(U)$ for each object U of $\mathcal{C}$, so that the function $f^*\colon G(V) \to G(U)$ associated with an arrow $f\colon U \to V$ in $\mathcal{C}$ is always a homomorphism of groups.

This can be restated using Yoneda's lemma.

PROPOSITION 2.12. To give a group object structure on an object G of $\mathcal{C}$ is equivalent to assigning three arrows $\mathrm{m}_G\colon G \times G \to G$ (the multiplication), $\mathrm{i}_G\colon G \to$

G (the inverse), and $\mathrm{e}_G\colon \mathrm{pt} \to G$ (the identity), such that the following diagrams commute.

(i) The identity is a left and right identity:

$$\begin{array}{ccc} \mathrm{pt} \times G & \xrightarrow{\mathrm{e}_G \times \mathrm{id}_G} & G \times G \\ & \searrow^{=} & \downarrow{\scriptstyle \mathrm{m}_G} \\ & & G \end{array} \quad \text{and} \quad \begin{array}{ccc} G \times \mathrm{pt} & \xrightarrow{\mathrm{id}_G \times \mathrm{e}_G} & G \times G \\ & \searrow^{=} & \downarrow{\scriptstyle \mathrm{m}_G} \\ & & G \end{array}$$

(ii) Multiplication is associative:

$$\begin{array}{ccc} G \times G \times G & \xrightarrow{\mathrm{m}_G \times \mathrm{id}_G} & G \times G \\ \downarrow{\scriptstyle \mathrm{id}_G \times \mathrm{m}_G} & & \downarrow{\scriptstyle \mathrm{m}_G} \\ G \times G & \xrightarrow{\mathrm{m}_G} & G \end{array}$$

(iii) The inverse is a left and right inverse:

$$\begin{array}{ccc} G & \xrightarrow{\langle \mathrm{i}_G, \mathrm{id}_G \rangle} & G \times G \\ \downarrow & & \downarrow{\scriptstyle \mathrm{m}_G} \\ \mathrm{pt} & \xrightarrow{\mathrm{e}_G} & G \end{array} \quad \text{and} \quad \begin{array}{ccc} G & \xrightarrow{\langle \mathrm{id}_G, \mathrm{i}_G \rangle} & G \times G \\ \downarrow & & \downarrow{\scriptstyle \mathrm{m}_G} \\ \mathrm{pt} & \xrightarrow{\mathrm{e}_G} & G \end{array}$$

PROOF. It is immediate to check that, if $\mathcal{C}$ is the category of sets, the commutativity of the diagram above gives the usual group axioms. Hence the result follows by evaluating the diagrams above (considered as diagrams of functors) at any object U of $\mathcal{C}$. □

Thus, for example, a topological group is simply a group, that has a structure of a topological space, such that the multiplication map and the inverse map are continuous (of course the map from a point giving the identity is automatically continuous).

Let us give examples of group schemes.

The first examples are the schemes $\mathbb{A}^n_S \to S$; these represent the functor $\mathcal{O}^n$ sending a scheme $U \to S$ to the set $\mathcal{O}(U)^n$, which has an evident additive group structure.

The group scheme $\mathbb{A}^1_S$ is often denote by $\mathbb{G}_{\mathrm{a},S}$.

Also, $\mathbb{G}_{\mathrm{m},S} = \mathbb{A}^1_S \setminus 0_S$ represents the functor $\mathcal{O}^*\colon (\mathrm{Sch}/S)^{\mathrm{op}} \to (\mathrm{Set})$, that sends each scheme $U \to S$ to the group $\mathcal{O}^*(U)$; this gives $\mathbb{G}_{\mathrm{m},S}$ an obvious structure of group scheme.

Now consider the functor $(\mathrm{Sch}/S)^{\mathrm{op}} \to (\mathrm{Set})$ that sends each scheme $U \to S$ to the set $\mathrm{M}_n\big(\mathcal{O}(U)\big)$ of $n \times n$ matrices with coefficients into the ring $\mathcal{O}(U)$. This is obviously represented by the scheme $\mathrm{M}_{n,S} \overset{\mathrm{def}}{=} \mathbb{A}^{n^2}_S$. Consider the determinant mapping as morphism of schemes $\det\colon \mathrm{M}_{n,S} \to \mathbb{A}^1_S$; denote by $\mathrm{GL}_{n,S}$ the inverse image of the open subscheme $\mathbb{G}_{\mathrm{m},S} \subseteq \mathbb{A}^1_S$. Then $\mathrm{GL}_{n,S}$ is an open subscheme of $\mathrm{M}_{n,S}$; the functor it represents is the functor sending each scheme $U \to S$ to the set of matrices in $\mathrm{M}_n\big(\mathcal{O}(U)\big)$ with invertible determinants. But these are the invertible matrices, and they form a group. This gives $\mathrm{GL}_{n,S}$ the structure of a group scheme on S.

There are various subschemes of $\mathrm{GL}_{n,S}$ that are group schemes. For example, $\mathrm{SL}_{n,S}$, the inverse image of the identity section $1_S\colon S \to \mathbb{G}_{\mathrm{m},S}$ via the morphism $\det\colon \mathrm{GL}_{n,S} \to \mathbb{G}_{\mathrm{m},S}$ represents the functor sending each scheme $U \to S$ to the group $\mathrm{SL}\bigl(\mathcal{O}(U)\bigr)$ of $n \times n$ matrices with determinant 1.

We leave it to the reader to define the orthogonal group scheme $\mathrm{O}_{n,S}$ and the symplectic group scheme $\mathrm{Sp}_{n,S}$.

DEFINITION 2.13. If G and H are group objects, we define a *homomorphism* of group objects as an arrow $G \to H$ in $\mathcal{C}$, such that for each object U of $\mathcal{C}$ the induced function $G(U) \to H(U)$ is a group homomorphism.

Equivalently, a homomorphism is an arrow $f\colon G \to H$ such that the diagram

$$\begin{array}{ccc} G \times G & \xrightarrow{\mathrm{m}_G} & G \\ {\scriptstyle f\times f}\downarrow & & \downarrow{\scriptstyle f} \\ H \times H & \xrightarrow{\mathrm{m}_H} & H \end{array}$$

commutes.

The identity is obviously a homomorphism from a group object to itself. Furthermore, the composite of homomorphisms of group objects is still a homomorphism; thus, group objects in a fixed category form a category, which we denote by $\mathrm{Grp}(\mathcal{C})$.

REMARK 2.14. Suppose that $\mathcal{C}$ and $\mathcal{D}$ are categories with products and terminal object $\mathrm{pt}_{\mathcal{C}}$ and $\mathrm{pt}_{\mathcal{D}}$. Suppose that $F\colon \mathcal{C} \to \mathcal{D}$ is a functor that preserves finite products, and G is a group object in $\mathcal{C}$. The arrow $\mathrm{e}_G\colon \mathrm{pt}_{\mathcal{C}} \to G$ yields an arrow $F\mathrm{e}_G\colon F\mathrm{pt}_{\mathcal{C}} \to FG$; this can be composed with the inverse of the unique arrow $F\mathrm{pt}_{\mathcal{C}} \to \mathrm{pt}_{\mathcal{D}}$, which is an isomorphism, because $F\mathrm{pt}_{\mathcal{C}}$ is a terminal object, to get an arrow $\mathrm{e}_{FG}\colon \mathrm{pt}_{\mathcal{D}} \to FG$. Analogously one uses $F\mathrm{m}_G\colon F(G \times G) \to FG$ and the inverse of the isomorphism $F(G \times G) \simeq FG \times FG$ to define an arrow $\mathrm{m}_{FG}\colon FG \times FG \to FG$. Finally we set $\mathrm{i}_{FG} \overset{\text{def}}{=} F\mathrm{i}_G\colon FG \to FG$.

We leave it to the reader to check that this gives FG the structure of a group object, and this induces a functor from the category of group objects on $\mathcal{C}$ to the category of group objects on $\mathcal{D}$.

2.2.1. Actions of group objects. There is an obvious notion of left action of a functor into groups on a functor into sets.

DEFINITION 2.15. A left action α of a functor $G\colon \mathcal{C}^{\mathrm{op}} \to (\mathrm{Grp})$ on a functor $F\colon \mathcal{C}^{\mathrm{op}} \to (\mathrm{Set})$ is a natural transformation $G \times F \to F$, such that for any object U of $\mathcal{C}$, the induced function $G(U) \times F(U) \to F(U)$ is an action of the group $G(U)$ on the set $F(U)$.

In the definition above, we denote by $G \times F$ the functor that sends an object U of $\mathcal{C}$ to the product of the set underlying the group GU with the set FU. In other words, $G \times F$ is the product $\widetilde{G} \times F$, where $\widetilde{G}$ is the composite of G with the forgetful functor $(\mathrm{Grp}) \to (\mathrm{Set})$.

Equivalently, a left action of G on F consists of an action of $G(U)$ on $F(U)$ for all objects U of $\mathcal{C}$, such that for any arrow $f\colon U \to V$ in $\mathcal{C}$, any $g \in G(V)$ and any $x \in F(V)$ we have

$$f^*g \cdot f^*x = f^*(g \cdot x) \in F(U).$$

Right actions are defined analogously.

We define an action of a group object G on an object X as an action of the functor $\mathrm{h}_G\colon \mathcal{C}^{\mathrm{op}} \to (\mathrm{Grp})$ on $\mathrm{h}_X : \mathcal{C}^{\mathrm{op}} \to (\mathrm{Set})$.

Again, we can reformulate this definition in terms of diagrams.

PROPOSITION 2.16. Giving a left action of a group object G on an object X is equivalent to assigning an arrow $\alpha\colon G \times X \to X$, such that the following two diagrams commute.

(i) The identity of G acts like the identity on X:

$$\begin{array}{ccc} \mathrm{pt} \times X & \xrightarrow{\mathrm{e}_G \times \mathrm{id}_X} & G \times X \\ & \searrow_{=} & \downarrow{\scriptstyle \alpha} \\ & & X \end{array}$$

(ii) The action is associative with respect to the multiplication on G:

$$\begin{array}{ccc} G \times G \times X & \xrightarrow{\mathrm{m}_G \times \mathrm{id}_X} & G \times X \\ \downarrow{\scriptstyle \mathrm{id}_G \times \alpha} & & \downarrow{\scriptstyle \alpha} \\ G \times X & \xrightarrow{\alpha} & X \end{array}$$

PROOF. It is immediate to check that, if $\mathcal{C}$ is the category of sets, the commutativity of the diagram above gives the usual axioms for a left action. Hence the result follows from Yoneda's lemma by evaluating the diagrams above (considered as diagrams of functors) on any object U of $\mathcal{C}$. $\square$

DEFINITION 2.17. Let X and Y be objects of $\mathcal{C}$ with an action of G, an arrow $f\colon X \to Y$ is called *G-equivariant* if for all objects U of $\mathcal{C}$ the induced function $X(U) \to Y(U)$ is $G(U)$-equivariant.

Equivalently, f is G-equivariant if the diagram

$$\begin{array}{ccc} G \times X & \longrightarrow & X \\ \downarrow{\scriptstyle \mathrm{id}_G \times f} & & \downarrow{\scriptstyle f} \\ G \times Y & \longrightarrow & Y \end{array}$$

where the rows are given by the actions, commutes (the equivalence of these two definitions follows from Yoneda's lemma).

There is yet another way to define the action of a functor $G\colon \mathcal{C}^{\mathrm{op}} \to (\mathrm{Grp})$ on an object X of $\mathcal{C}$. Given an object U of $\mathcal{C}$, we denote by $\mathrm{End}_U(U \times X)$ the set of arrows $U \times X \to U \times X$ that commute with the projection $\mathrm{pr}_1 : U \times X \to U$; this set has the structure of a monoid, the operation being the composition. In other words, $\mathrm{End}_U(U \times X)$ is the monoid of endomorphisms of $\mathrm{pr}_1 : U \times X \to U$ considered as an object of the comma category $(\mathcal{C}/U)$. We denote the group of automorphisms in $\mathrm{End}_U(U \times X)$ by $\mathrm{Aut}_U(U \times X)$.

Let us define a functor

$$\underline{\mathrm{Aut}}_{\mathcal{C}}(X)\colon \mathcal{C}^{\mathrm{op}} \longrightarrow (\mathrm{Grp})$$

sending each object U of $\mathcal{C}$ to the group $\underline{\mathrm{Aut}}_{\mathcal{C}}(X)(U) \overset{\mathrm{def}}{=} \mathrm{Aut}_U(U \times X)$. The group $\underline{\mathrm{Aut}}_{\mathcal{C}}(X)(\mathrm{pt})$ is canonically isomorphic to $\mathrm{Aut}_{\mathcal{C}}(X)$.

Consider an arrow $f\colon U \to V$ in $\mathcal{C}$; with this we need to associate a group homomorphism $f^*\colon \operatorname{Aut}_V(V \times X) \to \operatorname{Aut}_U(U \times X)$. The diagram

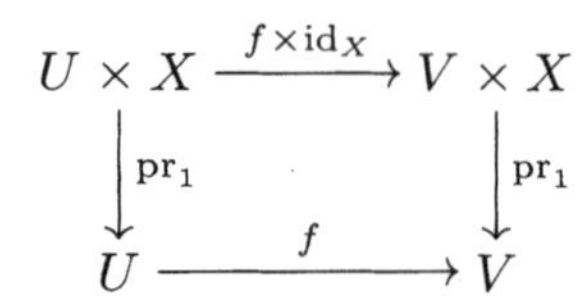

is cartesian; hence, given an arrow $\beta\colon V \times X \to V \times X$ over V, there is a unique arrow $\alpha\colon U \times X \to U \times X$ making the diagram

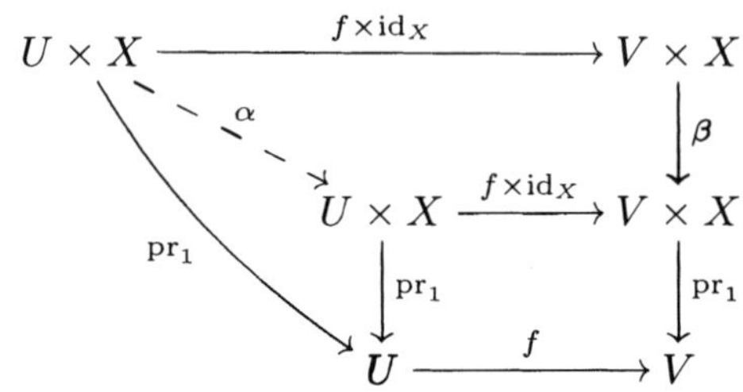

commute. This gives a function from the set $\operatorname{End}_V(V \times X)$ to $\operatorname{End}_U(U \times X)$, which is easily checked to be a homomorphism of monoids (that is, it sends the identity to the identity, and it preserves composition). It follows that it restricts to a homomorphism of groups $f^*\colon \operatorname{Aut}_V(V \times X) \to \operatorname{Aut}_U(U \times X)$. This gives $\underline{\operatorname{Aut}}_{\mathcal{C}}(X)$ the structure of a functor.

This construction is a very particular case of that of Section 3.7.

Proposition 2.18. Let $G\colon \mathcal{C}^{\mathrm{op}} \to (\mathrm{Grp})$ a functor, X an object of $\mathcal{C}$. To give an action of G on X is equivalent to giving a natural transformation $G \to \underline{\operatorname{Aut}}_{\mathcal{C}}(X)$ of functors $\mathcal{C}^{\mathrm{op}} \to (\mathrm{Grp})$.

Proof. Suppose that we are given a natural transformation $G \to \underline{\operatorname{Aut}}_{\mathcal{C}}(X)$. Then for each object U of $\mathcal{C}$ we have a group homomorphism $G(U) \to \operatorname{Aut}_U(U \times X)$. The set $X(U)$ is in bijective correspondence with the set of sections $U \to U \times X$ to the projection $\mathrm{pr}_1\colon U \times X \to U$, and if $s\colon U \to U \times X$ is a section, $\alpha \in \operatorname{Aut}_U(U \times X)$, then $\alpha \circ s\colon U \to U \times X$ is still a section. This induces an action of $\operatorname{Aut}_U(U \times X)$ on $X(U)$, and, via the given homomorphism $G(U) \to \operatorname{Aut}_U(U \times X)$, also an action of $G(U)$ on $X(U)$. It is easy to check that this defines an action of G on X.

Conversely, suppose that G acts on X, let U be an object of $\mathcal{C}$, and $g \in G(U)$. We need to associate with g an object of

$$\underline{\operatorname{Aut}}_{\mathcal{C}}(X)(U) = \operatorname{Aut}_U(U \times X).$$

We will use Yoneda's lemma once again, and consider $U \times X$ as a functor $U \times X\colon (\mathcal{C}/U)^{\mathrm{op}} \to (\mathrm{Set})$. For each arrow $V \to U$ in $\mathcal{C}$ there is a bijective correspondence between the set $X(V)$ and the set of arrows $V \to U \times X$ in $\mathcal{C}/U$, obtained by composing an arrow $V \to U \times X$ with the projection $\mathrm{pr}_2\colon U \times X \to X$. Now we are given an action of $G(V)$ on $X(V)$, and this induces an action of $G(V)$ on $\operatorname{Hom}_{(\mathcal{C}/U)}(V, U \times X) = (U \times X)(V)$. The arrow $V \to U$ induces a group homomorphism $G(U) \to G(V)$, so the element $g \in G(U)$ induces a permutation of $(U \times X)(V)$. There are several things to check: all of them are straightforward and left to the reader as an exercise.

(i) This construction associates with each g an automorphism of the functor $U \times X$, hence an automorphism of $U \times X$ in $(\mathcal{C}/U)$.
(ii) The resulting function $G(U) \to \underline{\mathrm{Aut}}_{\mathcal{C}}(X)(U)$ is a group homomorphism.
(iii) This defines a natural transformation $G \to \underline{\mathrm{Aut}}_{\mathcal{C}}(X)$.
(iv) The resulting functions from the set of actions of G on X and the set of natural transformations $G \to \underline{\mathrm{Aut}}_{\mathcal{C}}(X)$ are inverse to each other. □

2.2.2. Discrete groups. There is a standard notion of action of a group on an object of a category: a group Γ acts on an object X of $\mathcal{C}$ when there is given a group homomorphism $\Gamma \to \mathrm{Aut}_{\mathcal{C}}(X)$. With appropriate hypothesis, this action can be interpreted as the action of a *discrete group object* on U.

In many concrete cases, a category of geometric objects has objects that can be called *discrete.* For example in the category of topological spaces we have discrete spaces: these are spaces with the discrete topology, or, in other words, disjoint unions of point. In the category (Sch/S) of schemes over S an object should be called discrete when it is the disjoint union of copies of S. In categorical terms, a disjoint union is a coproduct; thus a discrete object of (Sch/S) is a scheme U over S, with the property that the functor $\mathrm{Hom}_S(U,-)\colon (\mathrm{Sch}/S) \to (\mathrm{Set})$ is the product of copies of $\mathrm{Hom}_S(S,-)$.

DEFINITION 2.19. Let $\mathcal{C}$ be a category. We say that $\mathcal{C}$ *has discrete objects* if it has a terminal object pt, and for any set I the coproduct $\coprod_{i\in I}\mathrm{pt}$ exist.

An object of $\mathcal{C}$ that is isomorphic to one of the form $\coprod_{i\in I}\mathrm{pt}$ for some set I is called a *discrete object.*

Suppose that $\mathcal{C}$ has discrete objects. If I and J are two sets and $\phi\colon I \to J$ is a function, we get a collection of arrows $\mathrm{pt} \to \coprod_{j\in J}\mathrm{pt}$ parametrized by I: with each $i \in I$ we associate the tautological arrow $\mathrm{pt} \to \coprod_{j\in J}\mathrm{pt}$ corresponding to the element $\phi(j) \in J$. In this way we have defined an arrow

$$\phi_*\colon \coprod_{i\in I}\mathrm{pt} \longrightarrow \coprod_{j\in J}\mathrm{pt}.$$

It is immediate to check that if $\phi\colon I \to J$ and $\psi\colon J \to K$ are functions, we have

$$(\psi\circ\phi)_* = \psi_*\circ\phi_*\colon \coprod_{i\in I}\mathrm{pt} \longrightarrow \coprod_{k\in K}\mathrm{pt}.$$

In this way we have defined a functor $\Delta\colon (\mathrm{Set}) \to \mathcal{C}$ that sends a set I to $\coprod_{i\in I}\mathrm{pt}$. This is called *the discrete object functor*. By construction, it is a left adjoint to the functor $\mathrm{Hom}_{\mathcal{C}}(\mathrm{pt},-)$. Recall that this means that for every set I and every object U of $\mathcal{C}$ one has a bijective correspondence between $\mathrm{Hom}_{\mathcal{C}}(\Delta I, U)$ and the set of functions $I \to \mathrm{Hom}_{\mathcal{C}}(\mathrm{pt}, U)$; furthermore this bijective correspondence is functorial in I and U.

Conversely, if we assume that $\mathcal{C}$ has a terminal object pt, and that $\Delta\colon (\mathrm{Set}) \to \mathcal{C}$ is a left adjoint to the functor $\mathrm{Hom}_{\mathcal{C}}(\mathrm{pt},-)$, then it is easy to see for each set I the object ΔI is a coproduct $\coprod_{i\in I}\mathrm{pt}$.

We are interested in constructing discrete group objects in a category $\mathcal{C}$; for this, we need to have discrete objects, and, according to Remark 2.14, we need to have that the discrete object functor $(\mathrm{Set}) \to \mathcal{C}$ preserves finite products. Here is a condition to ensure that this happens.

Suppose that $\mathcal{C}$ is a category with finite products. Assume furthermore that for any object U in $\mathcal{C}$ and any set I the coproduct $\coprod_{i\in I} U$ exists in $\mathcal{C}$; in particular, $\mathcal{C}$

has discrete objects. If U is an object of $\mathcal{C}$ and I is a set, we will set $I \times U \stackrel{\text{def}}{=} \coprod_{i\in I} U$. By definition, an arrow $I \times U \to V$ is defined by a collection of arrows $f_i \colon U \to V$ parametrized by I. In particular, $\emptyset \times U$ is an initial object of $\mathcal{C}$.

Notice the following fact. Let I be a set, U an object of $\mathcal{C}$. If $i \in I$ then $\Delta\{i\}$ is a terminal object of $\mathcal{C}$, hence there is a canonical isomorphism $U \simeq \Delta\{i\} \times U$ (the inverse of the projection $\Delta\{i\} \times U \to U$). On the other hand the embedding $\iota_i \colon \{i\} \hookrightarrow I$ induces an arrow $\Delta\iota_i \times \mathrm{id}_U \colon \Delta\{i\} \times U \to \Delta I \times U$. By composing these $\Delta\iota_i$ with the isomorphisms $U \simeq \Delta\{i\} \times U$ we obtain a set of arrows $U \to \Delta I \times U$ parametrized by I, hence an arrow $I \times U \to \Delta I \times U$.

DEFINITION 2.20. A category $\mathcal{C}$ *has discrete group objects* when the following conditions are satisfied.

(i) $\mathcal{C}$ has finite products.
(ii) For any object U in $\mathcal{C}$ and any set I the coproduct $I \times U \stackrel{\text{def}}{=} \coprod_{i\in I} U$ exists;
(iii) For any object U in $\mathcal{C}$ and any set I, the canonical arrow $I \times U \to \Delta I \times U$ is an isomorphism.

For example, in the category (Top) a terminal object is a point (in other words, a topological space with one element), while the coproducts are disjoint unions. The conditions of the definition are easily checked. This also applies to the category (Sch/S) of schemes over a fixed base scheme S; in this case a terminal object is S itself.

PROPOSITION 2.21. If $\mathcal{C}$ has discrete group objects, then the discrete object functor $\Delta \colon (\mathrm{Set}) \to \mathcal{C}$ preserves finite products.

So, by Remark 2.14, when the category $\mathcal{C}$ has discrete group object the functor $\Delta \colon (\mathrm{Set}) \to \mathcal{C}$ gives a functor, also denoted by $\Delta \colon (\mathrm{Grp}) \to \mathrm{Grp}(\mathcal{C})$, from the category of groups to the category of group objects in $\mathcal{C}$. A group object in $\mathcal{C}$ is called *discrete* when it isomorphic to one of the form $\Delta\Gamma$, where Γ is a group.

PROOF. Let $\mathcal{C}$ be a category with discrete group objects. To prove that Δ preserves finite products, it is enough to check that Δ sends a terminal object to a terminal object, and that it preserves products of two objects. The first fact follows immediately from the definition of Δ.

Let us show that, given two sets I and J, the natural arrow $\Delta(I \times J) \to \Delta I \times \Delta J$ is an isomorphism. By definition, $\Delta(I \times J) = (I \times J) \times \mathrm{pt}$. On the other hand there is a canonical well-known isomorphism of

$$(I \times J) \times \mathrm{pt} = \coprod_{(i,j)\in I\times J} \mathrm{pt}$$

with

$$\coprod_{i\in I}\Bigl(\coprod_{j\in J} \mathrm{pt}\Bigr) = I \times (J \times \mathrm{pt}) = I \times \Delta J.$$

If we compose this isomorphism $\Delta(I \times J) \simeq I \times \Delta J$ with the isomorphism $I \times \Delta J \simeq \Delta I \times \Delta J$ discussed above we obtain an isomorphism $\Delta(I \times J) \simeq \Delta I \times \Delta J$. It is easy to check that the projections $\Delta(I \times J) \to \Delta I$ and $\Delta(I \times J) \to \Delta J$ are induced by the projections $I \times J \to I$ and $I \times J \to J$; this finishes the proof. $\square$

An action of a group is the same as an action of the associated discrete group object.

PROPOSITION 2.22. Suppose that $\mathcal{C}$ has finite group objects. Let X be an object of $\mathcal{C}$, Γ a group, $\Delta\Gamma$ the associated discrete group object of $\mathcal{C}$. Then giving an action of Γ on X, that is, giving a group homomorphism $\Gamma \to \mathrm{Aut}_{\mathcal{C}}(X)$, is equivalent to giving an action of the group object $\Delta\Gamma$ on X.

PROOF. A function from Γ to the set $\mathrm{Hom}_{\mathcal{C}}(X, X)$ of arrows from X to itself corresponds, by definition, to an arrow $\Gamma \times X \to X$; the isomorphism $\Gamma \times X \simeq \Delta\Gamma \times X$ above gives a bijective correspondence between functions $\Gamma \to \mathrm{Hom}_{\mathcal{C}}(X, X)$ and arrows $\Delta\Gamma \times X \to X$. We have to check that a function $\Gamma \to \mathrm{Hom}_{\mathcal{C}}(X, X)$ gives an action of Γ on X if and only if the corresponding arrow $\Delta\Gamma \times X \to X$ gives an action of $\Delta\Gamma$ on X. This is straightforward and left to the reader. □

REMARK 2.23. The terminology "$\mathcal{C}$ has discrete group objects" is perhaps misleading; for $\mathcal{C}$ to have discrete group objects would be sufficient to have discrete objects, and that the functor Δ preserves finite products.

However, for discrete group objects to be well behaved we need more than their existence, we want Proposition 2.22 to hold: and for this purpose the conditions of Definition 2.20 seem to be optimal (except that one does not need to assume that $\mathcal{C}$ has all products; but this hypothesis is satisfied in all the examples I have in mind).

2.3. Sheaves in Grothendieck topologies

2.3.1. Grothendieck topologies. The reader is familiar with the notion of sheaf on a topological space. A presheaf on a topological space X can be considered as a functor. Denote by X_{cl} the category in which the objects are the open subsets of X, and the arrows are given by inclusions. Then a presheaf of sets on X is a functor ${X_{\mathrm{cl}}}^{\mathrm{op}} \to (\mathrm{Set})$; and this is a sheaf when it satisfies appropriate gluing conditions.

There are more general circumstances under which we can ask whether a functor is a sheaf. For example, consider a functor $F\colon (\mathrm{Top})^{\mathrm{op}} \to (\mathrm{Set})$; for each topological space X we can consider the restriction F_X to the subcategory X_{cl} of (Top). We say that F is a *sheaf* on (Top) if F_X is a sheaf on X for all X.

There is a very general notion of sheaf in a Grothendieck topology; in this Section we review this theory.

In a Grothendieck topology the "open sets" of a space are *maps* into this space; instead of intersections we have to look at fibered products, while unions play no role. The axioms do not describe the "open sets", but the coverings of a space.

DEFINITION 2.24. Let $\mathcal{C}$ be a category. A *Grothendieck topology* on $\mathcal{C}$ is the assignment to each object U of $\mathcal{C}$ of a collection of sets of arrows $\{U_i \to U\}$, called *coverings of* U, so that the following conditions are satisfied.

(i) If $V \to U$ is an isomorphism, then the set $\{V \to U\}$ is a covering.
(ii) If $\{U_i \to U\}$ is a covering and $V \to U$ is any arrow, then the fibered products $\{U_i \times_U V\}$ exist, and the collection of projections $\{U_i \times_U V \to V\}$ is a covering.
(iii) If $\{U_i \to U\}$ is a covering, and for each index i we have a covering $\{V_{ij} \to U_i\}$ (here j varies on a set depending on i), the collection of composites $\{V_{ij} \to U_i \to U\}$ is a covering of U.

A category with a Grothendieck topology is called a *site*.

Notice that from (ii) and (iii) it follows that if $\{U_i \to U\}$ and $\{V_j \to U\}$ are two coverings of the same object, then $\{U_i \times_U V_j \to U\}$ is also a covering.

REMARK 2.25. In fact what we have defined here is what is called a *pretopology* in [**SGA4**]; a pretopology defines a topology, and very different pretopologies can define the same topology. The point is that the sheaf theory only depends on the topology, and not on the pretopology. Two pretopologies induce the same topology if and only if they are equivalent, in the sense of Definition 2.47.

Despite its unquestionable technical advantages, I do not find the notion of topology, as defined in [**SGA4**], very intuitive, so I prefer to avoid its use (just a question of habit, undoubtedly).

However, *sieves*, the objects that intervene in the definition of a topology, are quite useful, and will be used extensively.

Here are some examples of Grothendieck topologies. In what follows, a set $\{U_i \to U\}$ of functions, or morphisms of schemes, is called *jointly surjective* when the set-theoretic union of their images equals U.

EXAMPLE 2.26 (The site of a topological space). Let X be a topological space; denote by X_{cl} the category in which the objects are the open subsets of X, and the arrows are given by inclusions. Then we get a Grothendieck topology on X_{cl} by associating with each open subset $U \subseteq X$ the set of open coverings of U.

In this case, if $U_1 \to U$ and $U_2 \to U$ are arrows, the fibered product $U_1 \times_U U_2$ is the intersection $U_1 \cap U_2$.

EXAMPLE 2.27 (The global classical topology). Here $\mathcal{C}$ is the category (Top) of topological spaces. If U is a topological space, then a covering of U will be a jointly surjective collection of open embeddings $U_i \to U$.

Notice here we must interpret "open embedding" as meaning an open continuous injective map $V \to U$; if by an open embedding we mean the inclusion of an open subspace, then condition (i) of Definition 2.24 is not satisfied.

EXAMPLE 2.28 (The global étale topology for topological spaces). Here $\mathcal{C}$ is the category (Top) of topological spaces. If U is a topological space, then a covering of U will be a jointly surjective collection of local homeomorphisms $U_i \to U$.

Here is an extremely important example from algebraic geometry.

EXAMPLE 2.29 (The small étale site of a scheme). Let X be a scheme. Consider the full subcategory $X_{\text{ét}}$ of (Sch/X), consisting of morphisms $U \to X$ locally of finite presentation, that are étale. If $U \to X$ and $V \to X$ are objects of $X_{\text{ét}}$, then an arrow $U \to V$ over X is necessarily étale.

A covering of $U \to X$ in the small étale topology is a jointly surjective collection of morphisms $U_i \to U$.

Here are topologies that one can put on the category (Sch/S) of schemes over a fixed scheme S. Several more have been used in different contexts.

EXAMPLE 2.30 (The global Zariski topology). Here a covering $\{U_i \to U\}$ is a collection of open embeddings covering U. As in the example of the global classical topology, an open embedding must be defined as a morphism $V \to U$ that gives an isomorphism of V with an open subscheme of U, and not simply as the embedding of an open subscheme.

EXAMPLE 2.31 (The global étale topology). A covering $\{U_i \to U\}$ is a jointly surjective collection of étale maps locally of finite presentation.

EXAMPLE 2.32 (The fppf topology). A covering $\{U_i \to U\}$ is a jointly surjective collection of flat maps locally of finite presentation.

The abbreviation fppf stands for "fidèlement plat et de présentation finie".

2.3.2. The fpqc topology. It is sometimes useful to consider coverings that are not locally finitely presented. One can define a topology on (Sch/S) simply by taking all collections of morphisms $\{U_i \to U\}$ such that the resulting morphism $\coprod_i U_i \to U$ is faithfully flat. Unfortunately, this topology is not well behaved (see Remarks 1.14 and 2.56). One needs some finiteness condition in order to get a reasonable topology.

For example, one could define a covering as a collection of morphisms $\{U_i \to U\}$ such that the resulting morphism $\{\coprod_i U_i \to U\}$ is faithfully flat and quasi-compact, as I did in the first version of these notes; but then Zariski covers would not be included, and the resulting topology would not be comparable with the Zariski topology. The definition of the fpqc topology that follows, suggested by Steve Kleiman, gives the correct sheaf theory.

PROPOSITION 2.33. Let $f\colon X \to Y$ be a surjective morphism of schemes. Then the following properties are equivalent.

(i) Every quasi-compact open subset of Y is the image of a quasi-compact open subset of X.
(ii) There exists a covering $\{V_i\}$ of Y by open affine subschemes, such that each V_i is the image of a quasi-compact open subset of X.
(iii) Given a point $x \in X$, there exists an open neighborhood U of x in X, such that the image fU is open in Y, and the restriction $U \to fU$ of f is quasi-compact.
(iv) Given a point $x \in X$, there exists a quasi-compact open neighborhood U of x in X, such that the image fU is open and affine in Y.

PROOF. It is obvious that (i) implies (ii). The fact that (iv) implies (iii) follows from the fact that a morphism from a quasi-compact scheme to an affine scheme is quasi-compact.

It is also easy to show that (iii) implies (iv): if U' is an open subset of X containing x, whose image fU' in Y is open, take an affine neighborhood V of $f(x)$ in fU', and set $U = f^{-1}V$.

Since f is surjective, we see that (iv) implies (ii).

Conversely, assuming (ii), take a point $x \in X$. Then $f(x)$ will be contained in some V_i. Let U' be a quasi-compact open subset of X with image V_i, and U'' an open neighborhood of x in $f^{-1}V_i$. Then $U = U' \cup U''$ is quasi-compact, contains x and has image V_i.

We only have left to prove that (ii) implies (i). Let V be a quasi-compact open subset of Y. The open affine subsets of Y that are contained in some $V \cap V_i$ form a covering of V, so we can choose finitely many of them, call them $W_1, \dots, W_r$. Given one of the W_j, choose an index i such that $W_j \subseteq V_i$ and a quasi-compact open subset U_i of X with image V_i; the restriction $U_i \to V_i$ is quasi-compact, so the inverse image W_j' of W_j in U_i is quasi-compact. Then $\bigcup_{j=1}^r W_j'$ is an open quasi-compact subscheme of X with image $\bigcup_{j=1}^r W_j = V$. □

DEFINITION 2.34. An *fpqc morphism of schemes* is a faithfully flat morphism that satisfies the equivalent conditions of Proposition 2.33.

The abbreviation fpqc stands for "fidèlement plat et quasi-compact".

Here are some properties of fpqc morphisms.

PROPOSITION 2.35.

(i) The composite of fpqc morphisms is fpqc.
(ii) If $f\colon X \to Y$ is a morphism of schemes, and there is an open covering V_i of Y, such that the restriction $f^{-1}V_i \to V_i$ is fpqc, then f is fpqc.
(iii) An open faithfully flat morphism is fpqc.
(iv) A faithfully flat morphism that is locally of finite presentation is fpqc.
(v) A morphism obtained by base change from an fpqc morphism is fpqc.
(vi) If $f\colon X \to Y$ is an fpqc morphism, a subset of Y is open in Y if and only if its inverse image is open in X.

PROOF. (i) follows from the definition, using the characterization (i) in Proposition 2.33. Also (ii) follows easily, using the characterization (ii), and (iii) follows from condition (iii). (iv) follows from (iii) and the fact that a faithfully flat morphism that is locally of finite presentation is open (Proposition 1.12).

For (iii), suppose that we are given a cartesian diagram of schemes

$$\begin{array}{ccc} X' & \longrightarrow & X \\ \downarrow & & \downarrow \\ Y' & \longrightarrow & Y \end{array}$$

such that $X \to Y$ is fpqc. Take a covering V_i of Y by open affine subschemes, and for each of them choose an open quasi-compact open subset U_i of X mapping onto V_i. If we denote by V_i' its inverse image of V_i in Y' and U_i' the inverse image of U_i in X, it is easy to check that $U_i' = V_i' \times_{V_i} U_i$. Since the morphism $U_i \to V_i$ is quasi-compact, it follows that $U_i' \to V_i'$ is also quasi-compact. Now take a covering $\{V_j''\}$ by open affine subschemes, such that each V_j'' is contained in some V_i'; then each V_j'' is the image of a quasi-compact open subset of X', its inverse image in some U_i.

Let us prove (vi). Let A be a subset of Y whose inverse image in X is open. Pick a covering $\{V_i\}$ of Y by open affine subsets, each of which is the image of a quasi-compact open subset U_i of X. Then the inverse image of A in each U_i will be open, and according to Proposition 1.13 this implies that each $A \cap V_i$ is open in V_i, so A is open in Y. □

The *fpqc topology* on the category (Sch/S) is the topology in which the coverings $\{U_i \to U\}$ are collections of morphisms, such that the induced morphism $\coprod U_i \to U$ is fpqc.

Let us verify that this is indeed a topology, by checking the three conditions of Definition 2.24. Condition (i) is obvious, because an isomorphism is fpqc.

Condition (ii) follows from Proposition 2.35 (v).

Condition (iii) is easy to prove, from parts (i) and (ii) of Proposition 2.35.

The fpqc topology is finer than the fppf topology, which is finer than the étale topology, which is in turn finer than the Zariski topology.

Many properties of morphisms are local on the codomain in the fpqc topology.

PROPOSITION 2.36. Let $X \to Y$ be a morphism of schemes, $\{Y_i \to Y\}$ an fpqc covering. Suppose that for each i the projection $Y_i \times_Y X \to Y_i$ has one of the following properties:

(i) is separated,
(ii) is quasi-compact,
(iii) is locally of finite presentation,
(iv) is proper,
(v) is affine,
(vi) is finite,
(vii) is flat,
(viii) is smooth,
(ix) is unramified,
(x) is étale,
(xi) is an embedding,
(xii) is a closed embedding.

Then $X \to Y$ has the same property.

PROOF. This follows easily from the fact that each of the properties above is local in the Zariski topology in the codomain, from the characterization (ii) in Proposition 2.33, and from Proposition 1.15. □

2.3.3. Sheaves. If X is a topological space, a presheaf of sets on X is a functor ${X_{\mathrm{cl}}}^{\mathrm{op}} \to (\mathrm{Set})$, where X_{cl} is the category of open subsets of X, as in Example 2.26. The condition that F be a sheaf can easily be generalized to any site, provided that we substitute intersections, which do not make sense, with fibered products. (Of course, fibered products in X_{cl} are just intersections.)

DEFINITION 2.37. Let $\mathcal{C}$ be a site, $F\colon \mathcal{C}^{\mathrm{op}} \to (\mathrm{Set})$ a functor.

(i) F is *separated* if, given a covering $\{U_i \to U\}$ and two sections a and b in FU whose pullbacks to each FU_i coincide, it follows that $a = b$.
(ii) F is a *sheaf* if the following condition is satisfied. Suppose that we are given a covering $\{U_i \to U\}$ in $\mathcal{C}$, and a set of elements $a_i \in FU_i$. Denote by $\mathrm{pr}_1\colon U_i \times_U U_j \to U_i$ and $\mathrm{pr}_2\colon U_i \times_U U_j \to U_j$ the first and second projection respectively, and assume that $\mathrm{pr}_1^* a_i = \mathrm{pr}_2^* a_j \in F(U_i \times_U U_j)$ for all i and j. Then there is a unique section $a \in FU$ whose pullback to FU_i is a_i for all i.

If F and G are sheaves on a site $\mathcal{C}$, a *morphism of sheaves* $F \to G$ is simply a natural transformation of functors.

A sheaf on a site is clearly separated.

Of course one can also define sheaves of groups, rings, and so on, as usual: a functor from $\mathcal{C}^{\mathrm{op}}$ to the category of groups, or rings, is a sheaf if its composite with the forgetful functor to the category of sets is a sheaf.

The reader might find our definition of sheaf pedantic, and wonder why we did not simply say "assume that the pullbacks of a_i and a_j to $F(U_i \times_U U_j)$ coincide". The reason is the following: when $i = j$, in the classical case of a topological space we have $U_i \times_U U_i = U_i \cap U_i = U_i$, so the two possible pullbacks from $U_i \times_U U_i \to U_i$ coincide; but if the map $U_i \to U$ is not injective, then the two projections $U_i \times_U U_i \to U_i$ will be different. So, for example, in the classical case coverings with one subset are not interesting, and the sheaf condition is automatically verified for them, while in the general case this is very far from being true.

An alternative way to state the condition that F is a sheaf is the following.

Let A, B and C be sets, and suppose that we are given a diagram

$$A \xrightarrow{f} B \overset{g}{\underset{h}{\rightrightarrows}} C.$$

(that is, we are given a function $f\colon A \to B$ and two functions $f, g\colon B \to C$). We say that the diagram is an *equalizer* if f is injective, and maps A surjectively onto the subset $\{b \in B \mid g(b) = h(b)\} \subseteq B$.

Equivalently, the diagram is an equalizer if $g \circ f = h \circ f$, and every function $p\colon D \to B$ such that $g \circ p = h \circ p$ factors uniquely through A.

Now, take a functor $F\colon \mathcal{C}^{\mathrm{op}} \to (\mathrm{Set})$ and a covering $\{U_i \to U\}$ in $\mathcal{C}$. There is a diagram

$$FU \longrightarrow \prod_i FU_i \overset{\mathrm{pr}_1^*}{\underset{\mathrm{pr}_2^*}{\rightrightarrows}} \prod_{i,j} F(U_i \times_U U_j) \tag{2.3.1}$$

where the function $FU \to \prod_i FU_i$ is induced by the restrictions $FU \to FU_i$, while

$$\mathrm{pr}_1^*\colon \prod_i FU_i \longrightarrow \prod_{i,j} F(U_i \times_U U_j)$$

sends an element $(a_i) \in \prod_i FU_i$ to the element $\mathrm{pr}_1^*(a_i) \in \prod_{i,j} F(U_i \times_U U_j)$ whose component in $F(U_i \times_U U_j)$ is the pullback $\mathrm{pr}_1^* a_i$ of a_i along the first projection $U_i \times_U U_j \to U_i$. The function

$$\mathrm{pr}_2^*\colon \prod_i FU_i \longrightarrow \prod_{i,j} F(U_i \times_U U_j)$$

is defined similarly.

One immediately sees that F is a sheaf if and only if the diagram (2.3.1) is an equalizer for all coverings $\{U_i \to U\}$ in $\mathcal{C}$.

2.3.4. Sieves. Given an object U in a category $\mathcal{C}$ and a set of arrows $\mathcal{U} = \{U_i \to U\}$ in $\mathcal{C}$, we define a subfunctor $h_{\mathcal{U}} \subseteq h_U$, by taking $h_{\mathcal{U}}(T)$ to be the set of arrows $T \to U$ with the property that for some i there is a factorization $T \to U_i \to U$. In technical terms, $h_{\mathcal{U}}$ is the *sieve* associated with the covering $\mathcal{U}$. The term is suggestive: think of the U_i as holes on U. Then an arrow $T \to U$ is in $h_{\mathcal{U}}T$ when it goes through one of the holes. So a sieve is determined by what goes through it.

Definition 2.38. Let U be an object of a category $\mathcal{C}$. A *sieve* on U is a subfunctor of $h_U\colon \mathcal{C}^{\mathrm{op}} \to (\mathrm{Set})$.

Given a subfunctor $S \subseteq h_U$, we get a collection $\mathcal{S}$ of arrows $T \to U$ (consisting of union of the ST with T running through all objects of $\mathcal{C}$), with the property that every time an arrow $T \to U$ is in $\mathcal{S}$, every composite $T' \to T \to U$ is in $\mathcal{S}$. Conversely, from such a collection we get a subfunctor $S \subseteq h_U$, in which ST is the set of all arrows $T \to U$ that are in $\mathcal{S}$.

Now, let $\mathcal{U} = \{U_i \to U\}$ be a set of arrows, $F\colon \mathcal{C}^{\mathrm{op}} \to (\mathrm{Set})$ a functor. We define $F\mathcal{U}$ to be the set of elements of $\prod_i FU_i$ whose images in $\prod_{i,j} F(U_i \times_U U_j)$ are equal. Then the restrictions $FU \to FU_i$ induce a function $FU \to F\mathcal{U}$; by definition, a sheaf is a functor F such that $FU \to F\mathcal{U}$ is a bijection for all coverings $\mathcal{U} = \{U_i \to U\}$.

The set $F\mathcal{U}$ can be defined in terms of sieves.

PROPOSITION 2.39. There is a canonical bijection

$$R\colon \operatorname{Hom}(\mathrm{h}_{\mathcal{U}}, F) \simeq F\mathcal{U}$$

such that the diagram

$$\begin{array}{ccc} \operatorname{Hom}(\mathrm{h}_U, F) & \longrightarrow & FU \\ \downarrow & & \downarrow \\ \operatorname{Hom}(\mathrm{h}_{\mathcal{U}}, F) & \xrightarrow{R} & F\mathcal{U} \end{array}$$

in which the top row is the Yoneda isomorphism, the left hand column is the restriction function induced by the embedding of $\mathrm{h}_{\mathcal{U}}$ in h_U and the right hand column is induced by the restriction functions $FU \to FU_i$, commutes.

PROOF. Take a natural transformation $\phi\colon \mathrm{h}_{\mathcal{U}} \to F$. For each i, the arrow $U_i \to U$ is an object of $\mathrm{h}_{\mathcal{U}}U_i$; from this we get an element $R\phi \overset{\text{def}}{=} \big(\phi(U_i \to U)\big) \in \prod_i FU_i$. The pullbacks $\mathrm{pr}_1^* \phi(U_i \to U)$ and $\mathrm{pr}_2^* \phi(U_j \to U)$ to FU_{ij} both coincide with $\phi(U_{ij} \to U)$, hence $R\phi$ is an element of $F\mathcal{U}$. This defines a function $R\colon \operatorname{Hom}(\mathrm{h}_{\mathcal{U}}, F) \to F\mathcal{U}$; the commutativity of the diagram is immediately checked.

We need to show that R is a bijection. Take two natural transformations $\phi, \psi\colon \mathrm{h}_{\mathcal{U}} \to F$ such that $R\phi = R\psi$. Consider an element $T \to U$ of some $\mathrm{h}_{\mathcal{U}}T$; by definition, this factors as $T \xrightarrow{f} U_i \to U$ for some arrow $f\colon T \to U$. Then, by definition of a natural transformation we have

$$\phi(T \to U) = f^*\phi(U_i \to U) = f^*\psi(U_i \to U) = \psi(T \to U),$$

hence $\phi = \psi$. This proves the injectivity of R.

For surjectivity, take an element $(\xi_i) \in F\mathcal{U}$; we need to define a natural transformation $\mathrm{h}_{\mathcal{U}} \to F$. If $T \to U$ is an element of $\mathrm{h}_{\mathcal{U}}T$, choose a factorization $T \xrightarrow{f} U_i \to U$; this defines an element $f^*\xi_i$ of FU. This element is independent of the factorization: two factorizations $T \xrightarrow{f} U_i \to U$ and $T \xrightarrow{g} U_j \to U$ give an arrow $T \to U_{ij}$, whose composites with $\mathrm{pr}_1\colon U_{ij} \to U_i$ and $\mathrm{pr}_2\colon U_{ij} \to U_j$ are equal to f and g. Since $\mathrm{pr}_1^*\xi_i = \mathrm{pr}_2^*\xi_j$, we see that $f^*\xi_i = g^*\xi_j$.

This defines a function $\mathrm{h}_{\mathcal{U}}T \to FT$ for each T. We leave it to the reader to show that this defines a natural transformation $\phi\colon \mathrm{h}_{\mathcal{U}} \to F$, and that $R\phi = (\xi_i)$. □

As an immediate corollary, we get the following characterization of sheaves.

COROLLARY 2.40. A functor $F\colon \mathcal{C}^{\mathrm{op}} \to (\mathrm{Set})$ is a sheaf if and only if for any covering $\mathcal{U} = \{U_i \to U\}$ in $\mathcal{C}$, the induced function

$$FU \simeq \operatorname{Hom}(\mathrm{h}_U, F) \longrightarrow \operatorname{Hom}(\mathrm{h}_{\mathcal{U}}, F)$$

is bijective. Furthermore, F is separated if and only if this function is always injective.

This characterization can be sharpened.

DEFINITION 2.41. Let $\mathcal{T}$ be a Grothendieck topology on a category $\mathcal{C}$. A sieve $S \subseteq \mathrm{h}_U$ on an object U of $\mathcal{C}$ is said to *belong to* $\mathcal{T}$ if there exists a covering $\mathcal{U}$ of U such that $\mathrm{h}_{\mathcal{U}} \subseteq S$.

If $\mathcal{C}$ is a site, we will talk about the *sieves of* $\mathcal{C}$ to mean the sieves belonging to the topology of $\mathcal{C}$.

The importance of the following characterization will be apparent after the proof of Proposition 2.49.

PROPOSITION 2.42. A functor $F\colon \mathcal{C}^{\mathrm{op}} \to (\mathrm{Set})$ is a sheaf in a topology $\mathcal{T}$ if and only if for any sieve S belonging to $\mathcal{T}$ the induced function

$$FU \simeq \mathrm{Hom}(\mathrm{h}_U, F) \longrightarrow \mathrm{Hom}(S, F)$$

is bijective. Furthermore, F is separated if and only if this function is always injective.

PROOF. The fact that this condition implies that F is a sheaf is an immediate consequence of Corollary 2.42.

To show the converse, let F be a sheaf, take a sieve $S \subseteq \mathrm{h}_U$ belonging to $\mathcal{C}$, and choose a covering $\mathcal{U}$ of U with $\mathrm{h}_{\mathcal{U}} \subseteq S$. The composite

$$\mathrm{Hom}(\mathrm{h}_U, F) \longrightarrow \mathrm{Hom}(S, F) \longrightarrow \mathrm{Hom}(\mathrm{h}_{\mathcal{U}}, F)$$

is a bijection, again because of Corollary 2.42, so the thesis follows from the next Lemma.

LEMMA 2.43. If F is separated, the restriction function

$$\mathrm{Hom}(S, F) \longrightarrow \mathrm{Hom}(\mathrm{h}_{\mathcal{U}}, F)$$

is injective.

PROOF. Let us take two natural transformations $\phi, \psi\colon S \to F$ with the same image in $\mathrm{Hom}(\mathrm{h}_{\mathcal{U}}, F)$, an element $T \to U$ of ST, and let us show that $\phi(T \to U) = \psi(T \to U) \in FT$.

Set $\mathcal{U} = \{U_i \to U\}$, and consider the fibered products $T \times_U U_i$ with their projections $p_i\colon T \times_U U_i \to T$. Since $T \times_U U_i \to U$ is in $\mathrm{h}_{\mathcal{U}}(T \times_U U_i)$ we have

$$p_i^*\phi(T \to U) = \phi(T \times_U U_i) = \psi(T \times_U U_i) = p_i^*\psi(T \to U).$$

Since $\{p_i\colon T \times_U U_i \to T\}$ is a covering and F is a presheaf, we conclude that $\phi(T \to U) = \psi(T \to U) \in FT$, as desired. □

This concludes the proof of Proposition 2.42. □

We conclude with a remark. Suppose that $\mathcal{U} = \{U_i \to U\}$ and $\mathcal{V} = \{V_j \to U\}$ are coverings. Then $\mathcal{U} \times_U \mathcal{V} \overset{\mathrm{def}}{=} \{U_i \times_U V_j \to U\}$ is a covering. An arrow $T \to U$ factors through $U_i \times_U V_j$ if and only if it factors through U_i and through U_j. This simple observation is easily seen to imply the following fact.

PROPOSITION 2.44.

(1) If $\mathcal{U} = \{U_i \to U\}$ and $\mathcal{V} = \{V_j \to U\}$ are coverings, then

$$\mathrm{h}_{\mathcal{U} \times_U \mathcal{V}} = \mathrm{h}_{\mathcal{U}} \cap \mathrm{h}_{\mathcal{V}} \subseteq \mathrm{h}_U.$$

(2) If S_1 and S_2 are sieves on U belonging to $\mathcal{T}$, the intersection $S_1 \cap S_2 \subseteq \mathrm{h}_U$ also belongs to $\mathcal{T}$.

2.3.5. Equivalence of Grothendieck topologies. Sometimes two different topologies on the same category define the same sheaves.

DEFINITION 2.45. Let $\mathcal{C}$ be a category, $\{U_i \to U\}_{i \in I}$ a set of arrows. A *refinement* $\{V_a \to U\}_{a \in A}$ is a set of arrows such that for each index $a \in A$ there is some index $i \in I$ such that $V_a \to U$ factors through $U_i \to U$.

Notice that the choice of factorizations $V_a \to U_i \to U$ is *not* part of the data, we simply require their existence.

This relation between sets of arrows is most easily expressed in terms of sieves. The following fact is immediate.

PROPOSITION 2.46. Let there be given two sets of arrows $\mathcal{U} = \{U_i \to U\}$ and $\mathcal{V} = \{V_a \to U\}$. Then $\mathcal{V}$ is a refinement of $\mathcal{U}$ if and only if $\mathrm{h}_{\mathcal{V}} \subseteq \mathrm{h}_{\mathcal{U}}$.

A refinement of a refinement is obviously a refinement. Also, any covering is a refinement of itself: thus, the relation of being a refinement is a pre-order on the set of coverings of an object U.

DEFINITION 2.47. Let $\mathcal{C}$ be a category, $\mathcal{T}$ and $\mathcal{T}'$ two topologies on $\mathcal{C}$. We say that $\mathcal{T}$ is *subordinate* to $\mathcal{T}'$, and write $\mathcal{T} \prec \mathcal{T}'$, if every covering in $\mathcal{T}$ has a refinement that is a covering in $\mathcal{T}'$.

If $\mathcal{T} \prec \mathcal{T}'$ and $\mathcal{T}' \prec \mathcal{T}$, we say that $\mathcal{T}$ and $\mathcal{T}'$ are *equivalent*, and write $\mathcal{T} \equiv \mathcal{T}'$.

Being a refinement is a relation between sets of arrows into U that is transitive and reflexive. Therefore being subordinate is a transitive and reflexive relation between topologies on $\mathcal{C}$, and being equivalent is an equivalence relation.

This relation between topologies is naturally expressed in terms of sieves.

PROPOSITION 2.48. Let $\mathcal{T}$ and $\mathcal{T}'$ be topologies on a category $\mathcal{C}$. Then $\mathcal{T} \prec \mathcal{T}'$ if and only if every sieve belonging to $\mathcal{T}$ also belongs to $\mathcal{T}'$.

In particular, two topologies are equivalent if and only if they have the same sieves.

This is clear from Proposition 2.46.

PROPOSITION 2.49. Let $\mathcal{T}$ and $\mathcal{T}'$ be two Grothendieck topologies on the same category $\mathcal{C}$. If $\mathcal{T}$ is subordinate to $\mathcal{T}'$, then every sheaf in $\mathcal{T}'$ is also a sheaf in $\mathcal{T}$.

In particular, two equivalent topologies have the same sheaves.

The proof is immediate from Propositions 2.42 and 2.48.

In Grothendieck's language what we have defined would be called a pretopology, and two equivalent pretopologies define the same topology.

EXAMPLE 2.50. The global classical topology on (Top) (Example 2.27), and the global étale topology of Example 2.28, are equivalent.

EXAMPLE 2.51. If S is a base scheme, there is another topology that we can define over the category (Sch/S), the *smooth topology*, in which a covering $\{U_i \to U\}$ is a jointly surjective set of smooth morphisms locally of finite presentation.

By [**EGAIV4**, Corollaire 17.16.3], given a smooth covering $\{U_i \to U\}$ we can find an étale surjective morphism $V \to U$ that factors through the disjoint union $\coprod_i U_i \to U$; given such a factorization, if V_i is the inverse image of U_i in V. we have that $\{V_i \to U\}$ is an étale covering that is a refinement of $\{U_i \to U\}$. This means that the smooth topology is subordinate to the étale topology. Since obviously every étale covering is a smooth cover, the two topologies are equivalent.

DEFINITION 2.52. A topology $\mathcal{T}$ on a category $\mathcal{C}$ is called *saturated* if, given a covering $\mathcal{U}$ in $\mathcal{T}$, every refinement of $\mathcal{U}$ is also in $\mathcal{T}$.

If $\mathcal{T}$ is a topology of $\mathcal{C}$, the *saturation* $\overline{\mathcal{T}}$ of $\mathcal{T}$ is the set of refinements of coverings in $\mathcal{T}$.

PROPOSITION 2.53. Let $\mathcal{T}$ be a topology on a category $\mathcal{C}$.

(i) The saturation $\overline{\mathcal{T}}$ of $\mathcal{T}$ is a saturated topology.
(ii) $\mathcal{T} \subseteq \overline{\mathcal{T}}$.
(iii) $\overline{\mathcal{T}}$ is equivalent to $\mathcal{T}$.
(iv) The topology $\mathcal{T}$ is saturated if and only if $\mathcal{T} = \overline{\mathcal{T}}$.
(v) A topology $\mathcal{T}'$ on $\mathcal{C}$ is subordinate to $\mathcal{T}$ if and only if $\mathcal{T}' \subseteq \overline{\mathcal{T}}$.
(vi) A topology $\mathcal{T}'$ on $\mathcal{C}$ is equivalent to $\mathcal{T}$ if and only if $\overline{\mathcal{T}'} = \overline{\mathcal{T}}$.
(vii) A topology on $\mathcal{C}$ is equivalent to a unique saturated topology.

We leave the easy proofs to the reader.

2.3.6. Sheaf conditions on representable functors.

PROPOSITION 2.54. A representable functor $(\mathrm{Top})^{\mathrm{op}} \to (\mathrm{Set})$ is a sheaf in the global classical topology.

This amounts to saying that, given two topological spaces U and X, an open covering $\{U_i \subseteq U\}$, and continuous functions $f_i \colon U_i \to X$, with the property that the restriction of f_i and f_j to $U_i \cap U_j$ coincide for all i and j, there exists a unique continuous function $U \to X$ whose restriction $U_i \to X$ is f_i. This is essentially obvious (it boils down to the fact that, for a function, the property of being continuous is local on the domain). For similar reasons, it is easy to show that a representable functor on the category (Sch/S) over a base scheme S is a sheaf in the Zariski topology.

On the other hand the following is not easy at all: a scheme is a topological space, together with a sheaf of rings in the Zariski topology. A priori, there does not seem to be a reason why we should be able to glue morphisms of schemes in a finer topology than the Zariski topology.

THEOREM 2.55 (Grothendieck). A representable functor on (Sch/S) is a sheaf in the fpqc topology.

So, in particular, it is also a sheaf in the étale and in the fppf topologies.

Here is another way of expressing this result. Recall that in a category $\mathcal{C}$ an arrow $f \colon V \to U$ is called an *epimorphism* if, whenever we have two arrows $U \rightrightarrows X$ with the property that the two composites $V \to U \rightrightarrows X$ coincide, then the two arrow are equal. In other words, we require that the function $\mathrm{Hom}_{\mathcal{C}}(U, X) \to \mathrm{Hom}_{\mathcal{C}}(V, X)$ be injective for any object X of $\mathcal{C}$.

On the other hand, $V \to U$ is called an *effective epimorphism* if for any object X of $\mathcal{C}$, any arrow $V \to X$ with the property that the two composites

$$V \times_U V \underset{\mathrm{pr}_2}{\overset{\mathrm{pr}_1}{\rightrightarrows}} V \longrightarrow X$$

coincide, factors uniquely through U. In other words, we require that the diagram

$$\mathrm{Hom}_{\mathcal{C}}(U, X) \longrightarrow \mathrm{Hom}_{\mathcal{C}}(V, X) \rightrightarrows \mathrm{Hom}_{\mathcal{C}}(V \times_U V, X)$$

be an equalizer.

Then Theorem 2.55 says that every fpqc morphism of schemes is an effective epimorphism in (Sch/S).

REMARK 2.56. As we have already observed at the beginning of §2.3.2, there is a "wild" flat topology in which the coverings are jointly surjective sets $\{U_i \to U\}$ of flat morphisms. However, this topology is very badly behaved; in particular, not all representable functors are sheaves.

Take an integral smooth curve U over an algebraically closed field, with quotient field K and let $V_p = \operatorname{Spec} \mathcal{O}_{U,p}$ for all closed points $p \in U(k)$, as in Remark 1.14. Then $\{V_p \to U\}$ is a covering in this wild flat topology.

Each V_p contains the closed point p, and $V_p \setminus \{p\} = \operatorname{Spec} K$ is the generic point of V_p; furthermore $V_p \times_U V_q = V_p$ if $p = q$, otherwise $V_p \times_U V_q = \operatorname{Spec} K$.

We can form a (very non-separated) scheme X by gluing together all the V_p along $\operatorname{Spec} K$; then the embeddings $V_p \hookrightarrow X$ and $V_q \hookrightarrow X$ agree when restricted to $V_p \times_U V_q$, so the give an element of $\prod_p \mathrm{h}_X V_p$ whose two images in $\prod_{p,q} \mathrm{h}_X(V_p \times_U V_q)$ agree. However, there is no morphism $U \to X$ whose restriction to each V_p is the natural morphism $V_p \to U$. In fact, such a morphism would have to send each closed point $p \in U$ into $p \in V_p \subseteq X$, and the generic point to the generic point; but the resulting set-theoretic function $U \to X$ is not continuous, since all subsets of X formed by closed points are closed, while only the finite sets are closed in U.

DEFINITION 2.57. A topology $\mathcal{T}$ on a category $\mathcal{C}$ is called *subcanonical* if every representable functor on $\mathcal{C}$ is a sheaf with respect to $\mathcal{T}$.

A *subcanonical site* is a category endowed with a subcanonical topology.

There are examples of sites that are not subcanonical (we have just seen one in Remark 2.56), but I have never had dealings with any of them.

The name "subcanonical" comes from the fact that on a category $\mathcal{C}$ there is a topology, known as the *canonical topology*, which is the finest topology in which every representable functor is a sheaf. We will not need this fact.

DEFINITION 2.58. Let $\mathcal{C}$ be a site, S an object of $\mathcal{C}$. We define the *comma topology* on the comma category $(\mathcal{C}/S)$ as the topology in which a covering of an object $U \to S$ of $(\mathcal{C}/S)$ is a collection of arrows

$$\begin{array}{ccc} U_i & \xrightarrow{f_i} & U \\ & \searrow \quad \swarrow & \\ & S & \end{array}$$

such that the collection $\{f_i \colon U_i \to U\}$ is a covering in $\mathcal{C}$. In other words, the coverings of $U \to S$ are simply the coverings of U.

It is very easy to check that the comma topology is in fact a topology.

For example, if $\mathcal{C}$ is the category of all schemes (or, equivalently, the category of schemes over $\mathbb{Z}$), then $(\mathcal{C}/S)$ is the category of schemes over S, and the comma topology induced by the fpqc topology on $(\mathcal{C}/S)$ is the fpqc topology. Analogous statements hold for the Zariski, étale and fppf topology.

PROPOSITION 2.59. If $\mathcal{C}$ is a subcanonical site and S is an object of $\mathcal{C}$, then $(\mathcal{C}/S)$ is also subcanonical.

PROOF. We need to show that for any covering $\{U_i \to U\}$ in $(\mathcal{C}/S)$ the sequence

$$\operatorname{Hom}_S(U,X) \longrightarrow \prod_i \operatorname{Hom}_S(U_i,X) \rightrightarrows \prod_{i,j} \operatorname{Hom}_S(U_i \times_U U_i, X)$$

is an equalizer. The injectivity of the function

$$\operatorname{Hom}_S(U,X) \longrightarrow \prod_i \operatorname{Hom}_S(U_i,X)$$

is clear, since $\operatorname{Hom}_S(U,X)$ injects into $\operatorname{Hom}_S(U,X)$, $\prod_i \operatorname{Hom}_S(U_i,X)$ injects into $\prod_i \operatorname{Hom}(U_i,X)$, and $\operatorname{Hom}_S(U,X)$ injects into $\prod_i \operatorname{Hom}(U_i,X)$, because h_X is a sheaf. On the other hand, let us suppose that we are given an element (a_i) of $\prod_i \operatorname{Hom}_S(U_i,X)$, with the property that for all pairs i, j of indices $\mathrm{pr}_1^* a_i = \mathrm{pr}_2^* a_j$ in $\operatorname{Hom}_S(U_i \times_U U_j, X)$. Then there exists a morphism $a \in \operatorname{Hom}(U,X)$ such that the composite $U_i \to U \xrightarrow{a} X$ coincides with a_i for all, and we only have to check that a is a morphism of S-schemes. But the composite $U_i \to U \xrightarrow{a} X \to S$ coincides with the structure morphism $U_i \to S$ for all i; since $\operatorname{Hom}(-,S)$ is a sheaf on the category of schemes. so that $\operatorname{Hom}(U,S)$ injects into $\prod_i \operatorname{Hom}(U_i,S)$, this implies that the composite $U \xrightarrow{a} X \to S$ is the structure morphism of U, and this completes the proof. □

PROOF OF THEOREM 2.55. We will use the following useful criterion.

LEMMA 2.60. Let S be a scheme, $F\colon (\mathrm{Sch}/S)^{\mathrm{op}} \to (\mathrm{Set})$ a functor. Suppose that F satisfies the following two conditions.

(i) F is a sheaf in the global Zariski topology.
(ii) Whenever $V \to U$ is a faithfully flat morphism of affine S-schemes, the diagram

$$FU \longrightarrow FV \underset{\mathrm{pr}_2^*}{\overset{\mathrm{pr}_1^*}{\rightrightarrows}} F(V \times_U V)$$

is an equalizer.

Then F is a sheaf in the fpqc topology.

PROOF. Take a covering $\{U_i \to U\}$ of schemes over S in the fpqc topology, and set $V = \coprod_i U_i$. The induced morphism $V \to U$ is fpqc. Since F is a Zariski sheaf, the function $FV \to \prod_i FU_i$ induced by restrictions is an isomorphism. We have a commutative diagram of sets

$$\begin{array}{ccccc}
FU & \longrightarrow & FV & \underset{\mathrm{pr}_2^*}{\overset{\mathrm{pr}_1^*}{\rightrightarrows}} & F(V \times_U V) \\
\| & & \downarrow & & \downarrow \\
FU & \longrightarrow & \prod_i FU_i & \underset{\mathrm{pr}_2^*}{\overset{\mathrm{pr}_1^*}{\rightrightarrows}} & \prod_{i,j} F(U_i \times_U U_j)
\end{array}$$

where the columns are bijections; hence to show that the bottom row is an equalizer it is enough to show that the top row is an equalizer. In other words, we have shown that it is enough to consider coverings $\{V \to U\}$ consisting of a single morphism. Similarly, to check that F is separated we may limit ourselves to considering coverings consisting of a single morphism.

This argument also shows that if $\{U_i \to U\}$ is a finite covering, such that U and the U_i are affine, then the diagram

$$FU \longrightarrow \prod_i FU_i \overset{\mathrm{pr}_1^*}{\underset{\mathrm{pr}_2^*}{\rightrightarrows}} \prod_{i,j} F(U_i \times_U U_j)$$

is an equalizer. In fact, in this case the finite disjoint union $\coprod_i U_i$ is also affine.

Now we are given an fpqc morphism $f\colon V \to U$; take an open covering $\{V_i\}$ of V by open quasi-compact subsets, whose image $U_i = fV_i$ is open and affine. Write each V_i as a union of finitely many open subschemes V_{ia}. Consider the diagram

$$\begin{array}{ccccc}
FU & \longrightarrow & FV & \rightrightarrows & F(V \times_U V) \\
\downarrow & & \downarrow & & \downarrow \\
\prod_i FU_i & \longrightarrow & \prod_i \prod_a FV_{ia} & \rightrightarrows & \prod_i \prod_{a,b} F(V_{ia} \times_U V_{ib}) \\
\downarrow\downarrow & & \downarrow\downarrow & & \\
\prod_{i,j} F(U_i \cap U_j) & \longrightarrow & \prod_{i,j} \prod_{a,b} F(V_{ia} \cap V_{jb}). & &
\end{array}$$

Its columns are equalizers, because F is a sheaf in the Zariski topology. On the other hand, the second row is an equalizer, because each diagram

$$FU_i \longrightarrow \prod_k FV_{ia} \rightrightarrows \prod_{a,b} F(V_{ia} \times_U V_{ib})$$

is an equalizer, and a product of equalizers is an equalizer. Hence the restriction function $FU \to FV$ is injective, so F is separated. But this implies that the bottom row is injective, and with an easy diagram chasing one shows that the top row is exact. $\square$

To prove Theorem 2.55 we need to check that if $F = \mathrm{h}_X$, where X is an S-scheme, then the second condition of Lemma 2.60 is satisfied. First of all, by Proposition 2.59 it is enough to prove the result in case $S = \operatorname{Spec} \mathbb{Z}$, that is, when (Sch/S) is simply the category of all schemes. So for the rest of the proof we only need to work with morphism of schemes, without worrying about base schemes.

We will assume at first that X is affine. Set $U = \operatorname{Spec} A$, $V = \operatorname{Spec} B$, $X = \operatorname{Spec} R$. In this case the result is an easy consequence of the following lemma. Consider the ring homomorphism $f\colon A \to B$ corresponding to the morphism $V \to U$, and the two homomorphisms of A-algebras $e_1, e_2\colon B \to B \otimes_A B$ defined by $e_1(b) = b \otimes 1$ and $e_2(b) = 1 \otimes b$; these correspond to the two projections $V \times_U V \to V$.

LEMMA 2.61. The sequence

$$0 \longrightarrow A \xrightarrow{f} B \xrightarrow{e_1 - e_2} B \otimes_A B$$

is exact.

PROOF. The injectivity of f is clear, because B is faithfully flat over A. Also, it is clear that the image of f is contained in the kernel of $e_1 - e_2$, so we have only to show that the kernel of $e_1 - e_2$ is contained in the image of f.

Assume that there exists a homomorphism of A-algebras $g\colon B \to A$ (in other words, assume that the morphism $V \to U$ has a section). Then the composite $g \circ f\colon A \to A$ is the identity. Take an element $b \in \ker(e_1 - e_2)$; by definition, this means that $b \otimes 1 = 1 \otimes b$ in $B \otimes_A B$. By applying the homomorphism $g \otimes \mathrm{id}_B\colon B \otimes_A$

$B \to A \otimes_A B = B$ to both members of the equality we obtain that $f(gb) = b$, hence $b \in \operatorname{im} f$.

In general, there will be no section $U \to V$; however, suppose that there exists a faithfully flat A algebra $A \to A'$, such that the homomorphism $f \otimes \mathrm{id}_{A'} \colon A' \to B \otimes A'$ obtained by base change has a section $B \otimes A' \to A'$ as before. Set $B' = B \otimes A'$. Then there is a natural isomorphism of A'-algebras $B' \otimes_{A'} B' \simeq (B \otimes_A B) \otimes_A A'$, making the diagram

$$\begin{array}{ccccccc} 0 & \longrightarrow & A' & \xrightarrow{f \otimes \mathrm{id}_{A'}} & B' & \xrightarrow{e_1' - e_2'} & B' \otimes_{A'} B' \\ & & \| & & \| & & \downarrow \\ 0 & \longrightarrow & A' & \xrightarrow{f \otimes \mathrm{id}_{A'}} & B' & \xrightarrow{(e_1 - e_2) \otimes \mathrm{id}_{A'}} & (B \otimes_A B) \otimes_A A' \end{array}$$

commutative. The top row is exact, because of the existence of a section, and so the bottom row is exact. The thesis follows, because A' is faithfully flat over A.

But to find such homomorphism $A \to A'$ it is enough to set $A' = B$; the product $B \otimes_A A' \to A'$ defined by $b \otimes b' \mapsto bb'$ gives the desired section. In geometric terms, the diagonal $V \to V \times_U V$ gives a section of the first projection $V \times_U V \to V$. $\square$

To finish the proof of Theorem 2.55 in the case that X is affine, recall that morphisms of schemes $U \to X$, $V \to X$ and $V \times_U V \to X$ correspond to ring homomorphisms $R \to A$, $R \to B$ and $R \to B \otimes_A B$; then the result is immediate from the lemma above. This proves that h_X is a sheaf when X is affine.

If X is not necessarily affine, write $X = \cup_i X_i$ as a union of affine open subschemes.

Let us show that h_X is separated. Given a covering $V \to U$, take two morphisms $f, g \colon U \to X$ such that the two composites $V \to U \to X$ are equal. Since $V \to U$ is surjective, f and g coincide set-theoretically, so we can set $U_i = f^{-1} X_i = g^{-1} X_i$, and call V_i the inverse images of U_i in V. The two composites

$$V_i \longrightarrow U_i \underset{g|_{U_i}}{\overset{f|_{U_i}}{\rightrightarrows}} X_i$$

coincide, and X_i is affine; hence $f|_{U_i} = g|_{U_i}$ for all i, so $f = g$, as desired.

To complete the proof, suppose that $g \colon V \to X$ is a morphism with the property that the two composites

$$V \times_U V \underset{\mathrm{pr}_2}{\overset{\mathrm{pr}_1}{\rightrightarrows}} V \xrightarrow{g} X$$

are equal; we need to show that g factors through U. The morphism $V \to U$ is surjective, so, from Lemma 2.62 below, g factors through U set-theoretically. Since U has the quotient topology induced by the morphism $V \to U$ (Proposition 1.13), we get that the resulting function $f \colon U \to X$ is continuous.

Set $U_i = f^{-1} X_i$ and $V_i = g^{-1} V_i$ for all i. The composites

$$V_i \times_U V_i \underset{\mathrm{pr}_2}{\overset{\mathrm{pr}_1}{\rightrightarrows}} V_i \xrightarrow{g|_{V_i}} V_i \longrightarrow X_i$$

coincide, and X_i is affine, so $g\,|_{V_i}\colon V_i \to X$ factors uniquely through a morphism $f_i\colon U_i \to X_i$. We have

$$f_i\,|_{U_i \cap U_j} = f_j\,|_{U_i \cap U_j}\colon U_i \cap U_j \longrightarrow X,$$

because h_X is separated; hence the f_i glue together to give the desired factorization $V \to U \to X$. □

LEMMA 2.62. *Let $f_1\colon X_1 \to Y$ and $f_2\colon X_2 \to Y$ be morphisms of schemes. If x_1 and x_2 are points of X_1 and X_2 respectively, and $f_1(x_1) = f_2(x_2)$, then there exists a point z in the fibered product $X_1 \times_Y X_2$ such that $\mathrm{pr}_1(z) = x_1$ and $\mathrm{pr}_2(z) = x_2$.*

PROOF. Set $y = f(x_1) = f_2(x_2) \in Y$. Consider the extensions $k(y) \subseteq k(x_1)$ and $k(y) \subseteq k(x_2)$; the tensor product $k(x_1) \otimes_{k(y)} k(x_2)$ is not 0, because the tensor product of two vector spaces over a field is never 0, unless one of the vector spaces is 0. Hence $k(x_1) \otimes_{k(y)} k(x_2)$ has a maximal ideal; the quotient field K is an extension of $k(y)$ containing both $k(x_1)$ and $k(x_2)$. The two composites $\operatorname{Spec} K \to \operatorname{Spec} k(x_1) \to X_1 \xrightarrow{f_1} U$ and $\operatorname{Spec} K \to \operatorname{Spec} k(x_2) \to X_2 \xrightarrow{f_2} U$ coincide, so we get a morphism $\operatorname{Spec} K \to X_1 \times_Y X_2$. We take z to be the image of $\operatorname{Spec} K$ in $X_1 \times_Y X_2$. □

The proof of Theorem 2.55 is now complete.

2.3.7. The sheafification of a functor. The usual construction of the sheafification of a presheaf of sets on a topological space carries over to this more general context.

DEFINITION 2.63. Let $\mathcal{C}$ be a site, $F\colon \mathcal{C}^{\mathrm{op}} \to (\mathrm{Set})$ a functor. A *sheafification* of F is a sheaf $F^{\mathrm{a}}\colon \mathcal{C}^{\mathrm{op}} \to (\mathrm{Set})$, together with a natural transformation $F \to F^{\mathrm{a}}$, such that:

(i) given an object U of $\mathcal{C}$ and two objects ξ and η of FU whose images ξ^{a} and η^{a} in $F^{\mathrm{a}}U$ are isomorphic, there exists a covering $\{\sigma_i\colon U_i \to U\}$ such that $\sigma_i^*\xi = \sigma_i^*\eta$, and
(ii) for each object U of $\mathcal{C}$ and each $\overline{\xi} \in F^{\mathrm{a}}(U)$, there exists a covering $\{\sigma_i\colon U_i \to U\}$ and elements $\xi_i \in F(U_i)$ such that $\xi_i^{\mathrm{a}} = \sigma_i^*\xi$.

THEOREM 2.64. *Let $\mathcal{C}$ be a site, $F\colon \mathcal{C}^{\mathrm{op}} \to (\mathrm{Set})$ a functor.*

(i) If $F^{\mathrm{a}}\colon \mathcal{C}^{\mathrm{op}} \to (\mathrm{Set})$ is a sheafification of F, any morphism from F to a sheaf factors uniquely through F^{a}.
(ii) There exists a sheafification $F \to F^{\mathrm{a}}$, which is unique up to a canonical isomorphism.
(iii) The natural transformation $F \to F^{\mathrm{a}}$ is injective (that is, each function $FU \to F^{\mathrm{a}}U$ is injective) if and only if F is separated.

SKETCH OF PROOF. For part (i), we leave to the reader to check uniqueness of the factorization.

For existence, let $\phi\colon F \to G$ be a natural transformation from F to a sheaf $G\colon \mathcal{C}^{\mathrm{op}} \to (\mathrm{Set})$. Given an element $\overline{\xi}$ of $F^{\mathrm{a}}U$, we want to define the image of ξ in GU. There exists a covering $\{\sigma_i\colon U_i \to U\}$ and elements $\xi_i \in FU_i$, such that the image of ξ_i in $F^{\mathrm{a}}U_i$ is $\sigma_i^*\overline{\xi}$. Set $\eta_i = \phi\xi_i \in GU_i$. The pullbacks $\mathrm{pr}_1^*\eta_i$ and $\mathrm{pr}_2^*\eta_j$ in GU_{ij} both have as their image in $F^{\mathrm{a}}U_{ij}$ the pullback of $\overline{\xi}$; hence there is a covering $\{U_{ij\alpha} \to U_{ij}\}$ such that the pullbacks of $\mathrm{pr}_1^*\xi_i$ and $\mathrm{pr}_2^*\xi_j$ in $FU_{ij\alpha}$ coincide for each α. By applying ϕ, and keeping in mind that it is a natural transformation, and that G is a sheaf, we see that the pullbacks of η_i and η_j to GU_{ij} are the same, for any pair of indices i and j. Hence there an element η of GU whose pullback to each GU_i is η_i.

We leave to the reader to verify that this $\eta \in GU$ only depends on ξ, and that by sending each ξ to the corresponding η we define a natural transformation $F^{\mathrm{a}} \to G$, whose composition with the given morphism $F \to F^{\mathrm{a}}$ is ϕ.

Let us prove part (ii). For each object U of $\mathcal{C}$, we define an equivalence relation $\sim$ on FU as follows. Given two elements a and b of FU, we write $a \sim b$ if there is a covering $U_i \to U$ such that the pullbacks of a and b to each U_i coincide. We check easily that this is an equivalence relation, and we define $F^{\mathrm{s}}U = FU/\sim$. We also verify that if $V \to U$ is an arrow in $\mathcal{C}$, the pullback $FU \to FV$ is compatible with the equivalence relations, yielding a pullback $F^{\mathrm{s}}U \to F^{\mathrm{s}}V$. This defines the functor F^{s} with the surjective morphism $F \to F^{\mathrm{s}}$. It is straightforward to verify that F^{s} is separated, and that every natural transformation from F to a separated functor factors uniquely through F^{s}.

To construct F^{a}, we take for each object U of $\mathcal{C}$ the set of pairs $(\{U_i \to U\}, \{a_i\})$, where $\{U_i \to U\}$ is a covering, and $\{a_i\}$ is a set of elements with $a_i \in F^{\mathrm{s}}U_i$ such that the pullback of a_i and a_j to $F^{\mathrm{s}}(U_i \times_U U_j)$, along the first and second projection respectively, coincide. On this set we impose an equivalence relation, by declaring $(\{U_i \to U\}, \{a_i\})$ to be equivalent to $(\{V_j \to U\}, \{b_j\})$ when the restrictions of a_i and b_j to $F^{\mathrm{s}}(U_i \times_U V_j)$, along the first and second projection respectively, coincide. To verify the transitivity of this relation we need to use the fact that the functor F^{s} is separated.

For each U, we denote by $F^{\mathrm{a}}U$ the set of equivalence classes. If $V \to U$ is an arrow, we define a function $F^{\mathrm{a}}U \to F^{\mathrm{a}}V$ by associating with the class of a pair $(\{U_i \to U\}, \{a_i\})$ in $F^{\mathrm{a}}U$ the class of the pair $(\{U_i \times_U V\}, p_i^* a_i)$, where $p_i \colon U_i \times_U V \to U_i$ is the projection. Once we have checked that this is well defined, we obtain a functor $F^{\mathrm{a}} \colon \mathcal{C}^{\mathrm{op}} \to (\mathrm{Set})$. There is also a natural transformation $F^{\mathrm{s}} \to F^{\mathrm{a}}$, obtained by sending an element $a \in F^{\mathrm{s}}U$ into $(\{U = U\}, a)$. Then one verifies that F^{a} is a sheaf, and that the composite of the natural transformations $F \to F^{\mathrm{s}}$ and $F^{\mathrm{s}} \to F^{\mathrm{a}}$ has the desired universal property.

The uniqueness up to a canonical isomorphism follows directly from part (i). Part (iii) follows easily from the definition. □

A slicker, but equivalent, definition is as follows. Consider the set S_i of sieves belonging to $\mathcal{T}$ on an object U of $\mathcal{C}$. These form a ordered set: we set $i \leq j$ if $S_j \subseteq S_i$. According to Proposition 2.44, this is a direct system, that is, given two indices i and j there is some k such that $k \geq i$ and $k \geq j$. Then $F^{\mathrm{a}}U$ is in a canonical bijective correspondence with the direct limit $\varinjlim_i \mathrm{Hom}_{\mathcal{C}}(S_i, F^{\mathrm{s}})$.

CHAPTER 3

Fibered categories

3.1. Fibered categories

3.1.1. Definition and first properties. In this Section we will fix a category $\mathcal{C}$; the topology will play no role. We will study categories over $\mathcal{C}$, that is, categories $\mathcal{F}$ equipped with a functor $p_{\mathcal{F}} \colon \mathcal{F} \to \mathcal{C}$.

We will draw several commutative diagrams involving objects of $\mathcal{C}$ and $\mathcal{F}$; an arrow going from an object ξ of $\mathcal{F}$ to an object U of $\mathcal{C}$ will be of type "$\xi \mapsto U$", and will mean that $p_{\mathcal{F}}\xi = U$. Furthermore the commutativity of the diagram

$$\begin{array}{ccc} \xi & \xrightarrow{\phi} & \eta \\ \downarrow & & \downarrow \\ U & \xrightarrow{f} & V \end{array}$$

will mean that $p_{\mathcal{F}}\phi = f$.

DEFINITION 3.1. Let $\mathcal{F}$ be a category over $\mathcal{C}$. An arrow $\phi \colon \xi \to \eta$ of $\mathcal{F}$ is *cartesian* if for any arrow $\psi \colon \zeta \to \eta$ in $\mathcal{F}$ and any arrow $h \colon p_{\mathcal{F}}\zeta \to p_{\mathcal{F}}\xi$ in $\mathcal{C}$ with $p_{\mathcal{F}}\phi \circ h = p_{\mathcal{F}}\psi$, there exists a unique arrow $\theta \colon \zeta \to \xi$ with $p_{\mathcal{F}}\theta = h$ and $\phi \circ \theta = \psi$, as in the commutative diagram

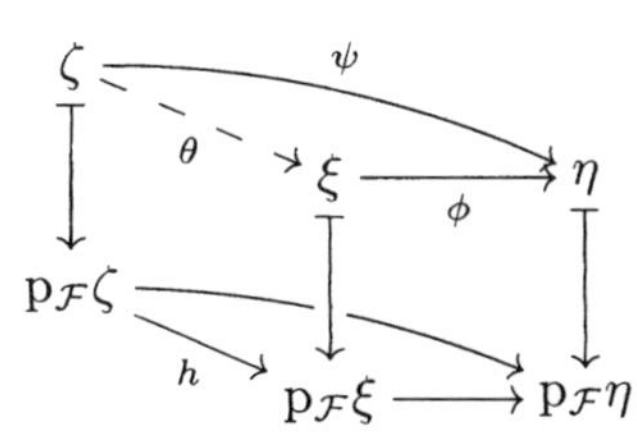

If $\xi \to \eta$ is a cartesian arrow of $\mathcal{F}$ mapping to an arrow $U \to V$ of $\mathcal{C}$, we also say that ξ is *a pullback of* η *to* U.

REMARK 3.2. The definition of cartesian arrow we give is more restrictive than the definition in [**SGA1**]; our cartesian arrows are called *strongly cartesian* in [**Gra66**]. However, the resulting notions of fibered category coincide.

REMARK 3.3. Given two pullbacks $\phi\colon \xi \to \eta$ and $\widetilde{\phi}\colon \widetilde{\xi} \to \eta$ of η to U, the unique arrow $\theta\colon \widetilde{\xi} \to \xi$ that fits into the diagram

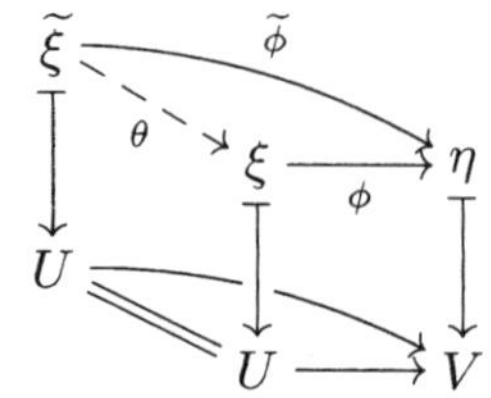

is an isomorphism; the inverse is the arrow $\widetilde{\xi} \to \xi$ obtained by exchanging ξ and $\widetilde{\xi}$ in the diagram above.

In other words, a pullback is unique, up to a unique isomorphism.

The following facts are easy to prove, and are left to the reader.

PROPOSITION 3.4.

(i) If $\mathcal{F}$ is a category over $\mathcal{C}$, the composite of cartesian arrows in $\mathcal{F}$ is cartesian.
(ii) An arrow in $\mathcal{F}$ whose image in $\mathcal{C}$ is an isomorphism is cartesian if and only if it is an isomorphism.
(iii) If $\xi \to \eta$ and $\eta \to \zeta$ are arrows in $\mathcal{F}$ and $\eta \to \zeta$ is cartesian, then $\xi \to \eta$ is cartesian if and only if the composite $\xi \to \zeta$ is cartesian.
(iv) Let $\mathrm{p}_{\mathcal{G}}\colon \mathcal{G} \to \mathcal{C}$ and $F\colon \mathcal{F} \to \mathcal{G}$ be functors, $\phi\colon \xi \to \eta$ an arrow in $\mathcal{F}$. If ϕ is cartesian over its image $F\phi\colon F\xi \to F\eta$ in $\mathcal{G}$ and $F\phi$ is cartesian over its image $\mathrm{p}_{\mathcal{G}}F\phi\colon \mathrm{p}_{\mathcal{G}}F\xi \to \mathrm{p}_{\mathcal{F}}F\eta$ in $\mathcal{C}$, then ϕ is cartesian over its image $\mathrm{p}_{\mathcal{G}}F\phi$ in $\mathcal{C}$.

DEFINITION 3.5. A *fibered category over* $\mathcal{C}$ is a category $\mathcal{F}$ over $\mathcal{C}$, such that given an arrow $f\colon U \to V$ in $\mathcal{C}$ and an object η of $\mathcal{F}$ mapping to V, there is a cartesian arrow $\phi\colon \xi \to \eta$ with $\mathrm{p}_{\mathcal{F}}\phi = f$.

In other words, in a fibered category $\mathcal{F} \to \mathcal{C}$ we can pull back objects of $\mathcal{F}$ along any arrow of $\mathcal{C}$.

DEFINITION 3.6. If $\mathcal{F}$ and $\mathcal{G}$ are fibered categories over $\mathcal{C}$, then a *morphism of fibered categories* $F\colon \mathcal{F} \to \mathcal{G}$ is a functor such that:

(i) F is *base-preserving*, that is, $\mathrm{p}_{\mathcal{G}} \circ F = \mathrm{p}_{\mathcal{F}}$;
(ii) F sends cartesian arrows to cartesian arrows.

Notice that in the definition above the equality $\mathrm{p}_{\mathcal{G}} \circ F = \mathrm{p}_{\mathcal{F}}$ must be interpreted as an actual equality. In other words, the existence of an isomorphism of functors between $\mathrm{p}_{\mathcal{G}} \circ F$ and $\mathrm{p}_{\mathcal{F}}$ is not enough.

PROPOSITION 3.7. Let there be given two functors $\mathcal{F} \to \mathcal{G}$ and $\mathcal{G} \to \mathcal{C}$. If $\mathcal{F}$ is fibered over $\mathcal{G}$ and $\mathcal{G}$ is fibered over $\mathcal{C}$, then $\mathcal{F}$ is fibered over $\mathcal{C}$.

PROOF. This follows from Proposition 3.4 (iv). □

3.1.2. Fibered categories as pseudo-functors.

DEFINITION 3.8. Let $\mathcal{F}$ be a fibered category over $\mathcal{C}$. Given an object U of $\mathcal{C}$, the *fiber* $\mathcal{F}(U)$ of $\mathcal{F}$ over U is the subcategory of $\mathcal{F}$ whose objects are the objects ξ of $\mathcal{F}$ with $\mathrm{p}_{\mathcal{F}}\xi = U$, and whose arrows are arrows ϕ in $\mathcal{F}$ with $\mathrm{p}_{\mathcal{F}}\phi = \mathrm{id}_U$.

By definition, if $F\colon \mathcal{F} \to \mathcal{G}$ is a morphism of fibered categories over $\mathcal{C}$ and U is an object of $\mathcal{C}$, the functor F sends $\mathcal{F}(U)$ to $\mathcal{G}(U)$, so we have a restriction functor $F_U\colon \mathcal{F}(U) \to \mathcal{G}(U)$.

Notice that formally we could give the same definition of a fiber for any functor $\mathrm{p}_{\mathcal{F}}\colon \mathcal{F} \to \mathcal{C}$, without assuming that $\mathcal{F}$ is fibered over $\mathcal{C}$. However, we would end up with a useless notion. For example, it may very well happen that we have two objects U and V of $\mathcal{C}$ which are isomorphic, but such that $\mathcal{F}(U)$ is empty while $\mathcal{F}(V)$ is not. This kind of pathology does not arise for fibered categories, and here is why.

Let $\mathcal{F}$ be a category fibered over $\mathcal{C}$, and $f\colon U \to V$ an arrow in $\mathcal{C}$. For each object η over V, we choose a pullback $\phi_\eta\colon f^*\eta \to \eta$ of η to U. We define a functor $f^*\colon \mathcal{F}(V) \to \mathcal{F}(U)$ by sending each object η of $\mathcal{F}(V)$ to $f^*\eta$, and each arrow $\beta\colon \eta \to \eta'$ of $\mathcal{F}(V)$ to the unique arrow $f^*\beta\colon f^*\eta \to f^*\eta'$ in $\mathcal{F}(U)$ making the diagram

$$\begin{array}{ccc} f^*\eta & \longrightarrow & \eta \\ {\scriptstyle f^*\beta}\downarrow & & \downarrow{\scriptstyle \beta} \\ f^*\eta' & \longrightarrow & \eta' \end{array}$$

commute.

DEFINITION 3.9. A *cleavage* of a fibered category $\mathcal{F} \to \mathcal{C}$ consists of a class K of cartesian arrows in $\mathcal{F}$ such that for each arrow $f\colon U \to V$ in $\mathcal{C}$ and each object η in $\mathcal{F}(V)$ there exists a unique arrow in K with target η mapping to f in $\mathcal{C}$.

By the axiom of choice, every fibered category has a cleavage. Given a fibered category $\mathcal{F} \to \mathcal{C}$ with a cleavage, we associate with each object U of $\mathcal{C}$ a category $\mathcal{F}(U)$, and to each arrow $f\colon U \to V$ a functor $f^*\colon \mathcal{F}(V) \to \mathcal{F}(U)$, constructed as above. It is very tempting to believe that in this way we have defined a functor from $\mathcal{C}$ to the category of categories; however, this is not quite correct. First of all, pullbacks $\mathrm{id}_U^*\colon \mathcal{F}(U) \to \mathcal{F}(U)$ are not necessarily identities. Of course we could just choose all pullbacks along identities to be identities on the fiber categories: this would certainly work, but it is not very natural, as there are often natural defined pullbacks where this does not happen (in Example 3.15 and many others). What happens in general is that, when U is an object of $\mathcal{C}$ and ξ an object of $\mathcal{F}(U)$, we have the pullback $\epsilon_U(\xi)\colon \mathrm{id}_U^*\xi \to \xi$ is an isomorphism, because of Proposition 3.4 (ii), and this defines an isomorphism of functors $\epsilon_U\colon \mathrm{id}_U^* \simeq \mathrm{id}_{\mathcal{F}(U)}$.

A more serious problem is the following. Suppose that we have two arrows $f\colon U \to V$ and $g\colon V \to W$ in $\mathcal{C}$, and an object ζ of $\mathcal{F}$ over W. Then $f^*g^*\zeta$ is a pullback of ζ to U; however, pullbacks are not unique, so there is no reason why $f^*g^*\zeta$ should coincide with $(gf)^*\zeta$. However, there is a canonical isomorphism $\alpha_{f,g}(\zeta)\colon f^*g^*\zeta \simeq (gf)^*\zeta$ in $\mathcal{F}(U)$, because both are pullbacks, and this gives an isomorphism $\alpha_{f,g}\colon f^*g^* \simeq (gf)^*$ of functors $\mathcal{F}(W) \to \mathcal{F}(U)$.

So, after choosing a cleavage a fibered category almost gives a functor from $\mathcal{C}$ to the category of categories, but not quite. The point is that the category of categories is not just a category, but what is known as a 2-category; that is, its arrows are functors, but two functors between the same two categories in turn form a category, the arrows being natural transformations of functors. Thus there are 1-arrows (functors) between objects (categories), but there are also 2-arrows (natural transformations) between 1-arrows.

What we get instead of a functor is what is called a *pseudo-functor*, or, in a more modern terminology, a *lax 2-functor*.

DEFINITION 3.10. A *pseudo-functor* Φ on $\mathcal{C}$ consists of the following data.

(i) For each object U of $\mathcal{C}$ a category ΦU.
(ii) For each arrow $f\colon U \to V$ a functor $f^*\colon \Phi V \to \Phi U$.
(iii) For each object U of $\mathcal{C}$ an isomorphism $\epsilon_U\colon \mathrm{id}_U^* \simeq \mathrm{id}_{\Phi U}$ of functors $\Phi U \to \Phi U$.
(iv) For each pair of arrows $U \xrightarrow{f} V \xrightarrow{g} W$ an isomorphism

$$\alpha_{f,g}\colon f^*g^* \simeq (gf)^*\colon \Phi W \longrightarrow \Phi U$$

of functors $\Phi W \to \Phi U$.

These data are required to satisfy the following conditions.

(a) If $f\colon U \to V$ is an arrow in $\mathcal{C}$ and η is an object of ΦV, we have

$$\alpha_{\mathrm{id}_U,f}(\eta) = \epsilon_U(f^*\eta)\colon \mathrm{id}_U^* f^*\eta \longrightarrow f^*\eta$$

and

$$\alpha_{f,\mathrm{id}_V}(\eta) = f^*\epsilon_V(\eta)\colon f^*\mathrm{id}_V^*\eta \longrightarrow f^*\eta.$$

(b) Whenever we have arrows $U \xrightarrow{f} V \xrightarrow{g} W \xrightarrow{h} T$ and an object θ of $\mathcal{F}(T)$, the diagram

$$\begin{array}{ccc} f^*g^*h^*\theta & \xrightarrow{\alpha_{f,g}(h^*\theta)} & (gf)^*h^*\theta \\ \downarrow{\scriptstyle f^*\alpha_{g,h}(\theta)} & & \downarrow{\scriptstyle \alpha_{gh,f}(\theta)} \\ f^*(hg)^*\theta & \xrightarrow{\alpha_{f,hg}(\theta)} & (hgf)^*\theta \end{array}$$

commutes.

In this definition we only consider (contravariant) pseudo-functor into the category of categories. Of course, there is a much more general notion of pseudo-functor with values in a 2-category, which we will not use at all.

A functor Φ from S into the category of categories can be considered as a pseudo-functor, in which every ϵ_U is the identity on ΦU, and every $\alpha_{f,g}$ is the identity on $f^*g^* = (gf)^*$.

We have seen how to associate with a fibered category over $\mathcal{C}$, equipped with a cleavage, the data for a pseudo-functor; we still have to check that the two conditions of the definition are satisfied.

PROPOSITION 3.11. A fibered category over $\mathcal{C}$ with a cleavage defines a pseudo-functor on $\mathcal{C}$.

PROOF. We have to check that the two conditions are satisfied. Let us do this for condition (b) (the argument for condition (a) is very similar). The point is that $f^*g^*h^*\zeta$ and $(hgf)^*\zeta$ are both pullbacks of ζ, and so, by the definition of cartesian arrow, there is a unique arrow $f^*g^*h^*\zeta \to (hgf)^*\zeta$ lying over the identity on U, and making the diagram

$$\begin{array}{ccc} f^*g^*h^*\zeta & \longrightarrow & (hgf)^*\zeta \\ & \searrow \quad \swarrow & \\ & \zeta & \end{array}$$

commutative. But one sees immediately that both $\alpha_{gh,f}(\zeta)\circ\alpha_{f,g}(h^*\zeta)$ and $\alpha_{f,hg}(\zeta)\circ f^*\alpha_{g,h}(\zeta)$ satisfy this condition. □

A functor $\Phi\colon \mathcal{C}^{\mathrm{op}} \to (\mathrm{Cat})$ from $\mathcal{C}$ into the category of categories gives rise to a pseudo-functor on $\mathcal{C}$, simply by defining all ϵ_U and all $\alpha_{f,g}$ to be identities.

It is easy to see when a cleavage defines a functor from $\mathcal{C}$ into the category of categories.

DEFINITION 3.12. A cleavage on a fibered category is a *splitting* if it contains all the identities, and it is closed under composition.

A fibered category endowed with a splitting is called *split*.

PROPOSITION 3.13. The pseudo-functor associated with a cleavage is a functor if and only if the cleavage is a splitting.

The proof is immediate.

In general a fibered category does not admit a splitting.

EXAMPLE 3.14. Every group G can be considered as a category with one object, where the set of arrows is exactly G, and the composition is given by the operation in G. A group homomorphism $G \to H$ can be considered as a functor. An arrow in G (that is, an element of G) is always cartesian; hence G is fibered over H if and only if $G \to H$ is surjective.

Given a surjective homomorphism $G \to H$, a cleavage K is a subset of G that maps bijectively onto H; and a cleavage is a splitting if and only if K is a subgroup of G. So, a splitting is a splitting $H \to G$ of the homomorphism $G \to H$, in the usual sense of a group homomorphism such that the composite $H \to G \to H$ is the identity on H. But of course such a splitting does not always exist.

Despite this, every fibered category is equivalent to a split fibered category (Theorem 3.45).

3.1.3. The fibered category associated with a pseudo-functor. So, every fibered category with a cleavage defines a pseudo-functor. Conversely, from a pseudo-functor on $\mathcal{C}$ one gets a fibered category over $\mathcal{C}$ with a cleavage. First of all, let us analyze the case that the pseudo-functor is simply a functor $\Phi\colon \mathcal{C}^{\mathrm{op}} \to (\mathrm{Cat})$ into the category of categories, considered as a 1-category. This means that with each object U of $\mathcal{C}$ we associate a category ΦU, and with each arrow $f\colon U \to V$ a functor $\Phi f\colon \Phi V \to \Phi U$, in such a way that $\Phi \mathrm{id}_U\colon \Phi U \to \Phi U$ is the identity, and $\Phi(g \circ f) = \Phi f \circ \Phi g$ every time we have two composable arrows f and g in $\mathcal{C}$.

With this Φ, we can associate a fibered category $\mathcal{F} \to \mathcal{C}$, such that for any object U in $\mathcal{C}$ the fiber $\mathcal{F}(U)$ is canonically equivalent to the category ΦU. An object of $\mathcal{F}$ is a pair (ξ, U) where U is an object of $\mathcal{C}$ and ξ is an object of $\mathcal{F}(U)$. An arrow $(a, f)\colon (\xi, U) \to (\eta, V)$ in $\mathcal{F}$ consists of an arrow $f\colon U \to V$ in $\mathcal{C}$, together with an arrow $a\colon \xi \to \Phi f(\eta)$ in ΦU.

The composition is defined as follows: if

$$(a, f)\colon (\xi, U) \longrightarrow (\eta, V) \quad \text{and} \quad (b, g)\colon (\eta, V) \longrightarrow (\zeta, W)$$

are two arrows, then

$$(b, g) \circ (a, f) = (\Phi b \circ a, g \circ f)\colon (\xi, U) \longrightarrow (\zeta, W).$$

There is an obvious functor $\mathcal{F} \to \mathcal{C}$ that sends an object (ξ, U) into U and an arrow (a, f) into f; I claim that this functor makes $\mathcal{F}$ into a fibered category over

$\mathcal{C}$. In fact, given an arrow $f\colon U \to V$ in $\mathcal{C}$ and an object (η, V) in $\mathcal{F}(V)$, then $(\Phi f(\eta), U)$ is an object of $\mathcal{F}(U)$, and it is easy to check that the pair $(\mathrm{id}_{\Phi f(\eta)}, f)$ gives a cartesian arrow $(\Phi f(\eta), U) \to (\eta, V)$.

The fiber of $\mathcal{F}$ is canonically isomorphic to the category ΦU: the isomorphism $\mathcal{F}(U) \to \Phi U$ is obtained at the level of objects by sending (ξ, U) to ξ, and at the level of arrows by sending (a, id_U) to a. The collection of all the arrows of type $(\mathrm{id}_{\Phi f(\eta)}, f)$ gives a splitting.

The general case is similar, only much more confusing. Consider a pseudo-functor Φ on $\mathcal{C}$. As before, we define the objects of $\mathcal{F}$ to be pairs (ξ, U) where U is an object of $\mathcal{C}$ and ξ is an object of $\mathcal{F}(U)$. Again, an arrow $(a, f)\colon (\xi, U) \to (\eta, V)$ in $\mathcal{F}$ consists of an arrow $f\colon U \to V$ in $\mathcal{C}$, together with an arrow $a\colon \xi \to f^*(\eta)$ in ΦU.

Given two arrows $(a, f)\colon (\xi, U) \to (\eta, V)$ and $(b, g)\colon (\eta, V) \to (\zeta, W)$, we define the composite $(b, g) \circ (a, f)$ as the pair $(b \cdot a, gf)$, where $b \cdot a = \alpha_{f,g}(\zeta) \circ f^*b \circ a$ is the composite

$$\xi \xrightarrow{a} f^*\zeta \xrightarrow{f^*b} f^*g^*\zeta \xrightarrow{\alpha_{f,g}(\zeta)} (gf)^*\zeta$$

in ΦU.

Let us check that composition is associative. Given three arrows

$$(\xi, U) \xrightarrow{(a,f)} (\eta, V) \xrightarrow{(b,g)} (\zeta, W) \xrightarrow{(c,h)} (\theta, T)$$

we have to show that

$$(c, h) \circ \bigl((b, g) \circ (a, f)\bigr) \stackrel{\text{def}}{=} \bigl(c \cdot (b \cdot a), hgf\bigr)$$

equals

$$\bigl((c, h) \circ (b, g)\bigr) \circ (a, f) \stackrel{\text{def}}{=} \bigl((c \cdot b) \cdot a, hgf\bigr).$$

By the definition of the composition, we have

$$\begin{aligned} c \cdot (b \cdot a) &= \alpha_{gf,h}(\theta) \circ (gf)^*c \circ (b \cdot a) \\ &= \alpha_{gf,h}(\theta) \circ (gf)^*c \circ \alpha_{f,g}(\zeta) \circ f^*b \circ a \end{aligned}$$

while

$$\begin{aligned} (c \cdot b) \cdot a &= \alpha_{f,hg}(\theta) \circ f^*(c \cdot b) \circ a \\ &= \alpha_{f,hg}(\zeta) \circ f^*\alpha_{g,h}(\theta) \circ f^*g^*c \circ f^*b \circ a; \end{aligned}$$

hence it is enough to show that the diagram

$$\begin{array}{ccccc} f^*g^*\zeta & \xrightarrow{f^*g^*c} & f^*g^*h^*\theta & \xrightarrow{f^*\alpha_{g,h}(\theta)} & f^*(hg)^*\theta \\ \downarrow{\scriptstyle \alpha_{f,g}(\zeta)} & & \downarrow{\scriptstyle \alpha_{gf,h}(\theta)} & & \downarrow{\scriptstyle \alpha_{f,hg}(\theta)} \\ (gf)^*\zeta & \xrightarrow{(gf)^*c} & (gf)^*h^*\theta & \xrightarrow{\alpha_{gf,h}(\theta)} & (hgf)^*\theta \end{array}$$

commutes. But the commutativity of the first square follows from the fact that $\alpha_{f,g}$ is a natural transformation of functor, while that of the second is condition (b) in Definition 3.10.

Given an object (ξ, U) of $\mathcal{F}$, we have the isomorphism $\epsilon_U(\xi)\colon \mathrm{id}_U^*\xi \to \xi$; we define the identity $\mathrm{id}_{(\xi,U)}\colon (\xi, U) \to (\xi, U)$ as $\mathrm{id}_{(\xi,U)} = \bigl(\epsilon_U(\xi)^{-1}, \mathrm{id}_U\bigr)$. To check that this is neutral with respect to composition, take an arrow $(a, f)\colon (\xi, U) \to (\eta, V)$; we have

$$(a, f) \circ \bigl(\epsilon_U(\xi)^{-1}, \mathrm{id}_U\bigr) = \bigl(a \cdot \epsilon_U(\xi)^{-1}, f\bigr)$$

and

$$a \cdot \epsilon_U(\xi)^{-1} = \alpha_{\mathrm{id}_U, f}(f^*\eta) \circ \mathrm{id}_U^* a \circ \epsilon_U(\xi)^{-1}.$$

But condition (a) of Definition 3.10 says that $\alpha_{\mathrm{id}_U, f}(f^*\eta)$ equals $\epsilon_U(f^*\eta)$, while the diagram

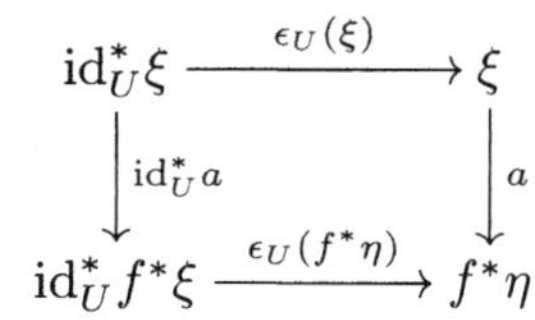

commutes, because ϵ_U is a natural transformation. This implies that $a \cdot \epsilon_U(\xi)^{-1} = a$, and therefore $(a, f) \circ \bigl(\epsilon_U(\xi)^{-1}, \mathrm{id}_U\bigr) = (a, f)$.

A similar argument shows that $\bigl(\epsilon_U(\xi)^{-1}, \mathrm{id}_U\bigr)$ is also a left identity.

Hence $\mathcal{F}$ is a category. There is an obvious functor $\mathrm{p}_{\mathcal{F}} \colon \mathcal{F} \to \mathcal{C}$ sending an object (ξ, U) to U and an arrow (a, f) to f. I claim that this makes $\mathcal{F}$ into a category fibered over $\mathcal{C}$.

Take an arrow $f \colon U \to V$ of $\mathcal{C}$, and an object (η, V) of $\mathcal{F}$ over V. I claim that the arrow

$$(\mathrm{id}_{f^*\eta}, f) \colon (f^*\eta, U) \longrightarrow (\eta, V)$$

is cartesian. To prove this, suppose that we are given a diagram

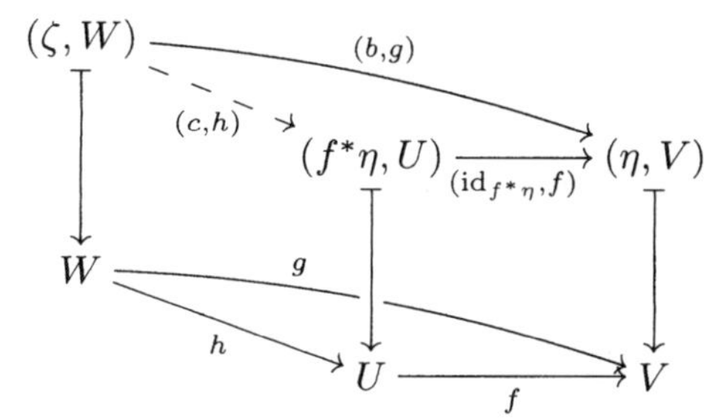

(without the dotted arrow); we need to show that there is a unique arrow (c, h) that can be inserted in the diagram. But it is easy to show that

$$(\mathrm{id}_{f^*\eta}, f) \circ (c, h) = (\alpha_{h,f}(\eta) \circ c, hg),$$

and this tells us that the one and only arrow that fits into the diagram is $(\alpha_{h,f}(\eta)^{-1} \circ b, h)$.

This shows that $\mathcal{F}$ is fibered over $\mathcal{C}$, and also gives us a cleavage.

Finally, let us notice that for all objects U of $\mathcal{C}$ there is functor $\mathcal{F}(U) \to \Phi U$, sending an object (ξ, U) to ξ and an arrow (a, f) into a. This is an isomorphism of categories.

The cleavage constructed above gives, for each arrow $f \colon U \to V$, functors $f^* \colon \mathcal{F}(V) \to \mathcal{F}(U)$. If we identify each $\mathcal{F}(U)$ with ΦU via the isomorphism above, then these functors correspond to the $f^* \colon \Phi V \to \Phi U$. Hence if we start with a pseudo-functor, we construct the associated fibered category with a cleavage, and then we take the associated pseudo-functor, this is isomorphic to the original pseudo-functor (in the obvious sense).

Conversely, it is easy to see that if we start from a fibered category with a cleavage, construct the associated pseudo-functor, and then take the associated fibered category with a cleavage, we get something isomorphic to the original fibered category with a cleavage (again in the obvious sense). So really giving a pseudo-functor is the same as giving a fibered category with a cleavage.

On the other hand, since cartesian pullbacks are unique up to a unique isomorphism (Remark 3.2), also cleavages are unique up to a unique isomorphism. This means that, in a sense that one could make precise, the theory of fibered categories is equivalent to the theory of pseudo-functors. On the other hand, as was already remarked in [**SGA1**, Remarque, pp. 193–194], often the choice of a cleavage hinders more than it helps.

3.2. Examples of fibered categories

EXAMPLE 3.15. Assume that $\mathcal{C}$ has fibered products. Let $\operatorname{Arr}\mathcal{C}$ be the category of arrows in $\mathcal{C}$; its objects are the arrows in $\mathcal{C}$, while an arrow from $f: X \to U$ to $g: Y \to V$ is a commutative diagram

$$\begin{array}{ccc} X & \longrightarrow & Y \\ \Big\downarrow{\scriptstyle f} & & \Big\downarrow{\scriptstyle g} \\ U & \longrightarrow & V \end{array}$$

The functor $\mathrm{p}_{\operatorname{Arr}\mathcal{C}}: \operatorname{Arr}\mathcal{C} \to \mathcal{C}$ sends each arrow $S \to U$ to its codomain U, and each commutative diagram to its bottom row.

I claim that $\operatorname{Arr}\mathcal{C}$ is a fibered category over $\mathcal{C}$. In fact, it easy to check that the cartesian diagrams are precisely the cartesian squares, so the statement follows from the fact that $\mathcal{C}$ has fibered products.

DEFINITION 3.16. A class $\mathcal{P}$ of arrows in a category $\mathcal{C}$ is *stable* if the following two conditions hold.

(a) If $f: X \to U$ is in $\mathcal{P}$, and $\phi: X' \simeq X$, $\psi: U \simeq U'$ are isomorphisms, the composite
$$\psi \circ f \circ \phi: X' \longrightarrow U'$$
is in $\mathcal{P}$.

(b) Given an arrow $Y \to V$ in $\mathcal{P}$ and any other arrow $U \to V$, then a fibered product $U \times_V Y$ exists, and the projection $U \times_V Y \to U$ is in $\mathcal{P}$.

EXAMPLE 3.17. As a variant of the example above, let $\mathcal{P}$ be a stable class of arrows. The arrows in $\mathcal{P}$ are the objects in a category, again denoted by $\mathcal{P}$, in which an arrow from $X \to U$ to $Y \to V$ is a commutative diagram

$$\begin{array}{ccc} X & \longrightarrow & Y \\ \Big\downarrow & & \Big\downarrow \\ U & \longrightarrow & V \end{array}$$

It is easy to see that this is a fibered category over $\mathcal{C}$; the cartesian arrows are precisely the cartesian diagrams.

EXAMPLE 3.18. Let G a topological group. The *classifying stack* of G is the fibered category $\mathcal{B}G \to (\mathrm{Top})$ over the category of topological spaces, whose objects are principal G bundles $P \to U$, and whose arrows (ϕ, f) from $P \to U$ to $Q \to V$ are commutative diagrams

$$\begin{array}{ccc} P & \xrightarrow{\phi} & Q \\ \downarrow & & \downarrow \\ U & \xrightarrow{f} & V \end{array}$$

where the function ϕ is G-equivariant. The functor $\mathcal{B}G \to (\mathrm{Top})$ sends a principal bundle $P \to U$ into the topological space U, and an arrow (ϕ, f) into f.

This fibered category $\mathcal{B}G \to (\mathrm{Top})$ has the property that each of its arrows is cartesian.

EXAMPLE 3.19. Here is an interesting example, suggested by one of the participants in the school. Consider the forgetful functor $F\colon (\mathrm{Top}) \to (\mathrm{Set})$ that associates with each topological space X it underlying set FX, and to each continuous function the function itself.

I claim that this makes (Top) fibered over (Set). Suppose that you have a topological space Y, a set U and a function $f\colon U \to FY$. Denote by X the set U with the *initial topology*, in which the open sets are the inverse images of the opens subsets of Y; this is the coarsest topology that makes f continuous. If T is a topological space, a function $T \to X$ is continuous if and only if the composite $T \to X \to Y$ is continuous; this means that $f\colon X \to Y$ is a cartesian arrow over the given arrow $f\colon U \to FY$.

The fiber of (Top) over a set U is the partially ordered set of topologies on U, made into a category in the usual way.

Notice that in this example the category (Top) has a canonical splitting over (Set).

We are interested in categories of sheaves. The simplest example is the fibered category of sheaves on objects of a site, defined as follows.

EXAMPLE 3.20. Let $\mathcal{C}$ be a site, $\mathcal{T}$ its topology. We will refer to a sheaf in the category $(\mathcal{C}/X)$, endowed with the comma topology (Definition 2.58) as a *sheaf on* X, and denote the category of sheaves on X by $\mathrm{Sh}\, X$.

If $f\colon X \to Y$ is an arrow in $\mathcal{C}$, there is a corresponding restriction functor $f^*\colon \mathrm{Sh}\, Y \to \mathrm{Sh}\, X$, defined as follows.

If G is a sheaf on Y and $U \to X$ is an object of $(\mathcal{C}/X)$, we define $f^*G(U \to Y) = G(U \to Y)$, where $U \to Y$ is the composite of $U \to X$ with f.

If $U \to X$ and $V \to X$ are objects of $(\mathcal{C}/X)$ and $\phi\colon U \to V$ is an arrow in $(\mathcal{C}/X)$, then ϕ is also an arrow from $U \to Y$ to $V \to Y$, hence it induces a function $\phi^*\colon F^*(V \to X) = F(U \to Y) \to F(V \to Y) = f^*F(V \to X)$. This gives f^*F the structure of a functor $(\mathcal{C}/X)^{\mathrm{op}} \to (\mathrm{Set})$. One sees easily that f^*F is a sheaf on $(\mathcal{C}/X)$.

If $\phi\colon F \to G$ is a natural transformation of sheaves on $(\mathcal{C}/Y)$, there is an induced natural transformation $f^*\phi\colon f^*F \to f^*G$ of sheaves on $(\mathcal{C}/X)$, defined in the obvious way. This defines a functor $f^*\colon \mathrm{Sh}\, Y \to \mathrm{Sh}\, X$.

It is immediate to check that, if $f\colon X \to Y$ and $g\colon Y \to Z$ are arrows in $\mathcal{C}$, we have an *equality* of functors $(gf)^* = f^*g^*\colon (\mathcal{C}/Z) \to (\mathcal{C}/X)$. Furthermore $\mathrm{id}_X^*\colon \mathrm{Sh}\, X \to \mathrm{Sh}\, X$ is the identity. This means that we have defined a functor from $\mathcal{C}$ to the category of categories, sending an object X into the category of categories.

According to the result of §3.1.3, this yields a category $(\mathrm{Sh}/\mathcal{C}) \to \mathcal{C}$, whose fiber over X is $\mathrm{Sh}\, X$.

There are many variants on this example, by considering sheaves in abelian groups, rings, and so on.

This example is particularly simple, because it is defined by a functor. In most of the cases that we are interested in, the sheaves on a given object will be defined in a site that is not the one inherited from the base category $\mathcal{C}$; this creates some difficulties, and forces one to use the unpleasant machinery of pseudo-functors. On the other hand, this discrepancy between the topology on the base and the topology in which the sheaves are defined is what makes descent theory for quasi-coherent sheaves so much more than an exercise in formalism.

Let us consider directly the example we are interested in, that is, fibered categories of quasi-coherent sheaves.

3.2.1. The fibered category of quasi-coherent sheaves. Here $\mathcal{C}$ will be the category (Sch/S) of schemes over a fixed base scheme S. For each scheme U we define $\mathrm{QCoh}(U)$ to be the category of quasi-coherent sheaves on U. Given a morphism $f\colon U \to V$, we have a functor $f^*\colon \mathrm{QCoh}(V) \to \mathrm{QCoh}(U)$. Unfortunately, given two morphisms $U \xrightarrow{f} V \xrightarrow{g} W$, the pullback $(gf)^*\colon \mathrm{QCoh}(W) \to \mathrm{QCoh}(U)$ does not coincide with the composite $f^*g^*\colon \mathrm{QCoh}(W) \to \mathrm{QCoh}(U)$, but it is only canonically isomorphic to it. This may induce one to suspect that we are in the presence of a pseudo-functor; and this is indeed the case.

The neatest way to prove this is probably by exploiting the fact that the push-forward $f_*\colon \mathrm{QCoh}(U) \to \mathrm{QCoh}(V)$ is functorial, that is, $(gf)_*$ equals g_*f_* on the nose, and f^* is a left adjoint to f_*. This means that, given quasi-coherent sheaves $\mathcal{N}$ on U and $\mathcal{M}$ on V, there is a canonical isomorphism of groups

$$\Theta_f(\mathcal{N},\mathcal{M})\colon \mathrm{Hom}_{\mathcal{O}_V}(\mathcal{N}, f_*\mathcal{M}) \simeq \mathrm{Hom}_{\mathcal{O}_U}(f^*\mathcal{N},\mathcal{M})$$

that is natural in $\mathcal{M}$ and $\mathcal{N}$. More explicitly, there are two functors

$$\mathrm{QCoh}(U)^{\mathrm{op}} \times \mathrm{QCoh}(V) \longrightarrow (\mathrm{Grp})$$

defined by

$$(\mathcal{M},\mathcal{N}) \mapsto \mathrm{Hom}_{\mathcal{O}_V}(G, f_*\mathcal{M})$$

and

$$(\mathcal{M},\mathcal{N}) \mapsto \mathrm{Hom}_{\mathcal{O}_U}(f^*\mathcal{N},\mathcal{M});$$

then Θ_f defines a natural isomorphism from the first to the second.

Equivalently, if $\alpha\colon \mathcal{M} \to \mathcal{M}'$ and $\beta\colon \mathcal{N} \to \mathcal{N}'$ are homomorphisms of quasi-coherent sheaves on U and V respectively, the diagrams

$$\begin{array}{ccc}
\mathrm{Hom}_{\mathcal{O}_V}(\mathcal{N}, f_*\mathcal{M}) & \xrightarrow{\Theta_f(\mathcal{N},\mathcal{M})} & \mathrm{Hom}_{\mathcal{O}_U}(f^*\mathcal{N},\mathcal{M}) \\
\big\downarrow{\scriptstyle f_*\alpha\circ -} & & \big\downarrow{\scriptstyle \alpha\circ -} \\
\mathrm{Hom}_{\mathcal{O}_V}(\mathcal{N}, f_*\mathcal{M}') & \xrightarrow{\Theta_f(\mathcal{N}',\mathcal{M})} & \mathrm{Hom}_{\mathcal{O}_U}(f^*\mathcal{N},\mathcal{M}')
\end{array}$$

and

$$\begin{array}{ccc}
\operatorname{Hom}_{\mathcal{O}_V}(\mathcal{N}', f_*\mathcal{M}) & \xrightarrow{\Theta_f(\mathcal{N},\mathcal{M}')} & \operatorname{Hom}_{\mathcal{O}_U}(f^*\mathcal{N}', \mathcal{M}) \\
\big\downarrow{\scriptstyle -\circ\beta} & & \big\downarrow{\scriptstyle -\circ f^*\beta} \\
\operatorname{Hom}_{\mathcal{O}_V}(\mathcal{N}, f_*\mathcal{M}) & \xrightarrow{\Theta_f(\mathcal{N},\mathcal{M})} & \operatorname{Hom}_{\mathcal{O}_U}(f^*\mathcal{N}, \mathcal{M})
\end{array}$$

commute.

If U is a scheme over S and $\mathcal{N}$ a quasi-coherent sheaf on U, then the pushforward functor $(\mathrm{id}_U)_*\colon \mathrm{QCoh}(U) \to \mathrm{QCoh}(U)$ is the identity (this has to be interpreted literally, I am not simply asserting the existence of a canonical isomorphism between $(\mathrm{id}_U)_*$ and the identity on $\mathrm{QCoh}(U)$). Now, if $\mathcal{M}$ is a quasi-coherent sheaf on U, there is a canonical adjunction isomorphism

$$\Theta_{\mathrm{id}_U}(\mathcal{M}, -)\colon \operatorname{Hom}_{\mathcal{O}_U}(\mathcal{M}, (\mathrm{id}_U)_*-) = \operatorname{Hom}_{\mathcal{O}_U}(\mathcal{M}, -) \simeq \operatorname{Hom}_{\mathcal{O}_U}(\mathrm{id}_U^*\mathcal{M}, -)$$

of functors from $\mathrm{QCoh}(U)$ to (Set). By the dual version of Yoneda's lemma (Remark 2.10) this corresponds to an isomorphism $\epsilon_U(\mathcal{M})\colon \mathrm{id}_U^*\mathcal{M} \simeq \mathcal{M}$. This is easily seen to be functorial, and therefore defines an isomorphism

$$\epsilon_U \colon \mathrm{id}_U^* \simeq \mathrm{id}_{\mathrm{QCoh}(U)}$$

of functors from $\mathrm{QCoh}(U)$ to itself. This isomorphism is the usual one: a section $s \in \mathcal{M}(A)$, for some open subset $A \subseteq U$, yields a section $\mathrm{id}_U^* s \in \mathcal{M}(\mathrm{id}_U^{-1}A) = \mathcal{M}(A)$, and $\epsilon_U(\mathcal{M})$ sends $\mathrm{id}_U^* s$ to s. This is the first piece of data that we need.

For the second, consider two morphisms $U \xrightarrow{f} V \xrightarrow{g} W$ and a quasi-coherent sheaf $\mathcal{P}$ on W. We have the chain of isomorphisms of functors $\mathrm{QCoh}(U) \to (\mathrm{Grp})$

$$\begin{aligned}
\operatorname{Hom}_{\mathcal{O}_U}\big((gf)^*\mathcal{P}, -\big) &\simeq \operatorname{Hom}_{\mathcal{O}_W}\big(\mathcal{P}, (gf)_*-\big) && (\text{this is } \Theta_{gf}(\mathcal{P}, -)^{-1}) \\
&= \operatorname{Hom}_{\mathcal{O}_W}\big(\mathcal{P}, g_*f_*-\big) \\
&\simeq \operatorname{Hom}_{\mathcal{O}_V}(g^*\mathcal{P}, f_*-) && (\text{this is } \Theta_g(\mathcal{P}, f_*-)) \\
&\simeq \operatorname{Hom}_{\mathcal{O}_U}(g^*f^*\mathcal{P}, -) && (\text{this is } \Theta_f(g^*\mathcal{P}, -));
\end{aligned}$$

the composite

$$\begin{aligned}
&\Theta_f(g^*\mathcal{P}, -) \circ \Theta_g(\mathcal{P}, f_*-) \circ \Theta_{gf}(\mathcal{P}, -)^{-1}\colon \\
&\qquad\qquad \operatorname{Hom}_{\mathcal{O}_U}\big((gf)^*\mathcal{P}, -\big) \simeq \operatorname{Hom}_{\mathcal{O}_U}(g^*f^*\mathcal{P}, -)
\end{aligned}$$

corresponds, again because of the covariant Yoneda lemma, to an isomorphism $\alpha_{f,g}(\mathcal{P})\colon f^*g^*\mathcal{P} \simeq (gf)^*\mathcal{P}$. These give an isomorphism $\alpha_{f,g}\colon f^*g^* \simeq (gf)^*$ of functors $\mathrm{QCoh}(W) \to \mathrm{QCoh}(U)$. Once again, this is the usual isomorphism: given a section $s \in \mathcal{P}(A)$ for some open subset $A \subseteq W$, there are two sections

$$(gf)^*s \in (gf)^*\mathcal{P}\big((gf)^{-1}A\big) = (gf)^*\mathcal{P}(f^{-1}g^{-1}A)$$

and

$$f^*g^*s \in f^*g^*\mathcal{P}(f^{-1}g^{-1}A);$$

the isomorphism $\alpha_{f,g}(\mathcal{P})$ sends f^*g^*s into $(gf)^*s$. Since the sections of type $(gf)^*s$ generate $(gf)^*\mathcal{P}$ as a sheaf of $\mathcal{O}_U$-modules, this characterizes $\alpha_{f,g}(\mathcal{P})$ uniquely.

We have to check that the ϵ_U and $\alpha_{f,g}$ satisfy the conditions of Definition 3.10. This can be done directly at the level of sections, or using the definition of the two isomorphisms via the covariant Yoneda lemma; we will follow the second route.

Take a morphism of schemes $f\colon U \to V$. We need to prove that for any quasi-coherent sheaf $\mathcal{N}$ on V we have the equality

$$\alpha_{\mathrm{id}_U, f}(\mathcal{N}) = \epsilon_U(f^*\mathcal{N})\colon \mathrm{id}_U^* f^*\mathcal{N} \longrightarrow f^*\mathcal{N}.$$

This is straightforward: by the covariant Yoneda lemma, it is enough to show that $\alpha_{id_U, f}(\mathcal{N})$ and $\epsilon_U(f^*\mathcal{N})$ induce the same natural transformation

$$\mathrm{Hom}_{\mathcal{O}_U}(f^*\mathcal{N}, -) \longrightarrow \mathrm{Hom}_{\mathcal{O}_U}(\mathrm{id}_U^* f^*\mathcal{N}, -).$$

But by definition the natural transformation induced by $\epsilon_U(f^*\mathcal{N})$ is

$$\Theta_{\mathrm{id}_U}(f^*\mathcal{N}, -),$$

while that induced by $\alpha_{\mathrm{id}_U, f}(\mathcal{N})$ is

$$\Theta_{\mathrm{id}_U}(f^*\mathcal{N}, -) \circ \Theta_f(\mathcal{N}, (\mathrm{id}_U)_* -) \circ \Theta_f(\mathcal{N}, -)^{-1} = \Theta_{\mathrm{id}_U}(f^*\mathcal{N}, -).$$

Similar arguments works for the second part of the first condition and for the second condition.

The fibered category on (Sch/S) associated with this pseudo-functor is the fibered category of quasi-coherent sheaves, and will be denoted by $(\mathrm{QCoh}/S) \to (\mathrm{Sch}/S)$.

There are many variants on this example. For example, one can define the fibered category of sheaves of $\mathcal{O}$-modules over the category of ringed topological spaces in exactly the same way.

3.3. Categories fibered in groupoids

DEFINITION 3.21. A *category fibered in groupoids over $\mathcal{C}$* is a category $\mathcal{F}$ fibered over $\mathcal{C}$, such that the category $\mathcal{F}(U)$ is a groupoid for any object U of $\mathcal{C}$.

In the literature one often finds a different definition of a category fibered in groupoids.

PROPOSITION 3.22. Let $\mathcal{F}$ be a category over $\mathcal{C}$. Then $\mathcal{F}$ is fibered in groupoids over $\mathcal{C}$ if and only if the following two conditions hold.

(i) Every arrow in $\mathcal{F}$ is cartesian.
(ii) Given an object η of $\mathcal{F}$ and an arrow $f\colon U \to \mathrm{p}_{\mathcal{F}}\eta$ of $\mathcal{C}$, there exists an arrow $\phi\colon \xi \to \eta$ of $\mathcal{F}$ with $\mathrm{p}_{\mathcal{F}}\phi = f$.

PROOF. Suppose that these two conditions hold: then clearly $\mathcal{F}$ is fibered over $\mathcal{C}$. Also, if $\phi\colon \xi \to \eta$ is an arrow of $\mathcal{F}(U)$ for some object U of $\mathcal{C}$, then we see from condition 3.22 (i) that there exists an arrow $\psi\colon \eta \to \xi$ with $\mathrm{p}_{\mathcal{F}}\psi = \mathrm{id}_U$ and $\phi\psi = \mathrm{id}_\eta$; that is, every arrow in $\mathcal{F}(U)$ has a right inverse. But this right inverse ψ also must also have a right inverse, and then the right inverse of ψ must be ϕ. This proves that every arrow in $\mathcal{F}(U)$ is invertible.

Conversely, assume that $\mathcal{F}$ is fibered over $\mathcal{C}$, and each $\mathcal{F}(U)$ is a groupoid. Condition (ii) is trivially verified. To check condition (i), let $\phi\colon \xi \to \eta$ be an arrow in $\mathcal{C}$ mapping to $f\colon U \to V$ in $\mathcal{C}$. Choose a pullback $\phi'\colon \xi' \to \eta$ of η to U; by definition there will be an arrow $\alpha\colon \xi \to \xi'$ in $\mathcal{F}(U)$ such that $\phi'\alpha = \phi$. Since $\mathcal{F}(U)$ is a a groupoid, α will be an isomorphism, and this implies that ϕ is cartesian. □

COROLLARY 3.23. Any base-preserving functor from a fibered category to a category fibered in groupoids is a morphism.

PROOF. This is clear, since every arrow in a category fibered in groupoids is cartesian. □

Of the examples of Section 3.1, 3.15 and 3.17 are not in general fibered in groupoids, while the classifying stack of a topological group introduced in 3.18 is always fibered in groupoids.

3.4. Functors and categories fibered in sets

The notion of category generalizes the notion of set: a set can be thought of as a category in which every arrow is an identity. Furthermore functors between sets are simply functions.

Similarly, fibered categories are generalizations of functors.

DEFINITION 3.24. A *category fibered in sets over* $\mathcal{C}$ is a category $\mathcal{F}$ fibered over $\mathcal{C}$, such that for any object U of $\mathcal{C}$ the category $\mathcal{F}(U)$ is a set.

Here is an useful characterization of categories fibered in sets.

PROPOSITION 3.25. Let $\mathcal{F}$ be a category over $\mathcal{C}$. Then $\mathcal{F}$ is fibered in sets if and only if for any object η of $\mathcal{F}$ and any arrow $f\colon U \to \mathrm{p}_{\mathcal{F}}\eta$ of $\mathcal{C}$, there is a unique arrow $\phi\colon \xi \to \eta$ of $\mathcal{F}$ with $\mathrm{p}_{\mathcal{F}}\phi = f$.

PROOF. Suppose that $\mathcal{F}$ is fibered in sets. Given η and $f\colon U \to \mathrm{p}_{\mathcal{F}}\eta$ as above, pick a cartesian arrow $\xi \to \eta$ over f. If $\xi' \to \eta$ is any other arrow over f, by definition there exists an arrow $\xi' \to \xi$ in $\mathcal{F}(U)$ making the diagram

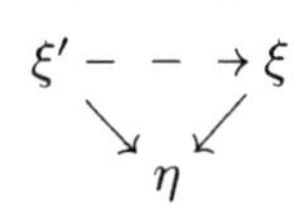

commutative. Since $\mathcal{F}(U)$ is a set, it follows that this arrow $\xi' \to \xi$ is the identity, so the two arrows $\xi \to \eta$ and $\xi' \to \eta$ coincide.

Conversely, assume that the condition holds. Given a diagram

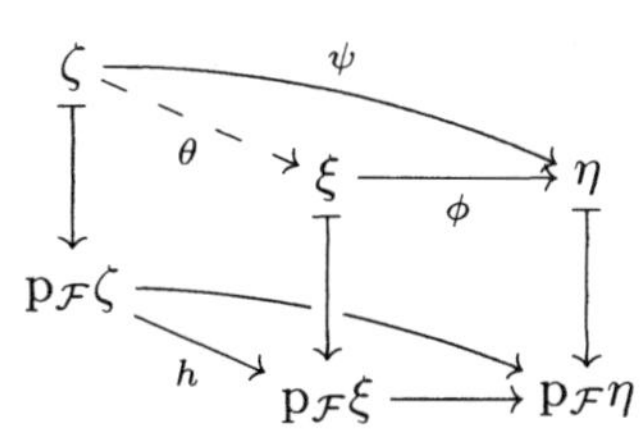

the condition implies that the only arrow $\theta\colon \zeta \to \xi$ over h makes the diagram commutative; so the category $\mathcal{F}$ is fibered.

It is obvious that the condition implies that $\mathcal{F}(U)$ is a set for all U. □

So, for categories fibered in sets the pullback of an object of $\mathcal{F}$ along an arrow of $\mathcal{C}$ is strictly unique. It follows from this that when $\mathcal{F}$ is fibered in sets over $\mathcal{C}$ and $f\colon U \to V$ is an arrow in $\mathcal{C}$, the pullback map $f^*\colon \mathcal{F}(V) \to \mathcal{F}(U)$ is uniquely defined, and the composition rule $f^*g^* = (gf)^*$ holds. Also for any object U of $\mathcal{C}$ we have that $\mathrm{id}_U^*\colon \mathcal{F}(U) \to \mathcal{F}(U)$ is the identity. This means that we have defined a functor $\Phi_{\mathcal{F}}\colon \mathcal{C}^{\mathrm{op}} \to (\mathrm{Set})$ by sending each object U of $\mathcal{C}$ to $\mathcal{F}(U)$, and each arrow $f\colon U \to V$ of $\mathcal{C}$ to the function $f^*\colon \mathcal{F}(V) \to \mathcal{F}(U)$.

Furthermore, if $F\colon \mathcal{F} \to \mathcal{G}$ is a morphism of categories fibered in sets, because of the condition that $\mathrm{p}_{\mathcal{G}} \circ F = \mathrm{p}_{\mathcal{F}}$, then every arrow in $\mathcal{F}(U)$, for some object U

of $\mathcal{C}$, will be send to $\mathcal{F}(U)$ itself. So we get a function $F_U : \mathcal{F}(U) \to \mathcal{G}(U)$. It is immediate to check that this gives a natural transformation $\phi_F : \Phi_{\mathcal{F}} \to \Phi_{\mathcal{G}}$.

There is a category of categories fibered in sets over $\mathcal{C}$, where the arrows are morphisms of fibered categories; the construction above gives a functor from this category to the category of functors $\mathcal{C}^{\mathrm{op}} \to (\mathrm{Set})$.

PROPOSITION 3.26. This is an equivalence of the category of categories fibered in sets over $\mathcal{C}$ and the category of functors $\mathcal{C}^{\mathrm{op}} \to (\mathrm{Set})$.

PROOF. The inverse functor is obtained by the construction of §3.1.3. If $\Phi\colon \mathcal{C}^{\mathrm{op}} \to (\mathrm{Set})$ is a functor, we construct a category fibered in sets $\mathcal{F}_\Phi$ as follows. The objects of $\mathcal{F}_\Phi$ will be pairs (U, ξ), where U is an object of $\mathcal{C}$, and $\xi \in \Phi U$. An arrow from (U, ξ) to (V, η) is an an arrow $f\colon U \to V$ of $\mathcal{C}$ with the property that $\Phi f \eta = \xi$. It follows from Proposition 3.25 that $\mathcal{F}_\Phi$ is fibered in sets over $\mathcal{C}$.

With each natural transformation of functors $\phi\colon \Phi \to \Phi'$ we associate a morphism $F_\phi\colon \mathcal{F}_\Phi \to \mathcal{F}_{\Phi'}$. An object (U, ξ) of $\mathcal{F}_\Phi$ will be sent to $(U, \phi_U \xi)$. If $f\colon (U, \xi) \to (V, \eta)$ is an arrow in $\mathcal{F}_\Phi$, then f is simply an arrow $f\colon U \to V$ in $\mathcal{C}$, with the property that $\Phi f(\eta) = \xi$. This implies that $\Phi'(f)\phi_V(\eta) = \phi_U \Phi(f)(\eta) = \phi_V \xi$, so the same f will yield an arrow $f\colon (U, \phi_U \xi) \to (V, \phi_V \eta)$.

We leave it the reader to check that this defines a functor from the category of functors to the category of categories fibered in sets. □

So, any functor $\mathcal{C}^{\mathrm{op}} \to (\mathrm{Set})$ will give an example of a fibered category over $\mathcal{C}$.

REMARK 3.27. It is interesting to notice that if $F\colon \mathcal{C}^{\mathrm{op}} \to (\mathrm{Set})$ is a functor and $\mathcal{F} \to \mathcal{C}$ the associated category fibered in sets, then an object (X, ξ) of $\mathcal{F}$ is universal pair for the functor F if and only if it is a terminal object for $\mathcal{F}$. Hence F is representable if and only if $\mathcal{F}$ has a terminal object.

In particular, given an object X of $\mathcal{C}$, we have the representable functor

$$\mathrm{h}_X : \mathcal{C}^{\mathrm{op}} \longrightarrow (\mathrm{Set}),$$

defined on objects by the rule $\mathrm{h}_X U = \mathrm{Hom}_{\mathcal{C}}(U, X)$. The category in sets over $\mathcal{C}$ associated with this functor is the comma category $(\mathcal{C}/X)$, and the functor $(\mathcal{C}/X) \to \mathcal{C}$ is the functor that forgets the arrow into X.

So the situation is the following. From Yoneda's lemma we see that the category $\mathcal{C}$ is embedded into the category of functors $\mathcal{C}^{\mathrm{op}} \to (\mathrm{Set})$, while the category of functors is embedded into the category of fibered categories.

From now we will identify a functor $F\colon \mathcal{C}^{\mathrm{op}} \to (\mathrm{Set})$ with the corresponding category fibered in sets over $\mathcal{C}$, and will (inconsistently) call a category fibered in sets simply "a functor".

3.4.1. Categories fibered over an object.

PROPOSITION 3.28. Let $\mathcal{G}$ be a category fibered in sets over $\mathcal{C}$, $\mathcal{F}$ another category, $F\colon \mathcal{F} \to \mathcal{G}$ a functor. Then $\mathcal{F}$ is fibered over $\mathcal{G}$ if and only if it is fibered over $\mathcal{C}$ via the composite $\mathrm{p}_{\mathcal{G}} \circ F\colon \mathcal{F} \to \mathcal{C}$.

Furthermore, $\mathcal{F}$ is fibered in groupoids over $\mathcal{G}$ if and only if it fibered in groupoids over $\mathcal{C}$, and is fibered in sets over $\mathcal{G}$ if and only if it fibered in sets over $\mathcal{C}$.

PROOF. One sees immediately that an arrow of $\mathcal{G}$ is cartesian over its image in $\mathcal{F}$ if and only if it is cartesian over its image in $\mathcal{C}$, and the first statement follows from this.

Furthermore, one sees that the fiber of $\mathcal{F}$ over an object U of $\mathcal{C}$ is the disjoint union, as a category, of the fibers of $\mathcal{F}$ over all the objects of $\mathcal{G}$ over U; these fiber are groupoids, or sets, if and only if their disjoint union is. □

This can be used as follows. Suppose that S is an object of $\mathcal{C}$, and consider the category fibered in sets $(\mathcal{C}/S) \to \mathcal{C}$, corresponding to the representable functor $\mathrm{h}_S \colon \mathcal{C}^{\mathrm{op}} \to (\mathrm{Set})$. By Proposition 3.28, a fibered category $\mathcal{F} \to (\mathcal{C}/S)$ is the same as a fibered category $\mathcal{F} \to \mathcal{C}$, together with a morphism $\mathcal{F} \to (\mathcal{C}/S)$ of categories fibered over $\mathcal{C}$.

In particular, categories fibered in sets correspond to functors; hence we get that giving a functor $(\mathcal{C}/S)^{\mathrm{op}} \to (\mathrm{Set})$ is equivalent to assigning a functor $\mathcal{C}^{\mathrm{op}} \to (\mathrm{Set})$ together with a natural transformation $F \to \mathrm{h}_S$. Describing this process for functors seems less natural than for fibered categories in general.

Given a functor $F \colon (\mathcal{C}/S)^{\mathrm{op}} \to (\mathrm{Set})$, this corresponds to a category fibered in sets $F \to (\mathcal{C}/S)$, which can be composed with the forgetful functor $(\mathcal{C}/S) \to \mathcal{C}$ to get a category fibered in sets $F \to \mathcal{C}$, which in turn corresponds to a functor $F' \colon \mathcal{C}^{\mathrm{op}} \to (\mathrm{Set})$. What is this functor? One minute's thought will convince you that it can be described as follows: $F'(U)$ is the disjoint union of the $F(U \to S)$ for all the arrows $U \to S$ in $\mathcal{C}$. The action of F' on arrows is the obvious one.

3.4.2. Fibered subcategories.

DEFINITION 3.29. Let $\mathcal{F} \to \mathcal{C}$ be a fibered category. A *fibered subcategory* $\mathcal{G}$ of $\mathcal{F}$ is a subcategory of $\mathcal{F}$, such that the composite $\mathcal{G} \hookrightarrow \mathcal{F} \to \mathcal{C}$ makes $\mathcal{G}$ into a fibered category over $\mathcal{C}$, and such that any cartesian arrow in $\mathcal{G}$ is also cartesian in $\mathcal{F}$.

The last condition is equivalent to requiring that the inclusion $\mathcal{G} \hookrightarrow \mathcal{F}$ is a morphism of fibered categories.

EXAMPLE 3.30. Let $\mathcal{F} \to \mathcal{C}$ be a fibered category, $\mathcal{G}$ a full subcategory of $\mathcal{F}$, with the property that if η is an object of $\mathcal{G}$ and $\xi \to \eta$ is a cartesian arrow in $\mathcal{F}$, then ξ is also is $\mathcal{G}$. Then $\mathcal{G}$ is a fibered subcategory of $\mathcal{F}$; the cartesian arrows in $\mathcal{G}$ are the cartesian arrows in $\mathcal{F}$ whose target is is in $\mathcal{G}$.

So, for example, the category of locally free sheaves is a fibered subcategory of the fibered category (QCoh/S) over (Sch/S).

Here is an interesting example.

DEFINITION 3.31. Let $\mathcal{F} \to \mathcal{C}$ be a fibered category. The *category fibered in groupoids associated with* $\mathcal{F}$ is the subcategory $\mathcal{F}_{\mathrm{cart}}$ of $\mathcal{F}$, whose objects are all the objects of $\mathcal{F}$, and whose arrows are the cartesian arrows of $\mathcal{F}$.

PROPOSITION 3.32. If $\mathcal{F} \to \mathcal{C}$ is a fibered category, then $\mathcal{F}_{\mathrm{cart}} \to \mathcal{C}$ is fibered in groupoids.

Furthermore, if $\mathcal{G} \to \mathcal{F}$ is a morphism of fibered categories and $\mathcal{G}$ is fibered in groupoids, then the image of F is in $\mathcal{F}_{\mathrm{cart}}$.

The proof is immediate from Proposition 3.22 and from Proposition 3.4 (iii).

3.5. Equivalences of fibered categories

3.5.1. Natural transformations of functors. The fact that fibered categories are categories, and not functors, has strong implications, and does cause difficulties. As usual, the main problem is that functors between categories can be isomorphic without being equal; in other words, functors between two fixed categories form a category, the arrows being given by natural transformations.

DEFINITION 3.33. Let $\mathcal{F}$ and $\mathcal{G}$ be two categories fibered over $\mathcal{C}$, $F, G: \mathcal{F} \to \mathcal{G}$ two morphisms. A *base-preserving natural transformation* $\alpha: F \to G$ is a natural transformation such that for any object ξ of $\mathcal{F}$, the arrow $\alpha_\xi: F\xi \to G\xi$ is in $\mathcal{G}(U)$, where $U \stackrel{\text{def}}{=} \mathrm{p}_{\mathcal{F}}\xi = \mathrm{p}_{\mathcal{G}}(F\xi) = \mathrm{p}_{\mathcal{G}}(G\xi)$.

An *isomorphism of* F *with* G is a base-preserving natural transformation $F \to G$ which is an isomorphism of functors.

It is immediate to check that the inverse of a base-preserving isomorphism is also base-preserving.

There is a category whose objects are the morphism from a $\mathcal{F}$ to $\mathcal{G}$, and the arrows are base-preserving natural transformations; we denote it by $\mathrm{Hom}_{\mathcal{C}}(\mathcal{F}, \mathcal{G})$.

3.5.2. Equivalences.

DEFINITION 3.34. Let $\mathcal{F}$ and $\mathcal{G}$ be two fibered categories over $\mathcal{C}$. An *equivalence*, of $\mathcal{F}$ with $\mathcal{G}$ is a morphism $F: \mathcal{F} \to \mathcal{G}$, such that there exists another morphism $G: \mathcal{G} \to \mathcal{F}$, together with isomorphisms of $G \circ F$ with $\mathrm{id}_{\mathcal{F}}$ and of $F \circ G$ with $\mathrm{id}_{\mathcal{G}}$.

We call G simply an *inverse* to F.

PROPOSITION 3.35. Suppose that $\mathcal{F}$, $\mathcal{F}'$, $\mathcal{G}$ and $\mathcal{G}'$ are categories fibered over $\mathcal{C}$. Suppose that $F: \mathcal{F}' \to \mathcal{F}$ and $G: \mathcal{G} \to \mathcal{G}'$ are equivalences. Then there an equivalence of categories

$$\mathrm{Hom}_{\mathcal{C}}(\mathcal{F}, \mathcal{G}) \longrightarrow \mathrm{Hom}_{\mathcal{C}}(\mathcal{F}', \mathcal{G}')$$

that sends each $\Phi: \mathcal{F} \to \mathcal{G}$ into the composite

$$G \circ \Phi \circ F: \mathcal{F}' \longrightarrow \mathcal{G}'.$$

The proof is left as an exercise to the reader.

The following is the basic criterion for checking whether a morphism of fibered categories is an equivalence.

PROPOSITION 3.36. Let $F: \mathcal{F} \to \mathcal{G}$ be a morphism of fibered categories. Then F is an equivalence if and only if the restriction $F_U: \mathcal{F}(U) \to \mathcal{G}(U)$ is an equivalence of categories for any object U of $\mathcal{C}$.

PROOF. Suppose that $G: \mathcal{G} \to \mathcal{F}$ is an inverse to F; the two isomorphisms $F \circ G \simeq \mathrm{id}_{\mathcal{G}}$ and $G \circ F \simeq \mathrm{id}_{\mathcal{F}}$ restrict to isomorphisms $F_U \circ G_U \simeq \mathrm{id}_{\mathcal{G}(U)}$ and $G_U \circ F_U \simeq \mathrm{id}_{\mathcal{F}(U)}$, so G_U is an inverse to F_U.

Conversely, we assume that $F_U: \mathcal{F}(U) \to \mathcal{G}(U)$ is an equivalence of categories for any object U of $\mathcal{C}$, and construct an inverse $G: \mathcal{G} \to \mathcal{F}$. Here is the main fact that we are going to need.

LEMMA 3.37. Let $F: \mathcal{F} \to \mathcal{G}$ a morphism of fibered categories such that every restriction $F_U: \mathcal{F}(U) \to \mathcal{G}(U)$ is fully faithful. Then the functor F is fully faithful.

PROOF. We need to show that, given two objects ξ' and η' of $\mathcal{F}$ and an arrow $\phi\colon F\xi' \to F\eta'$ in $\mathcal{G}$, there is a unique arrow $\phi'\colon \xi' \to \eta'$ in $\mathcal{F}$ with $F\phi' = \phi$. Set $\xi = F\xi'$ and $\eta = F\eta'$. Let $\eta_1' \to \eta'$ be a pullback of η' to U, $\eta_1 = F\eta_1'$. Then the image $\eta_1 \to \eta$ of $\eta_1' \to \eta'$ is cartesian, so every morphism $\xi \to \eta$ factors uniquely as $\xi \to \eta_1 \to \eta$, where the arrow $\xi \to \xi_1$ is in $\mathcal{G}(U)$. Analogously all arrows $\xi' \to \eta'$ factor uniquely through through η_1; since every arrow $\xi \to \eta_1$ in $\mathcal{G}(U)$ lifts uniquely to an arrow $\xi' \to \eta_1'$ in $\mathcal{F}(U)$, we have proved the Lemma. □

For any object ξ of $\mathcal{G}$ pick an object $G\xi$ of $\mathcal{F}(U)$, where $U = p_{\mathcal{G}}\xi$, together with an isomorphism $\alpha_\xi\colon \xi \simeq F(G\xi)$ in $\mathcal{G}(U)$; these $G\xi$ and α_ξ exist because $F_U\colon \mathcal{F}(U) \to \mathcal{G}(U)$ is an equivalence of categories.

Now, if $\phi\colon \xi \to \eta$ is an arrow in $\mathcal{G}$, by the Lemma there is a unique arrow $G\phi\colon G\xi \to G\eta$ such that $F(G\phi) = \alpha_\eta \circ \phi \circ \alpha_\xi^{-1}$, that is, such that the diagram

$$\begin{array}{ccc} \xi & \xrightarrow{\phi} & \eta \\ \downarrow{\scriptstyle \alpha_\xi} & & \downarrow{\scriptstyle \alpha_\eta} \\ F(G\xi) & \xrightarrow{F(G\phi)} & F(G\eta) \end{array}$$

commutes.

These operations define a functor $G\colon \mathcal{G} \to \mathcal{F}$. It is immediate to check that by sending each object ξ to the isomorphism $\alpha_\xi\colon \xi \simeq F(G\xi)$ we define an isomorphism of functors $\mathrm{id}_{\mathcal{F}} \simeq F \circ G\colon \mathcal{G} \to \mathcal{G}$.

We only have left to check that $G \circ F\colon \mathcal{F} \to \mathcal{F}$ is isomorphic to the identity $\mathrm{id}_{\mathcal{F}}$.

Fix an object ξ' of $\mathcal{F}$ over an object U of $\mathcal{C}$; we have a canonical isomorphism $\alpha_{F\xi'}\colon F\xi' \simeq F(G(F\xi'))$ in $\mathcal{G}(U)$. Since F_U is fully faithful there is a unique isomorphism $\beta_{\xi'}\colon \xi' \simeq G(F\xi')$ in $\mathcal{F}(U)$ such that $F\beta_{\xi'} = \alpha_{F\xi'}$; one checks easily that this defines an isomorphism of functors $\beta\colon G \circ F \simeq \mathrm{id}_{\mathcal{G}}$. □

3.5.3. Categories fibered in equivalence relations. As we remarked in §3.4, the notion of category generalizes the notion of set.

It is also possible to characterize the categories that are equivalent to a set: these are the equivalence relations.

Suppose that $R \subseteq X \times X$ is an equivalence relation on a set X. We can produce a category (X, R) in which X is the set of objects, R is the set of arrows, and the source and target maps $R \to X$ are given by the first and second projection. Then given x and y in X, there is precisely one arrow (x, y) if x and y are in the same equivalence class, while there is none if they are not. Then transitivity assures us that we can compose arrows, while reflexivity tell us that over each object $x \in X$ there is a unique arrow (x, x), which is the identity. Finally symmetry tells us that any arrow (x, y) has an inverse (y, x). So, (X, R) is groupoid such that from a given object to another there is at most one arrow.

Conversely, given a groupoid such that from a given object to another there is at most one arrow, if denote by X the set of objects and by R the set of arrows, the source and target maps induce an injective map $R \to X \times X$, which gives an equivalence relation on X.

So an equivalence relation can be thought of as a groupoid such that from a given object to another there is at most one arrow. Equivalently, an equivalence relation is a groupoid in which the only arrow from an object to itself is the identity.

PROPOSITION 3.38. A category is equivalent to a set if and only if it is an equivalence relation.

PROOF. If a category is equivalent to a set, it is immediate to see that it is an equivalence relation. If (X, R) is an equivalence relation and X/R is the set of isomorphism classes of objects, that is, the set of equivalence classes, one checks easily that the function $X \to X/R$ gives a functor that is fully faithful and essentially surjective, so it is an equivalence. □

There is an analogous result for fibered categories.

DEFINITION 3.39. A category $\mathcal{F}$ over $\mathcal{C}$ is a *quasi-functor*, or that it is *fibered in equivalence relations*, if it is fibered, and each fiber $\mathcal{F}(U)$ is an equivalence relation.

We have the following characterization of quasi-functors.

PROPOSITION 3.40. A category $\mathcal{F}$ over $\mathcal{C}$ is a quasi-functor if and only if the following two conditions hold.

(i) Given an object η of $\mathcal{F}$ and an arrow $f\colon U \to \mathrm{p}_{\mathcal{F}}\eta$ of $\mathcal{C}$, there exists an arrow $\phi\colon \xi \to \eta$ of $\mathcal{F}$ with $\mathrm{p}_{\mathcal{F}}\phi = f$.
(ii) Given two objects ξ and η of $\mathcal{F}$ and an arrow $f\colon \mathrm{p}_{\mathcal{F}}\xi \to \mathrm{p}_{\mathcal{F}}\eta$ of $\mathcal{C}$, there exists at most one arrow $\xi \to \eta$ over f.

The easy proof is left to the reader.

PROPOSITION 3.41. A fibered category over $\mathcal{C}$ is a quasi-functor if and only if it is equivalent to a functor.

PROOF. This is an application of Proposition 3.36.

Suppose that a fibered category $\mathcal{F}$ is equivalent to a functor Φ; then every category $\mathcal{F}(U)$ is equivalent to the set ΦU, so $\mathcal{F}$ is fibered in equivalence relations over $\mathcal{C}$ by Proposition 3.38.

Conversely, assume that $\mathcal{F}$ is fibered in equivalence relations. In particular it is fibered in groupoids, so every arrow in $\mathcal{F}$ is cartesian, by Proposition 3.22. For each object U of $\mathcal{C}$, denote by ΦU the set of isomorphism classes of elements in $\mathcal{F}(U)$. Given an arrow $f\colon U \to V$ in $\mathcal{C}$, two isomorphic objects η and η' of $\mathcal{F}(V)$, and two pullbacks ξ and ξ' of η and η' to $\mathcal{F}(U)$, we have that ξ and ξ' are isomorphic in $\mathcal{F}(U)$; this gives a well defined function $f^*\colon \Phi V \to \Phi U$ that sends an isomorphism class $[\eta]$ in $\mathcal{F}(V)$ into the isomorphism class of pullbacks of η. It is easy to see that this gives Φ the structure of a functor $\mathcal{C}^{\mathrm{op}} \to (\mathrm{Set})$. If we think of Φ as a category fibered in sets, we get by construction a morphism $\mathcal{F} \to \Phi$. Its restriction $\mathcal{F}(U) \to \Phi U$ is an equivalence for each object U of $\mathcal{C}$, so by Proposition 3.36 the morphism $\mathcal{F} \to \Phi$ is an equivalence. □

Here are a few useful facts.

PROPOSITION 3.42.

(i) If $\mathcal{G}$ is fibered in groupoids, then $\mathrm{Hom}_{\mathcal{C}}(\mathcal{F}, \mathcal{G})$ is a groupoid.
(ii) If $\mathcal{G}$ is a quasi-functor, then $\mathrm{Hom}_{\mathcal{C}}(\mathcal{F}, \mathcal{G})$ is an equivalence relation.
(iii) If $\mathcal{G}$ is a functor, then $\mathrm{Hom}_{\mathcal{C}}(\mathcal{F}, \mathcal{G})$ is a set.

We leave the easy proofs to the reader.

In 2-categorical terms, part (iii) says that the 2-category of categories fibered in sets is in fact just a 1-category, while part (ii) says that the 2-category of quasi-functors is equivalent to a 1-category.

3.6. Objects as fibered categories and the 2-Yoneda Lemma

3.6.1. Representable fibered categories. In §2.1 we have seen how we can embed a category $\mathcal{C}$ into the functor category $\mathrm{Hom}(\mathcal{C}^{\mathrm{op}}, (\mathrm{Set}))$, while in §3.4 we have seen how to embed the category $\mathrm{Hom}(\mathcal{C}^{\mathrm{op}}, (\mathrm{Set}))$ into the 2-category of fibered categories over $\mathcal{C}$. By composing these embeddings we have embedded $\mathcal{C}$ into the 2-category of fibered categories: an object X of $\mathcal{C}$ is sent to the fibered category $(\mathcal{C}/X) \to \mathcal{C}$. Furthermore, an arrow $f\colon X \to Y$ goes to the morphism of fibered categories $(\mathcal{C}/f)\colon (\mathcal{C}/X) \to (\mathcal{C}/Y)$ that sends an object $U \to X$ of $(\mathcal{C}/X)$ to the composite $U \to X \xrightarrow{f} Y$. The functor $(\mathcal{C}/f)$ sends an arrow

$$\begin{array}{ccc} U & \longrightarrow & V \\ & \searrow \quad \swarrow & \\ & X & \end{array}$$

of $(\mathcal{C}/X)$ to the commutative diagram obtained by composing both sides with $f\colon X \to Y$.

This is the 2-categorical version of the weak Yoneda lemma.

THE WEAK 2-YONEDA LEMMA. *The function that sends each arrow $f\colon X \to Y$ to the functor $(\mathcal{C}/f)\colon (\mathcal{C}/X) \to (\mathcal{C}/Y)$ is a bijection.*

DEFINITION 3.43. A fibered category over $\mathcal{C}$ is *representable* if it is equivalent to a category of the form $(\mathcal{C}/X)$.

So a representable category is necessarily a quasi-functor, by Proposition 3.41. However, we should be careful: if $\mathcal{F}$ and $\mathcal{G}$ are fibered categories, equivalent to $(\mathcal{C}/X)$ and $(\mathcal{C}/Y)$ for two objects X and Y of $\mathcal{C}$, then

$$\mathrm{Hom}(X, Y) = \mathrm{Hom}\big((\mathcal{C}/X), (\mathcal{C}/Y)\big),$$

and according to Proposition 3.35 we have an equivalence of categories

$$\mathrm{Hom}\big((\mathcal{C}/X), (\mathcal{C}/Y)\big) \simeq \mathrm{Hom}_{\mathcal{C}}(\mathcal{F}, \mathcal{G});$$

but $\mathrm{Hom}_{\mathcal{C}}(\mathcal{F}, \mathcal{G})$ need not be a set, it could very well be an equivalence relation.

3.6.2. The 2-categorical Yoneda lemma. As in the case of functors, we have a stronger version of the 2-categorical Yoneda lemma. Suppose that $\mathcal{F}$ is a category fibered over $\mathcal{C}$, and that X is an object of $\mathcal{C}$. Let there be given a morphism $F\colon (\mathcal{C}/X) \to \mathcal{F}$; with this we can associate an object $F(\mathrm{id}_X) \in \mathcal{F}(X)$. Also, to each base-preserving natural transformation $\alpha\colon F \to G$ of functors $F, G\colon (\mathcal{C}/X) \to \mathcal{F}$ we associate the arrow $\alpha_{\mathrm{id}_X}\colon F(\mathrm{id}_X) \to G(\mathrm{id}_X)$. This defines a functor

$$\mathrm{Hom}_{\mathcal{C}}\big((\mathcal{C}/X), \mathcal{F}\big) \longrightarrow \mathcal{F}(X).$$

Conversely, given an object $\xi \in \mathcal{F}(X)$ we get a functor $F_\xi\colon (\mathcal{C}/X) \to \mathcal{F}$ as follows. Given an object $\phi\colon U \to X$ of $(\mathcal{C}/X)$, we define $F_\xi(\phi) = \phi^*\xi \in \mathcal{F}(U)$; with an arrow

$$\begin{array}{ccc} U & \xrightarrow{\quad f \quad} & V \\ & {\scriptstyle\phi}\searrow \quad \swarrow{\scriptstyle\psi} & \\ & X & \end{array}$$

in $(\mathcal{C}/X)$ we associate the only arrow $\theta\colon \phi^*\xi \to \psi^*\xi$ in $\mathcal{F}(U)$ making the diagram

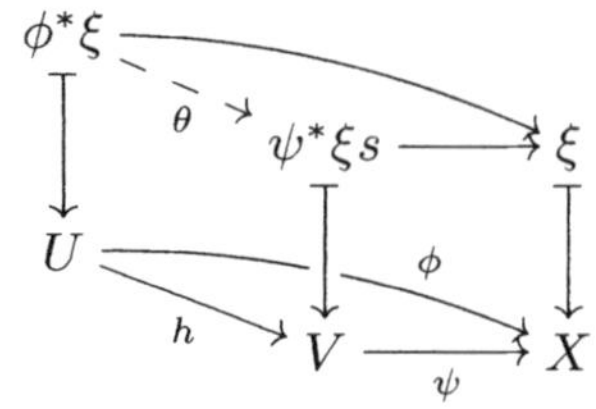

commutative. We leave it to the reader to check that F_ξ is indeed a functor.

2-YONEDA LEMMA. *The two functors above define an equivalence of categories*

$$\operatorname{Hom}_{\mathcal{C}}\bigl((\mathcal{C}/X), \mathcal{F}\bigr) \simeq \mathcal{F}(X).$$

PROOF. To check that the composite

$$\mathcal{F}(X) \longrightarrow \operatorname{Hom}_{\mathcal{C}}\bigl((\mathcal{C}/X), \mathcal{F}\bigr) \longrightarrow \mathcal{F}(X)$$

is isomorphic to the identity, notice that for any object $\xi \in \mathcal{F}(X)$, the composition applied to ξ yields $F_\xi(\xi) = \mathrm{id}_X^*\xi$, which is canonically isomorphic to ξ. It is easy to see that this defines an isomorphism of functors.

For the composite

$$\operatorname{Hom}_{\mathcal{C}}\bigl((\mathcal{C}/X), \mathcal{F}\bigr) \longrightarrow \mathcal{F}(X) \longrightarrow \operatorname{Hom}_{\mathcal{C}}\bigl((\mathcal{C}/X), \mathcal{F}\bigr)$$

take a morphism $F\colon (\mathcal{C}/X) \to \mathcal{F}$ and set $\xi = F(\mathrm{id}_X)$. We need to produce a base-preserving isomorphism of functors of F with F_ξ. The identity id_X is a terminal object in the category $(\mathcal{C}/X)$, hence for any object $\phi\colon U \to X$ there is a unique arrow $\phi\colon \mathrm{id}_X$, which is clearly cartesian. Hence it will remain cartesian after applying F, because F is a functor: this means that $F(\phi)$ is a pullback of $\xi = F(\mathrm{id}_X)$ along $\phi\colon U \to X$, so there is a canonical isomorphism $F_\xi(\phi) = \phi^*\xi \simeq F(\phi)$ in $\mathcal{F}(U)$. It is easy to check that this defines a base-preserving isomorphism of functors, and this ends the proof. $\square$

We have identified an object X with the functor $\mathrm{h}_X : \mathcal{C}^{\mathrm{op}} \to (\mathrm{Set})$ it represents, and we have identified the functor h_X with the corresponding category $(\mathcal{C}/X)$: so, to be consistent, we have to identify X and $(\mathcal{C}/X)$. So, we will write X for $(\mathcal{C}/X)$.

As for functors, the strong form of the 2-Yoneda Lemma can be used to reformulate the condition of representability. A morphism $(\mathcal{C}/X) \to \mathcal{F}$ corresponds to an object $\xi \in \mathcal{F}(X)$, which in turn defines the functor $F'\colon (\mathcal{C}/X) \to \mathcal{F}$ described above; this is isomorphic to the original functor F. Then F' is an equivalence if and only if for each object U of $\mathcal{C}$ the restriction

$$F'_U\colon \operatorname{Hom}_{\mathcal{C}}(U, X) = (\mathcal{C}/X)(U) \longrightarrow \mathcal{F}(U)$$

that sends each $f\colon U \to X$ to the pullback $f^*\xi \in \mathcal{F}(U)$, is an equivalence of categories. Since $\operatorname{Hom}_{\mathcal{C}}(U, X)$ is a set, this is equivalent to saying that $\mathcal{F}(U)$ is a groupoid, and each object of $\mathcal{F}(U)$ is isomorphic to the image of a unique element of $\operatorname{Hom}_{\mathcal{C}}(U, X)$ via a unique isomorphism. Since the isomorphisms $\rho \simeq f^*\xi$ in U correspond to cartesian arrows $\rho \to \xi$, and in a groupoid all arrows are cartesian, this means that $\mathcal{F}$ is fibered in groupoids, and for each $\rho \in \mathcal{F}(U)$ there exists a unique arrow $\rho \to \xi$. We have proved the following.

PROPOSITION 3.44. A fibered category $\mathcal{F}$ over $\mathcal{C}$ is representable if and only if $\mathcal{F}$ is fibered in groupoids, and there is an object U of $\mathcal{C}$ and an object ξ of $\mathcal{F}(U)$, such that for any object ρ of $\mathcal{F}$ there exists a unique arrow $\rho \to \xi$ in $\mathcal{F}$.

3.6.3. Splitting a fibered category. As we have seen in Example 3.14, a fibered category does not necessarily admit a splitting. However, a fibered category is always equivalent to a split fibered category.

THEOREM 3.45. Let $\mathcal{F} \to \mathcal{C}$ be a fibered category. Then there exists a canonically defined split fibered category $\widetilde{\mathcal{F}} \to \mathcal{C}$ and an equivalence of fibered categories of $\widetilde{\mathcal{F}}$ with $\mathcal{F}$.

PROOF. In this proof, if U is an object of $\mathcal{C}$, we will identify the functor h_U with the comma category $(\mathcal{C}/U)$. We have a functor $\mathrm{Hom}(-, \mathcal{F})\colon \mathcal{C}^{\mathrm{op}} \to (\mathrm{Cat})$ from the category $\mathcal{C}^{\mathrm{op}}$ into the category of all categories. If U is an object of $\mathcal{C}$ this functor will send U into the category $\mathrm{Hom}_{\mathcal{C}}(\mathrm{h}_U, \mathcal{F})$ of base-preserving natural transformations. An arrow $U \to V$ corresponds to a natural transformation $\mathrm{h}_U \to \mathrm{h}_V$, and this induces a functor $\mathrm{Hom}_{\mathcal{C}}(\mathrm{h}_V, \mathcal{F}) \to \mathrm{Hom}_{\mathcal{C}}(\mathrm{h}_U, \mathcal{F})$.

Let us denote by $\widetilde{\mathcal{F}}$ the fibered category associated with this functor: by definition, $\widetilde{\mathcal{F}}$ comes with a splitting. There is an obvious morphism $\widetilde{\mathcal{F}} \to \mathcal{F}$, sending an object $\phi\colon \mathrm{h}_U \to \mathcal{F}$ into $\phi(\mathrm{id}_U) \in \mathcal{F}(U)$. According to the 2-Yoneda Lemma, and the criterion of Proposition 3.36, this is an equivalence. □

It is an interesting exercise to figure out what this construction yields in the case of a surjective group homomorphism $G \to H$, as in Example 3.14.

3.7. The functors of arrows of a fibered category

Suppose that $\mathcal{F} \to \mathcal{C}$ is a fibered category; if U is an object in $\mathcal{C}$ and ξ, η are objects of $\mathcal{F}(U)$, we denote by $\mathrm{Hom}_U(\xi, \eta)$ the set of arrow from ξ to η in $\mathcal{F}(U)$.

Let ξ and η be two objects of $\mathcal{F}$ over the same object S of $\mathcal{C}$. Let $u_1\colon U_1 \to S$ and $u_2\colon U_2 \to S$ be arrows in $\mathcal{C}$; these are objects of the comma category $(\mathcal{C}/S)$. Suppose that $\xi_i \to \xi$ and $\eta_i \to \eta$ are pullbacks along $u_i\colon U_i \to S$ for $i = 1, 2$. For each arrow $f\colon U_1 \to U_2$ in $(\mathcal{C}/S)$, by definition of pullback there are two arrows, each unique, $\alpha_f\colon \xi_1 \to \xi_2$ and $\beta_f\colon \eta_1 \to \eta_2$, such that and the two diagrams

$$\begin{array}{ccc} \xi_1 & \xrightarrow{\alpha_f} & \xi_2 \\ & \searrow \quad \swarrow & \\ & \xi & \end{array} \qquad \text{and} \qquad \begin{array}{ccc} \eta_1 & \xrightarrow{\beta_f} & \eta_2 \\ & \searrow \quad \swarrow & \\ & \xi & \end{array}$$

commute. By Proposition 3.4 (iii) the arrows α_f and β_f are cartesian; we define a pullback function

$$f^*\colon \mathrm{Hom}_{U_2}(\xi_2, \eta_2) \longrightarrow \mathrm{Hom}_{U_1}(\xi_1, \eta_1)$$

in which $f^*\phi$ is defined as the only arrow $f^*\phi\colon \xi_1 \to \eta_1$ in $\mathcal{F}(U_1)$ making the diagram

$$\begin{array}{ccc} \xi_1 & \xrightarrow{f^*\phi} & \eta_1 \\ \big\downarrow{\scriptstyle \alpha_f} & & \big\downarrow{\scriptstyle \beta_f} \\ \xi_2 & \xrightarrow{\phi} & \eta_2 \end{array}$$

commute. If we are given a third arrow $g\colon U_2 \to U_3$ in $(\mathcal{C}/S)$ with pullbacks $\xi_3 \to \xi$ and $\eta_3 \to \eta$, we have arrows $\alpha_g\colon \xi_2 \to \xi_3$ and $\beta_g\colon \eta_2 \to \eta_3$; it is immediate to check that

$$\alpha_{gf} = \alpha_g \circ \alpha_f\colon \xi_1 \longrightarrow \xi_3 \quad \text{and} \quad \beta_{gf} = \beta_g \circ \beta_f\colon \eta_1 \longrightarrow \eta_3$$

and this implies that

$$(gf)^* = f^*g^*\colon \operatorname{Hom}_{U_3}(\xi_3, \eta_3) \longrightarrow \operatorname{Hom}_{U_1}(\xi_1, \eta_1).$$

After choosing a cleavage for $\mathcal{F}$, we can define a functor

$$\underline{\operatorname{Hom}}_S(\xi, \eta)\colon (\mathcal{C}/S)^{\mathrm{op}} \longrightarrow (\mathrm{Set})$$

by sending each object $u\colon U \to S$ into the set $\operatorname{Hom}_U(u^*\xi, u^*\eta)$ of arrows in the category $\mathcal{F}(U)$. An arrow $f\colon U_1 \to U_2$ from $u_1\colon U_1 \to S$ to $u_2\colon U_2 \to S$ yields a function

$$f^*\colon \operatorname{Hom}_{U_2}(u_2^*\xi, u_2^*\eta) \longrightarrow \operatorname{Hom}_{U_1}(u_1^*\xi, u_1^*\eta);$$

and this defines the effect of $\underline{\operatorname{Hom}}_S(\xi, \eta)$ on arrows.

It is easy to check that the functor $\underline{\operatorname{Hom}}_S(\xi, \eta)$ is independent of the choice of a cleavage, in the sense that cleavages give canonically isomorphic functors. Suppose that we have chosen for each $f\colon U \to V$ and each object ζ in $\mathcal{F}(V)$ another pullback $f^\vee\zeta \to \zeta$: then there is a canonical isomorphism $u^*\eta \simeq u^\vee\eta$ in $\mathcal{F}(U)$ for each arrow $u\colon U \to S$, and this gives a bijective correspondence

$$\operatorname{Hom}_U(u^*\xi, u^*\eta) \simeq \operatorname{Hom}_S(u^\vee\xi, u^\vee\eta),$$

yielding an isomorphism of the functors of arrows defined by the two pullbacks.

In fact, $\underline{\operatorname{Hom}}_S(\xi, \eta)$ can be more naturally defined as a quasi-functor

$$\mathcal{H}om_S(\xi, \eta) \longrightarrow (\mathcal{C}/S);$$

this does not require any choice of cleavages.

From this point of view, the objects of $\mathcal{H}om_S(\xi, \eta)$ over some object $u\colon U \to S$ of $(\mathcal{C}/S)$ are triples

$$(\xi_1 \longrightarrow \xi, \eta_1 \longrightarrow \eta, \phi),$$

where $\xi_1 \to \xi$ and $\eta_1 \to \eta$ are cartesian arrows of $\mathcal{F}$ over u, and $\phi\colon \xi_1 \to \eta_1$ is an arrow in $\mathcal{F}(U)$. An arrow from $(\xi_1 \to \xi, \eta_1 \to \eta, \phi_1)$ over $u_1\colon U_1 \to S$ and $(\xi_2 \to \xi, \eta_2 \to \eta, \phi_2)$ over $u_2\colon U_2 \to S$ is an arrow $f\colon U_1 \to U_2$ in $(\mathcal{C}/S)$ such that $f^*\phi_2 = \phi_1$.

From Proposition 3.40 we see that $\mathcal{H}om_S(\xi, \eta)$ is a quasi-functor over $\mathcal{C}$, and therefore, by Proposition 3.41, it is equivalent to a functor: of course this is the functor $\underline{\operatorname{Hom}}_S(\xi, \eta)$ obtained by the previous construction.

This can be proved as follows: the objects of $\underline{\operatorname{Hom}}_S(\xi, \eta)$, thought of as a category fibered in sets over $(\mathcal{C}/S)$ are pairs $(\phi, u\colon U \to S)$, where $u\colon U \to S$ is an object of $(\mathcal{C}/S)$ and $\phi\colon u^*\xi \to u^*\eta$ is an arrow in $\mathcal{F}(U)$; this also gives an object $(u^*\xi \to \xi, u^*\eta \to \eta, \phi)$ of $\mathcal{H}om_S(\xi, \eta)$ over U. The arrows between objects of $\underline{\operatorname{Hom}}_S(\xi, \eta)$ are precisely the arrows between the corresponding objects of $\mathcal{H}om_S(\xi, \eta)$, so we have an embedding of $\underline{\operatorname{Hom}}_S(\xi, \eta)$ into $\mathcal{H}om_S(\xi, \eta)$. But every object of $\mathcal{H}om_S(\xi, \eta)$ is isomorphic to an object of $\underline{\operatorname{Hom}}_S(\xi, \eta)$, hence the two fibered categories are equivalent.

3.8. Equivariant objects in fibered categories

The notion of an equivariant sheaf of modules on a scheme with the action of a group scheme, as defined in [**MFK94**, Chapter 1, § 3], or in [**Tho87**]), is somewhat involved and counterintuitive. The intuition is that if we are given the action of a group scheme G on a scheme X, an equivariant sheaf should be a sheaf $\mathcal{F}$, together with an action of G on the *pair* $(X, \mathcal{F})$, which is compatible with the action of G on X. Since the pair $(X, \mathcal{F})$ is an object of the fibered category of sheaves of modules, the language of fibered categories is very well suited for expressing this concept.

Let $G\colon \mathcal{C} \to (\mathrm{Grp})$ be a functor, $\mathcal{F} \to \mathcal{C}$ a fibered category, X an object of $\mathcal{C}$ with an action of G (see §2.2.1).

DEFINITION 3.46. A *G-equivariant object* of $\mathcal{F}(X)$ is an object ρ of $\mathcal{F}(X)$, together with an action of $G(U)$ on the set $\mathrm{Hom}_{\mathcal{F}}(\xi, \rho)$ for any $\xi \in \mathcal{F}(U)$, such that the following two conditions are satisfied.

(i) For any arrow $\phi\colon \xi \to \eta$ of $\mathcal{F}$ mapping to an arrow $f\colon U \to V$, the induced function $\phi^*\colon \mathrm{Hom}_{\mathcal{F}}(\eta, \rho) \to \mathrm{Hom}_{\mathcal{F}}(\xi, \rho)$ is equivariant with respect to the group homomorphism $f^*\colon G(V) \to G(U)$.

(ii) The function $\mathrm{Hom}_{\mathcal{F}}(\xi, \rho) \to \mathrm{Hom}_{\mathcal{C}}(U, X)$ induced by $\mathrm{p}_{\mathcal{F}}$ is $G(U)$-equivariant.

An arrow $u\colon \rho \to \sigma$ in $\mathcal{F}(X)$ is *G-equivariant* if it has the property that the induced function $\mathrm{Hom}_{\mathcal{F}}(\xi, \sigma) \to \mathrm{Hom}_{\mathcal{F}}(\xi, \rho)$ is $G(U)$-equivariant for all U and all $\xi \in \mathcal{F}(U)$.

The first condition can be expressed by saying that the data define an action of $G \circ \mathrm{p}_{\mathcal{F}}\colon \mathcal{F}^{\mathrm{op}} \to (\mathrm{Grp})$ on the object ρ of $\mathcal{F}$. In other words, for any object ξ of $\mathcal{F}$, the action defines a set theoretic action

$$(G \circ \mathrm{p}_{\mathcal{F}})(\xi) \times \mathrm{h}_\rho(\xi) \longrightarrow \mathrm{h}_\rho(\xi)$$

and this action is required to give a natural transformation of functors $\mathcal{F} \to (\mathrm{Set})$

$$(G \circ \mathrm{p}_{\mathcal{F}}) \times \mathrm{h}_\rho \longrightarrow \mathrm{h}_\rho.$$

The second condition can be thought of as saying that the action of G on ξ is compatible with the action of G on X.

The G-equivariant objects over X are the objects of a category $\mathcal{F}^G(X)$, in which the arrows are the equivariant arrows in $\mathcal{F}(X)$.

It is not hard to define the fibered category of G-equivariant objects of $\mathcal{F}$ over the category of G-equivariant objects of $\mathcal{C}$, but we will not do this.

Now assume that G is a group object in $\mathcal{C}$ acting on an object X of $\mathcal{C}$, corresponding to an arrow $\alpha\colon G \times X \to X$, as in Proposition 2.16. Take a fibered category $\mathcal{F} \to \mathcal{C}$: the category $\mathcal{F}^G(X)$ of equivariant objects over X has a different description. Choose a cleavage for $\mathcal{F}$.

Let ρ be an object of $\mathcal{F}^G(X)$. Consider the pullback $\mathrm{pr}_2^* \rho \in \mathcal{F}(G \times X)$, and the functor $\mathrm{h}_{\mathrm{pr}_2^* \rho}\colon \mathcal{F} \to (\mathrm{Set})$ it represents. If $\phi\colon \xi \to \mathrm{pr}_2^* \rho$ is an arrow in $\mathcal{F}$, we obtain an arrow $\xi \to \rho$ by composing ϕ with the given cartesian arrow $\mathrm{pr}_2^* \rho \to \rho$, and an arrow $\mathrm{p}_{\mathcal{F}}\xi \to G$ by composing $\mathrm{p}_{\mathcal{F}}\phi\colon \mathrm{p}_{\mathcal{F}}\xi \to G \times X$ with the projection $G \times X \to G$. This defines a natural transformation $\mathrm{h}_{\mathrm{pr}_2^* \rho} \to (G \circ \mathrm{p}_{\mathcal{F}}) \times \mathrm{h}_\rho$.

The fact that the canonical arrow $p\colon \mathrm{pr}_2^* \rho \to \rho$ is cartesian implies that each pair consisting of an arrow $\xi \to \rho$ in $\mathcal{F}$ and an arrow $\mathrm{p}_{\mathcal{F}}\xi \to G$ in $\mathcal{C}$ comes from a unique arrow $\xi \to \mathrm{pr}_2^* \rho$. This means that the natural transformation above is in fact an isomorphism $\mathrm{h}_{\mathrm{pr}_2^* \rho} \simeq (G \circ \mathrm{p}_{\mathcal{F}}) \times \mathrm{h}_\xi$ of functors $\mathcal{F}^{\mathrm{op}} \to (\mathrm{Set})$. Hence, by Yoneda's lemma, a morphism $(G \circ \mathrm{p}_{\mathcal{F}}) \times \mathrm{h}_\rho \to \mathrm{h}_\rho$ corresponds to an arrow

$\beta\colon \mathrm{pr}_2^*\,\rho \to \rho$. Condition (ii) of Definition 3.46 can be expressed as saying that $\mathrm{p}_{\mathcal{F}}\beta = \alpha$.

There are two conditions that define an action. First consider the natural transformation $\mathrm{h}_\rho \to (G\circ \mathrm{p}_{\mathcal{F}})\times \mathrm{h}_\rho$ that sends an object $u \in \mathrm{h}_\rho(\xi)$ to the pair $(1,u) \in G(\mathrm{p}_{\mathcal{F}}\xi)\times \mathrm{h}_\rho(\xi)$; this corresponds to an arrow $\epsilon_\rho\colon \rho \to \mathrm{pr}_2^*\,\rho$, whose composite with $p\colon \mathrm{pr}_2^*\,\rho \to \rho$ is the identity id_ρ, and whose image in $\mathcal{C}$ is the arrow $\epsilon_X\colon X = \mathrm{pt}\times X \to G\times X$ induced by $\mathrm{e}_G\colon \mathrm{pt}\to G$. Since p is cartesian, these two conditions characterize ϵ_ρ uniquely.

The first condition that defines an action of $(G\circ \mathrm{p}_{\mathcal{F}})$ on ρ (see Proposition 2.16) is that the composite $\mathrm{h}_\rho \to (G\circ \mathrm{p}_{\mathcal{F}})\times \mathrm{h}_\rho \to \mathrm{h}_\rho$ be the identity; and this is equivalent to saying that the composite $\beta\circ\epsilon_\rho\colon \rho\to\rho$ is the identity id_ρ.

The second conditions can be expressed similarly. The functor

$$(G\circ \mathrm{p}_{\mathcal{F}})\times (G\circ \mathrm{p}_{\mathcal{F}})\times \mathrm{h}_\rho\colon \mathcal{F} \longrightarrow (\mathrm{Grp})$$

is represented by the pullback $\mathrm{pr}_3^*\,\rho$ of ρ along the third projection $\mathrm{pr}_3\colon G\times G\times X \to X$. Now, given any arrow $f\colon G\times G\times X \to G\times X$ whose composite with $\mathrm{pr}_2\colon F\times X \to X$ equals $\mathrm{pr}_2\colon G\times G\times X \to X$, there is a unique arrow $\widetilde{f}\colon\ \mathrm{pr}_3^*\,\rho \to \mathrm{p}_2^*\rho$ mapping to f, such that the composite $p\circ\widetilde{f}\colon\ \mathrm{pr}_3^*\,\rho\to\rho$ equals the canonical arrow $q\colon\ \mathrm{pr}_3^*\,\rho \to \rho$. Then it is an easy matter to convince oneself that the second condition that defines an action is equivalent to the commutativity of the diagram

$$\begin{array}{ccc} \mathrm{pr}_3^*\,\rho & \xrightarrow{\widetilde{\mathrm{m}_G\times\mathrm{id}_X}} & \mathrm{pr}_2^*\,\rho \\ \Big\downarrow{\scriptstyle \widetilde{\mathrm{id}_G\times\alpha}} & & \Big\downarrow{\scriptstyle\beta} \\ \mathrm{pr}_2^*\,\rho & \xrightarrow{\ \ \beta\ \ } & \rho \end{array} \tag{3.8.1}$$

This essentially proves the following fact (we leave the easy details to the reader).

PROPOSITION 3.47. *Let ρ be an object of $\mathcal{F}(X)$. To give ρ the structure of a G-equivariant object is the same as assigning an arrow $\beta\colon\ \mathrm{pr}_2^*\,\rho\to\rho$ with $\mathrm{p}_{\mathcal{F}}\beta = \alpha$, satisfying the following two conditions.*

(i) $\beta\circ\epsilon_\rho = \mathrm{id}_\rho$.
(ii) *The diagram* (3.8.1) *commutes.*

Furthermore, if ρ and ρ' are G-equivariant objects, and we denote by $\beta\colon\ \mathrm{pr}_2^*\,\rho\to \rho$ and $\beta'\colon\ \mathrm{pr}_2^*\,\rho'\to\rho'$ the corresponding arrows, then $u\colon \rho\to\rho'$ is G-equivariant if and only if the diagram

$$\begin{array}{ccc} \mathrm{pr}_2^*\,\rho & \xrightarrow{\ \ \beta\ \ } & \rho \\ \Big\downarrow{\scriptstyle \mathrm{pr}_2^*\,u} & & \Big\downarrow{\scriptstyle u} \\ \mathrm{pr}_2^*\,\rho' & \xrightarrow{\ \ \beta'\ \ } & \rho' \end{array}$$

commutes.

This can be restated further, to make it look more like the classical definition of an equivariant sheaf. First of all, let us notice that if an arrow $\beta\colon\ \mathrm{pr}_2^*\,\rho\to\rho$ corresponds to a G-equivariant structure on ρ, then it is cartesian. This can be shown as follows.

There is an automorphism i of $G\times X$, defined in functorial terms by the equation $i(g,x) = (g^{-1}, gx)$ whenever U is an object of $\mathcal{C}$, $g \in G(U)$ and $x \in X(U)$; this has the property that

$$\mathrm{pr}_2 \circ i = \alpha\colon G \times X \longrightarrow X.$$

Analogously one can use the action of $(G \circ \mathrm{p}_{\mathcal{F}})$ on ρ to define an automorphism I of $(G \circ \mathrm{p}_{\mathcal{F}}) \times \mathrm{h}_\rho$, hence an automorphism of $\mathrm{pr}_2^* \rho$, whose composite with the canonical arrow $p\colon \mathrm{pr}_2^* \rho \to \rho$ equals β. Since I is an isomorphism, hence is cartesian, the canonical arrow p is cartesian, and the composite of cartesian arrow is cartesian, it follows that β is cartesian.

Now, start from a *cartesian* arrow $\beta\colon \mathrm{pr}_2^* \rho \to \rho$ with $\mathrm{p}_{\mathcal{F}}\beta = \alpha$. Assume that the diagram (3.8.1) is commutative. I claim that in this case we also have $\beta \circ \epsilon_\rho = \mathrm{id}_\rho$. This can be checked in several ways: here is one.

The arrow β corresponds to a natural transformation $(G \circ \mathrm{p}_{\mathcal{F}}) \times \mathrm{h}_\rho \to \mathrm{h}_\rho$. The commutativity of the diagram (3.8.1) expresses the fact that $(g_1 g_2)x = g_1(g_2 x)$ for any object ξ of $\mathcal{F}$ and any $g_1, g_2 \in G(\mathrm{p}_{\mathcal{F}}\xi)$ and $x \in \mathrm{h}_\rho \xi$. The arrow $\beta \circ \epsilon_\rho\colon \rho \to \rho$ corresponds to the natural transformation $\mathrm{h}_\rho \to \mathrm{h}_\rho$ given by multiplication by the identity: and, because of the previous identity, this is an idempotent endomorphism of h_ρ. Hence $\beta \circ \epsilon_\rho$ is an idempotent arrow in $\mathrm{Hom}_{\mathcal{F}}(\rho, \rho)$.

On the other hand, Proposition 3.4 (iii) implies that ϵ_ρ is a cartesian arrow, so $\beta \circ \epsilon_\rho$ is also cartesian. But $\beta \circ \epsilon_\rho$ maps to id_X in $\mathcal{C}$, hence is an isomorphism: and the only idempotent isomorphism is the identity.

This allows us to rewrite the conditions as follows.

PROPOSITION 3.48. Let ρ be an object of $\mathcal{F}(X)$. To give ρ the structure of a G-equivariant object is the same as assigning a cartesian arrow $\beta\colon \mathrm{pr}_2^* \rho \to \rho$ with $\mathrm{p}_{\mathcal{F}}\beta = \alpha$, such that the diagram (3.8.1) commutes.

Furthermore, let ρ and ρ' be G-equivariant objects, and denote by $\beta\colon \mathrm{pr}_2^* \rho \to \rho$ and $\beta'\colon \mathrm{pr}_2^* \rho' \to \rho'$ the corresponding arrows, Then $u\colon \rho \to \rho'$ is G-equivariant if and only if the diagram

$$\begin{array}{ccc}
\mathrm{pr}_2^* \rho & \xrightarrow{\beta} & \rho \\
\downarrow{\scriptstyle \mathrm{pr}_2^* u} & & \downarrow{\scriptstyle u} \\
\mathrm{pr}_2^* \rho' & \xrightarrow{\beta'} & \rho'
\end{array}$$

commutes.

A final restatement is obtained via a cleavage, in the language of pseudo-functors. Recall that an arrow $\mathrm{pr}_2^* \rho \to \rho$ mapping to α in $\mathcal{C}$ corresponds to an arrow $\phi\colon \mathrm{pr}_2^* \rho \to \alpha^* \rho$ in $\mathcal{F}(G \times X)$, and that β is cartesian if and only if ϕ is an isomorphism.

We also have the equalities

$$\mathrm{pr}_3 = \mathrm{pr}_2 \circ (\mathrm{m}_G \times \mathrm{id}_X) = \mathrm{pr}_2 \circ \mathrm{pr}_{23}\colon G \times G \times X \to X,$$

$$A \overset{\text{def}}{=} \alpha \circ (\mathrm{m}_G \times \mathrm{id}_X) = \alpha \circ (\mathrm{id}_G \times \alpha)\colon G \times G \times X \to X.$$

and

$$B \overset{\text{def}}{=} \mathrm{pr}_2 \circ (\mathrm{id}_G \times \alpha) = \alpha \circ \mathrm{pr}_{23}\colon G \times G \times X \longrightarrow X.$$

We leave it to the reader to unwind the various definitions and check that the following is equivalent to the previous statement.

PROPOSITION 3.49. Let ρ be an object of $\mathcal{F}(X)$. To give ρ the structure of a G-equivariant object is the same as assigning an isomorphism $\phi\colon \mathrm{pr}_2^*\rho \cong \alpha^*\rho$ in $\mathcal{F}(G \times X)$, such that the diagram

$$\begin{array}{ccc} \mathrm{pr}_3^*\rho & \xrightarrow{(\mathrm{m}_G\times\mathrm{id}_X)^*} & A^*\rho \\ & {\scriptstyle \mathrm{pr}_{23}^*\phi}\searrow \qquad \nearrow{\scriptstyle (\mathrm{id}_G\times\alpha)^*\phi} & \\ & B^*\rho & \end{array}$$

commutes.

When applied to the fibered category of sheaves of some kind (for example, quasi-coherent sheaves) one gets precisely the usual definition of an equivariant sheaf.

CHAPTER 4

Stacks

4.1. Descent of objects of fibered categories

4.1.1. Gluing continuous maps and topological spaces. The following is the archetypal example of descent. Take (Cont) to be the category of continuous maps (that is, the category of arrows in (Top), as in Example 3.15); this category is fibered over (Top) via the functor $\mathrm{p}_{(\mathrm{Cont})}\colon (\mathrm{Cont}) \to (\mathrm{Top})$ sending each continuous map to its codomain. Now, suppose that $f\colon X \to U$ and $g\colon Y \to U$ are two objects of (Cont) mapping to the same object U in (Top); we want to construct a continuous map $\phi\colon X \to Y$ over U, that is, an arrow in $(\mathrm{Cont})(U) = (\mathrm{Top}/U)$. Suppose that we are given an open covering $\{U_i\}$ of U, and continuous maps $\phi_i\colon f^{-1}U_i \to g^{-1}U_i$ over U_i; assume furthermore that the restriction of ϕ_i and ϕ_j to $f^{-1}(U_i \cap U_j) \to g^{-1}(U_i \cap U_j)$ coincide. Then there is a unique continuous map $\phi\colon X \to Y$ over U whose restriction to each $f^{-1}U_i$ coincides with f_i.

This can be written as follows. The category (Cont) is fibered over (Top), and if $f\colon V \to U$ is a continuous map, $X \to U$ an object of $(\mathrm{Cont})(U) = (\mathrm{Top}/U)$, then a pullback of $X \to U$ to V is given by the projection $V \times_U X \to V$. The functor $f^*\colon (\mathrm{Cont})(U) \to (\mathrm{Cont})(V)$ sends each object $X \to U$ to $V \times_U X \to V$, and each arrow in (Top/U), given by continuous function $\phi\colon X \to Y$ over U, to the continuous function $f^*\phi = \mathrm{id}_V \times_U f\colon V \times_U X \to V \times_U Y$.

Suppose that we are given two topological spaces X and Y with continuous maps $X \to S$ and $Y \to S$. Consider the functor

$$\underline{\mathrm{Hom}}_S(X,Y)\colon (\mathrm{Top}/S) \longrightarrow (\mathrm{Set})$$

from the category of topological spaces over S, defined in Section 3.7. This sends each arrow $U \to S$ to the set of continuous maps $\mathrm{Hom}_U(U \times_S X, U \times_S Y)$ over U. The action on arrows is obtained as follows: given a continuous function $f\colon V \to U$, we send each continuous function $\phi\colon U \times_S X \to U \times_S Y$ to the function

$$f^*\phi = \mathrm{id}_V \times \phi\colon V \times_S X = V \times_U (U \times_S X) \longrightarrow V \times_U (U \times_S Y) = V \times_S Y.$$

Then the fact that continuous functions can be constructed locally and then glued together can be expressed by saying that the functor

$$\underline{\mathrm{Hom}}_S(X,Y)\colon (\mathrm{Top}/S)^{\mathrm{op}} \longrightarrow (\mathrm{Set})$$

is a sheaf in the classical topology of (Top).

But there is more: not only we can construct continuous functions locally: we can also do this for spaces, although this is more complicated.

PROPOSITION 4.1. *Suppose that we are given a topological space U with an open covering $\{U_i\}$; for each triple of indices i, j and k choose fibered products $U_{ij} = U_i \cap U_j$ and $U_{ijk} = U_i \cap U_j \cap U_k$. Assume that for each i we have a* continuous map $u_i\colon X_i \to U_i$, and that for each pair of indices i and j we have a homeomorphism $\phi_{ij}\colon u_j^{-1}U_{ij} \simeq u_i^{-1}U_{ij}$ over U_{ij}, satisfying the cocycle condition

$$\phi_{ik} = \phi_{ij} \circ \phi_{jk}\colon u_k^{-1}U_{ijk} \longrightarrow u_j^{-1}U_{ijk} \longrightarrow u_i^{-1}U_{ijk}.$$

Then there exists a continuous map $u\colon X \to U$, together with isomorphisms $\phi_i\colon u^{-1}U_i \simeq X_i$, such that $\phi_{ij} = \phi_i \circ \phi_j^{-1}\colon u_j^{-1}U_{ij} \to u^{-1}U_{ij} \to u_iU_{ij}$ for all i and j.

PROOF. Consider the disjoint union U' of the U_i; the fibered product $U' \times_U U'$ is the disjoint union of the U_{ij}. The disjoint union X' of the X_i, maps to U'; consider the subset $R \subseteq X' \times X'$ consisting of pairs $(x_i, x_j) \in X_i \times X_j \subseteq X' \times X'$ such that $x_i = \phi_{ij}x_j$. I claim that R is an equivalence relation in X'. Notice that the cocycle condition $\phi_{ii} = \phi_{ii} \circ \phi_{ii}$ implies that ϕ_{ii} is the identity on X_i, and this show that the equivalence relation is reflexive. The fact that $\phi_{ii} = \phi_{ij} \circ \phi_{ji}$, and therefore $\phi_{ji} = \phi_{ij}^{-1}$, prove that it is symmetric; and transitivity follows directly from the general cocycle condition. We define X to be the quotient X'/R.

If two points of X' are equivalent, then their images in U coincide; so there is an induced continuous map $u\colon X \to U$. The restriction to $X_i \subseteq X'$ of the projection $X' \to X$ gives a continuous map $\phi_i\colon X_i \to u^{-1}U_i$, which is easily checked to be a homeomorphism. One also sees that $\phi_{ij} = \phi_i \circ \phi_j^{-1}$, and this completes the proof. □

The fact that we can glue continuous maps and topological spaces says that (Cont) is a *stack* over (Top).

4.1.2. The category of descent data. Let $\mathcal{C}$ be a site. We have seen that a fibered category over $\mathcal{C}$ should be thought of as a functor from $\mathcal{C}$ to the category of categories, that is, as a presheaf of categories over $\mathcal{C}$. A stack is, morally, a sheaf of categories over $\mathcal{C}$.

Let $\mathcal{F}$ be a category fibered over $\mathcal{C}$. We fix a cleavage; but we will also indicate how the definitions can be given without resorting to the choice of a cleavage.

Given a covering $\{\sigma\colon U_i \to U\}$, set $U_{ij} = U_i \times_U U_j$ and $U_{ijk} = U_i \times_U U_j \times_U U_k$ for each triple of indices i, j and k.

DEFINITION 4.2. Let $\mathcal{U} = \{\sigma_i\colon U_i \to U\}$ be a covering in $\mathcal{C}$. An *object with descent data* $(\{\xi_i\}, \{\phi_{ij}\})$ on $\mathcal{U}$, is a collection of objects $\xi_i \in \mathcal{F}(U_i)$, together with isomorphisms $\phi_{ij}\colon \operatorname{pr}_2^*\xi_j \simeq \operatorname{pr}_1^*\xi_i$ in $\mathcal{F}(U_i \times_U U_j)$, such that the following cocycle condition is satisfied.

For any triple of indices i, j and k, we have the equality

$$\operatorname{pr}_{13}^*\phi_{ik} = \operatorname{pr}_{12}^*\phi_{ij} \circ \operatorname{pr}_{23}^*\phi_{jk}\colon \operatorname{pr}_3^*\xi_k \longrightarrow \operatorname{pr}_1^*\xi_i$$

where the pr_{ab} and pr_a are projections on the ath and b^{th} factor, or the ath factor respectively.

The isomorphisms ϕ_{ij} are called *transition isomorphisms* of the object with descent data.

An arrow between objects with descent data

$$\{\alpha_i\}\colon (\{\xi_i\}, \{\phi_{ij}\}) \longrightarrow (\{\eta_i\}, \{\psi_{ij}\})$$

is a collection of arrows $\alpha_i \colon \xi_i \to \eta_i$ in $\mathcal{F}(U_i)$, with the property that for each pair of indices i, j, the diagram

$$\begin{array}{ccc} \mathrm{pr}_2^* \xi_j & \xrightarrow{\mathrm{pr}_2^* \alpha_j} & \mathrm{pr}_2^* \eta_j \\ \downarrow{\scriptstyle \phi_{ij}} & & \downarrow{\scriptstyle \psi_{ij}} \\ \mathrm{pr}_1^* \xi_i & \xrightarrow{\mathrm{pr}_1^* \alpha_i} & \mathrm{pr}_1^* \eta_i \end{array}$$

commutes.

In understanding the definition above it may be useful to contemplate the cube

(4.1.1)

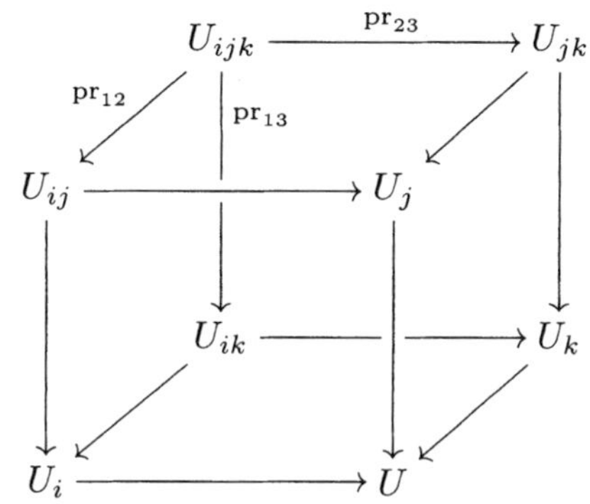

in which all arrows are given by projections, and every face is cartesian.

There is an obvious way of composing morphisms, which makes objects with descent data the objects of a category, denoted by $\mathcal{F}(\mathcal{U}) = \mathcal{F}(\{U_i \to U\})$.

REMARK 4.3. This category does not depend on the choice of fibered products U_{ij} and U_{ijk}, in the sense that with different choices we get isomorphic categories.

For each object ξ of $\mathcal{F}(U)$ we can construct an object with descent data on a covering $\{\sigma_i \colon U_i \to U\}$ as follows. The objects are the pullbacks $\sigma_i^*\xi$; the isomorphisms $\phi_{ij} \colon \mathrm{pr}_2^* \sigma_j^* \xi \simeq \mathrm{pr}_1^* \sigma_i^* \xi$ are the isomorphisms that come from the fact that both $\mathrm{pr}_2^* \sigma_j^* \xi$ and $\mathrm{pr}_1^* \sigma_i^* \xi$ are pullbacks of ξ to U_{ij}. If we identify $\mathrm{pr}_2^* \sigma_j^* \xi$ with $\mathrm{pr}_1^* \sigma_i^* \xi$, as is commonly done, then the ϕ_{ij} are identities.

Given an arrow $\alpha \colon \xi \to \eta$ in $\mathcal{F}(U)$, we get arrows $\sigma_i^* \colon \sigma_i^* \xi \to \sigma_i^* \eta$, yielding an arrow from the object with descent associated with ξ to the one associated with η. This defines a functor $\mathcal{F}(U) \to \mathcal{F}(\{U_i \to U\})$.

It is important to notice that these construction do not depend on the choice of a cleavage, in the following sense. Given a different cleavage, for each covering $\{U_i \to U\}$ there is a canonical isomorphism of the resulting categories $\mathcal{F}(\{U_i \to U\})$; and the functors $\mathcal{F}(U) \to \mathcal{F}(\{U_i \to U\})$ commute with these equivalences.

Here is a definition of the category of descent data that does not depend on choosing of a cleavage. Let $\{U_i \to U\}_{i \in I}$ be a covering. We define an object with descent data to be a triple of sets

$$(\{\xi_i\}_{i \in I}, \{\xi_{ij}\}_{i,j \in I}, \{\xi_{ijk}\}_{i,j,k \in I}),$$

where each ξ_α is an object of $\mathcal{F}(U_\alpha)$, plus, for each triple of indices i, j and k, a commutative diagram

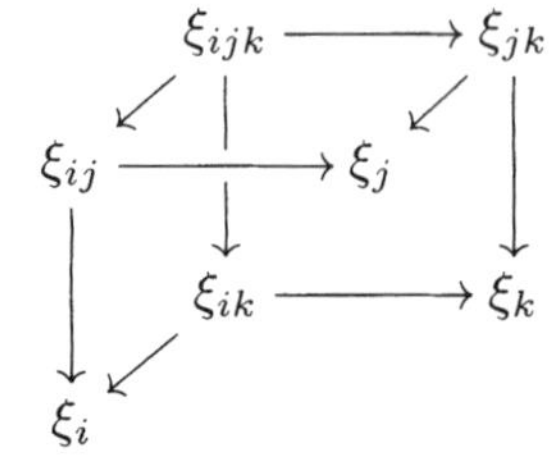

in which every arrow is cartesian, and such that when applying $\mathrm{p}_{\mathcal{F}}$ every arrow maps to the appropriate projection in the diagram (4.1.1). These form the objects of a category $\mathcal{F}_{\mathrm{desc}}(\{U_i \to U\})$.

An arrow

$$\{\phi_i\}_{i\in I}\colon (\{\xi_i\}, \{\xi_{ij}\}, \{\xi_{ijk}\}) \longrightarrow (\{\eta_i\}, \{\eta_{ij}\}, \{\eta_{ijk}\})$$

consists of set of arrows with $\phi_i : \xi_i \to \eta_i$ in $\mathcal{F}(U_i)$, such that for every pair of indices i and j we have

$$\mathrm{pr}_1^* \phi_i = \mathrm{pr}_2^* \phi_j \colon \xi_{ij} \longrightarrow \eta_{ij}.$$

Alternatively, and perhaps more naturally, we could define an arrow as a triple $((\{\phi_i\}_{i\in I}, \{\phi_{ij}\}_{i,j\in I}, \{\phi_{ijk}\}_{i,j,k\in I}))$, where $\phi_\alpha\colon \xi_\alpha \to \eta_\alpha$ is an arrow in $\mathcal{F}(U_\alpha)$ for each α in I, $I \times I$ or $I \times I \times I$, with the obvious compatibility conditions with the various arrows involved in the definition of an object. We leave it to the reader to check that these two definitions of an arrow are equivalent.

Once we have chosen a cleavage, there is a functor from $\mathcal{F}_{\mathrm{desc}}(\{U_i \to U\})$ to $\mathcal{F}(\{U_i \to U\})$. Given an object $(\{\xi_i\}, \{\xi_{ij}\}, \{\xi_{ijk}\})$ of $\mathcal{F}_{\mathrm{desc}}(\{U_i \to U\})$, the arrows $\xi_{ij} \to \xi_i$ and $\xi_{ij} \to \xi_j$ induce isomorphisms $\xi_{ij} \simeq \mathrm{pr}_1^* \xi_i$ and $\xi_{ij} \simeq \mathrm{pr}_2^* \xi_j$; the resulting isomorphism $\mathrm{pr}_2^* \xi_j \simeq \mathrm{pr}_1^* \xi_i$ is easily seen to satisfy the cocycle condition, thus defining an object of $\mathcal{F}(\{U_i \to U\})$. An arrow $\{\phi_i\}$ in $\mathcal{F}_{\mathrm{desc}}(\{U_i \to U\})$ is already an arrow in $\mathcal{F}(\{U_i \to U\})$.

It is not hard to check that this functor is an equivalence of categories.

We can not define a functor $\mathcal{F}(U) \to \mathcal{F}_{\mathrm{desc}}(\{U_i \to U\})$ directly, without the choice of a cleavage. However, let us define another category

$$\mathcal{F}_{\mathrm{comp}}(\{U_i \to U\}),$$

in which the objects are quadruples $(\xi, \{\xi_i\}, \{\xi_{ij}\}, \{\xi_{ijk}\}))$, where ξ is an object of $\mathcal{F}(U)$ and each ξ_α is an object of $\mathcal{F}(U_\alpha)$, plus a commutative cube

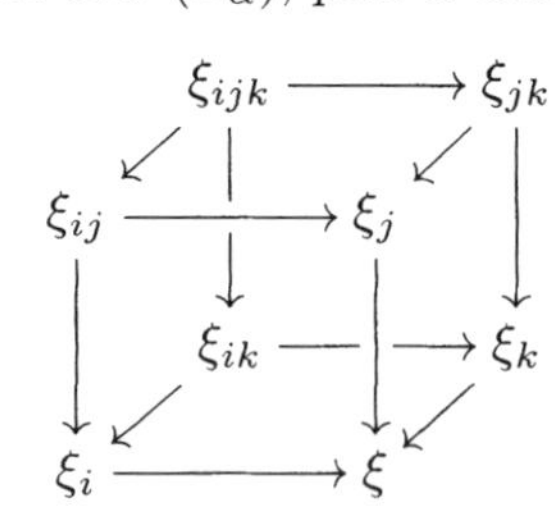

in $\mathcal{F}$ for all the triples of indices, in which all the arrows are cartesian, and whose image in $\mathcal{C}$ is the cube (4.1.1) above. An arrow from $(\xi, \{\xi_i\}, \{\xi_{ij}\}, \{\xi_{ijk}\}))$ to $(\eta, \{\eta_i\}, \{\eta_{ij}\}, \{\eta_{ijk}\}))$ can be indifferently defined as an arrow $\phi\colon \xi \to \eta$ in $\mathcal{F}(U)$, or as collections of arrows $\xi \to \eta$, $\xi_i \to \eta_i$, $\xi_{ij} \to \eta_{ij}$ and $\xi_{ijk} \to \eta_{ijk}$ satisfying the obvious commutativity conditions.

There is a functor from $\mathcal{F}_{\mathrm{comp}}(\{U_i \to U\})$ to $\mathcal{F}(U)$ that sends a whole object $(\xi, \{\xi_i\}, \{\xi_{ij}\}, \{\xi_{ijk}\}))$ to ξ, and is easily seen to be an equivalence. There is also a functor from $\mathcal{F}_{\mathrm{comp}}(\{U_i \to U\})$ to $\mathcal{F}_{\mathrm{desc}}(\{U_i \to U\})$ that forgets the object of $\mathcal{F}(U)$. This takes the place of the functor from $\mathcal{F}(U)$ to $\mathcal{F}(\{U_i \to U\})$ defined using cleavages.

REMARK 4.4. Of course, if one really wants to be consistent, one should not assume that the category $\mathcal{C}$ has a canonical choice of fibered products, and not suppose that the U_{ij} and the U_{ijk} are given a priori, but allow them to be arbitrary fibered products.

The most elegant definition of objects with descent data is one that uses sieves; it does not require choosing anything. Let $\mathcal{U} = \{U_i \to U\}$ be a covering in $\mathcal{C}$. The sieve $\mathrm{h}_{\mathcal{U}} \colon \mathcal{C}^{\mathrm{op}} \to (\mathrm{Set})$ is a functor, whose associated category fibered in sets is the full subcategory of $(\mathcal{C}/U)$, whose objects are arrows $T \to U$ that factor through some $U_i \to U$. According to our principle that functors and categories fibered in sets should be identified, we denote by $\mathrm{h}_{\mathcal{U}}$ this category. By the same principle, we also denote by h_U the category $(\mathcal{C}/U)$.

There is a functor $\mathrm{Hom}_{\mathcal{C}}(\mathrm{h}_{\mathcal{U}}, \mathcal{F}) \to \mathcal{F}_{\mathrm{desc}}(\mathcal{U})$, defined as follows. Suppose that we are given a morphism $F \colon \mathrm{h}_{\mathcal{U}} \to (\mathrm{Set})$. For any triple of indices i, j and k we have objects $U_i \to U$, $U_{ij} \to U$ and $U_{ijk} \to U$ of $\mathrm{h}_{\mathcal{U}}$, and each of the projections of (4.1.1) not landing in U is an arrow in $\mathrm{h}_{\mathcal{U}}$. Hence we can apply F and get a diagram

$$
\begin{array}{ccccccc}
 & & F(U_{ijk}) & \longrightarrow & F(U_{jk}) & & \\
 & \swarrow & \downarrow & & \downarrow & \swarrow & \\
F(U_{ij}) & & \longrightarrow & & F(U_j) & & \\
\downarrow & & F(U_{ik}) & \longrightarrow & F(U_k) & & \\
 & \swarrow & & & & & \\
F(U_i) & & & & & &
\end{array}
$$

giving an object of $\mathcal{F}_{\mathrm{desc}}(\mathcal{U})$. This extends to a functor

$$\mathrm{Hom}(\mathrm{h}_{\mathcal{U}}, \mathcal{F}) \longrightarrow \mathcal{F}_{\mathrm{desc}}(\mathcal{U})$$

in the obvious way.

Also, consider the functor

$$\mathrm{Hom}_{\mathcal{C}}(\mathrm{h}_U, \mathcal{F}) \longrightarrow \mathrm{Hom}_{\mathcal{C}}(\mathrm{h}_{\mathcal{U}}, \mathcal{F})$$

induced by the embedding $\mathrm{h}_{\mathcal{U}} \subseteq \mathrm{h}_U$; after choosing a cleavage, it is easy to verify that the composite of functors

$$\mathrm{Hom}_{\mathcal{C}}(\mathrm{h}_U, \mathcal{F}) \longrightarrow \mathrm{Hom}_{\mathcal{C}}(\mathrm{h}_{\mathcal{U}}, \mathcal{F}) \simeq \mathcal{F}_{\mathrm{desc}}(\mathcal{U}) \simeq \mathcal{F}(\mathcal{U})$$

is isomorphic to the composite

$$\mathrm{Hom}_{\mathcal{C}}(\mathrm{h}_U, \mathcal{F}) \simeq \mathcal{F}(U) \longrightarrow \mathcal{F}(\mathcal{U})$$

where the first functor is the equivalence of the 2-Yoneda Lemma.

The following generalizes Proposition 2.39.

PROPOSITION 4.5. *The functor $\mathrm{Hom}(\mathrm{h}_{\mathcal{U}}, \mathcal{F}) \to \mathcal{F}_{\mathrm{desc}}(\mathcal{U})$ is an equivalence.*

PROOF. Let us construct a functor $\mathcal{F}_{\mathrm{desc}}(\mathcal{U}) \to \mathrm{Hom}(\mathrm{h}_{\mathcal{U}}, \mathcal{F})$. Set $\mathcal{U} = \{U_i \to U\}$, and, for each $T \to U$ in $\mathrm{h}_{\mathcal{U}} T$, choose a factorization $T \to U_i \to U$. Assume that we given an object $(\{\xi_i\}, \{\xi_{ij}\}, \{\xi_{ijk}\})$ of $\mathcal{F}_{\mathrm{desc}}(U)$; for each arrow $T \to U$ in the category $\mathrm{h}_{\mathcal{U}}$ we get an object ξ_T of T by pulling back ξ_i along the chosen arrow $T \to U_i$. This defines a function from the set of objects of $\mathrm{h}_{\mathcal{U}}$ to $\mathcal{F}$.

Given an arrow $T' \to T \to U$ in $\mathrm{h}_{\mathcal{U}}$, chose a factorization $T' \to U_j \to U$ of the composite $T' \to U$. This, together with the composite $T' \to T \to U_i$ yields an arrow $T' \to U_{ij}$ fitting into a diagram

$$\begin{array}{ccccc} T' & \longrightarrow & U_{ij} & \longrightarrow & U_j \\ \downarrow & & \downarrow & & \downarrow \\ T & \longrightarrow & U_i & \longrightarrow & U. \end{array}$$

Since the given arrow $\xi_{ij} \to \xi_j$ is cartesian, the canonical arrow $\xi_{T'} \to \xi_j$, that is given by definition, because $\xi_{T'}$ is a pullback of ξ_j, will factor uniquely as $\xi_{T'} \to \xi_{ij} \to \xi_j$, in such a way that $\xi_{T'} \to \xi_{ij}$ maps to $T' \to U_{ij}$. Now the composite $\xi_{T'} \to \xi_{ij} \to \xi_i$ will factor as $\xi_{T'} \to \xi_T \to \xi_i$ for a unique arrow $\xi_{T'} \to \xi_T$ mapping to the given arrow $T' \to T$ in $\mathcal{C}$. According to Proposition 3.4 (iii), the arrow $\xi_{T'} \to \xi_T$ is cartesian.

These two functions, on objects and on arrows, define a morphism $\mathrm{h}_{\mathcal{U}} \to \mathcal{F}$. We need to check that the composites

$$\mathcal{F}_{\mathrm{desc}}(\mathcal{U}) \longrightarrow \mathrm{Hom}(\mathrm{h}_{\mathcal{U}}, \mathcal{F}) \longrightarrow \mathcal{F}_{\mathrm{desc}}(\mathcal{U})$$

and

$$\mathrm{Hom}(\mathrm{h}_{\mathcal{U}}, \mathcal{F}) \longrightarrow \mathcal{F}_{\mathrm{desc}}(\mathcal{U}) \longrightarrow \mathrm{Hom}(\mathrm{h}_{\mathcal{U}}, \mathcal{F})$$

are isomorphic to the identities. This is straightforward, and left to the reader. □

If we choose a cleavage, the composite of functors

$$\mathrm{Hom}_{\mathcal{C}}(\mathrm{h}_U, \mathcal{F}) \longrightarrow \mathrm{Hom}_{\mathcal{C}}(\mathrm{h}_{\mathcal{U}}, \mathcal{F}) \simeq \mathcal{F}_{\mathrm{desc}}(\mathcal{U}) \simeq \mathcal{F}(\mathcal{U})$$

is isomorphic to the composite

$$\mathrm{Hom}_{\mathcal{C}}(\mathrm{h}_U, \mathcal{F}) \simeq \mathcal{F}(U) \longrightarrow \mathcal{F}(\mathcal{U})$$

where the first functor is the equivalence of the 2-Yoneda Lemma.

4.1.3. Fibered categories with descent.

DEFINITION 4.6. Let $\mathcal{F} \to \mathcal{C}$ be a fibered category on a site $\mathcal{C}$.

(i) $\mathcal{F}$ is a *prestack* over $\mathcal{C}$ if for each covering $\{U_i \to U\}$ in $\mathcal{C}$, the functor $\mathcal{F}(U) \to \mathcal{F}(\{U_i \to U\})$ is fully faithful.

(ii) $\mathcal{F}$ is a *stack* over $\mathcal{C}$ if for each covering $\{U_i \to U\}$ in $\mathcal{C}$, the functor $\mathcal{F}(U) \to \mathcal{F}(\{U_i \to U\})$ is an equivalence of categories.

Concretely, for $\mathcal{F}$ to be a prestack means the following. Let U be an object of $\mathcal{C}$, ξ and η object of $\mathcal{F}(U)$, $\{U_i \to U\}$ a covering, ξ_i and η_i pullbacks of ξ and η to U_i, ξ_{ij} and η_{ij} pullbacks of ξ and η to U_{ij}. Suppose that there are arrows $\alpha_i \colon \xi_i \to \eta_i$ in $\mathcal{F}(U_i)$, such that $\mathrm{pr}_1^* \alpha_i = \mathrm{pr}_2^* \alpha_j \colon \xi_{ij} \to \eta_{ij}$ for all i and j. Then there is a unique arrow $\alpha \colon \xi \to \eta$ in $\mathcal{F}(U)$, whose pullback to $\xi_i \to \eta_i$ is α_i for all i.

This condition can be restated using the functor of arrows of Section 3.7, and the comma topology on the category $(\mathcal{C}/S)$(Definition 2.58).

PROPOSITION 4.7. Let $\mathcal{F}$ be a fibered category over a site $\mathcal{C}$. Then $\mathcal{F}$ is a prestack if and only if for any object S of $\mathcal{C}$ and any two objects ξ and η in $\mathcal{F}(S)$, the functor $\underline{\mathrm{Hom}}_S(\xi,\eta)\colon (\mathcal{C}/S) \to (\mathrm{Set})$ is a sheaf in the comma topology.

PROOF. Let us prove the first part. Assume that for any object S of $\mathcal{C}$ and any two objects ξ and η in $\mathcal{F}(S)$, the functor $\underline{\mathrm{Hom}}_S(\xi,\eta)\colon (\mathcal{C}/S) \to (\mathrm{Set})$ is a sheaf. Take an object U of $\mathcal{C}$, a covering $\{U_i \to U\}$, and two objects ξ and η of $\mathcal{F}(U)$. If we denote by $(\{\xi_i\}, (\alpha_{ij}))$ and $(\{\eta_i\}, (\beta_{ij}))$ the descent data associated with ξ and η respectively, we see easily that the arrows in $\mathcal{F}(\{U_i \to U\})$ are the collections of arrows $\{\phi_i\colon \xi_i \to \eta_i\}$ such that the restrictions of ϕ_i and ϕ_j to the pullbacks of ξ and η to U_{ij} coincide. The fact that $\underline{\mathrm{Hom}}_U(\xi,\eta)$ is a sheaf ensures that this comes from a unique arrow $\xi \to \eta$ in $\mathcal{F}(U)$; but this means precisely that the functor $\mathcal{F}(U) \to \mathcal{F}(\{U_i \to U\})$ is fully faithful.

The proof of the opposite implication is similar, and left to the reader. □

DEFINITION 4.8. An object with descent data $(\{\xi_i\}, \{\phi_{ij}\})$ in $\mathcal{F}(\{U_i \to U\})$ is *effective* if it is isomorphic to the image of an object of $\mathcal{F}(U)$.

Here is another way of saying this: an object with descent data $(\{\xi_i\}, \{\phi_{ij}\})$ in $\mathcal{F}(\{U_i \to U\})$ is effective if there exists an object ξ of $\mathcal{F}(U)$, together with cartesian arrows $\xi_i \to \xi$ over $\sigma_i\colon U_i \to U$, such that the diagram

$$\begin{array}{ccccc}
\mathrm{pr}_2^*\,\xi_j & & \xrightarrow{\ \phi_{ij}\ } & & \mathrm{pr}_1^*\,\xi_i \\
\downarrow & & & & \downarrow \\
\xi_j & & & & \xi_i \\
& \searrow & & \swarrow & \\
& & \xi & &
\end{array}$$

commutes for all i and j. In fact, the cartesian arrows $\xi_i \to \xi$ correspond to isomorphisms $\xi_i \simeq \sigma_i^*\xi$ in $\mathcal{F}(U_i)$; and the commutativity of the diagram above is easily seen to be equivalent to the cocycle condition.

Clearly, $\mathcal{F}$ is a stack if and only if it is a prestack, and all objects with descent data in $\mathcal{F}$ are effective.

Stacks are the correct generalization of sheaves, and give the right notion of "sheaf of categories". We should of course prove the following statement.

PROPOSITION 4.9. Let $\mathcal{C}$ be a site, $F\colon \mathcal{C}^{\mathrm{op}} \to (\mathrm{Set})$ a functor; we can also consider it as a category fibered in sets $F \to \mathcal{C}$.

(i) F is a prestack if and only if it is a separated functor.
(ii) F is stack if and only if it is a sheaf.

PROOF. Consider a covering $\{U_i \to U\}$. The fiber of the category $F \to \mathcal{C}$ over U is precisely the set $F(U)$, while the category $F(\{U_i \to U\})$ is the set of elements $(\xi_i) \in \prod_i F(U_i)$ such that the pullbacks of ξ_i and ξ_j to $F(U_i \times_U U_j)$, via the first and second projections $U_i \times_U U_j \to U_i$ and $U_i \times_U U_j \to U_i$, coincide. The functor $F(U) \to F(\{U_i \to U\})$ is the function that sends each element $\xi \in F(U)$ to the collection of restrictions $(\xi\,|_{U_i})$.

Now, to say that a function, thought of as a functor between discrete categories, is fully faithful is equivalent to saying that it is injective; while to say that it is an equivalence means that it is a bijection. From this both statements follow. □

REMARK 4.10. The terminology here, due to Grothendieck, is a little unfortunate. Fibered categories are a generalization of functors: however, a presheaf is simply a functor, and thus, by analogy, a prestack should be simply a fibered category. What we call a prestack should be called a separated prestack.

I have decided to stick with Grothendieck's terminology, mostly because there is a notion of "separated stack" in the theory of algebraic stacks, and using the more rational term "separated prestack" would make "separated stack" pleonastic.

EXAMPLE 4.11. Let $\mathcal{C}$ be a site. Then I claim that the fibered category $(\mathrm{Sh}/\mathcal{C}) \to \mathcal{C}$, defined in Example 3.20, is a stack.

Here is a sketch of proof. Let F and G be two sheaves on an object X of $\mathcal{C}$: to show that $(\mathrm{Sh}/\mathcal{C})$ is a prestack we want to show that $\underline{\mathrm{Hom}}_X \colon (\mathcal{C}/X)^{\mathrm{op}} \to (\mathrm{Set})$ is a sheaf.

For each arrow $U \to X$, let us denote by F_U and G_U the restrictions of F and G to U. Let $\{U_i \to U\}$ be a covering, $\phi_i \colon F_{U_i} \to G_{U_i}$ a morphism of sheaves on $(\mathcal{C}/U_i)$, such that the restrictions of ϕ_i and ϕ_j to $(\mathcal{C}/U_{ij})$ coincide. Denote by $\phi_{ij} \colon F_{U_{ij}} \to G_{U_{ij}}$ this restriction. If $T \to U$ is an arrow, set $T_i = U_i \times_U T$, and consider the covering $\{T_i \to T\}$. Each $T_i \to U$ factors through U_i, so ϕ_i defines a function $\phi_i \colon FT_i \to GT_i$, and analogously ϕ_{ij} defines functions $\phi_{ij} \colon FT_{ij} \to GT_{ij}$. There is commutative diagram of sets with rows that are equalizers

$$\begin{array}{ccccc} FT & \longrightarrow & \prod_i FT_i & \rightrightarrows & \prod_{ij} FT_{ij} \\ \downarrow & & \downarrow{\scriptstyle \prod_i \phi_i} & & \downarrow{\scriptstyle \prod_{ij} \phi_{ij}} \\ GT & \longrightarrow & \prod_i GT_i & \rightrightarrows & \prod_{ij} GT_{ij}. \end{array}$$

There is a unique function $\phi_T \colon FT \to GT$ that one can insert in the diagram while keeping it commutative. This proves uniqueness. Also, it is easy to check that the collection of the ϕ_T defines a natural transformation $\phi_U \colon F_U \to G_U$, whose restriction $F_{U_i} \to G_{U_i}$ is ϕ_i.

Now let us show that every object with descent data $(\{F_i\}, \{\phi_{ij}\})$ is effective. Here F_i is a sheaf on $(\mathcal{C}/U_i)$, and ϕ_{ij} is an isomorphism of sheaves on $(\mathcal{C}/U_{ij})$ between the restrictions $(F_j)_{U_{ij}} \simeq (F_i)_{U_{ij}}$.

For each object $T \to U$ of $(\mathcal{C}/U)$, set $T_i = U_i \times_U T$ as before, and define FT to be the subset of $\prod_i F_i T_i$ consisting of objects $(s_i) \in \prod_i FT_i$, with the property that ϕ_{ij} carries the restriction $(s_j)_{T_{ij}}$ to $(s_i)_{T_{ij}}$. In other words, FT is the equalizer of two functions $\prod_i F_i T_i \to \prod_{ij} F_i T_{ij}$, where the first sends (s_i) to the collections of restrictions $\bigl((s_i)_{U_{ij}}\bigr)$, and the second sends it to $\bigl(\phi_{ij}(s_j)_{U_{ij}}\bigr)$.

For any arrow $T' \to T$ in $(\mathcal{C}/U)$, it easy to see that the product of the restriction functions $\prod_i F_i T_i \to \prod_i F_i T_i'$ carries FT to FT'; this gives F the structure of a functor $(\mathcal{C}/U)^{\mathrm{op}} \to (\mathrm{Set})$. We leave it to the reader to check that F is a sheaf.

Now we have to show that the image of F into $(\mathrm{Sh}/S)(\{U_i \to U\})$ is isomorphic to $(\{F_i\}, \{\phi_{ij}\})$.

For each index k let us construct an isomorphism of the restriction F_{U_k} with F_k as sheaves on $(\mathcal{C}/U_k)$. Let $T \to U_k$ be an object of $(\mathcal{C}/U_k)$, s an element of $F_k T$. Each T_i maps into U_{ik}, so we produce an element $\bigl(\phi_{ik}(s_{T_i})\bigr) \in \prod F_i T_i$; the cocycle condition ensures that this is an element of FT. This defines a natural transformation $F_k \to F_{U_k}$.

In the other direction, let $T \to U_k$ be an object of $(\mathcal{C}/U_k)$, (s_i) an element of $FT \subseteq \prod_i F_i T_i$. Factor each $T_i \to U_i$ through the projection $U_{ki} \to U_i$. Then $\phi_{ki}(s_i)$ is an element of $F_k T_i$, and the cocycle condition implies that the restrictions of $\phi_{ki}(s_i)$ and $\phi_{kj}(s_j)$ to $F_k T_{ij}$ coincide. Hence there a unique element of $F_k T$ that restricts to $\phi_{ki}(s_i) \in F_k T_i$ for each i. This construction defines a function $FT \to F_k T$ for each T, which is easily seen to give a natural transformation $F_{U_k} \to F_k$.

We leave it to the reader to check that these two natural transformations are inverse to one another, so they define an isomorphism of sheaves $F_{U_k} \simeq F_k$; and that this collection of isomorphisms constitutes an isomorphism in $(\mathrm{Sh}/S)(\{U_i \to U\})$ between the object associated with F and the given object $(\{F_i\}, \{\phi_{ij}\})$, which is therefore effective.

4.1.4. The functorial behavior of descent data. Descent data have three kinds of functorial properties: they are functorial for morphisms of fibered categories, functorial on the objects, and functorial under refinement.

Let $F\colon \mathcal{F} \to \mathcal{G}$ a morphism of categories fibered over $\mathcal{C}$. For any covering $\mathcal{U} = \{U_i \to U\}$ we get a functor $F_{\mathcal{U}}\colon \mathcal{F}(\mathcal{U}) \to \mathcal{G}(\mathcal{U})$ defined at the level of objects by the obvious rule

$$F_{\mathcal{U}}(\{\xi_i\}, \{\phi_{ij}\}) = (\{F\xi_i\}, \{F\phi_{ij}\})$$

and at the level of arrows by the equally obvious rule

$$F_{\mathcal{U}}\{\alpha_i\} = \{F\alpha_i\}.$$

Furthermore, if $\rho\colon F \to G$ is a base-preserving natural transformation of morphisms, there is an induced natural transformation of functors $\rho_{\mathcal{U}}\colon F_{\mathcal{U}} \to G_{\mathcal{U}}$, defined by

$$(\rho_{\mathcal{U}})_{(\{\xi_i\}, \{\phi_{ij}\})} = \{\rho_{\xi_i}\}.$$

Therefore, if F is an equivalence of fibered categories, $F_{\mathcal{U}}$ is also an equivalence.

We leave it to the reader to check that the diagram

$$\begin{array}{ccc} \mathcal{F}(U) & \longrightarrow & \mathcal{F}(\mathcal{U}) \\ \downarrow & & \downarrow \\ \mathcal{G}(U) & \longrightarrow & \mathcal{G}(\mathcal{U}) \end{array}$$

commutes, in the sense that the two composites $\mathcal{F}(U) \to \mathcal{G}(\mathcal{U})$ are isomorphic. From this we obtain the following useful fact.

PROPOSITION 4.12.

(a) If F is an equivalence of fibered categories and $\mathcal{F}(U) \to \mathcal{F}(\mathcal{U})$ is an equivalence of categories, then $\mathcal{G}(U) \to \mathcal{G}(\mathcal{U})$ is also an equivalence of categories.

(b) If two fibered categories over a site are equivalent, and one of them is a stack, or a prestack, the other is also a stack, or a prestack.

All this can be restated more elegantly using sieves. The morphism $F\colon \mathcal{F} \to \mathcal{G}$ induces a functor $F_*\colon \operatorname{Hom}(h_{\mathcal{U}}, \mathcal{F}) \to \operatorname{Hom}(h_{\mathcal{U}}, \mathcal{G})$, that is the composite with F at the level of objects. In this case the diagram becomes

$$\begin{array}{ccc} \operatorname{Hom}(h_U, \mathcal{F}) & \longrightarrow & \operatorname{Hom}(h_{\mathcal{U}}, \mathcal{F}) \\ \downarrow & & \downarrow \\ \operatorname{Hom}(h_U, \mathcal{G}) & \longrightarrow & \operatorname{Hom}(h_{\mathcal{U}}, \mathcal{G}) \end{array}$$

which strictly commutative, that is, the two composites are equal, not simply isomorphic. We leave the easy details to the reader.

We are not going to need the functoriality of descent data for the objects, so we will only sketch the idea: if $\{U_i \to U\}$ is a covering and $V \to U$ is an arrow, then there is a functor $\mathcal{F}(\{U_i \to U\}) \to \mathcal{F}(\{V \times_U U_i \to V\})$. If $(\{\xi_i\}, \{\phi_{ij}\})$ is an object of $\mathcal{F}(\{U_i \to U\})$, its image in $\mathcal{F}(\{V \times_U U_i \to V\})$ is obtained by pulling back the ξ_i and the ϕ_{ij} along the projection $V \times_U U_i \to U_i$.

Now suppose that $\mathcal{F}$ is a category fibered over $\mathcal{C}$, $\mathcal{U} = \{U_i \to U\}_{i \in I}$ a covering, $\mathcal{V} = \{V_{i'} \to U\}_{i' \in I'}$ a refinement of $\mathcal{U}$. For each index i' choose a factorization $V_{i'} \xrightarrow{f_{i'}} U_{\mu(i')} \to U$ for a certain $\mu(i') \in I$; this defines a function $\mu\colon I' \to I$.

This induces a functor $\mathcal{F}\mathcal{U} \to \mathcal{F}\mathcal{V}$, as follows. An object $(\{\xi_i\}, \{\phi_{ij}\})$ is sent to $(\{f_{i'}^* \xi_{\mu(i')}\}, \{f_{i'}^* \phi_{\mu(i')\mu(j')}\})$; we leave to the reader to check that this is also an object with descent data. An arrow

$$\{\alpha_i\}\colon (\{\xi_i\}, \{\phi_{ij}\}) \longrightarrow (\{\eta_i\}, \{\psi_{ij}\})$$

is a collection of arrows $\alpha_i\colon \xi_i \to \eta_i$ in $\mathcal{F}(U_i)$, and these can be pulled back to arrows $f_{i'}^* \alpha_{\mu(i')}\colon f_{i'}^* \xi_{\mu(i')} \to f_{i'}^* \eta_{\mu(i')}$. We leave it the reader to verify that the collection $\{f_{i'}^* \alpha_{\mu(i')}\}$ yields an arrow

$$(\{f_{i'}^* \xi_{\mu(i')}, \{f_{i'}^* \phi_{\mu(i')\mu(j')}\}) \longrightarrow (\{f_{i'}^* \eta\mu(i'), \{f_{i'}^* \psi_{\mu(i')\mu(j')}\}),$$

and that this defines a functor.

This functor is essentially independent of the function $\mu\colon I' \to I$, that is, if we change the function we get isomorphic functors. This is seen as follows. Suppose that $\nu\colon I' \to I$ is another function, and that there are factorizations $V_{i'} \xrightarrow{g_{i'}} U_{\nu(i')} \to U$. The two arrows $f_{i'}$ and $g_{i'}$ induce arrows $(f_{i'}, g_{i'})\colon U_{\mu(i')\nu(i')}$. We define an isomorphism $f_{i'}^* \xi_{\mu(i')} \simeq g_{i'}^* \xi_{\nu(i')}$ by composing the isomorphisms in the following diagram

$$\begin{array}{ccc} f_{i'}^* \xi_{\mu(i')} & \dashrightarrow & g_{i'}^* \xi_{\nu(i')} \\ \downarrow \simeq & & \uparrow \simeq \\ (f_{i'}, g_{i'})^* \operatorname{pr}_1^* \xi_{\mu(i')} & \xrightarrow{(f_{i'}, g_{i'})^* \phi_{ij}} & (f_{i'}, g_{i'})^* \operatorname{pr}_2^* \xi_{\nu(i')} \end{array}$$

The cocycle condition ensures that this gives an isomorphism of descent data between $(\{f_{i'}^* \xi_{\mu(i')}\}, \{f_{i'}^* \phi_{\mu(i')\mu(j')}\})$ and $(\{g_{i'}^* \xi_{\nu(i')}\}, \{g_{i'}^* \phi_{\nu(i')\nu(j')}\})$ (we leave the details to the reader). It is easy to check that this gives an isomorphism of functors.

Also, if $\mathcal{W} = \{W_{i''} \to U\}_{i'' \in I''}$ is a refinement of $\mathcal{V}$, it is also a refinement of $\mathcal{U}$. After choosing functions $\mu\colon I' \to I$, $\nu\colon I'' \to I'$ and $\rho\colon I'' \to I$, and factorizations $V_{i'} \xrightarrow{f_{i'}} U_{\mu(i')} \to U$, $W_{i''} \xrightarrow{g_{i''}} U_{\nu(i'')} \to U$ and $W_{i''} \xrightarrow{h_{i''}} U_{\rho(i'')} \to U$ we get functors $\mathcal{F}\mathcal{U} \to \mathcal{F}\mathcal{V}$, $\mathcal{F}\mathcal{V} \to \mathcal{F}\mathcal{W}$ and $\mathcal{F}\mathcal{U} \to \mathcal{F}\mathcal{W}$. I claim that the composite of the first two is isomorphic to the third.

To check this, we may change the factorizations $W_{i''} \xrightarrow{h_{i''}} U_{\rho(i'')} \to U$, because, as we have just seen, this does not change the isomorphism class of the functor $\mathcal{F}\mathcal{U} \to \mathcal{F}\mathcal{W}$; hence we may assume that $\rho = \mu \circ \nu\colon I'' \to I$, and that $h_{i''}\colon W_{i''} \to U_{\rho(i'')}$ equals the composite $f_{\nu(i'')} \circ g_{i''} \to I$. Given an object $(\{\xi_i\}, \{\phi_{ij}\})$ of $\mathcal{F}\mathcal{U}$, its image in $\mathcal{F}\mathcal{W}$ under the functor $\mathcal{F}\mathcal{U} \to \mathcal{F}\mathcal{W}$ is the object

$$(\{(f_{\nu(i'')} g_{i''})^* \xi_{i''}\}, \{(f_{\nu(i'')} g_{i''})^* \phi_{\nu\mu(i'')\,\nu\mu(j'')}\})$$

of $\mathcal{FW}$, while its image under the composite $\mathcal{FU} \to \mathcal{FV} \to \mathcal{FV} \to \mathcal{FW}$ is the object

$$\left(\{g_{i''}^* f_{\nu(i'')}^* \xi_{i''}\}, \{g_{i''}^* f_{\nu(i'')}^* \phi_{\nu\mu(i'')\,\nu\mu(j'')}\}\right).$$

The canonical isomorphisms $(f_{\nu(i'')} g_{i''})^* \xi_{i''} \simeq g_{i''}^* f_{\nu(i'')}^* \xi_{i''}$ give an isomorphism of the two objects of $\mathcal{FW}$, and this defines the desired isomorphism of functors.

Once again, in the language of sieves everything is much easier: if $\mathcal{V}$ is a refinement of $\mathcal{U}$, then $\mathrm{h}_{\mathcal{V}}$ is a subfunctor of $\mathrm{h}_{\mathcal{U}}$, and the embedding $\mathrm{h}_{\mathcal{V}} \hookrightarrow \mathrm{h}_{\mathcal{U}}$ induces a functor

$$\mathrm{Hom}_{\mathcal{C}}(\mathrm{h}_{\mathcal{U}}, \mathcal{F}) \longrightarrow \mathrm{Hom}_{\mathcal{C}}(\mathrm{h}_{\mathcal{V}}, \mathcal{F})$$

with no choice required.

Also, in this language the composite

$$\mathrm{Hom}_{\mathcal{C}}(\mathrm{h}_{\mathcal{U}}, \mathcal{F}) \longrightarrow \mathrm{Hom}_{\mathcal{C}}(\mathrm{h}_{\mathcal{V}}, \mathcal{F}) \longrightarrow \mathrm{Hom}_{\mathcal{C}}(\mathrm{h}_{\mathcal{W}}, \mathcal{F})$$

equals the functor $\mathrm{Hom}_{\mathcal{C}}(\mathrm{h}_{\mathcal{U}}, \mathcal{F}) \to \mathrm{Hom}_{\mathcal{C}}(\mathrm{h}_{\mathcal{W}}, \mathcal{F})$ on the nose.

4.1.5. Stacks and sieves. Using the description of the category of objects with descent data in Proposition 4.5 we can give the following very elegant characterization of stacks, which generalizes the characterization of sheaves given in Corollary 2.42.

COROLLARY 4.13. *A fibered category $\mathcal{F} \to \mathcal{C}$ is a stack if and only if for any covering $\mathcal{U}$ in $\mathcal{C}$ the functor*

$$\mathrm{Hom}_{\mathcal{C}}(\mathrm{h}_U, \mathcal{F}) \longrightarrow \mathrm{Hom}_{\mathcal{C}}(\mathrm{h}_{\mathcal{U}}, \mathcal{F})$$

induced by the embedding $\mathrm{h}_{\mathcal{U}} \subseteq \mathrm{h}_U$ is an equivalence.

This can sharpened, as in Proposition 2.42.

PROPOSITION 4.14. *A fibered category $\mathcal{F} \to \mathcal{C}$ is a stack if and only if for any object $\mathcal{U}$ of $\mathcal{C}$ and sieve S on U belonging to $\mathcal{T}$, the functor*

$$\mathrm{Hom}_{\mathcal{C}}(\mathrm{h}_U, \mathcal{F}) \longrightarrow \mathrm{Hom}_{\mathcal{C}}(S, \mathcal{F})$$

induced by the embedding $\mathrm{h}_U \subseteq S$ is an equivalence.

Furthermore, $\mathcal{F}$ is a prestack if and only if the functor above is fully faithful for all U and S.

PROOF. The fact that if the functor is an equivalence then $\mathcal{F}$ is a stack follows from Corollary 4.13, so we only need to prove the converse (and similarly for the second statement).

Let S be a sieve belonging to $\mathcal{T}$ on an object U of $\mathcal{C}$. Choose a covering $\mathcal{U} = \{U_i \to U\}$ of U such that $\mathrm{h}_{\mathcal{U}} \subseteq S$: the restriction functor $\mathrm{Hom}_{\mathcal{C}}(\mathrm{h}_U, \mathcal{F}) \to \mathrm{Hom}_{\mathcal{C}}(\mathrm{h}_{\mathcal{U}}, \mathcal{F})$ is an equivalence, and it factors as

$$\mathrm{Hom}_{\mathcal{C}}(\mathrm{h}_U, \mathcal{F}) \longrightarrow \mathrm{Hom}_{\mathcal{C}}(S, \mathcal{F}) \longrightarrow \mathrm{Hom}_{\mathcal{C}}(\mathrm{h}_{\mathcal{U}}, \mathcal{F}).$$

Again by Corollary 4.13, $\mathrm{Hom}_{\mathcal{C}}(\mathrm{h}_U, \mathcal{F}) \to \mathrm{Hom}_{\mathcal{C}}(\mathrm{h}_{\mathcal{U}}, \mathcal{F})$ is fully faithful, and it is an equivalence when $\mathcal{F}$ is a stack: hence $\mathrm{Hom}_{\mathcal{C}}(S, \mathcal{F}) \to \mathrm{Hom}_{\mathcal{C}}(\mathrm{h}_{\mathcal{U}}, \mathcal{F})$ is full, and it essentially surjective whenever $\mathcal{F}$ is a stack. So we see that the following lemma suffices.

LEMMA 4.15. Let $\mathcal{F}$ be a prestack over a site $\mathcal{C}$, S and S' be sieves belonging to the topology of $\mathcal{C}$ with $S' \subseteq S$. Then the induced restriction functor

$$\mathrm{Hom}_{\mathcal{C}}(S, \mathcal{F}) \longrightarrow \mathrm{Hom}_{\mathcal{C}}(S', \mathcal{F})$$

is faithful.

This is a generalization of Lemma 2.43.

PROOF. The proof is very similar to that of Lemma 2.43. Let F and G be two morphisms $S \to \mathcal{F}$, ϕ and ψ two base-preserving natural transformations, inducing the same natural transformations from the restriction of F to $h_{\mathcal{U}}$ to that of G. Let $T \to U$ be an arrow in S; we need to prove that

$$\phi_{T \to U} = \psi_{T \to U} \colon F(T \to U) \longrightarrow G(T \to U).$$

Consider the fibered products $T \times_U U_i$, with the first projections $p_i \colon T \times_U U_i$. Since F and G are morphisms, and S is a functor, so that every arrow in S is cartesian, the arrows

$$F(p_i)\colon F(T \times_U U_i \to U) \longrightarrow F(T \to U)$$

and

$$G(p_i)\colon G(T \times_U U_i \to U) \longrightarrow G(T \to U)$$

are cartesian. Consider the covering $\{T \times_U U_i \to T\}$: since the composite $T \times_U U_i \to T \to U$ is in $h_{\mathcal{U}}T$, we have

$$\phi_{T \times_U U_i \to U} = \psi_{T \times_U U_i \to U} \colon F(T \times_U U_i \to U) \longrightarrow G(T \times_U U_i \to U).$$

Hence the commutativity of the diagrams

$$\begin{array}{ccc} F(T \times_U U_i \to U) & \xrightarrow{\phi_{T \times_U U_i \to U}} & G(T \times_U U_i \to U) \\ \Big\downarrow{\scriptstyle F(p_i} & & \Big\downarrow{\scriptstyle G(p_i)} \\ F(T \to U) & \xrightarrow{\phi_{T \to U}} & G(T \to U) \end{array}$$

can be interpreted as saying that the pullbacks of $\phi_{T \to U}$ and $\psi_{T \to U}$ to $T \times_U U_i$ are the same. Since the functors of arrows of $\mathcal{F}$ form a sheaf, since $\mathcal{F}$ is a prestack, this implies that $\phi_{T \to U}$ and $\psi_{T \to U}$ are equal, as claimed. □

This ends the proof of Proposition 4.14. □

Since two equivalent topologies on the same category have the same sieves, we obtain the following generalization of Proposition 2.49.

PROPOSITION 4.16. Let $\mathcal{C}$ a category, $\mathcal{T}$ and $\mathcal{T}'$ two topologies on $\mathcal{C}$, $\mathcal{F} \to \mathcal{C}$ a fibered category. Suppose that $\mathcal{T}'$ is subordinate to $\mathcal{T}$. If $\mathcal{F}$ is a prestack, or a stack, relative to $\mathcal{T}$, then it is also a prestack, or a stack, relative to $\mathcal{T}'$.

In particular, if $\mathcal{T}$ and $\mathcal{T}'$ are equivalent, then $\mathcal{F}$ is a stack relative to $\mathcal{T}$ if and only if it is also a stack relative to $\mathcal{T}'$.

For later use, we note the following consequence of Lemma 4.15.

LEMMA 4.17. If $\mathcal{F}$ is a prestack on a site, $\mathcal{U}$ and $\mathcal{V}$ two coverings of an object U of $\mathcal{C}$, such that $\mathcal{V}$ is a refinement of $\mathcal{U}$, and $\mathcal{F}(U) \to \mathcal{F}(\mathcal{V})$ is an equivalence, then $\mathcal{F}(U) \to \mathcal{F}(\mathcal{U})$ is also an equivalence.

4.1.6. Substacks.

DEFINITION 4.18. Let $\mathcal{C}$ be a site, $\mathcal{F} \to \mathcal{C}$ a stack. A *substack* of $\mathcal{F}$ is a fibered subcategory that is a stack.

EXAMPLE 4.19. Let $\mathcal{C}$ be a site, $\mathcal{F} \to \mathcal{C}$ a stack, $\mathcal{G}$ a full subcategory of $\mathcal{F}$ satisfying the following two conditions.

(i) Any cartesian arrow in $\mathcal{F}$ whose target is in $\mathcal{G}$ is also in $\mathcal{G}$.
(ii) Let $\{U_i \to U\}$ be a covering in $\mathcal{C}$, ξ an object of $\mathcal{F}(U)$, ξ_i pullbacks of ξ to U_i. If ξ_i is in $\mathcal{G}$ for all i, then ξ is in $\mathcal{G}$
Then $\mathcal{G}$ is a substack.

There are many examples of the situation above: for example, as we shall see (Theorem 4.23) the fibered category (QCoh/S) is a stack over (Sch/S) with the fpqc topology. Then the full subcategory of (QCoh/S) consisting of locally free sheaves of finite rank satisfies the two conditions, hence it is a substack.

PROPOSITION 4.20. Let $\mathcal{C}$ be a site, $\mathcal{F} \to \mathcal{C}$ a fibered category. Recall that $\mathcal{F}_{\mathrm{cart}}$ is the associated category fibered in groupoids (Definition 3.31).

(i) If $\mathcal{F}$ is a stack, so is $\mathcal{F}_{\mathrm{cart}}$.
(ii) If $\mathcal{F}$ is a prestack and $\mathcal{F}_{\mathrm{cart}}$ is a stack, then $\mathcal{F}$ is also a stack.

PROOF. The isomorphisms in $\mathcal{F}_{\mathrm{cart}}$ are all cartesian; hence, given a covering $\{U_i \to U\}$ in $\mathcal{C}$, the categories $\mathcal{F}(\{U_i \to U\})$ and $\mathcal{F}_{\mathrm{cart}}(\{U_i \to U\})$ have the same objects, and the effective objects with descent data are the same. So is enough to prove that if $\mathcal{F}$ is a prestack then $\mathcal{F}_{\mathrm{cart}}$ is a prestack.

Let ξ and η be two objects in some $\mathcal{F}(U)$. Let $\{U_i \to U\}$ be a covering, ξ_i and η_i pullbacks of ξ and η to U_i, ξ_{ij} and η_{ij} pullbacks to U_{ij}, $\alpha_i \colon \xi_i \to \eta_i$ arrows in $\mathcal{F}_{\mathrm{cart}}(U_i)$, such that $\mathrm{pr}_1^* \alpha_i = \mathrm{pr}_2^* \alpha_j \colon \xi_{ij} \to \eta_{ij}$. Then there is unique arrow $\alpha \colon \xi \to \eta$ that restricts to α_i for each i; and it is enough to show that α is cartesian. But the cartesian arrows in $\mathcal{F}(U)$ and in each $\mathcal{F}(U_i)$ are the isomorphisms; hence the α_i are isomorphisms, and the arrows $\alpha_i^{-1} \colon \eta_i \to \xi$ come from a unique arrow $\beta \colon \eta \to \xi$. The composites $\beta \circ \alpha$ and $\alpha \circ \beta$ pull back to identities in each $\mathcal{F}(U_i)$, and so they must be identities in $\mathcal{F}(U)$. This shows that α is in $\mathcal{F}_{\mathrm{cart}}(U)$, and completes the proof. □

4.2. Descent theory for quasi-coherent sheaves

4.2.1. Descent for modules over commutative rings. Here we develop an affine version of the descent theory for quasi-coherent sheaves. It is only needed to prove Theorem 4.23 below, so it may be a good idea to postpone reading it until after reading the next section on descent for quasi-coherent sheaves.

If A is a commutative ring, we will denote by Mod_A the category of modules over A.

Consider a ring homomorphism $f \colon A \to B$. If M is an A-module, we denote by $\iota_M \colon M \otimes_A B \simeq B \otimes_A M$ the usual isomorphism of A-modules defined by $\iota_M(n \otimes b) = b \otimes n$. Furthermore, we denote by $\alpha_M \colon M \to B \otimes_A M$ the homomorphism defined by $\alpha_M(m) = 1 \otimes m$.

For each $r \geq 0$ set

$$B^{\otimes r} = \overbrace{B \otimes_A B \otimes_A \cdots \otimes_A B}^{r \text{ times}}.$$

A B-module N becomes a module over $B^{\otimes 2}$ in two different ways, as $N \otimes_A B$ and $B \otimes_A N$; in both cases the multiplication is defined by the formula $(b_1 \otimes b_2)(x_1 \otimes x_2) = b_1x_1 \otimes b_2x_2$. Analogously, N becomes a module over $B^{\otimes 3}$ as $N \otimes_A B \otimes_A B$, $B \otimes_A N \otimes_A B$ and $B \otimes_A B \otimes_A N$ (more generally, N becomes a module over $B^{\otimes r}$ in r different ways; but we will not need this).

Let us assume that we have a homomorphism of $B^{\otimes 2}$-modules $\psi\colon N \otimes_A B \to B \otimes_A N$. Then there are three associated homomorphism of $B^{\otimes 3}$-modules

$$\begin{aligned}
\psi_1 &\colon B \otimes_A N \otimes_A B \to B \otimes_A B \otimes_A N,\\
\psi_2 &\colon N \otimes_A B \otimes_A B \to B \otimes_A B \otimes_A N,\\
\psi_3 &\colon N \otimes_A B \otimes_A B \to B \otimes_A N \otimes B
\end{aligned}$$

by inserting the identity in the first, second and third position, respectively. More explicitly, we have $\psi_1 = \mathrm{id}_B \otimes \psi$, $\psi_2 = \psi \otimes \mathrm{id}_B$, while we have $\psi_2(x_1 \otimes x_2 \otimes x_3) = \sum_i y_i \otimes x_2 \otimes z_i$ if $\psi(x_1 \otimes x_3) = \sum_i y_i \otimes z_i \in B \otimes_A N$. Alternatively, $\psi_2 = (\mathrm{id}_B \otimes \iota_N) \circ \psi \circ (\mathrm{id}_N \otimes \iota_B)$.

Let us define a category $\mathrm{Mod}_{A\to B}$ as follows. Its objects are pairs (N,ψ), where N is a B-module and $\psi\colon N \otimes_A B \simeq B \otimes_A N$ is an isomorphism of $B^{\otimes 2}$-modules such that

$$\psi_2 = \psi_1 \circ \psi_3 \colon N \otimes_A B \otimes_A B \longrightarrow B \otimes_A B \otimes_A N.$$

An arrow $\beta\colon (N,\psi) \to (N',\psi')$ is a homomorphism of B-modules $\beta\colon N \to N'$, making the diagram

$$\begin{array}{ccc}
N \otimes_A B & \xrightarrow{\ \psi\ } & B \otimes_A N\\
\Big\downarrow{\scriptstyle \beta\otimes \mathrm{id}_B} & & \Big\downarrow{\scriptstyle \mathrm{id}_B \otimes \beta}\\
N' \otimes_A B & \xrightarrow{\ \psi'\ } & B \otimes_A N'
\end{array}$$

commutative.

We have a functor $F\colon \mathrm{Mod}_A \to \mathrm{Mod}_{A\to B}$, sending an A-module M to the pair $(B \otimes_A M, \psi_M)$, where

$$\psi_M\colon (B \otimes_A M) \otimes_A B \longrightarrow B \otimes_A (B \otimes_A M)$$

is defined by the rule

$$\psi_M(b \otimes m \otimes b') = b \otimes b' \otimes m.$$

In other words, $\psi_M = \mathrm{id}_B \otimes \iota_M$.

It is easily checked that ψ_M is an isomorphism of $B^{\otimes 2}$-modules, and that $(M \otimes_A B, \psi_M)$ is in fact an object of $\mathrm{Mod}_{A\to B}$.

If $\alpha\colon M \to M'$ is a homomorphism of A-modules, one sees immediately that $\mathrm{id}_B \otimes \alpha\colon B \otimes_A M \to B \otimes_A M'$ is an arrow in $\mathrm{Mod}_{A\to B}$. This defines the desired functor F.

THEOREM 4.21. *If B is faithfully flat over A, the functor*

$$F\colon \mathrm{Mod}_A \longrightarrow \mathrm{Mod}_{A\to B}$$

defined above is an equivalence of categories.

PROOF. Let us define a functor $G\colon \mathrm{Mod}_{A\to B} \to \mathrm{Mod}_A$. We send an object (N,ψ) to the A-submodule $GN \subseteq N$ consisting of elements $n \in N$ such that $1 \otimes n = \psi(n \otimes 1)$.

Given an arrow $\beta\colon (N,\psi) \to (N',\psi')$ in $\mathrm{Mod}_{A\to B}$, it follows from the definition of an arrow that β takes GN to GN'; this defines the functor G.

We need to check that the composites GF and FG are isomorphic to the identity. For this we need the following generalization of Lemma 2.61. Recall that we have defined the two homomorphisms of A-algebras

$$e_1, e_2 \colon B \longrightarrow B \otimes_A B$$

by $e_1(b) = b \otimes 1$ and $e_2(b) = 1 \otimes b$.

LEMMA 4.22. *Let M be an A-module. Then the sequence*

$$0 \longrightarrow M \xrightarrow{\alpha_M} B \otimes_A M \xrightarrow{(e_1 - e_2) \otimes \mathrm{id}_M} B^{\otimes 2} \otimes M$$

is exact.

The proof is a simple variant of the proof of Lemma 2.61.

Now notice that

$$\begin{aligned}\big((e_1 - e_2) \otimes \mathrm{id}_M\big)(b \otimes m) &= b \otimes 1 \otimes m - 1 \otimes b \otimes m \\ &= \psi_M(b \otimes m \otimes 1) - 1 \otimes b \otimes m\end{aligned}$$

for all m and b; and this implies that

$$\big((e_1 - e_2) \otimes \mathrm{id}_M\big)(x) = \psi_M(x \otimes 1) - 1 \otimes x$$

for all $x \in B \otimes_A M$. Hence $G(B \otimes_A M, \psi_M)$ is the kernel of $(e_1 - e_2) \otimes \mathrm{id}_M$, and the homomorphism $M \to B \otimes M$ establishes a natural isomorphism between M and $G(M \otimes_A B) = GF(M)$, showing that GF is isomorphic to the identity.

Now take an object (N, ψ) of $\mathrm{Mod}_{A \to B}$, and set $M = G(N, \psi)$. The fact that M is an A-submodule of the B-module N induces a homomorphism of B-modules $\theta \colon B \otimes_A M \to N$ with the usual rule $\theta(b \otimes m) = bm$. Let us check that θ is an arrow in $\mathrm{Mod}_{A \to B}$, that is, that the diagram

$$\begin{array}{ccc} B \otimes_A B \otimes_A M & \xrightarrow{\mathrm{id}_B \otimes \theta} & B \otimes_A N \\ \Big\downarrow{\scriptstyle \mathrm{id}_B \otimes \iota_M} & & \Big\downarrow{\scriptstyle \psi} \\ B \otimes_A M \otimes_A B & \xrightarrow{\theta \otimes \mathrm{id}_B} & N \otimes_A B \end{array}$$

commutes. The calculation is as follows:

$$\begin{aligned}\psi(\mathrm{id}_B \otimes \theta)(b \otimes b' \otimes m) &= \psi(b \otimes b'm) \\ &= \psi\big((b \otimes b')(1 \otimes m)\big) \\ &= (b \otimes b')\psi(1 \otimes m) \\ &= (b \otimes b')(1 \otimes m) \qquad \text{(because } m \in M) \\ &= (bm \otimes b') \\ &= (\theta \otimes \mathrm{id}_B)(b \otimes m \otimes b') \\ &= (\theta \otimes \mathrm{id}_B)(\mathrm{id}_B \otimes \iota_M)(b \otimes b' \otimes m).\end{aligned}$$

So this θ defines a natural transformation $\mathrm{id} \to FG$. We have to check that θ is an isomorphism.

Consider the homomorphisms $\alpha, \beta \colon N \to B \otimes N$ defined by $\alpha(n) = 1 \otimes n$ and $\beta(n) = \psi(n \otimes 1)$; by definition, M is the kernel of $\alpha - \beta$. There is a diagram with exact rows

$$\begin{array}{ccccccccc} 0 & \longrightarrow & M \otimes B & \xrightarrow{i \otimes \mathrm{id}_B} & N \otimes_A B & \xrightarrow{(\alpha-\beta)\otimes_A \mathrm{id}_B} & B \otimes_A N \otimes_A B \\ & & \downarrow{\scriptstyle \theta\circ\iota_M} & & \downarrow{\scriptstyle \psi} & & \downarrow{\scriptstyle \psi_1} \\ 0 & \longrightarrow & N & \xrightarrow{\alpha_M} & B \otimes_A N & \xrightarrow{(e_2-e_1)\otimes \mathrm{id}_N} & B \otimes_A B \otimes_A N \end{array}$$

where $i \colon M \hookrightarrow N$ denotes the inclusion. Let us show that it is commutative. For the first square, we have

$$\alpha_M \theta \iota_M (m \otimes b) = 1 \otimes bm$$

while

$$\begin{aligned} \psi(i \otimes \mathrm{id}_B)(m \otimes b) &= \psi(m \otimes b) \\ &= \psi\big((1 \otimes b)(m \otimes 1)\big) \\ &= (1 \otimes b)\psi(m \otimes 1) \\ &= (1 \otimes b)(1 \otimes m) \\ &= 1 \otimes bm. \end{aligned}$$

For the second square, it is immediate to check that $\psi_1 \circ (\alpha \otimes \mathrm{id}_B) = (e_2 \otimes \mathrm{id}_N) \circ \psi$. On the other hand

$$\begin{aligned} \psi_1(\beta \otimes \mathrm{id}_B)(n \otimes b) &= \psi_1\big(\psi(n \otimes 1) \otimes b\big) \\ &= \psi_1 \psi_3 (n \otimes 1 \otimes b) \\ &= \psi_2(n \otimes 1 \otimes b) \\ &= (e_1 \otimes \mathrm{id}_N)\psi(n \otimes b). \end{aligned}$$

Both ψ and ψ_1 are isomorphisms; hence $\theta \circ \iota_N$ is an isomorphism, so θ is an isomorphism, as desired.

This finishes the proof of Theorem 4.21. □

4.2.2. Descent for quasi-coherent sheaves. Here is the main result of descent theory for quasi-coherent sheaves. It states that quasi-coherent sheaves satisfy descent with respect to the fpqc topology; in other words, they form a stack with respect to either topology. This is quite remarkable, because quasi-coherent sheaves are sheaves in that Zariski topology, which is much coarser, so a priori one would not expect this to happen.

Given a scheme S, recall that in §3.2.1 we have constructed the fibered category (QCoh/S) of quasi-coherent sheaves, whose fiber of a scheme U over S is the category $\mathrm{QCoh}(U)$ of quasi-coherent sheaves on U.

THEOREM 4.23. *Let S be a scheme. The fibered category (QCoh/S) over (Sch/S) is stack with respect to the fpqc topology.*

REMARK 4.24. This would fail in the "wild" flat topology of Remark 2.56: in this topology (QCoh/S) is not even a prestack.

Take the covering $\{V_p \to U\}$ defined there, and the quasi-coherent sheaf $\oplus_p \mathcal{O}_p$, the direct sum of the structure sheaves of all the closed points. The restriction of $\oplus_p \mathcal{O}_p$ to each V_p is the structure sheaf $\mathcal{O}_p$ of the closed point, since pullbacks commute with direct sums, and the restriction of each $\mathcal{O}_q$ to V_p is zero for $p \neq q$.

For each p consider the projection $\pi_p \colon \mathcal{O}_{V_p} \to \mathcal{O}_p$; it easy to see that $\mathrm{pr}_1^* \pi_p = \mathrm{pr}_2^* \pi_q \colon \mathcal{O}_{V_p \times_U V_q} \to f_{p,q}^*(\oplus_p \mathcal{O}_p)$, where $f_{p,q} \colon V_p \times_U V_q \to U$ is the obvious morphism.

On the other hand there is no homomorphism $\mathcal{O}_U \to \oplus_p \mathcal{O}_p$ that pulls back to π_p for each p. In fact, such a homomorphism would correspond to a section of $\oplus_p \mathcal{O}_p$ that is 1 at each closed point, and this does not exist because of the definition of direct sum.

For the proof of the theorem we will use the following criterion, a generalization of that of Lemma 2.60.

LEMMA 4.25. Let S be a scheme, $\mathcal{F}$ be a fibered category over the category (Sch/S). Suppose that the following conditions are satisfied.

(i) $\mathcal{F}$ is a stack with respect to the Zariski topology.
(ii) Whenever $V \to U$ is a flat surjective morphism of affine S-schemes, the functor
$$\mathcal{F}(U) \longrightarrow \mathcal{F}(V \to U)$$
is an equivalence of categories.

Then $\mathcal{F}$ is a stack with respect to the the fpqc topology.

PROOF. The proof is a little long and complicates, so for clarity we will divide it in several steps. According to Theorem 3.45 and Proposition 4.12 we may assume that $\mathcal{F}$ is split: this will only be used in the last two steps.

Step 1: $\mathcal{F}$ is a prestack. Given an S scheme $T \to S$ and two objects ξ and η of $\mathcal{F}(T)$, consider the functor
$$\underline{\mathrm{Hom}}_T(\xi, \eta) \colon (\mathrm{Sch}/T)^{\mathrm{op}} \longrightarrow (\mathrm{Set}).$$
We see immediately that the two conditions of Lemma 2.60 are satisfied, so the functor $\underline{\mathrm{Hom}}_T(\xi, \eta)$ is a sheaf, and $\mathcal{F}$ is a prestack in the fpqc topology.

Now we have to check that every object with descent data is effective.

Step 2: reduction to the case of a single morphism. We start by analyzing the sections of $\mathcal{F}$ over the empty scheme $\emptyset$.

LEMMA 4.26. The category $\mathcal{F}(\emptyset)$ is equivalent to a category with one object and one morphism.

Equivalently, between any two objects of $\mathcal{F}(\emptyset)$ there is a unique arrow.

PROOF. The scheme $\emptyset$ has the empty Zariski covering $\mathcal{U} = \emptyset$. By this I really mean the empty set, consisting of no morphisms at all, and not the set consisting of the embedding of $\emptyset \subseteq \emptyset$. There is only one object with descent data $(\emptyset, \emptyset)$ in $\mathcal{F}(\mathcal{U})$, and one morphism from $(\emptyset, \emptyset)$ to itself. Hence $\mathcal{F}(\mathcal{U})$ is equivalent to the category with one object and one morphism; but $\mathcal{F}(\emptyset)$ is equivalent to $\mathcal{F}(\mathcal{U})$, because $\mathcal{F}$ is a stack in the Zariski topology. □

LEMMA 4.27. If a scheme U is a disjoint union of open subschemes $\{U_i\}_{i \in I}$, then the functor $\mathcal{F}(U) \to \prod_{i \in I} \mathcal{F}(U_i)$ obtained from the various restriction functors $\mathcal{F}(U) \to \mathcal{F}(U_i)$ is an equivalence of categories.

PROOF. Let ξ and η be objects of $\mathcal{F}(U)$; denote by ξ_i and η_i their restrictions to U_i. The fact that $\underline{\mathrm{Hom}}_U(\xi, \eta) \colon (\mathrm{Sch}/T)^{\mathrm{op}} \to (\mathrm{Set})$ is a sheaf ensures that the function
$$\mathrm{Hom}_{\mathcal{F}(U)}(\xi, \eta) \longrightarrow \prod_i \mathrm{Hom}_{\mathcal{F}(U_i)}(\xi_i, \eta_i)$$

is a bijection; but this means precisely that the functor is fully faithful.

To check that it is essentially surjective, take an object (ξ_i) in $\prod_{i \in I} \mathcal{F}(U_i)$. We have $U_{ij} = \emptyset$ when $i \neq j$, and $U_{ij} = U_i$ when $i = j$; we can define transition isomorphisms $\phi_{ij}\colon \operatorname{pr}_2^* \xi_j \simeq \operatorname{pr}_1^* \xi_i$ as the identity when $i = j$, and as the only arrow from $\operatorname{pr}_2^* \xi_j$ to $\operatorname{pr}_1^* \xi_i$ in $\mathcal{F}(U_{ij}) = \mathcal{F}(\emptyset)$ when $i \neq j$. These satisfy the cocycle condition; hence there is an object ξ of $\mathcal{F}(U)$ whose restriction to each U_i is isomorphic to ξ_i. Then the image of ξ into $\prod_{i \in I} \mathcal{F}(U_i)$ is isomorphic to (ξ_i), and the functor is essentially surjective. $\square$

Given an arbitrary covering $\{U_i \to U\}$, set $V = \coprod_i U_i$, and denote by $f\colon V \to U$ the induced morphism. I claim that the functor $\mathcal{F}(U) \to \mathcal{F}(V \to U)$ is an equivalence if and only if $\mathcal{F}(U) \to \mathcal{F}(\{U_i \to U\})$ is. In fact, we will show that there is an equivalence of categories

$$\mathcal{F}(V \to U) \longrightarrow \mathcal{F}(\{U_i \to U\})$$

such that the composite

$$\mathcal{F}(U) \longrightarrow \mathcal{F}(V \to U) \longrightarrow \mathcal{F}(\{U_i \to U\})$$

is isomorphic to the functor

$$\mathcal{F}(U) \longrightarrow \mathcal{F}(\{U_i \to U\}).$$

This is obtained as follows. We have a natural isomorphism of U-schemes

$$V \times_U V \simeq \coprod_{i,j} U_i \times_U U_j,$$

so Lemma 4.27 gives us equivalences of categories

$$\mathcal{F}(V) \longrightarrow \prod_i \mathcal{F}(U_i) \tag{4.2.1}$$

and

$$\mathcal{F}(V \times_U V) \longrightarrow \prod_{i,j} \mathcal{F}(U_i \times_U U_j). \tag{4.2.2}$$

An object of $\mathcal{F}(V \to U)$ is a pair (η, ϕ), where η is an object of $\mathcal{F}(V)$ and $\phi\colon \operatorname{pr}_2^* \eta \simeq \operatorname{pr}_1^* \eta$ in $\mathcal{F}(V \times_U V)$ satisfying the cocycle condition. If η_i denotes the restriction of η to U_i for all i and $\phi_{ij}\colon \operatorname{pr}_2^* \eta \simeq \operatorname{pr}_1^* \eta$ the arrow pulled back from ϕ, the image of ϕ in $\prod_{i,j} \mathcal{F}(U_i \times_U U_j)$ is precisely the collection (ϕ_{ij}); it is immediate to see that (ϕ_{ij}) satisfies the cocycle condition.

In this way we associate with each object (η, ϕ) of $\mathcal{F}(V \to U)$ an object $(\{\eta_i\}, \{\phi_{ij}\})$ of $\mathcal{F}(\{U_i \to U\})$. An arrow $\alpha\colon (\eta, \phi) \to (\eta', \phi')$ is an arrow $\alpha\colon \eta \to \eta'$ in $\mathcal{F}(V)$ such that

$$\operatorname{pr}_1^* \alpha \circ \phi = \phi \circ \operatorname{pr}_2^* \alpha\colon \operatorname{pr}_2^* \eta \longrightarrow \operatorname{pr}_1^* \eta';$$

then one checks immediately that the collection of restrictions $\{\alpha_i\colon \eta_i \to \eta'_i\}$ gives an arrow $\{\alpha_i\}\colon (\{\eta_i\}, \{\phi_{ij}\}) \to (\{\eta'_i\}, \{\phi'_{ij}\})$.

Conversely, one can use the inverses of the functors (4.2.1) and (4.2.2) to define the inverse of the functor constructed above, thus showing that it is an equivalence (we leave the details to the reader). This equivalence has the desired properties.

This means that to check that descent data in $\mathcal{F}$ are effective we can restrict consideration to coverings consisting of one arrow.

Step 3: the case of a quasi-compact morphism with affine target. Consider the case that $V \to U$ is a flat surjective morphism of S-schemes, with U affine and V quasi-compact. Let $\{V_i\}$ be a finite covering of V by open affine subschemes, V' the disjoint union of the V_i. Then $\mathcal{F}(U) \to \mathcal{F}(V' \to U)$ is an equivalence, by hypothesis, so by Lemma 4.17 $\mathcal{F}(U) \to \mathcal{F}(V \to U)$ is also an equivalence.

Step 4: the case of a morphism with affine target. Now U is affine and $V \to U$ is an arbitrary fpqc morphism. By hypothesis, there is an open covering $\{V_i\}$ of V by quasi-compact open subschemes, all of which surject onto U. We will use the fact that $\mathcal{F}$ has a splitting. We need to show that $\mathcal{F}(U) \to \mathcal{F}(V \to U)$ is essentially surjective.

Choose an index i; $V_i \to U$ is also an fpqc cover, with V_i quasi-compact. We have a strictly commutative diagram of functors

$$\begin{array}{ccc} \mathcal{F}(U) & \longrightarrow & \mathcal{F}(V \to U) \\ & \searrow \quad \swarrow & \\ & \mathcal{F}(V_i \to U) & \end{array}$$

in which $\mathcal{F}(U) \to \mathcal{F}(V_i \to U)$ is an equivalence, because of the previous step, and $\mathcal{F}(U) \to \mathcal{F}(V \to U)$ is fully faithful. From this we see that to show that $\mathcal{F}(U) \to \mathcal{F}(V \to U)$ is essentially surjective it is enough to prove that $\mathcal{F}(V \to U) \to \mathcal{F}(V_i \to U)$ is fully faithful.

It is clear that $\mathcal{F}(V \to U) \to \mathcal{F}(V_i \to U)$ is full (because $\mathcal{F}(U) \to \mathcal{F}(V_i \to U)$ is), so it is enough to show that it is faithful. Let α and β be two arrows in $\mathcal{F}(V \to U)$ with the same image in $\mathcal{F}(V_i \to U)$. For any other index j we have that the restriction functor $\mathcal{F}(V_i \cup V_j \to U) \to \mathcal{F}(V_i \to U)$ is an equivalence, because in the strictly commutative diagram

$$\begin{array}{ccc} \mathcal{F}(U) & \longrightarrow & \mathcal{F}(V_i \cup V_j \to U) \\ & \searrow \quad \swarrow & \\ & \mathcal{F}(V_i \to U) & \end{array}$$

the top and left arrows are equivalences, so the restrictions of α and β to $V_i \cup V_j$ are the same. Hence the restrictions of α and β to each V_j are the same: since $\mathcal{F}$ is a prestack we can conclude that $\alpha = \beta$.

Step 5: the general case. Now consider a general fpqc morphism $f\colon V \to U$, with no restrictions on U or V. Take an open covering $\{U_i\}$ of U by affine subschemes, and let V_i be the inverse image of U_i in V. We need to show that any object (η, ϕ) of $\mathcal{F}(V \to U)$ comes from an object of $\mathcal{F}(U)$.

For each open subset $U' \subseteq U$, denote by $\Phi_{U'}\colon \mathcal{F}(U') \to \mathcal{F}(f^{-1}U' \to U')$ the functor that sends objects to objects with descent data, defined via the splitting. We will use the obvious fact that if U'' is an open subset of U', the diagram

$$\begin{array}{ccc} \mathcal{F}(f^{-1}U') & \longrightarrow & \mathcal{F}(f^{-1}U'') \\ \Big\downarrow{\scriptstyle \Phi_{U'}} & & \Big\downarrow{\scriptstyle \Phi_{U''}} \\ \mathcal{F}(f^{-1}U' \to U') & \longrightarrow & \mathcal{F}(f^{-1}U'' \to U'') \end{array}$$

where the rows are given by restrictions, is strictly commutative.

For each index i, let (η_i, ϕ_i) be the restriction of (η, ϕ) to V_i. This is an object of $\mathcal{F}(V_i \to U_i)$, hence by the previous step it there exists an object ξ_i of $\mathcal{F}(U_i)$ with an isomorphism $\alpha_i \colon \Phi_{U_i}\xi_i \simeq (\eta_i, \phi_i)$ in $\mathcal{F}(\{V_i \to U_i\})$. Now we want to glue together the ξ_i to a global object ξ of $\mathcal{F}(U)$; for this we need Zariski descent data $\phi_{ij} \colon \xi_j \mid_{U_{ij}} \simeq \xi_i \mid_{U_{ij}}$.

For each pair of indices i and j, set $V_{ij} = V_i \cap V_j$, so that V_{ij} is the inverse image of U_{ij} in V. By restricting the isomorphisms $\alpha_i \colon \Phi_{U_i}\xi_i \simeq (\eta_i, \phi_i)$ to V_{ij} we get isomorphisms

$$\Phi_{U_{ij}}(\xi_i \mid_{U_{ij}}) = (\Phi_{U_i}\xi_i) \mid_{V_{ij}} \overset{\alpha_i|_{V_{ij}}}{\simeq} (\eta \mid_{V_{ij}}, \phi \mid_{V_{ij}})$$

and from these isomorphisms

$$\alpha_i^{-1}\alpha_j \colon \Phi_{U_{ij}}(\xi_j \mid_{U_{ij}}) \simeq \Phi_{U_{ij}}(\xi_i \mid_{U_{ij}}).$$

Since $\Phi_{U_{ij}}$ is an equivalence of categories, there is one and only one isomorphism $\phi_{ij} \colon \xi_j \mid_{V_{ij}} \simeq \xi_i \mid_{V_{ij}}$ such that $\Phi_{U_{ij}}\phi_{ij} = \alpha_i^{-1}\alpha_j$.

By applying $\Phi_{U_{ijk}}$ we see easily that the cocycle condition $\phi_{ik} = \phi_{ij}\phi_{jk}$ is satisfied; hence there exists an object ξ of $\mathcal{F}(U)$, with isomorphisms $\xi \mid_{U_i} \simeq \xi_i$. If we denote by $f_i \colon V_i \to U$ the restriction of $: V \to U$, we obtain an isomorphism of $f_i^*\xi = f^*\xi \mid_{V_i}$ with the pullback of ξ_i to V_i. We also have an isomorphism of the pullback of ξ_i to V_i with η_i, obtained by pulling back α_i along $V_i \to U_i$. These isomorphisms $f^*\xi \mid_{V_i} \simeq \eta_i$ coincide when pulled back to V_{ij}, so they glue together to give an isomorphism $f^*\xi \simeq \eta$; and this is the desired isomorphism of $\Phi_U\xi$ with (η, ϕ).

This completes the proof of Lemma 4.25. □

It is a standard fact that (QCoh/S) is a stack in the Zariski topology; so we only need to check that the second condition of Lemma 4.25 is satisfied; for this, we use the theory of §4.2.1. Take a flat surjective morphism $V \to U$, corresponding to a faithfully flat ring homomorphism $f \colon A \to B$. We have the standard equivalence of categories $\mathrm{QCoh}(U) \simeq \mathrm{Mod}_A$; I claim that there is also an equivalence of categories $\mathrm{QCoh}(V \to U) \simeq \mathrm{Mod}_{A \to B}$. A quasi-coherent sheaf $\mathcal{M}$ on U corresponds to an A-module M. The inverse images $\mathrm{pr}_1^* \mathcal{M}$ and $\mathrm{pr}_2^* \mathcal{M}$ in $V \times_U B = \operatorname{Spec} B \otimes_A B$ correspond to the modules $M \otimes_A B$ and $B \otimes_A M$, respectively; hence an isomorphism $\phi \colon \mathrm{pr}_2^* \mathcal{M} \simeq \mathrm{pr}_1^* \mathcal{M}$ corresponds to an isomorphism $\psi \colon M \otimes_A B \simeq B \otimes_A M$. It is easy to see that ϕ satisfies the cocycle condition, so that $(\mathcal{M}, \phi)$ is an object of $\mathrm{QCoh}(V \to U)$, if and only if ψ satisfies the condition $\psi_1\psi_3 = \psi_2$; this gives us the equivalence $\mathrm{QCoh}(V \to U) \simeq \mathrm{Mod}_{A \to B}$. The functor $\mathrm{QCoh}(U) \to \mathrm{QCoh}(V \to U)$ corresponds to the functor $\mathrm{Mod}_A \to \mathrm{Mod}_{A \to B}$ defined in §4.2.1, in the sense that the composites

$$\mathrm{QCoh}(U) \longrightarrow \mathrm{QCoh}(V \longrightarrow U) \simeq \mathrm{Mod}_{A \to B}$$

and

$$\mathrm{QCoh}(U) \simeq \mathrm{Mod}_A \simeq \mathrm{Mod}_{A \to B}$$

are isomorphic. Since $\mathrm{Mod}_A \to \mathrm{Mod}_{A \to B}$ is an equivalence, this finishes the proof of Theorem 4.23.

Here is an interesting question. Let us call a morphism of schemes $V \to U$ a *descent morphism* if the functor $\mathrm{QCoh}(U) \to \mathrm{QCoh}(V \to U)$ is an equivalence.

Suppose that a morphism of schemes $V \to U$ has local sections in the fpqc topology, that is, there exists an fpqc covering $U_i \to U$ with sections $U_i \to V$.

This is equivalent to saying that $V \to U$ is a covering in the saturation of the fpqc topology, so, by Theorem 4.23 and Proposition 4.16, it is a descent morphism.

OPEN QUESTION 4.28. Do all descent morphisms have local sections in the fpqc topology? If not, is there an interesting characterization of descent morphisms?

4.2.3. Descent for sheaves of commutative algebras. There are many variants of Theorem 4.23. The general principle is that one has descent in the fpqc topology for quasi-coherent sheaves with an additional structure, as long as this structure is defined by homomorphisms of sheaves, satisfying conditions that are expressed by the commutativity of certain diagrams.

Here is a typical example. If U is a scheme, we may consider quasi-coherent sheaves of commutative algebras on U, that is, sheaves of commutative $\mathcal{O}_U$-algebras that are quasi-coherent as sheaves of $\mathcal{O}_U$-modules. The quasi-coherent sheaves of commutative algebras on a scheme U form a category, denoted by $\operatorname{QCohComm} U$.

We get a pseudo-functor on the category (Sch/S) by sending each $U \to S$ to the category $\operatorname{QCohComm} U$; we denote the resulting fibered category on (Sch/S) by $(\mathrm{QCohComm}/S)$.

THEOREM 4.29. $(\mathrm{QCohComm}/S)$ is a stack over (Sch/S).

Here is the key fact.

LEMMA 4.30. Let $\{\sigma_i \colon U_i \to U\}$ be an fpqc covering of schemes.

(i) If $\mathcal{A}$ and $\mathcal{B}$ are quasi-coherent sheaves of algebras over U, $\phi \colon \mathcal{A} \to \mathcal{B}$ is a homomorphism of quasi-coherent sheaves, such that each pullback $\sigma_i^*\phi \colon \sigma_i^*\mathcal{A} \to \sigma_i^*\mathcal{B}$ is a homomorphism of algebras for all i, then ϕ is a homomorphism of algebras.

(ii) Let $\mathcal{A}$ be a quasi-coherent sheaf on U. Assume that each pullback $\sigma_i^*\mathcal{A}$ has a structure of sheaf of commutative algebras, and that the canonical isomorphism of quasi-coherent sheaves $\mathrm{pr}_2^*\sigma_j^*\mathcal{A} \simeq \mathrm{pr}_1^*\sigma_i^*\mathcal{A}$ is an isomorphism of sheaves of algebras for each i and j. Then there exists a unique structure of sheaf of commutative algebras on $\mathcal{A}$ inducing the given structure on each $\sigma_i^*\mathcal{A}$.

PROOF. For part (i), we need to check that the two composites

$$\mathcal{A} \otimes_{\mathcal{O}_U} \mathcal{A} \xrightarrow{\mu_{\mathcal{A}}} \mathcal{A} \longrightarrow \mathcal{B}$$

and

$$\mathcal{A} \otimes_{\mathcal{O}_U} \mathcal{A} \xrightarrow{\phi\otimes\phi} \mathcal{B} \otimes_{\mathcal{O}_U} \mathcal{B} \xrightarrow{\mu_{\mathcal{B}}} \mathcal{B}$$

coincide. However, the composites

$$\sigma_i^*\mathcal{A} \otimes_{\mathcal{O}_{U_i}} \sigma_i^*\mathcal{A} \simeq \sigma_i^*(\mathcal{A} \otimes_{\mathcal{O}_U} \mathcal{A}) \xrightarrow{\sigma_i^*\mu_{\mathcal{A}}} \sigma_i^*\mathcal{A} \xrightarrow{\sigma_i^*\phi} \sigma_i^*\mathcal{B}$$

and

$$\sigma_i^*\mathcal{A} \otimes_{\mathcal{O}_{U_i}} \sigma_i^*\mathcal{A} \simeq \sigma_i^*(\mathcal{A} \otimes_{\mathcal{O}_U} \mathcal{A}) \xrightarrow{\sigma_i^*\phi\otimes\phi} \sigma_i^*(\mathcal{B} \otimes_{\mathcal{O}_U} \mathcal{B}) \xrightarrow{\sigma_i^*\mu_{\mathcal{B}}} \sigma_i^*\mathcal{B}$$

coincide, because $\sigma_i^*\phi$ is a homomorphism of sheaves of algebras; and we know that two homomorphism of quasi-coherent sheaves on a scheme that are locally equal in the fpqc topology, are in fact equal, because (QCoh/S) is a stack over (Sch/S).

Let us prove part (ii). From the algebra structure on each $\sigma_i^*\mathcal{A}$ we get homomorphisms of quasi-coherent sheaves

$$\mu_i \colon \sigma_i^*(\mathcal{A} \otimes_{\mathcal{O}_U} \mathcal{A}) \simeq \sigma_i^*\mathcal{A} \otimes_{\mathcal{O}_{U_i}} \sigma_i^*\mathcal{A} \longrightarrow \sigma_i^*\mathcal{A}.$$

We need to show that these homomorphisms are pulled back from a homomorphism $\mathcal{A} \otimes_{\mathcal{O}_U} \mathcal{A} \to \mathcal{A}$. Denote by $\sigma_{ij} : U_{ij} \to U$ the obvious morphism; since (QCoh/S) is a stack, and in particular the functors of arrows are sheaves, this is equivalent to proving that the pullbacks $\sigma_{ij}^*(\mathcal{A} \otimes_{\mathcal{O}_U} \mathcal{A}) \to \mathcal{A}$ of μ_i and μ_j coincide for all i and j. This is most easily checked at the level of sections.

Similarly, the homomorphisms $\mathcal{O}_{U_i} \to \mathcal{A}_i$ corresponding to the identity come from a homomorphism $\mathcal{O}_U \to \mathcal{A}$.

We also need to show that the resulting homomorphism $\mathcal{A} \otimes_{\mathcal{O}_U} \mathcal{A} \to \mathcal{A}$ gives $\mathcal{A}$ the structure of a sheaf of commutative algebras, that is, we need to prove that the product is associative and commutative, and that the homomorphism $\mathcal{O}_U \to \mathcal{A}$ gives an identity. Once again, this is easily done by looking at sections, and is left to the reader. □

From this it is easy to deduce Theorem 4.29. If $\{\sigma_i : U_i \to U\}$ is an fpqc covering, and $(\{\mathcal{A}_i\}, \{\phi_{ij}\})$ is an object of $(\mathrm{QCohComm}/S)(\{\sigma_i : U_i \to U\})$, we can forget the algebra structure on the $\mathcal{A}_i$, and simply consider it as an object of $(\mathrm{QCoh}/S)(\{\sigma_i : U_i \to U\})$; then it will come from a quasi-coherent sheaf $\mathcal{A}$ on U. However, each $\sigma_i^*\mathcal{A}$ is isomorphic to $\mathcal{A}_i$, thus it inherits a commutative algebra structure: and Lemma 4.30 implies that this comes from a structure of sheaf of commutative algebras on $\mathcal{A}$. This finishes the proof of Theorem 4.29.

Exactly in the same way one can defined fibered categories of sheaves of (not necessarily commutative) associative algebras, sheaves of Lie algebras, and so on, and prove that all these structures give stacks.

4.3. Descent for morphisms of schemes

Consider a site $\mathcal{C}$, a stable class $\mathcal{P}$ of arrows, and the associate fibered category $\mathcal{P} \to \mathcal{C}$, as in Example 3.17.

The following fact is often useful.

PROPOSITION 4.31. *Let $\mathcal{C}$ be a subcanonical site, $\mathcal{P}$ a stable class of arrows. Then $\mathcal{P} \to \mathcal{C}$ is a prestack.*

Recall (Definition 2.57) that a site is subcanonical when every representable functor is a sheaf. The site (Sch/S) with the fpqc topology is subcanonical (Theorem 2.55).

PROOF. Let $\{U_i \to U\}$ be a covering, $X \to U$ and $Y \to U$ two arrows in $\mathcal{P}$. The arrows in $\mathcal{P}(U)$ are the arrows $X \to Y$ in $\mathcal{C}$ that commute with the projections to U. Set $X_i = U_i \times_U X$ and $X_{ij} = U_{ij} \times_U X = X_i \times_X X_j$, and analogously for Y_i and Y_{ij}. Suppose that we have arrows $f_i : X_i \to Y_i$ in $\mathcal{P}(U_i)$, such that the arrows $X_{ij} \to Y_{ij}$ induced by f_i and f_j coincide; we need to show that there is a unique arrow $f : X \to Y$ in $\mathcal{P}(U)$ whose restriction $X_i \to Y_i$ coincides with f_i for each i.

The composites $X_i \xrightarrow{f_i} Y_i \to Y$ give sections $g_i \in \mathrm{h}_Y(X_i)$, such that the pullbacks of g_i and g_j to X_{ij} coincide. Since h_Y is a sheaf, $\{X_i \to X\}$ is a covering, and $X_{ij} = X_i \times_X X_j$ for any i and j, there is a unique arrow $f : X \to Y$ in $\mathcal{C}$, such the composite $X_i \to X \xrightarrow{f} Y$ is g_i, so that the diagram

$$\begin{array}{ccc} X_i & \xrightarrow{f_i} & Y_i \\ \downarrow & & \downarrow \\ X & \xrightarrow{f} & Y \end{array}$$

commutes for all i. It is also clear that the arrows $X \to U$ and $X \xrightarrow{f} Y \to U$ coincide, since they coincide when composed with $U_i \to U$ for all i, and since h_U is a sheaf, and in particular a separated functor. Hence the diagram

$$\begin{array}{ccc} X & \xrightarrow{f} & Y \\ \downarrow & & \downarrow \\ U & = & U \end{array}$$

commutes, and f is the only arrow in $\mathcal{P}(U)$ whose restriction to each U_i coincides with f_i. □

However, in general $\mathcal{P}$ will not be a stack. It is easy to see that $\mathcal{P}$ cannot be a stack unless it satisfies the following condition.

DEFINITION 4.32. A class of arrows $\mathcal{P}$ in $\mathcal{C}$ is *local* if it is stable (Definition 3.16), and the following condition holds. Suppose that you are given a covering $\{U_i \to U\}$ in $\mathcal{C}$ and an arrow $X \to U$. Then, if the projections $U_i \times_U X \to U_i$ are in $\mathcal{P}$ for all i, $X \to U$ is also in $\mathcal{P}$.

Still, a local class of arrow does not form a stack in general, effectiveness of descent data is not guaranteed, not even when $\mathcal{P}$ is the class of all arrows. Consider the following example. Take $\mathcal{C}$ to be the class of all schemes locally of finite type over a field k, of bounded dimension, with the arrows being morphisms of schemes over k. Let us equip it with the Zariski topology, and let $\mathcal{P}$ be the class of all arrows. Call $U = U_1 \coprod U_2 \coprod U_3 \coprod U_4 \coprod \dots$ the union of countably many copies of $U_1 = \operatorname{Spec} k$. The collection of inclusions $\{U_i \hookrightarrow U\}$ forms a covering. Over each U_i consider the scheme $\mathbb{A}^i_k$. Obviously $U_{ij} = \emptyset$ if $i \neq j$, and $U_{ij} = \operatorname{Spec} k$ if $i = j$, so we define transition isomorphisms in the only possible way as the identity $\phi_{ii} = \mathrm{id}_{\mathbb{A}^i_k} : \mathbb{A}^i_k \to \mathbb{A}^i_k$, and as the identity $\mathrm{id}_\emptyset : \emptyset \to \emptyset$ when $i \neq j$. These obviously satisfy the cocycle condition, being all identities. On the other hand there cannot be a scheme of bounded dimension over U, whose pullback to each U_i is $\mathbb{A}^i_k$.

This is an artificial example; obviously if we want to glue together infinitely many algebraic varieties, we shouldn't ask for the dimension to be bounded. And in fact, morphisms of schemes form a stack in the Zariski topology, and therefore a local category of arrows also forms a stack in the Zariski topology.

On the other hand, most of the interesting properties of morphisms of schemes are local in the fpqc topology on the codomain, such as for example being flat, being of finite presentation, being quasi-compact, being proper, being smooth, being affine, and so on (Proposition 2.36). For each of this properties we get a prestack of morphisms of schemes over (Sch/S), and we can ask if this is a stack in the fpqc topology.

The issue of effectiveness of descent data is rather delicate, however. We will give an example to show that it can fail even for proper and smooth morphisms, in the étale topology (see 4.4.2). In this section we will prove some positive results.

4.3.1. Descent for affine morphisms. Let $\mathcal{P}$ be the class of affine arrows in (Sch/S), and denote by $(\mathrm{Aff}/S) \to (\mathrm{Sch}/S)$ the resulting fibered category. The objects of (Aff/S) are affine morphisms $X \to U$, where $U \to S$ is an S-scheme.

THEOREM 4.33. *The fibered category (Aff/S) is a stack over (Sch/S) in the fpqc topology.*

First of all, (Aff/S) is a prestack, because of Proposition 4.31, so the only issue is effectiveness of descent data. By Proposition 4.20 (ii) it is enough to check that $(\mathrm{Aff}/S)_{\mathrm{cart}}$ is a stack.

Let $\mathcal{A}$ be a quasi-coherent sheaf of algebras on a scheme U. Then we denote by $\mathcal{S}pec_X \mathcal{A}$ the relative spectrum of $\mathcal{A}$; this is an affine scheme over U, and if $V \subseteq U$ is an open affine subscheme of U, the inverse image of V in $\mathcal{S}pec_U \mathcal{A}$ is the spectrum of the ring $\mathcal{A}(V)$.

A homomorphism of sheaves of commutative rings $\mathcal{A} \to \mathcal{B}$ induces a homomorphism of U-schemes $\mathcal{S}pec_U \mathcal{B} \to \mathcal{S}pec_U \mathcal{A}$; this is a contravariant functor from $\mathrm{QCohComm}\, U$ to the category $\mathrm{Aff}\, U$ of affine schemes over U, which is well-known to be an equivalence of categories $\mathrm{QCohComm}\, U^{\mathrm{op}} \simeq \mathrm{Aff}\, U$. The inverse functor sends an affine morphism $h\colon X \to U$ to the quasi-coherent sheaf of commutative algebras $h_*\mathcal{O}_X$.

There is a morphism of fibered categories

$$(\mathrm{QCohComm}/S)_{\mathrm{cart}} \longrightarrow (\mathrm{Aff}/S)_{\mathrm{cart}}$$

that sends a an object $\mathcal{A}$ of $\mathrm{QCohComm}\, U$ to the affine morphism $\mathcal{S}pec\,\mathcal{A} \to U$. Let $(f, \alpha)\colon (U, \mathcal{A}) \to (V, \mathcal{B})$ be an arrow in $(\mathrm{QCohComm}/S)_{\mathrm{cart}}$; $f\colon U \to V$ is a morphism of S-schemes, $\alpha\colon \mathcal{A} \simeq f^*\mathcal{B}$ an isomorphism of sheaves of $\mathcal{O}_U$-modules. Then α^{-1} gives an isomorphism $\mathcal{S}pec_U \mathcal{A} \simeq \mathcal{S}pec_U f^*\mathcal{B} = U \times_V \mathcal{S}pec_V \mathcal{B}$ of schemes over U, and the composite of this isomorphism with the projection $U \times_V \mathcal{S}pec_V \mathcal{B} \to \mathcal{S}pec_V \mathcal{B}$ gives an arrow from $\mathcal{S}pec_U \mathcal{A} \to U$ to $\mathcal{S}pec_V \mathcal{B} \to V$ in $(\mathrm{Aff}/S)_{\mathrm{cart}}$.

If we restrict the morphism to a functor

$$(\mathrm{QCohComm}/S)_{\mathrm{cart}}(U) \longrightarrow (\mathrm{Aff}/S)_{\mathrm{cart}}(U)$$

for some S-scheme U we obtain an equivalence of categories; hence this morphism is an equivalence of fibered categories over (Sch/S), by Proposition 3.36. Since $(\mathrm{QCoh}/S)_{\mathrm{cart}}$ is a stack, by Theorem 4.29 and Proposition 4.20 (i), we see from Proposition 4.12 that $(\mathrm{Aff}/S)_{\mathrm{cart}}$ is also a stack, and this concludes the proof of Theorem 4.33.

The following corollary will be used in §4.3.3.

COROLLARY 4.34. *Let $P \to U$ be a morphism of schemes, $\{U_i \to U\}$ an fpqc cover. For each i set $P_i \overset{\mathrm{def}}{=} U_i \times_U P$ and $P_{ij} \overset{\mathrm{def}}{=} U_{ij} \times_U P$. Suppose that for each i we have a closed subscheme X_i of P_i, with the property that for each pair of indices i and j the inverse images of X_i and X_j in P_{ij}, through the first and second projection respectively, coincide. Then there is a unique closed subscheme of P whose inverse image in P_i coincides with X_i for each i.*

PROOF. We have that $\{P_i \to P\}$ is an fpqc cover, and $P_{ij} = P_i \times_P P_j$. The pullbacks $\mathrm{pr}_2^* X_j$ and $\mathrm{pr}_1^* X_i$ to P_{ij} coincide as subschemes of P_{ij}, and this yields a canonical isomorphism $\phi_{ij}\,\mathrm{pr}_2^* X_j \simeq \mathrm{pr}_1^* X_i$. The cocycle condition is automatically satisfied, because any two morphisms of P_{ijk}-schemes that are embedded in P_{ijk} automatically coincide. Hence there is an affine morphism $X \to P$ that pulls back to $X_i \to P_i$ for each i; and this morphism is a closed embedding, because of Proposition 2.36.

Uniqueness is clear, because two closed subschemes of P that are isomorphic as P-schemes are in fact equal. □

4.3.2. The base change theorem. For the next result we are going to need a particular case of the base change theorem for quasi-coherent sheaves.

Suppose that we have a commutative diagram of schemes

$$\begin{array}{ccc} X & \xrightarrow{f} & Y \\ \downarrow{\scriptstyle \xi} & & \downarrow{\scriptstyle \eta} \\ U & \xrightarrow{\phi} & V \end{array} \tag{4.3.1}$$

and a sheaf of $\mathcal{O}_Y$-modules $\mathcal{L}$. Then there exists a natural base change homomorphism of $\mathcal{O}_U$-modules

$$\beta_{f,\phi}(\mathcal{L})\colon \phi^*\eta_*\mathcal{L} \longrightarrow \xi_* f^*\mathcal{L}$$

that is defined as follows. First of all, start from the natural adjunction homomorphism $\mathcal{L} \to f_* f^*\mathcal{L}$ (this is the homomorphism that corresponds to $\mathrm{id}_{f^*\mathcal{L}}$ in the natural adjunction isomorphism $\mathrm{Hom}_Y(\mathcal{L}, f_*\phi^*\mathcal{L}) \simeq \mathrm{Hom}_X(f^*\mathcal{L}, f^*\mathcal{L})$). This gives a homomorphism of $\mathcal{O}_V$-modules

$$\eta_*\mathcal{L} \longrightarrow \eta_* f_* f^*\mathcal{L} = \phi_*\xi_* f^*\mathcal{L}.$$

Then $\beta_{f,\phi}(\mathcal{L})$ corresponds to this homomorphism under the adjunction isomorphism

$$\mathrm{Hom}_U(\phi^*\eta_*\mathcal{L}, \xi_* f^*\mathcal{L}) \simeq \mathrm{Hom}_V(\eta_*\mathcal{L}, \phi_*\xi_* f^*\mathcal{L}).$$

The homomorphism $\beta_{f,\phi}(\mathcal{L})$ has the following useful characterization at the level of sections. If V_1 is an open subset of V, and $s \in \mathcal{L}(\eta^{-1}V_1) = \eta_*\mathcal{L}(V_1)$, then there is a pullback section $\phi^* s \in \phi^*\eta_*\mathcal{L}(\phi^{-1}V_1)$. The sections of this form generate $\phi^*\eta_*\mathcal{L}$ as an $\mathcal{O}_U$-module. The section

$$f^* s \in \mathcal{L}(f^{-1}\eta^{-1}V_1) = \mathcal{L}(\xi^{-1}\phi^{-1}s)$$

can be considered as an element of $\xi_* f^*\mathcal{L}(\phi^{-1}V_1)$; and then $\beta_{f,\phi}(\mathcal{L})$ is characterized as the only $\mathcal{O}_U$-linear homomorphism of sheaves such that

$$\beta_{f,\phi}(\mathcal{L})(\phi^* s) = f^* s \in \xi_* f^*\mathcal{L}(\phi^{-1}V_1)$$

for all s as above.

The base change homomorphism is functorial in $\mathcal{L}$. That is, there are two functors $\phi^*\eta_*$ and $\xi_* f^*$ from (QCoh/Y) to (QCoh/U), and $\beta_{f,\phi}$ gives a natural transformation $\phi^*\eta_* \to \xi_* f^*$.

The base change homomorphism also satisfies a compatibility condition.

PROPOSITION 4.35. Let

$$\begin{array}{ccccc} X & \xrightarrow{f} & Y & \xrightarrow{g} & Z \\ \downarrow{\scriptstyle \xi} & & \downarrow{\scriptstyle \eta} & & \downarrow{\scriptstyle \zeta} \\ U & \xrightarrow{\phi} & V & \xrightarrow{\psi} & W \end{array}$$

be a commutative diagram of schemes, $\mathcal{L}$ a sheaf of $\mathcal{O}_Z$-modules. Then the diagram of $\mathcal{O}_U$-modules

$$\begin{array}{ccccc} \phi^*\psi^*\zeta_*\mathcal{L} & & \xrightarrow{\alpha_{\phi,\psi}(\zeta_*\mathcal{L})} & & (\psi\phi)^*\zeta_*\mathcal{L} \\ \downarrow{\scriptstyle \phi^*\beta_{g,\psi}(\mathcal{L})} & & & & \downarrow{\scriptstyle \beta_{gf,\psi\phi}(\mathcal{L})} \\ \phi^*\eta_*g^*\mathcal{L} & & & & \xi_*(gf)^*\mathcal{L} \\ & \searrow{\scriptstyle \beta_{f,\phi}(g^*\mathcal{L})} & & \swarrow{\scriptstyle \xi_*\alpha_{f,g}(\mathcal{L})} & \\ & & \xi_*f^*g^*\mathcal{L} & & \end{array}$$

commutes.

PROOF. This is immediately proved by taking an open subset W_1 of W, a section $s \in \zeta_*\mathcal{L}(W_1) = \mathcal{L}(\zeta^{-1}W_1)$, and following $\phi^*\psi^*s$ in the diagram above. □

Since in Proposition 4.35 the homomorphisms $\alpha_{\phi,\psi}(\zeta_*\mathcal{L})$ and $\xi_*\alpha_{f,g}(\mathcal{L})$ are always isomorphism, we get the following corollary.

COROLLARY 4.36. In the situation of Proposition 4.35, assume that the base change homomorphism $\beta_{g,\psi}(\mathcal{L})$ is an isomorphism. Then $\beta_{gf,\psi\phi}(\mathcal{L})$ is an isomorphism if and only if $\beta_{f,\phi}(g^*\mathcal{L})$ is an isomorphism

Here is the base change theorem, in the form in which we are going to need it. This is completely standard in the noetherian case; the proof reduces to this case with reduction techniques that are also standard.

PROPOSITION 4.37. Suppose that the diagram (4.3.1) is cartesian, that η is proper and of finite presentation, and that $\mathcal{L}$ is quasi-coherent and of finite presentation, and flat over V. For any point $v \in V$ denote by Y_v the fiber of η over v, and by $\mathcal{L}_v$ the restriction of $\mathcal{L}$ to Y_v.

If $\mathrm{H}^1(Y_v, \mathcal{L}_v) = 0$ for all $v \in V$, then $\eta_*\mathcal{L}$ is locally free over V, and the base change homomorphism $\beta_{f,\phi}(\mathcal{L})\colon \phi^*\eta_*\mathcal{L} \to \xi_*f^*\mathcal{L}$ is an isomorphism.

PROOF. If V is noetherian, then the result follows from [**EGAIII2**, 7.7] (see also [**Har77**, III 12]).

In the general case, the base change homomorphism is easily seen to localize in the Zariski topology on U; hence we may assume that V is affine. Set $V = \operatorname{Spec} A$. According to [**EGAIV3**, Proposition 8.9.1, Théorème 8.10.5 and Théorème 11.2.6] there exists a subring $A_0 \subseteq A$ that is of finite type over $\mathbb{Z}$, hence noetherian, a scheme Y_0' that is proper over $\operatorname{Spec} A_0$, and a coherent sheaf $\mathcal{L}_0'$ on Y_0' that is flat over A_0, together with an isomorphism of $(Y, \mathcal{L})$ with the pullback of $(Y_0', \mathcal{L}_0')$ to $\operatorname{Spec} A$.

By semicontinuity ([**EGAIII2**, Théorème 7.6.9]), the set of points $v_0 \in \operatorname{Spec} A_0$ such that the restriction of $\mathcal{L}_0'$ to the fiber of Y_0' over v_0 has nontrivial H^1 is closed in $\operatorname{Spec} A_0$; obviously, it does not contain the image of $\operatorname{Spec} A$. Denote by V_0

the open subscheme that is the complement of this closed subset, by Y_0 and $\mathcal{L}_0$ the restrictions of Y_0' and $\mathcal{L}_0'$ to V_0. Then $\operatorname{Spec} A$ maps into V_0, and $(Y, \mathcal{L})$ is isomorphic to the pullback of $(Y_0, \mathcal{L}_0)$ to $\operatorname{Spec} A$; hence the result follows from Corollary 4.36 and from the noetherian case. □

4.3.3. Descent via ample invertible sheaves. Descent for affine morphism can be very useful, but is obviously limited in scope. One is more easily interested in projective morphisms, rather than in affine ones. Descent works in this case, as long as the projective morphisms are equipped with ample invertible sheaves, and these also come with descent data.

THEOREM 4.38. Let S be a scheme, $\mathcal{F}$ be a class of flat proper morphisms of finite presentation in (Sch/S) that is local in the fpqc topology (Definition 4.32). Suppose that for each object $\xi\colon X \to U$ of $\mathcal{F}$ one has given an invertible sheaf $\mathcal{L}_\xi$ on X that is ample relative to the morphism $X \to U$, and for each cartesian diagram

$$\begin{array}{ccc} X & \xrightarrow{f} & Y \\ \downarrow{\scriptstyle\xi} & & \downarrow{\scriptstyle\eta} \\ U & \xrightarrow{\phi} & V \end{array}$$

an isomorphism $\rho_{f,\phi}\colon f^*\mathcal{L}_\eta \simeq \mathcal{L}_\xi$ of invertible sheaves on X. These isomorphisms are required to satisfy the following condition: whenever we have a cartesian diagram of schemes

$$\begin{array}{ccccc} X & \xrightarrow{f} & Y & \xrightarrow{g} & Z \\ \downarrow{\scriptstyle\xi} & & \downarrow{\scriptstyle\eta} & & \downarrow{\scriptstyle\zeta} \\ U & \xrightarrow{\phi} & V & \xrightarrow{\psi} & W \end{array}$$

whose columns are in $\mathcal{F}$, then the diagram

$$\begin{array}{ccc} f^*g^*\mathcal{L}_\zeta & \xrightarrow{\alpha_{f,g}(\mathcal{L}_\zeta)} & (gf)^*\mathcal{L}_\zeta \\ \downarrow{\scriptstyle f^*\rho_{g,\psi}} & & \downarrow{\scriptstyle\rho_{gf,\psi\phi}} \\ f^*\mathcal{L}_\eta & \xrightarrow{\rho_{f,\phi}} & \mathcal{L}_\xi \end{array}$$

of quasi-coherent sheaves on X commutes. Here $\alpha_{f,g}(\mathcal{L}_\zeta)$ is the canonical isomorphism of §3.2.1.

Then $\mathcal{F}$ is a stack in the fpqc topology.

Another less cumbersome way to state the compatibility condition is using the formalism of fibered categories, which will be freely used in the proof: since $(gf)^*\mathcal{L}_\zeta$ is a pullback of $g^*\mathcal{L}_\zeta$ to Y, we can consider the pullback $f^*\rho_{g,\psi}\colon (gf)^*\mathcal{L}_\zeta \to f^*\mathcal{L}_\eta$, and then the condition is simply the equality

$$\rho_{gf,\psi\phi} = \rho_{f,\phi} \circ f^*\rho_{g,\psi}\colon (gf)^*\mathcal{L}_\zeta \longrightarrow \mathcal{L}_\xi.$$

EXAMPLE 4.39. For any fixed base scheme S and any non-negative integer g we can consider the class $\mathcal{F}_{g,S}$ of proper smooth morphisms, whose geometric fibers are connected curves of genus g. These morphisms form a local class in (Sch/S).

If $g \neq 1$ then the theorem applies. For $g \geq 2$ we can take $\mathcal{L}_{X\to U}$ to be the relative cotangent sheaf $\Omega^1_{X/U}$, or one of its powers, while for $g = 0$ we can take its dual. So $\mathcal{F}_{g,S}$ is a stack. The stack $(\mathcal{F}_{g,S})_{\mathrm{cart}}$ (Definition 3.31) is usually denoted by $\mathcal{M}_{g,S}$, and plays an important role in algebraic geometry.

There is no natural ample sheaf on families of curves of genus 1, so this theorem does not apply. In fact, $\mathcal{F}_{1,S}$, as we have defined it here, is not a stack: this follows from the counterexample of Raynaud in [**Ray70**, XIII 3.2].

See Remark 4.48 for further discussion.

PROOF OF THEOREM 4.38. The fact that $\mathcal{F}$ is a prestack in the fpqc topology follows from Proposition 4.31. It is also easy to check that $\mathcal{F}$ is a stack in the Zariski topology.

For each object $\xi\colon X \to U$ of $\mathcal{F}$ we define a quasi-coherent finitely presented sheaf $\mathcal{M}_\xi$ on U as $\mathcal{M}_\xi \overset{\mathrm{def}}{=} \xi_*\mathcal{L}_X$. Given a cartesian square

$$\begin{array}{ccc} X & \xrightarrow{f} & Y \\ \downarrow{\scriptstyle\xi} & & \downarrow{\scriptstyle\eta} \\ U & \xrightarrow{\phi} & V \end{array}$$

we get a homomorphism of quasi-coherent sheaves

$$\sigma_{f,\phi}\colon \phi^*\mathcal{M}_\eta = \phi^*\eta_*\mathcal{L}_\eta \longrightarrow \xi_*\mathcal{L}_X = \mathcal{M}_\xi$$

by composing the base change homomorphism

$$\beta_{f,\phi}\colon \phi^*\eta_*\mathcal{L}_\eta \longrightarrow \xi_*f^*\mathcal{L}_\eta$$

with the isomorphism

$$\xi_*\rho_{f,\phi}\colon \xi_*f^*\mathcal{L}_\eta \simeq \xi_*\mathcal{L}_\xi.$$

This homomorphism satisfies the following compatibility condition.

PROPOSITION 4.40. Given a cartesian diagram of schemes

$$\begin{array}{ccccc} X & \xrightarrow{f} & Y & \xrightarrow{g} & Z \\ \downarrow{\scriptstyle\xi} & & \downarrow{\scriptstyle\eta} & & \downarrow{\scriptstyle\zeta} \\ U & \xrightarrow{\phi} & V & \xrightarrow{\psi} & W \end{array}$$

whose columns are in $\mathcal{F}$, the composite

$$(\psi\phi)^*\mathcal{M}_\zeta \xrightarrow{\phi^*\sigma_{g,\psi}} \phi^*\mathcal{M}_\eta \xrightarrow{\sigma_{f,\phi}} \mathcal{M}_\xi$$

equals $\sigma_{gf,\psi\phi}$.

PROOF. Take a section s of $\mathcal{M}_\zeta$ on some open subset of W, and let us check that

$$\sigma_{f,\phi} \circ \phi^*\sigma_{g,\psi}\bigl((\psi\phi)^*s\bigr) = \sigma_{gf,\psi\phi}\bigl((\psi\phi)^*s\bigr)$$

(since the sections of the form $(gf)^*s$ generate $(\psi\phi)^*\mathcal{M}_\zeta$ as a sheaf of $\mathcal{O}_U$-modules, this is enough). The section s is a section of $\mathcal{L}_\zeta$ on some open subset of Z, and we have

$$\begin{aligned} \phi^*\sigma_{g,\psi}\bigl((\psi\phi)^*s\bigr) &= \phi^*(\eta_*\rho_{g,\psi} \circ \beta_{g,\psi})\bigl((\psi\phi)^*s\bigr) \\ &= \phi^*\bigl(\rho_{g,\psi} \circ \beta_{g,\psi}(\psi^*s)\bigr) \\ &= \phi^*\bigl(\rho_{g,\psi}(g^*s)\bigr); \end{aligned}$$

hence

$$\begin{aligned}
\sigma_{f,\phi} \circ \phi^* \sigma_{g,\psi}\big((\psi\phi)^* s\big) &= (\xi_* \rho_{f,\phi} \circ \beta_{f,\phi}) \phi^* \big(\rho_{g,\psi}(g^* s)\big) \\
&= \rho_{f,\phi}\big(f^* \rho_{g,\psi}(g^* s)\big) \\
&= \phi_{gf,\psi\phi}\big((gf)^* s\big) \\
&= \xi_* \phi_{gf,\psi\phi} \circ \beta_{gf,\psi\phi}\big((\psi\phi)^* s\big) \\
&= \sigma_{gf,\psi\phi}\big((\psi\phi)^* s\big).
\end{aligned}$$ □

We use once again the criterion of Lemma 4.25. Consider a flat surjective morphism $V \to U$ of affine S-schemes and object $\eta\colon Y \to V$ of $\mathcal{F}(V)$. Notice that, given a positive integer N, the isomorphisms $\rho_{f,\phi}\colon f^*\mathcal{L}_\eta \simeq \mathcal{L}_\xi$ as in the statement of the theorem induce isomorphisms $\rho_{f,\phi}^{\otimes N}\colon f^*(\mathcal{L}_\eta^{\otimes N}) \simeq \mathcal{L}_\xi^{\otimes N}$. These also satisfy the conditions of the theorem: whenever we have a diagram of schemes

$$\begin{array}{ccccc}
X & \xrightarrow{f} & Y & \xrightarrow{g} & Z \\
\downarrow{\scriptstyle \xi} & & \downarrow{\scriptstyle \eta} & & \downarrow{\scriptstyle \zeta} \\
U & \xrightarrow{\phi} & V & \xrightarrow{\psi} & W
\end{array}$$

whose columns are in $\mathcal{F}$, then the diagram

$$\begin{array}{ccc}
f^*g^*\mathcal{L}_\zeta^{\otimes N} & \xrightarrow{\alpha_{f,g}(\mathcal{L}_\zeta^{\otimes N})} & (gf)^*\mathcal{L}_\zeta^{\otimes N} \\
\downarrow{\scriptstyle f^*\rho_{g,\psi}^{\otimes N}} & & \downarrow{\scriptstyle \rho_{gf,\psi\phi}^{\otimes N}} \\
f^*\mathcal{L}_\eta & \xrightarrow{\rho_{f,\phi}^{\otimes N}} & \mathcal{L}_\xi^{\otimes N}
\end{array}$$

commutes (this is easily checked by following the action of the arrows on sections of the form $f^*g^*s^{\otimes N}$, where s is a section of $\mathcal{L}_\zeta$ over some open subset of W: since those generate $f^*g^*\mathcal{L}_\zeta^{\otimes N}$, if the two composites $f^*g^*\mathcal{L}_\zeta \to \mathcal{L}_\xi$ agree on them, they must be equal).

By substituting $\mathcal{L}_\eta$ with $\mathcal{L}_\eta^{\otimes N}$ for a sufficiently large integer N, we may assume that $\mathcal{L}_\eta$ is very ample on Y, and for any point $v \in V$ we have $\mathrm{H}^1(Y_v, \mathcal{L}_\eta\mid_{Y_v}) = 0$, where Y_v is the fiber of Y over v. This will have the consequence that all the base change homomorphisms that intervene in the following discussion are isomorphisms.

The two diagrams

$$\begin{array}{ccc}
Y \times_U V & \xrightarrow{\mathrm{pr}_1} & Y \\
{\scriptstyle \eta\times \mathrm{id}_V}\downarrow & & \downarrow{\scriptstyle \eta} \\
V \times_U V & \xrightarrow{\mathrm{pr}_1} & V
\end{array}
\qquad \text{and} \qquad
\begin{array}{ccc}
V \times_U Y & \xrightarrow{\mathrm{pr}_2} & Y \\
{\scriptstyle \mathrm{id}_V\times \eta}\downarrow & & \downarrow{\scriptstyle \eta} \\
V \times_U V & \xrightarrow{\mathrm{pr}_2} & V
\end{array}$$

are cartesian; therefore we can take $\eta \times \mathrm{id}_V\colon Y \times_U V \to V \times_U V$ as the pullback $\mathrm{pr}_1^*\eta \in \mathcal{F}(V \times_U V)$, and analogously, $\mathrm{id}_V \times f\colon V \times_U V \to V \times_U V$ as $\mathrm{pr}_2^*\eta \in \mathcal{F}(V \times_U V)$.

Analogously, the pullbacks of f along the three projections $V \times_U V \times_U V \to V$ are

$$\begin{aligned} &\eta \times \mathrm{id}_V \times \mathrm{id}_V \colon Y \times_U V \times_U V \to V \times_U V \times_U V, \\ &\mathrm{id}_V \times \eta \times \mathrm{id}_V \colon V \times_U Y \times_U V \to V \times_U V \times_U V \quad \text{and} \\ &\mathrm{id}_V \times \mathrm{id}_V \times \eta \colon V \times_U V \times_U Y \to V \times_U V \times_U V. \end{aligned}$$

Suppose that we are given an object $\eta\colon Y \to V$ in $\mathcal{F}(V)$ with descent data $\phi\colon V \times_U Y \simeq Y \times_U V$, that consists of an isomorphism of schemes over $V \times_U V$ satisfying the cocycle condition, that is the commutativity of the diagram

$$\begin{array}{ccc} V \times_U V \times_U Y & \xrightarrow{\mathrm{pr}_{23}^* \phi \,=\, \mathrm{id}_V \times \phi} & V \times_U Y \times_U V \\ & {\scriptstyle \mathrm{pr}_{13}^* \phi} \searrow \quad \swarrow {\scriptstyle \mathrm{pr}_{12}^* \phi \,=\, \phi \times \mathrm{id}_V} & \\ & Y \times_U V \times_U V & \end{array}$$

We will use ϕ to construct descent data for the quasi-coherent sheaf of finite presentation $\mathcal{M}_\eta$ on V. From the two cartesian diagrams

$$\begin{array}{ccc} V \times_U Y & \xrightarrow{\mathrm{pr}_2} & Y \\ {\scriptstyle \mathrm{id}_V \times \eta} \downarrow & & \downarrow {\scriptstyle \eta} \\ V \times_U V & \xrightarrow{\mathrm{pr}_2} & V \end{array} \qquad \text{and} \qquad \begin{array}{ccc} Y \times_U V & \xrightarrow{\mathrm{pr}_1} & Y \\ {\scriptstyle \eta \times \mathrm{id}_V} \downarrow & & \downarrow {\scriptstyle \eta} \\ V \times_U V & \xrightarrow{\mathrm{pr}_1} & V \end{array}$$

we get isomorphisms

$$\sigma_{\mathrm{pr}_2,\mathrm{pr}_2} \colon \mathrm{pr}_2^* \mathcal{M}_\eta \simeq \mathcal{M}_{\mathrm{id}_V \times \eta} \quad \text{and} \quad \sigma_{\mathrm{pr}_1,\mathrm{pr}_1} \colon \mathrm{pr}_1^* \mathcal{M}_\eta \simeq \mathcal{M}_{\eta \times \mathrm{id}_V};$$

and from the cartesian diagram

$$\begin{array}{ccc} V \times_U Y & \xrightarrow{\phi} & Y \times_U V \\ \downarrow {\scriptstyle \mathrm{id}_V \times \eta} & & \downarrow {\scriptstyle \eta \times \mathrm{id}_V} \\ V \times_U V & = & V \times_U V \end{array}$$

another isomorphism

$$\sigma_{\phi, \mathrm{id}_{V \times_U V}} \colon \mathcal{M}_{\eta \times \mathrm{id}_V} \simeq \mathcal{M}_{\mathrm{id}_V \times \eta}.$$

With these we define an isomorphism

$$\psi \overset{\text{def}}{=} \sigma_{\mathrm{pr}_1,\mathrm{pr}_1}^{-1} \circ \sigma_{\phi, \mathrm{id}_{V \times_U V}}^{-1} \circ \sigma_{\mathrm{pr}_2,\mathrm{pr}_2} \colon \mathrm{pr}_2^* \mathcal{M}_\eta \simeq \mathrm{pr}_1^* \mathcal{M}_\eta$$

of quasi-coherent sheaves on $V \times_U V$.

Let us check that ψ satisfies the cocycle condition. We will use our customary notation pr_1 and pr_2 to denote the projection onto the first and second factor of a product $X_1 \times_Y X_2$, and p_1, p_2 and p_3 for the first, second and third projection onto the factors of the triple product $X_1 \times_Y X_2 \times_Y X_3$. (Previously we have also denoted these by pr_1, pr_2 and pr_3, but here the risk of confusion seems more real.)

Consider the cartesian diagram

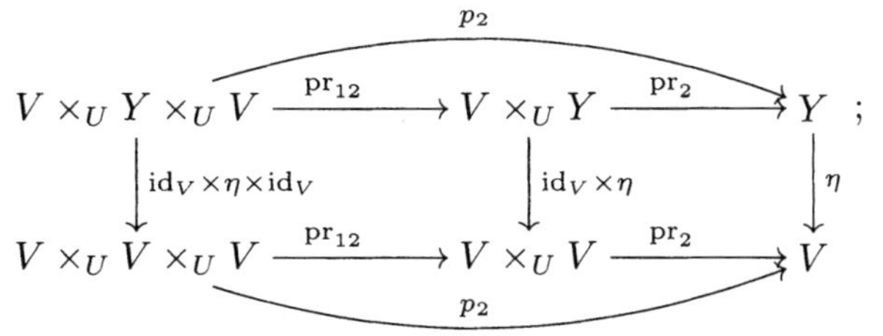

$$\begin{array}{ccccc} V \times_U Y \times_U V & \xrightarrow{\mathrm{pr}_{12}} & V \times_U Y & \xrightarrow{\mathrm{pr}_2} & Y \\ \Big\downarrow{\scriptstyle \mathrm{id}_V \times \eta \times \mathrm{id}_V} & & \Big\downarrow{\scriptstyle \mathrm{id}_V \times \eta} & & \Big\downarrow{\scriptstyle \eta} \\ V \times_U V \times_U V & \xrightarrow{\mathrm{pr}_{12}} & V \times_U V & \xrightarrow{\mathrm{pr}_2} & V \end{array} \quad ;$$

(with the composites of the top and bottom rows labelled p_2) according to Proposition 4.40, we have that $\sigma_{p_2,p_2} : p_2^* \mathcal{M}_\eta \to \mathcal{M}_{\mathrm{id}_V \times \eta \times \mathrm{id}_V}$ is the composite

$$p_2^* \mathcal{M}_\eta \xrightarrow{\mathrm{pr}_{12}^* \sigma_{\mathrm{pr}_2,\mathrm{pr}_2}} \mathrm{pr}_{12}^* \mathcal{M}_{\mathrm{id}_V \times \eta} \xrightarrow{\sigma_{\mathrm{pr}_{12} \cdot \mathrm{pr}_{12}}} \mathcal{M}_{\mathrm{id}_V \times \eta \times \mathrm{id}_V},$$

so we have the equality

$$\mathrm{pr}_{12}^* \sigma_{\mathrm{pr}_2,\mathrm{pr}_2} = \sigma_{\mathrm{pr}_{12},\mathrm{pr}_{12}}^{-1} \circ \sigma_{p_2,p_2} : p_2^* \mathcal{M}_\eta \longrightarrow \mathcal{M}_{\mathrm{id}_V \times \eta \times \mathrm{id}_V}.$$

In a completely analogous fashion we get the equalities

$$\begin{aligned} \mathrm{pr}_{12}^* \sigma_{\mathrm{pr}_1,\mathrm{pr}_1} &= \sigma_{\mathrm{pr}_{12},\mathrm{pr}_{12}}^{-1} \circ \sigma_{p_1,p_1}, \\ \mathrm{pr}_{23}^* \sigma_{\mathrm{pr}_1,\mathrm{pr}_1} &= \sigma_{\mathrm{pr}_{23},\mathrm{pr}_{23}}^{-1} \circ \sigma_{p_2,p_2}, \\ \mathrm{pr}_{23}^* \sigma_{\mathrm{pr}_2,\mathrm{pr}_2} &= \sigma_{\mathrm{pr}_{23},\mathrm{pr}_{23}}^{-1} \circ \sigma_{p_3,p_3}, \\ \mathrm{pr}_{13}^* \sigma_{\mathrm{pr}_1,\mathrm{pr}_1} &= \sigma_{\mathrm{pr}_{13},\mathrm{pr}_{13}}^{-1} \circ \sigma_{p_1,p_1} \quad \text{and} \\ \mathrm{pr}_{13}^* \sigma_{\mathrm{pr}_2,\mathrm{pr}_2} &= \sigma_{\mathrm{pr}_{13},\mathrm{pr}_{13}}^{-1} \circ \sigma_{p_3,p_3}. \end{aligned}$$

We need to prove the equality $\mathrm{pr}_{12}^* \psi \circ \mathrm{pr}_{23}^* \psi = \mathrm{pr}_{13}^* \psi$; using definition of ψ and the identities above, and doing some simplifications, the reader can check that this equality is equivalent to the equality

$$\begin{aligned} (\sigma_{\mathrm{pr}_{12},\mathrm{pr}_{12}} \circ \mathrm{pr}_{12}^* \sigma_\phi^{-1} \circ \sigma_{\mathrm{pr}_{12},\mathrm{pr}_{12}}^{-1}) \circ (\sigma_{\mathrm{pr}_{23},\mathrm{pr}_{23}} \circ \mathrm{pr}_{23}^* \sigma_\phi^{-1} \circ \sigma_{\mathrm{pr}_{23},\mathrm{pr}_{23}}^{-1}) \\ = (\sigma_{\mathrm{pr}_{23},\mathrm{pr}_{23}} \circ \mathrm{pr}_{23}^* \sigma_\phi^{-1} \circ \sigma_{\mathrm{pr}_{23},\mathrm{pr}_{23}}^{-1}) \end{aligned}$$

that we are going to prove as follows.

From the cartesian diagram

$$\begin{array}{ccccc} V \times_U Y \times_U V & \xrightarrow{\mathrm{pr}_{12}} & V \times_U Y & \xrightarrow{\phi} & Y \times_U V \\ \Big\downarrow{\scriptstyle \mathrm{id}_V \times \eta \times \mathrm{id}_V} & & \Big\downarrow{\scriptstyle \mathrm{id}_V \times \eta} & & \Big\downarrow{\scriptstyle \eta \times \mathrm{id}_V} \\ V \times_U V \times_U V & \xrightarrow{\mathrm{pr}_{12}} & V \times_U V & = & V \times_U V \end{array}$$

we get the equality

$$\mathrm{pr}_{12}^* \sigma_\phi^{-1} \circ \sigma_{\mathrm{pr}_{12},\mathrm{pr}_{12}}^{-1} = \sigma_{\phi \circ \mathrm{pr}_{12},\mathrm{pr}_{12}}^{-1};$$

from the other cartesian diagram

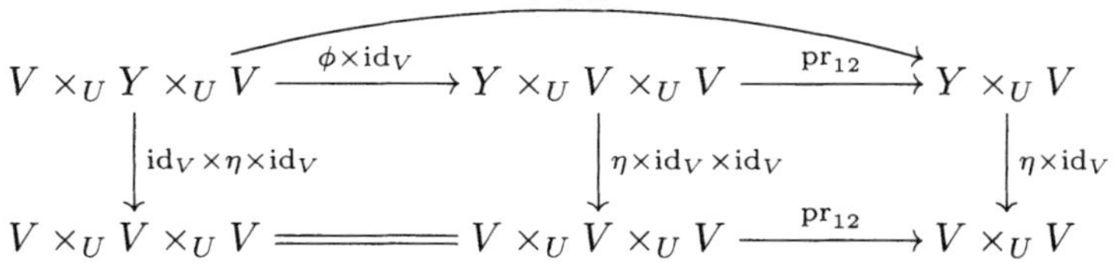

$$\begin{array}{ccccc} V \times_U Y \times_U V & \xrightarrow{\phi \times \mathrm{id}_V} & Y \times_U V \times_U V & \xrightarrow{\mathrm{pr}_{12}} & Y \times_U V \\ \Big\downarrow{\scriptstyle \mathrm{id}_V \times \eta \times \mathrm{id}_V} & & \Big\downarrow{\scriptstyle \eta \times \mathrm{id}_V \times \mathrm{id}_V} & & \Big\downarrow{\scriptstyle \eta \times \mathrm{id}_V} \\ V \times_U V \times_U V & = & V \times_U V \times_U V & \xrightarrow{\mathrm{pr}_{12}} & V \times_U V \end{array}$$

(with the composite of the top row labelled $\phi \circ p_{12}$)

we obtain that

$$\sigma_{\mathrm{pr}_{12},\mathrm{pr}_{12}} \circ \sigma^{-1}_{\phi\circ\mathrm{pr}_{12},\mathrm{pr}_{12}} = \sigma^{-1}_{\phi\times\mathrm{id}_V,\mathrm{id}_{V\times_U V\times_U V}} = \sigma^{-1}_{\mathrm{pr}_{12}^*\phi,\mathrm{id}_{V\times_U V\times_U V}}.$$

With analogous arguments we get the equalities

$$\sigma_{\mathrm{pr}_{23},\mathrm{pr}_{23}} \circ \sigma^{-1}_{\phi\circ\mathrm{pr}_{23},\mathrm{pr}_{23}} = \sigma^{-1}_{\mathrm{pr}_{23}^*\phi,\mathrm{id}_{V\times_U V\times_U V}}$$

and

$$\sigma_{\mathrm{pr}_{13},\mathrm{pr}_{13}} \circ \sigma^{-1}_{\phi\circ\mathrm{pr}_{13},\mathrm{pr}_{13}} = \sigma^{-1}_{\mathrm{pr}_{13}^*\phi,\mathrm{id}_{V\times_U V\times_U V}},$$

from which we see that the cocycle condition for ψ is equivalent to the equality

$$\sigma_{\mathrm{pr}_{23}^*\phi,\mathrm{id}_{V\times_U V\times_U V}} \circ \sigma_{\mathrm{pr}_{12}^*\phi,\mathrm{id}_{V\times_U V\times_U V}} = \sigma_{\mathrm{pr}_{13}^*\phi,\mathrm{id}_{V\times_U V\times_U V}}.$$

But this follows immediately, once again thanks to Proposition 4.40, from the cocycle condition on ϕ.

So $(\mathcal{M}_\eta, \psi)$ is a quasi-coherent sheaf with descent data, hence it will come from some quasi-coherent sheaf of finite presentation $\mathcal{M}$ on U.

Now let us go back to the general case. Given an arrow $\xi\colon X \to U$ in $\mathcal{F}$, we have an adjunction homomorphism

$$\tau_\xi\colon \xi^*\mathcal{M}_\xi = \xi^*\xi_*\mathcal{L}_\xi \longrightarrow \mathcal{L}_\xi$$

that is characterized at the level of sections by the equality $\tau_\xi(\xi^*s) = s$ for any section s of $\mathcal{L}_\xi$ over the inverse image of an open subset of U.

PROPOSITION 4.41. Given a cartesian square

$$\begin{array}{ccc} X & \xrightarrow{f} & Y \\ \downarrow{\scriptstyle\xi} & & \downarrow{\scriptstyle\eta} \\ U & \xrightarrow{\phi} & V \end{array}$$

the two composites

$$(\phi\xi)^*\mathcal{M}_\eta = (\eta f)^*\mathcal{M}_\eta \xrightarrow{f^*\tau_\eta} f^*\mathcal{L}_\eta \xrightarrow{\rho_{f,\phi}} \mathcal{L}_\xi$$

and

$$(\phi\xi)^*\mathcal{M}_\eta \xrightarrow{\xi^*\sigma_{f,\phi}} \xi^*\mathcal{M}_\xi \xrightarrow{\tau_\xi} \mathcal{L}_\xi$$

coincide.

PROOF. Let s be a section of $\mathcal{M}_\eta$ over an open subset of V, that is, a section of $\mathcal{L}_\eta$ over the inverse image in Y of an open subset of V. Then both composites are characterized by the property of sending $(\phi\xi)^*s = (\eta f)^*s$ to $\rho_{f,\phi}(f^*s)$. □

In the situation of the Proposition above, assume that $\mathcal{L}_\eta$ is very ample on Y relative to η, and that the base change homomorphism $\beta_{f,\phi}\colon \phi^*\eta_*\mathcal{L}_\eta \to \xi_*f^*\mathcal{L}_\eta$ is an isomorphism. Then $\sigma_{f,\phi}\colon \phi^*\mathcal{M}_\eta \to \mathcal{M}_\xi$ is an isomorphism, and this induces an isomorphism

$$U \times_V \mathbb{P}(\mathcal{M}_\eta) = \mathbb{P}(\phi^*\mathcal{M}_\eta) \simeq \mathbb{P}(\mathcal{M}_\xi)$$

of schemes over U, hence a cartesian square

$$\begin{array}{ccc} \mathbb{P}(\mathcal{M}_\xi) & \longrightarrow & U \\ \downarrow & & \downarrow{\scriptstyle\phi} \\ \mathbb{P}(\mathcal{M}_\eta) & \longrightarrow & V \end{array}$$

Also, since $\mathcal{L}_\xi$ and $\mathcal{L}_\eta$ are very ample, the base change homomorphisms $\tau_\xi \colon \xi^* \mathcal{M}_\xi \to \mathcal{L}_\xi$ and $\tau_\eta \colon \eta^* \mathcal{M}_\eta \to \mathcal{L}_\eta$ are surjective, and the corresponding morphisms of schemes $X \to \mathbb{P}(\mathcal{M}_\xi)$ and $Y \to \mathbb{P}(\mathcal{M}_\eta)$ are closed embeddings. Proposition 4.41 implies that the diagram

$$\begin{array}{ccccc}
X & \hookrightarrow & \mathbb{P}(\mathcal{M}_\xi) & \longrightarrow & U \\
\downarrow{\scriptstyle f} & & \downarrow & & \downarrow{\scriptstyle \phi} \\
Y & \hookrightarrow & \mathbb{P}(\mathcal{M}_\eta) & \longrightarrow & V
\end{array}$$

commutes, and is cartesian.

Going back to our covering $V \to U$, we have that the two diagrams

$$\begin{array}{ccccc}
V \times_U Y & \hookrightarrow & \mathbb{P}(\mathcal{M}_{\mathrm{id}_V \times \eta}) & \longrightarrow & V \times_U V \\
\downarrow{\scriptstyle \mathrm{pr}_1} & & \downarrow & & \downarrow{\scriptstyle \mathrm{pr}_1} \\
Y & \hookrightarrow & \mathbb{P}(\mathcal{M}_\eta) & \longrightarrow & V
\end{array}$$

and

$$\begin{array}{ccccc}
Y \times_U V & \hookrightarrow & \mathbb{P}(\mathcal{M}_{\eta \times \mathrm{id}_V}) & \longrightarrow & V \times_U V \\
\downarrow{\scriptstyle \mathrm{pr}_2} & & \downarrow & & \downarrow{\scriptstyle \mathrm{pr}_2} \\
Y & \hookrightarrow & \mathbb{P}(\mathcal{M}_\eta) & \longrightarrow & V
\end{array}$$

are cartesian; these, together with the diagram

$$\begin{array}{ccccc}
V \times_U Y & \hookrightarrow & \mathbb{P}(\mathcal{M}_{\mathrm{id}_V \times \eta}) & \longrightarrow & V \times_U V \\
\downarrow{\scriptstyle \phi} & & \downarrow & & \| \\
Y \times_U V & \hookrightarrow & \mathbb{P}(\mathcal{M}_\eta \times \mathrm{id}_V) & \longrightarrow & V \times_U V
\end{array}$$

and the definition of ψ, show that the two inverse images of $Y \subseteq \mathbb{P}(\mathcal{M}_\eta)$ in $\mathbb{P}(\mathrm{pr}_1^* \mathcal{M}_\eta)$ and in $\mathbb{P}(\mathrm{pr}_2^* \mathcal{M}_\eta)$ coincide. On the other hand, the quasi-coherent sheaf with descent data $(\mathcal{M}_\eta, \psi)$ is isomorphic to the pullback of the quasi-coherent sheaf $\mathcal{M}$. If $f \colon V \to U$ is the given morphism, $F \colon V \times_U V \to U$ the composite with the projections $V \times_U V \to V$, this implies that the two pullbacks of $Y \subseteq \mathbb{P}(\mathcal{M}_\eta) \simeq \mathbb{P}(f^* \mathcal{M})$ coincide. This implies that there is a unique closed subscheme $X \subseteq \mathbb{P}(\mathcal{M})$ that pulls back to $Y \subseteq \mathbb{P}(\mathcal{M}_\eta)$; and, since $\mathcal{F}$ is a local class in the fpqc topology, the morphism $X \to U$ is in $\mathcal{F}$. The scheme with descent data associated with $X \to U$ is precisely $(Y \to V, \phi)$; and this completes the proof of Theorem 4.38. □

4.4. Descent along torsors

One of the most interesting examples of descent is descent for quasi-coherent sheaves along fpqc torsors. This can be considered as a vast generalization of the well-known equivalence between the category of real vector spaces and the category of complex vector spaces with an anti-linear involution. Torsors are generalizations of principal fiber bundles in topology; and I always find it striking that among the simplest examples of torsors are Galois field extensions (see Example 4.45).

Here we only introduce the bare minimum of material that allows us to state and prove the main theorem. For a fuller treatment, see [**DG70**].

In this section we will work with a subcanonical site $\mathcal{C}$ with fibered products, and a group object G in $\mathcal{C}$. We will assume that $\mathcal{C}$ has a terminal object pt.

The examples that we have in mind are $\mathcal{C} = (\mathrm{Top})$, endowed with the global classical topology, where G is any topological group, and $\mathcal{C} = (\mathrm{Sch}/S)$, with the fpqc topology, where $G \to S$ is a group scheme.

4.4.1. Torsors. Torsors are what in other fields of mathematics are called *principal bundles.* Suppose that we have an object X of $\mathcal{C}$, with a left action $\alpha\colon G \times X \to X$ of G. An arrow $X \to Y$ is called *invariant* if for each object U of $\mathcal{C}$ the induced function $X(U) \to Y(U)$ is invariant with respect to the action of $G(U)$ on $X(U)$. Another way of saying this is that the composites of $X \to Y$ with the two arrows α and pr_2 from $G \times X$ to X are equal (the equivalence with the definition above follows from Yoneda's lemma).

Yet another equivalent definition is that the arrow f is G-equivariant, when Y is given the trivial G-action $\mathrm{pr}_2\colon G \times Y \to Y$.

If $\pi\colon X \to Y$ is an invariant arrow and $f\colon Y' \to Y$ is an arrow, there is an induced action of G on $Y' \times_Y X$; this is the unique actions that makes the first projection $\mathrm{pr}_1\colon Y' \times_Y X \to Y'$ invariant, and the second projection $\mathrm{pr}_2\colon Y' \times_Y X \to X$ G-equivariant. In functorial terms, if U is an object of $\mathcal{C}$, $g \in G(U)$, $x \in X(U)$ and $y' \in Y'(U)$ are elements with the same image in $Y(U)$, we have $g \cdot (y', x) = (y', g \cdot x)$.

The first example of a torsor is the *trivial torsor.* For each object Y of $\mathcal{C}$, consider the product $G \times Y$. This has an action of G, defined by the obvious formula

$$g \cdot (h, x) = (gh, x)$$

for all objects U of $\mathcal{C}$, all g and h in $G(U)$ and all x in $X(U)$.

More generally, a *trivial torsor* consists of an object X of $\mathcal{C}$ with a left action of G, together with an invariant arrow $f\colon X \to Y$, such that there is a G-equivariant isomorphism $\phi\colon G \times Y \simeq X$ making the diagram

$$\begin{array}{ccc} G \times Y & \xrightarrow{\ \phi\ } & X \\ & \searrow^{\mathrm{pr}_2} \quad \swarrow_{f} & \\ & Y & \end{array}$$

commute. (The isomorphism itself is not part of the data, only its existence is required.)

A G-torsor is an object X of $\mathcal{C}$ with an action of G and an invariant arrow $\pi\colon X \to Y$ that locally on Y is a trivial torsor. Here is the precise definition.

DEFINITION 4.42. A *G-torsor* in $\mathcal{C}$ consists of an object X of $\mathcal{C}$ with an action of G and an invariant arrow $\pi\colon X \to Y$, such that there exists a covering $\{Y_i \to Y\}$ of Y with the property that for each i the arrow $\mathrm{pr}_1\colon Y_i \times_Y X \to Y_i$ is a trivial torsor.

Here is an important characterization of torsors. Notice that every time we have an action $\alpha\colon G \times X \to X$ of G on an object X and an invariant arrow $f\colon X \to Y$, we get an arrow $\delta_\alpha\colon G \times X \to X \times_Y X$, defined as a natural transformation by the formula $(g, x) \mapsto (gx, x)$ for any object U of $\mathcal{C}$ and any $g \in G(U)$ and $x \in X(U)$.

PROPOSITION 4.43. Let X be an object of $\mathcal{C}$ with an action of G. An invariant arrow $\pi\colon X \to Y$ is a G-torsor if and only if

(i) There exists a covering $\{Y_i \to Y\}$ such that every arrow $Y_i \to Y$ factors through $\pi\colon X \to Y$, and
(ii) the arrow $\delta_\alpha\colon G \times X \to X \times_Y X$ is an isomorphism.

Notice that part (i) says that $X \to Y$ is a covering in the saturation of the topology of $\mathcal{C}$ (Definition 2.52).

PROOF. Assume that the two conditions are satisfied. The arrow δ_α is immediately checked to be G-equivariant; hence the pullback $X \times_Y X \to X$ of π through the covering $\pi\colon X \to Y$ is a trivial torsor, and therefore $\pi\colon X \to Y$ is a G-torsor.

Conversely, take a torsor $\pi\colon X \to Y$. First of all, assume that $\pi\colon X \to Y$ is a trivial torsor, and fix a G-equivariant isomorphism $\phi\colon G \times Y \simeq X$ over Y. There is a section $Y \to X$ of $\pi\colon X \to Y$, so condition (i) is satisfied for the covering $\{Y = Y\}$.

To verify condition (ii), notice that δ_α can be written as the composite of isomorphisms

$$G \times X \xrightarrow{\mathrm{id}_G \times \phi^{-1}} G \times G \times Y \simeq (G \times Y) \times_Y (G \times Y) \xrightarrow{\phi \times \phi} X \times_Y X,$$

where the isomorphism is in the middle is defined as a natural transformation by the rule $(g, h, y) \mapsto \big((g, y), (h, y)\big)$ for any object U of $\mathcal{C}$ and any $g, h \in G(U)$ and $y \in Y(U)$.

In the general case, when $\pi\colon X \to Y$ is not necessarily trivial, the result follows from the previous case and the following lemma.

LEMMA 4.44. Let $\mathcal{C}$ be a subcanonical site,

$$\begin{array}{ccc} X & \xrightarrow{f} & Y \\ & \searrow \quad \swarrow & \\ & S & \end{array}$$

a commutative diagram in $\mathcal{C}$. Suppose that there is a covering $\{S_i \to S\}$ such that the induced arrows

$$\mathrm{id}_{S_i} \times f\colon S_i \times_S X \longrightarrow S_i \times_S Y$$

are isomorphisms. Then f is also an isomorphism.

PROOF. The site $(\mathcal{C}/S)$ is subcanonical (Proposition 2.59): this means that we can substitute $(\mathcal{C}/S)$ for $\mathcal{C}$, and suppose that S is a terminal object of $\mathcal{C}$.

By Yoneda's lemma, it is enough to show that for any object U of $\mathcal{C}$ the function $f_U\colon X(U) \to Y(U)$ induced by f is a bijection. First of all, assume that the arrow $U \to S$ factors through some S_i. By hypothesis $\mathrm{id}_{S_i} \times f\colon S_i \times X \to S_i \times Y$ is an isomorphism, hence $\mathrm{id}_{S_i(U)} \times f_U\colon S_i(U) \times X(U) \to S_i(U) \times Y(U)$ is a bijection. If $S_i(U) \neq \emptyset$ it follows that f_U is a bijection.

For the general case we use the hypothesis that $\mathcal{C}$ is subcanonical. If U is arbitrary, and we set $U_i \overset{\mathrm{def}}{=} S_i \times U$, then $\{U_i \to U\}$ is a covering. Hence we have a diagram of sets

$$\begin{array}{ccccc} X(U) & \longrightarrow & \prod_i X(U_i) & \rightrightarrows & \prod_{ij} Y(U_{ij}) \\ \downarrow{\scriptstyle f_U} & & \downarrow{\scriptstyle \prod_i f_{U_i}} & & \downarrow{\scriptstyle \prod_{ij} f_{U_{ij}}} \\ Y(U) & \longrightarrow & \prod_i X(U_i) & \rightrightarrows & \prod_{ij} Y(U_{ij}) \end{array}$$

in which the rows are equalizers, because $\mathcal{C}$ is subcanonical. On the other hand each arrow $U_i \to S$ and $U_{ij} \to S$ factors through S_i, so f_{U_i} and $f_{U_{ij}}$ are bijections. It follows that f_U is a bijection, as required. □

Consider the arrow $\delta_\alpha \colon G \times X \to X \times_Y X$, and choose a covering $\{Y_i \to Y\}$ such that for each i the pullbacks $X_i \overset{\text{def}}{=} Y_i \times_Y X$ are trivial as torsors over Y_i. Denote by $\alpha_i \colon G \times X_i \to X_i$ the induced action; then $\delta_{\alpha_i} \colon G \times X_i \to X_i \times_{Y_i} X_i$ is an isomorphism. On the other hand there are standard isomorphisms $(G \times X) \times_Y Y_i \simeq G \times X_i$ and $(X \times_Y X) \times_Y Y_i \simeq X_i \times_{Y_i} X_i$, and the diagram

$$\begin{array}{ccc} (G \times X) \times_Y Y_i & \xrightarrow{\delta_\alpha \times \mathrm{id}_{Y_i}} & (X \times_Y X) \times_Y Y_i \\ \downarrow{\scriptstyle \mathrm{pr}_1} & & \downarrow{\scriptstyle \mathrm{pr}_1} \\ G \times X_i & \xrightarrow{\delta_{\alpha_i}} & X_i \times_{Y_i} X_i \end{array}$$

commutes. Hence $\delta_\alpha \times \mathrm{id}_{Y_i}$ is an isomorphism for all i, and it follows that δ_α is an isomorphism. □

EXAMPLE 4.45. Let $K \subseteq L$ be a finite Galois extensions, with Galois group G. Denote by G_K the discrete group scheme $G \times \operatorname{Spec} K \to \operatorname{Spec} K$ associated with G, as in §2.2.2. The action of G on $\operatorname{Spec} L$ defines an action $\alpha \colon G_K \times_{\operatorname{Spec} K} \operatorname{Spec} L = G \times \operatorname{Spec} L \to \operatorname{Spec} L$ of G_K on $\operatorname{Spec} L$ (Proposition 2.22), which leaves the morphism $\operatorname{Spec} L \to \operatorname{Spec} K$ invariant. (For convenience we will write the action of G on L on the right, so that the resulting action of G on $\operatorname{Spec} L$ is naturally written as a left action.)

By the primitive element theorem, L is generated as an extension of K by a unique element u; denote by $f \in K[x]$ its minimal polynomial. Then $L = K[x]/\big(f(x)\big)$. The group G acts on the roots of f simply transitively, so $f(x) = \prod_{g \in G}(x - ug) \in L[x]$.

The morphism

$$\delta_\alpha \colon G \times \operatorname{Spec} L \longrightarrow \operatorname{Spec} L \times_{\operatorname{Spec} K} \operatorname{Spec} L = \operatorname{Spec}(L \otimes_K L)$$

corresponds to the homomorphism of K-algebras $L \otimes_K L \to L^G$ defined as

$$a \otimes b \mapsto \big((ag)b\big)_{g \in G},$$

where by L^G we mean the product of copies of L indexed by G. We have an isomorphism

$$L \otimes_K L = K[x]/\big(\prod_{g \in G}(x - ug)\big) \otimes_K L \simeq L[x]/\big(\prod_{g \in G}(x - ug)\big);$$

by the chinese remainder theorem, the projection

$$L[x]/\big(\prod_{g \in G}(x - ug)\big) \longrightarrow \prod_{g \in G} L[x]/(x - ug) \simeq L^G$$

is an isomorphism. Thus we get an isomorphism $L \otimes L \simeq L^G$, that is easily seen to coincide the homomorphism corresponding to δ_α. Thus δ_α is an isomorphism; and since $\operatorname{Spec} L \to \operatorname{Spec} K$ is étale, this shows that $\operatorname{Spec} L$ is G_K-torsor over $\operatorname{Spec} K$.

Here is our main result.

THEOREM 4.46. Let $X \to Y$ be a G-torsor, and $\mathcal{F} \to \mathcal{C}$ a stack. Then there exists a canonical equivalence of categories between $\mathcal{F}(Y)$ and the category of G-equivariant objects $\mathcal{F}^G(X)$ defined in §3.8.

PROOF. Because of Proposition 4.16, we have an equivalence of $\mathcal{F}(Y)$ with $\mathcal{F}(X \to Y)$, so it is enough to produce an equivalence between $\mathcal{F}(X \to Y)$ and $\mathcal{F}^G(X)$.

For this we need the isomorphism $\delta_\alpha\colon G \times X \simeq X \times_Y X$ defined above, and also the one defined in the next Lemma.

LEMMA 4.47. If $X \to Y$ is a G-torsor, the arrow

$$\delta'_\alpha\colon G \times G \times X \longrightarrow X \times_Y X \times_Y X$$

defined in functorial terms by the rule

$$\delta'_\alpha(g, h, x) = (ghx, hx, x)$$

is an isomorphism.

Once again, one reduces to the case of a trivial torsor using Lemma 4.44. We leave the proof of this case to the reader.

Since the category $\mathcal{F}(X \to Y)$ does not depend on the choice of the fibered products $X \times_Y X$ and $X \times_Y X \times_Y X$, we can make the choice $X \times_Y X = G \times X$ and $X \times_Y X \times_Y X = G \times G \times X$, in such a way that δ_α and δ'_α become the identity. Then we have

$$\begin{aligned} \mathrm{pr}_1 &= \alpha\colon G \times X \to X, \\ \mathrm{pr}_{13} &= \mathrm{m}_G \times \mathrm{id}_X\colon G \times G \times X \to G \times X, \\ \mathrm{pr}_{12} &= \mathrm{id}_G \times \alpha\colon G \times G \times X \to G \times X, \end{aligned}$$

while $\mathrm{pr}_{23} = X \times_Y X \times_Y X \to X \times_Y X$ coincides with the projection $G \times G \times X \to G \times X$ on the second and third factor.

Then an object (ρ, ϕ) of $\mathcal{F}(X \to Y)$ is an object ρ of $\mathcal{F}(X)$, together with an isomorphism $\phi\colon \mathrm{pr}_2^* \rho \simeq \alpha^* \rho$ satisfying the cocycle condition: and the cocycle condition is precisely the condition for ϕ to define a G-equivariant structure on ρ, according to Proposition 3.49. Hence the category $\mathcal{F}(X \to Y)$ is canonically isomorphic to $\mathcal{F}^G(X)$ (for this we need to check what happens to arrows, but this is easy and left to the reader), and this concludes the proof of the theorem. □

4.4.2. Failure of descent for morphisms of schemes. Now we construct an example to show how descent can fail for proper smooth morphisms of proper schemes of finite type over a field.

The starting point is a variant on Hironaka's famous example of a nonprojective threefold ([**Hir62**], [**MFK94**, Chapter 3, § 3]), [**Har77**, Appendix B, Example 3.4.1]); this has been already been used to give examples of a smooth three-dimensional algebraic space over a field that is not a scheme ([**Knu71**, p. 14]).

Fix an algebraically closed field κ. Then one constructs a smooth proper connected three dimensional scheme M over κ, with an action of a cyclic group of order two $\mathrm{C}_2 = \{1, \sigma\}$, containing two copies L_1 of L_2 of $\mathbb{P}^1$ that are interchanged by σ, with the property that the 1-cycle $L_1 + L_2$ is algebraically equivalent to 0. This implies that there is no open affine subscheme U of M that intersects L_1 and L_2 simultaneously: if not, the complement S of U would be a surface in M that intersects both L_1 and L_2 in a finite number of points. But since $S \cdot (L_1 + L_2) = 0$

this finite number of points would have to be zero, and this would mean that L_1 and L_2 are entirely contained in U. This is impossible, because U is affine.

Now take a C_2-torsor $V \to U$ (a Galois étale cover with group C_2) with V irreducible, and set $Y = M \times_\kappa V$. The projection $\pi\colon Y \to V$ is smooth and proper. We need descent data for the covering $V \to U$; these are given by the diagonal action of C_2 on Y, obtained from the two actions on M and V. More precisely, the action $C_2 \times Y \to Y$ gives a cartesian diagram

$$\begin{array}{ccc} C_2 \times Y & \longrightarrow & Y \\ \downarrow & & \downarrow \\ C_2 \times V & \longrightarrow & V \end{array}$$

yielding an isomorphism of $C_2 \times Y$ with the pullback of $Y \to V$ to $C_2 \times V$, and defines an object with descent data on the covering $V \to V$ (keeping in mind that $V \times_U V = C_2 \times V$, since $V \to U$ is a C_2-torsor).

I claim that these descent data are not effective. Suppose that it is not so: then there is a cartesian diagram of schemes

$$\begin{array}{ccc} Y & \xrightarrow{f} & X \\ \downarrow{\scriptstyle \pi} & & \downarrow \\ V & \longrightarrow & U \end{array}$$

such that f is invariant under the action of C_2 on Y. Take an open affine subscheme $W \subseteq X$ that intersects $f(L_1 \times V) = f(L_2 \times V)$; then its inverse image $f^{-1}W \subseteq Y$ is affine, and if p is a generic closed point of V the intersection $\pi^{-1}p \cap f^{-1}W \subseteq M$ is an affine open subscheme of M that intersects both L_1 and L_2. As we have seen, this is impossible.

Another counterexample, already mentioned in Example 4.39, is in [**Ray70**, XIII 3.2].

REMARK 4.48. There is an extension of the theory of schemes, the theory of *algebraic spaces*, due to Michael Artin (see [**Art71**], [**Art73**] and [**Knu71**]). An algebraic space over a scheme S is an étale sheaf $(\mathrm{Sch}/S)^{\mathrm{op}} \to (\mathrm{Set})$, that is, in some sense, étale locally a scheme. The category of algebraic stacks contains the category of schemes over S with quasi-compact diagonal; furthermore, by a remarkable result of Artin, it is a stack in the fppf topology (it is probably also a stack in the fpqc topology, but I do not know this for sure: however, for most applications fppf descent is what is needed). Also, most of the concepts and techniques that apply to schemes extend to algebraic spaces. This is obvious for properties of schemes, and morphisms of schemes, such us being Cohen–Macaulay, smooth, or flat, that are local in the étale topology (on the domain). Global properties, such as properness, require more work.

So, in many contexts, when some descent data in the fppf topology fail to define a scheme, an algebraic space appears as a result. Also, algebraic spaces can be used to define stacks in situations when descent for schemes fails. For example, if we redefine that stack $\mathcal{F}_{1,S}$ of Example 4.39 so that the objects are proper smooth morphisms $X \to U$ whose fibers are curves of genus 1, where U is an S-scheme and X is an algebraic space, then $\mathcal{F}_{1,S}$ is a stack in the fppf topology.

Part 2

Construction of Hilbert and Quot schemes

Nitin Nitsure

CHAPTER 5

Construction of Hilbert and Quot schemes

Introduction

Any scheme X defines a contravariant functor h_X (called the functor of points of the scheme X) from the category of schemes to the category of sets, which associates to any scheme T the set $Mor(T, X)$ of all morphisms from T to X. The scheme X can be recovered (up to a unique isomorphism) from h_X by the Yoneda lemma. In fact, it is enough to know the restriction of this functor to the full subcategory consisting of affine schemes, in order to recover the scheme X.

It is often easier to directly describe the functor h_X than to give the scheme X. Such is typically the case with various parameter schemes and moduli schemes, or with various group-schemes over arbitrary bases, where we can directly define a contravariant functor F from the category of schemes to the category of sets which would be the functor of points of the scheme in question, without knowing in advance whether such a scheme indeed exists.

This raises the problem of representability of contravariant functors from the category of schemes to the category of sets. An important necessary condition for representability come from the fact that the functor h_X satisfies descent under faithfully flat quasi-compact coverings.

(Recall that descent for a set-valued functor F is the sheaf condition, which says that if $(f_i : U_i \to U)$ is an open cover of U in the fpqc topology, then the sequence $F(U) \to \prod_i F(U_i) \rightrightarrows \prod_{i,j} F(U_i \times_U U_j)$ is exact.)

The descent condition is often easy to verify for a given functor F, but it is not a sufficient condition for representability.

It is therefore a subtle and technically difficult problem in Algebraic Geometry to construct schemes which represent various important functors, such as moduli functors. Grothendieck addressed the issue by proving the representability of certain basic functors, namely, the Hilbert and Quot functors. The representing schemes that he constructed, known as Hilbert schemes and Quot schemes, are the foundation for proving representability of most moduli functors (whether as schemes or as algebraic stacks).

The techniques used by Grothendieck are based on the theories of descent and cohomology developed by him. In a sequence of talks in the Bourbaki seminar, collected under the title 'Fondements de la Géométrie Algébriques' (see [FGA]), he gave a sketch of the theory of descent, the construction of Hilbert and Quot schemes, and its application to the construction of Picard schemes (and also a sketch of formal schemes and some quotient techniques).

The following notes give an expository account of the construction of Hilbert and Quot schemes. We assume that the reader is familiar with the basics of the language of schemes and cohomology, say at the level of chapters 2 and 3 of Hartshorne's 'Algebraic Geometry' [**Har77**]. Some more advanced facts about flat morphisms (including the local criterion for flatness) that we need are available in Altman and Kleiman's 'Introduction to Grothendieck Duality Theory' [**AK70**]. Part 1 of this book on the theory of descent in this summer school contains in particular the background we need on descent. Certain advanced techniques of projective geometry, namely Castelnuovo–Mumford regularity and flattening stratification (to each of which we devote one lecture) are nicely given in Mumford's 'Lectures on Curves on an Algebraic Surface' [**Mum66**]. The book 'Neron Models' by Bosch, Lütkebohmert, Raynaud [**BLR90**] contains a quick exposition of descent, Quot schemes, and Picard schemes. The reader of these lecture notes is strongly urged to read Grothendieck's original presentation in [FGA].

5.1. The Hilbert and Quot functors

5.1.1. The functors $\mathfrak{Hilb}_{\mathbb{P}^n}$. The main problem addressed in Part 2, in its simplest form, is as follows. If S is a locally noetherian scheme, a *family of subschemes of* $\mathbb{P}^n$ *parametrised by* S will mean a closed subscheme $Y \subset \mathbb{P}^n_S = \mathbb{P}^n_{\mathbb{Z}} \times S$ such that Y is flat over S. If $f : T \to S$ is any morphism of locally noetherian schemes, then by pull-back we get a family $f^*(Y) = (\mathrm{id} \times f)^{-1}(Y) \subset \mathbb{P}^n_T$ parametrised by T, from a family Y parametrised by S. This defines a contravariant functor $\mathfrak{Hilb}_{\mathbb{P}^n}$ from the category of all locally noetherian schemes to the category of sets, which associates to any S the set of all such families

$$\mathfrak{Hilb}_{\mathbb{P}^n}(S) = \{Y \subset \mathbb{P}^n_S \mid Y \text{ is flat over } S\}.$$

Question: Is the functor $\mathfrak{Hilb}_{\mathbb{P}^n}$ representable?

Grothendieck proved that this question has an affirmative answer, that is, there exists a locally noetherian scheme $\mathrm{Hilb}_{\mathbb{P}^n}$ together with a family $Z \subset \mathbb{P}^n_{\mathbb{Z}} \times \mathrm{Hilb}_{\mathbb{P}^n}$ parametrised by $\mathrm{Hilb}_{\mathbb{P}^n}$, such that any family Y over S is obtained as the pull-back of Z by a uniquely determined morphism $\varphi_Y : S \to \mathrm{Hilb}_{\mathbb{P}^n}$. In other words, $\mathfrak{Hilb}_{\mathbb{P}^n}$ is isomorphic to the functor $Mor(-, \mathrm{Hilb}_{\mathbb{P}^n})$.

5.1.2. The functors $\mathfrak{Quot}_{\oplus^r \mathcal{O}_{\mathbb{P}^n}}$. A family Y of subschemes of $\mathbb{P}^n$ parameterised by S is the same as a coherent quotient sheaf $q : \mathcal{O}_{\mathbb{P}^n_S} \to \mathcal{O}_Y$ on $\mathbb{P}^n_S$, such that $\mathcal{O}_Y$ is flat over S. This way of looking at the functor $\mathfrak{Hilb}_{\mathbb{P}^n}$ has the following fruitful generalisation.

Let r be any positive integer. A *family of quotients of* $\oplus^r \mathcal{O}_{\mathbb{P}^n}$ *parametrised by a locally noetherian scheme* S will mean a pair $(\mathcal{F}, q)$ consisting of

(i) a coherent sheaf $\mathcal{F}$ on $\mathbb{P}^n_S$ which is flat over S, and

(ii) a surjective $\mathcal{O}_{\mathbb{P}^n_S}$-linear homomorphism of sheaves $q : \oplus^r \mathcal{O}_{\mathbb{P}^n_S} \to \mathcal{F}$.

Two such families $(\mathcal{F}, q)$ and $(\mathcal{F}, q)$ parametrised by S will be regarded as equivalent if there exists an isomorphism $f : \mathcal{F} \to \mathcal{F}'$ which takes q to q', that is, the following diagram commutes.

$$\begin{array}{ccc} \oplus^r \mathcal{O}_{\mathbb{P}^n} & \xrightarrow{q} & \mathcal{F} \\ \| & & \downarrow f \\ \oplus^r \mathcal{O}_{\mathbb{P}^n} & \xrightarrow{q'} & \mathcal{F}' \end{array}$$

This is the same as the condition $\ker(q) = \ker(q')$. We will denote by $\langle \mathcal{F}, q \rangle$ an equivalence class. If $f : T \to S$ is a morphism of locally noetherian schemes, then pulling back the quotient $q : \oplus^r \mathcal{O}_{\mathbb{P}^n_S} \to \mathcal{F}$ under $\mathrm{id} \times f : \mathbb{P}^n_T \to \mathbb{P}^n_S$ defines a family

$f^*(q) : \oplus^r \mathcal{O}_{\mathbb{P}^n_T} \to f^*(\mathcal{F})$ over T, which makes sense as tensor product is right-exact and preserves flatness. The operation of pulling back respects equivalence of families, therefore it gives rise to a contravariant functor $\mathfrak{Quot}_{\oplus^r \mathcal{O}_{\mathbb{P}^n}}$ from the category of all locally noetherian schemes to the category of sets, by putting

$$\mathfrak{Quot}_{\oplus^r \mathcal{O}_{\mathbb{P}^n}}(S) = \{ \text{ All } \langle \mathcal{F}, q \rangle \text{ parametrised by } S\}.$$

It is immediate that the functor $\mathfrak{Quot}_{\oplus^r \mathcal{O}_{\mathbb{P}^n}}$ satisfies faithfully flat descent. Grothendieck proved that in fact the above functor is representable on the category of all locally noetherian schemes by a scheme $\mathrm{Quot}_{\oplus^r \mathcal{O}_{\mathbb{P}^n}}$.

5.1.3. The functors $\mathfrak{Hilb}_{X/S}$ and $\mathfrak{Quot}_{E/X/S}$. The above functors $\mathfrak{Hilb}_{\mathbb{P}^n}$ and $\mathfrak{Quot}_{\oplus^r \mathcal{O}_{\mathbb{P}^n}}$ admit the following simple generalisations. Let S be a noetherian scheme and let $X \to S$ be a finite type scheme over it. Let E be a coherent sheaf on X. Let Sch_S denote the category of all locally noetherian schemes over S. For any $T \to S$ in Sch_S, a *family of quotients of E parametrised by T* will mean a pair $(\mathcal{F}, q)$ consisting of

(i) a coherent sheaf $\mathcal{F}$ on $X_T = X \times_S T$ such that the schematic support of $\mathcal{F}$ is proper over T and $\mathcal{F}$ is flat over T, together with

(ii) a surjective $\mathcal{O}_{X_T}$-linear homomorphism of sheaves $q : E_T \to \mathcal{F}$ where E_T is the pull-back of E under the projection $X_T \to X$.

Two such families $(\mathcal{F}, q)$ and $(\mathcal{F}, q)$ parametrised by T will be regarded as equivalent if $\ker(q) = \ker(q')$, and $\langle \mathcal{F}, q \rangle$ will denote an equivalence class. Then as properness and flatness are preserved by base-change, and as tensor product is right exact, the pull-back of $\langle \mathcal{F}, q \rangle$ under an S-morphism $T' \to T$ is well-defined, which gives a set-valued contravariant functor $\mathfrak{Quot}_{E/X/S} : Sch_S \to Sets$ under which

$$T \mapsto \{ \text{All } \langle \mathcal{F}, q \rangle \text{ parametrised by } T \}.$$

When $E = \mathcal{O}_X$, the functor $\mathfrak{Quot}_{\mathcal{O}_X/X/S} : Sch_S \to Sets$ associates to T the set of all closed subschemes $Y \subset X_T$ that are proper and flat over T. We denote this functor by $\mathfrak{Hilb}_{X/S}$.

Note in particular that we have

$$\mathfrak{Hilb}_{\mathbb{P}^n} = \mathfrak{Hilb}_{\mathbb{P}^n_{\mathbb{Z}}/\operatorname{Spec}\mathbb{Z}} \text{ and } \mathfrak{Quot}_{\oplus^r \mathcal{O}_{\mathbb{P}^n}} = \mathfrak{Quot}_{\oplus^r \mathcal{O}_{\mathbb{P}^n_{\mathbb{Z}}}/\mathbb{P}^n_{\mathbb{Z}}/\operatorname{Spec}\mathbb{Z}}.$$

It is clear that the functors $\mathfrak{Quot}_{E/X/S}$ and $\mathfrak{Hilb}_{X/S}$ satisfy faithfully flat descent, so it makes sense to pose the question of their representability.

5.1.4. Stratification by Hilbert polynomials. Let X be a scheme of finite type over a field k, together with a line bundle L. Recall that if F is a coherent sheaf on X whose support is proper over k, then the *Hilbert polynomial* $\Phi \in \mathbb{Q}[\lambda]$ of F is defined by the function

$$\Phi(m) = \chi(F(m)) = \sum_{i=0}^{n} (-1)^i \dim_k H^i(X, F \otimes L^{\otimes m})$$

where the dimensions of the cohomologies are finite because of the coherence and properness conditions. The fact that $\chi(F(m))$ is indeed a polynomial in m under the above assumption is a special case of what is known as Snapper's Lemma (see Part 5, B.7 for a proof).

Let $X \to S$ be a finite type morphism of noetherian schemes, and let L be a line bundle on X. Let $\mathcal{F}$ be any coherent sheaf on X whose schematic support is proper over S. Then for each $s \in S$, we get a polynomial $\Phi_s \in \mathbb{Q}[\lambda]$ which is the Hilbert polynomial of the restriction $\mathcal{F}_s = F|_{X_s}$ of $\mathcal{F}$ to the fiber X_s over s, calculated with respect to the line bundle $L_s = L|_{X_s}$. If $\mathcal{F}$ is flat over S then the function $s \mapsto \Phi_s$ from the set of points of S to the polynomial ring $\mathbb{Q}[\lambda]$ is known to be locally constant on S.

This shows that the functor $\mathfrak{Quot}_{E/X/S}$ naturally decomposes as a co-product

$$\mathfrak{Quot}_{E/X/S} = \coprod_{\Phi \in \mathbb{Q}[\lambda]} \mathfrak{Quot}^{\Phi,L}_{E/X/S}$$

where for any polynomial $\Phi \in \mathbb{Q}[\lambda]$, the functor $\mathfrak{Quot}^{\Phi,L}_{E/X/S}$ associates to any T the set of all equivalence classes of families $\langle \mathcal{F}, q \rangle$ such that at each $t \in T$ the Hilbert polynomial of the restriction $\mathcal{F}_t$, calculated using the pull-back of L, is Φ. Correspondingly, the representing scheme $\mathrm{Quot}_{E/X/S}$, when it exists, naturally decomposes as a co-product

$$\mathrm{Quot}_{E/X/S} = \coprod_{\Phi \in \mathbb{Q}[\lambda]} \mathrm{Quot}^{\Phi,L}_{E/X/S}.$$

Note. We will generally take X to be (quasi-)projective over S, and L to be a relatively very ample line bundle. Then indeed the Hilbert and Quot functors are representable by schemes, but not in general.

5.1.5. Elementary examples, exercises. (1) $\mathbb{P}^n_{\mathbb{Z}}$ as a Quot scheme. Show that the scheme $\mathbb{P}^n_{\mathbb{Z}} = \mathrm{Proj}\, \mathbb{Z}[x_0, \ldots, x_n]$ represents the functor φ from schemes to sets, which associates to any S the set of all equivalence classes $\langle \mathcal{F}, q \rangle$ of quotients $q : \oplus^{n+1} \mathcal{O}_S \to \mathcal{F}$, where $\mathcal{F}$ is an invertible $\mathcal{O}_S$-module. As coherent sheaves on S which are $\mathcal{O}_S$-flat with each fiber 1-dimensional are exactly the locally free sheaves on S of rank 1, it follows that φ is the functor $\mathfrak{Quot}^{1,\mathcal{O}_{\mathbb{Z}}}_{\oplus^{n+1}\mathcal{O}_{\mathbb{Z}}/\mathbb{Z}/\mathbb{Z}}$ (where in some places we write just $\mathbb{Z}$ for $\mathrm{Spec}\, \mathbb{Z}$ for simplicity). This shows that $\mathrm{Quot}^{1,\mathcal{O}_{\mathbb{Z}}}_{\oplus^{n+1}\mathcal{O}_{\mathbb{Z}}/\mathbb{Z}/\mathbb{Z}} = \mathbb{P}^n_{\mathbb{Z}}$. Under this identification, show that the universal family on $\mathrm{Quot}^{1,\mathcal{O}_{\mathbb{Z}}}_{\oplus^{n+1}\mathcal{O}_{\mathbb{Z}}/\mathbb{Z}/\mathbb{Z}}$ is the tautological quotient $\oplus^{n+1}\mathcal{O}_{\mathbb{P}^n_{\mathbb{Z}}} \to \mathcal{O}_{\mathbb{P}^n_{\mathbb{Z}}}(1)$.

More generally, show that if E is a locally free sheaf on a noetherian scheme S, the functor $\mathfrak{Quot}^{1,\mathcal{O}_S}_{E/S/S}$ is represented by the S-scheme $\mathbb{P}(E) = \mathbf{Proj}\, \mathrm{Sym}_{\mathcal{O}_S} E$, with the tautological quotient $\pi^*(E) \to \mathcal{O}_{\mathbb{P}(E)}(1)$ as the universal family.

(2) Grassmannian as a Quot scheme. For any integers $r \geq d \geq 1$, an explicit construction of the Grassmannian scheme $\mathrm{Grass}(r, d)$ over $\mathbb{Z}$, together with the tautological quotient $u : \oplus^r \mathcal{O}_{\mathrm{Grass}(r,d)} \to \mathcal{U}$ where $\mathcal{U}$ is a rank d locally free sheaf on $\mathrm{Grass}(r, d)$, has been given at the end of this section. A proof of the properness of $\pi : \mathrm{Grass}(r, d) \to \mathrm{Spec}\, \mathbb{Z}$ is given there, together with a closed embedding $\mathrm{Grass}(r, d) \hookrightarrow \mathbb{P}(\pi_* \det \mathcal{U}) = \mathbb{P}^m_{\mathbb{Z}}$ where $m = \binom{r}{d} - 1$.

Show that $\mathrm{Grass}(r, d)$ together with the quotient $u : \oplus^r \mathcal{O}_{\mathrm{Grass}(r,d)} \to \mathcal{U}$ represents the contravariant functor

$$\mathfrak{Grass}(r, d) = \mathfrak{Quot}^{d,\mathcal{O}_{\mathbb{Z}}}_{\oplus^r \mathcal{O}_{\mathbb{Z}}/\mathbb{Z}/\mathbb{Z}}$$

from schemes to sets, which associates to any T the set of all equivalence classes $\langle \mathcal{F}, q \rangle$ of quotients $q : \oplus^r \mathcal{O}_T \to \mathcal{F}$ where $\mathcal{F}$ is a locally free sheaf on T of rank d. Therefore, $\mathrm{Quot}^{d,\mathcal{O}_{\mathbb{Z}}}_{\oplus^r \mathcal{O}_{\mathbb{Z}}/\mathbb{Z}/\mathbb{Z}}$ exists, and equals $\mathrm{Grass}(r, d)$.

Grassmannian of a vector bundle. Show that for any ring A, the action of the group $GL_r(A)$ on the free module $\oplus^r A$ induces an action of $GL_r(A)$ on the set $\mathfrak{Grass}(r,d)(A)$, such that for any ring homomorphism $A \to B$, the set-map $\mathfrak{Grass}(r,d)(A) \to \mathfrak{Grass}(r,d)(B)$ is equivariant with respect to the group homomorphism $GL_r(A) \to GL_r(B)$. (In schematic terms, this means we have an action of the group-scheme $GL_{r,\mathbb{Z}}$ on $\mathrm{Grass}(r,d)$.)

Using the above show that, more generally, if S is a scheme and E is a locally free $\mathcal{O}_S$-module of rank r, the functor $\mathfrak{Grass}(E,d) = \mathfrak{Quot}^{d,\mathcal{O}_S}_{E/S/S}$ on all S-schemes which by definition associates to any T the set of all equivalence classes $\langle \mathcal{F}, q\rangle$ of quotients $q : E_T \to \mathcal{F}$ where $\mathcal{F}$ is a locally free sheaf on T of rank d, is representable. The representing scheme is denoted by $\mathrm{Grass}(E,d)$ and is called the rank d relative Grassmannian of E over S. It parametrises a universal quotient $\pi^*E \to \mathcal{F}$ where $\pi : \mathrm{Grass}(E,d) \to S$ is the projection. Show that the determinant line bundle $\bigwedge^d \mathcal{F}$ on $\mathrm{Grass}(E,d)$ is relatively very ample over S, and it gives a closed embedding $\mathrm{Grass}(E,d) \hookrightarrow \mathbb{P}(\pi_* \bigwedge^d \mathcal{F}) \subset \mathbb{P}(\bigwedge^d E)$. (The properness of the embedding follows from the properness of $\pi : \mathrm{Grass}(E,d) \to S$, which follows locally over S by base-change from properness of $\mathrm{Grass}(r,d)$ over $\mathbb{Z}$ – see Exercise **(5)** or **(7)** below.)

Grassmannian of a coherent sheaf. If E is a coherent sheaf on S, not necessarily locally free, then by definition the functor $\mathfrak{Grass}(E,d) = \mathfrak{Quot}^{d,\mathcal{O}_S}_{E/S/S}$ on all S-schemes associates to any T the set of all equivalence classes $\langle \mathcal{F}, q\rangle$ of quotients $q : E_T \to \mathcal{F}$ where $\mathcal{F}$ is locally free on T of rank d. If $r : E' \to E$ is a surjection of coherent sheaves on S, then show that the induced morphism of functors $\mathfrak{Grass}(E,d) \to \mathfrak{Grass}(E',d)$, which sends $\langle \mathcal{F}, q\rangle \mapsto \langle \mathcal{F}, q \circ r\rangle$, is a closed embedding. ¿From this, by locally expressing a coherent sheaf as a quotient of a vector bundle, show that $\mathfrak{Grass}(E,d)$ is representable even when E is a coherent sheaf on S which is not necessarily locally free. The representing scheme $\mathrm{Grass}(E,d)$ is proper over S, as locally over S it is a closed subscheme of the Grassmannian of a vector bundle. Show by arguing locally over S that the line bundle $\bigwedge^d \mathcal{F}$ on $\mathrm{Grass}(E,d)$ is relatively very ample over S, and therefore by using properness conclude that $\mathrm{Grass}(E,d)$ is projective over S.

(3) Grassmannian as a Hilbert scheme. Let $\Phi = 1 \in \mathbb{Q}[\lambda]$. Then the Hilbert scheme $\mathrm{Hilb}^{1,\mathcal{O}(1)}_{\mathbb{P}^n}$ is $\mathbb{P}^n_{\mathbb{Z}}$ itself. More generally, let $\Phi_r = \binom{r+\lambda}{r} \in \mathbb{Q}[\lambda]$ where $r \geq 0$. The Hilbert scheme $\mathrm{Hilb}^{\Phi_r,\mathcal{O}(1)}_{\mathbb{P}^n}$ is isomorphic to the Grassmannian scheme $\mathrm{Grass}(n+1, r+1)$ over $\mathbb{Z}$. This can be seen via the following steps, whose detailed verification is left to the reader as an exercise.

(i) The Grassmannian scheme $\mathrm{Grass}(n+1, r+1)$ over $\mathbb{Z}$ parametrises a tautological family of subschemes of $\mathbb{P}^n$ with Hilbert polynomial Φ_r. Therefore we get a natural transformation $h_{\mathrm{Grass}(n+1,r+1)} \to \mathfrak{Hilb}^{\Phi_r,\mathcal{O}(1)}_{\mathbb{P}^n}$.

(ii) Any closed subscheme $Y \subset \mathbb{P}^n_k$ with Hilbert polynomial Φ_r, where k is any field, is isomorphic to $\mathbb{P}^r_k$ embedded linearly in $\mathbb{P}^n_k$ over k. If V is a vector bundle over a noetherian base S, and if $Y \subset \mathbb{P}(V)$ is a closed subscheme flat over S with each schematic fiber Y_s an r-dimensional linear subspace of the projective space $\mathbb{P}(V_s)$, then Y defines a rank $r+1$ quotient vector bundle $V = \pi_*\mathcal{O}_{\mathbb{P}(V)}(1) \to \pi_*\mathcal{O}_Y(1)$ where $\pi : \mathbb{P}(V) \to S$ denotes the projection. This gives a natural transformation $\mathfrak{Hilb}^{\Phi_r,\mathcal{O}(1)}_{\mathbb{P}^n} \to h_{\mathrm{Grass}(n+1,r+1)}$.

(iii) The above two natural transformations are inverses of each other.

(4) Hilbert scheme of hypersurfaces in $\mathbb{P}^n$. Let $\Phi_d = \binom{n+\lambda}{n} - \binom{n-d+\lambda}{n} \in \mathbb{Q}[\lambda]$ where $d \geq 1$. The Hilbert scheme $\mathrm{Hilb}^{\Phi_d,\mathcal{O}(1)}_{\mathbb{P}^n}$ is isomorphic to $\mathbb{P}^m_{\mathbb{Z}}$ where

$m = \binom{n+d}{d} - 1$. This can be seen from the following steps, which are left as exercises.

(i) Any closed subscheme $Y \subset \mathbb{P}^n_k$ with Hilbert polynomial Φ_d, where k is any field, is a hypersurface of degree d in $\mathbb{P}^n_k$. *Hint:* If $Y \subset \mathbb{P}^n_k$ is a closed subscheme with Hilbert polynomial of degree $n-1$, then show that the schematic closure Z of the height 1 primary components is a hypersurface in $\mathbb{P}^n_k$ with $\deg(Z) = \deg(Y)$.

(ii) Any family $Y \subset \mathbb{P}^n_S$ is a Cartier divisor in $\mathbb{P}^n_S$. It gives rise to a line subbundle $\pi_*(I_Y \otimes \mathcal{O}_{\mathbb{P}^n_S}(d)) \subset \pi_*\mathcal{O}_{\mathbb{P}^n_S}(d)$, which defines a natural morphism $f_Y : S \to \mathbb{P}^m_{\mathbb{Z}}$ where $m = \binom{n+d}{d} - 1$. This gives a morphism of functors $\mathfrak{Hilb}^{\Phi_d}_{\mathbb{P}^n} \to \mathbb{P}^m$ where we denote $h_{\mathbb{P}^m_{\mathbb{Z}}}$ simply by $\mathbb{P}^m$.

(iii) The scheme $\mathbb{P}^m_{\mathbb{Z}}$ parametrises a tautological family of hypersurfaces of degree d, which gives a morphism of functors $\mathbb{P}^m \to \mathfrak{Hilb}^{\Phi_d, \mathcal{O}(1)}_{\mathbb{P}^n}$ in the reverse direction. These are inverses of each other.

(5) Base-change property of Hilbert and Quot schemes. Let S be a noetherian scheme, X a finite-type scheme over S, and E a coherent sheaf on X. If $T \to S$ is a morphism of noetherian schemes, then show that there is a natural isomorphism of functors $\mathfrak{Quot}_{E_T/X_T/T} \to \mathfrak{Quot}_{E/X/S} \times_{h_S} h_T$. Consequently, if $\mathrm{Quot}_{E/X/S}$ exists, then so does $\mathrm{Quot}_{E_T/X_T/T}$, which is naturally isomorphic to $\mathrm{Quot}_{E/X/S} \times_S T$. One can prove a similar statement involving $\mathfrak{Quot}^{\Phi,L}_{E/X/S}$. In particular, $\mathrm{Hilb}_{X/S}$ and $\mathrm{Hilb}^{\Phi,L}_{X/S}$, when they exist, base-change correctly..

(6) Descent condition in the fpqc topology. If U is an S-scheme and $(f_i : U_i \to U)$ is an open cover of U in the fpqc topology, then show that the following sequence of sets is exact:

$$\mathfrak{Quot}_{E/X/S}(U) \to \prod_i \mathfrak{Quot}_{E/X/S}(U_i) \rightrightarrows \prod_{i,j} \mathfrak{Quot}_{E/X/S}(U_i \times_U U_j).$$

(7) Valuative criterion for properness. When $X \to S$ is proper, show that the morphism of functors $\mathfrak{Quot}_{E/X/S} \to h_S$ satisfies the valuative criterion of properness with respect to discrete valuation rings, that is, if R is a discrete valuation ring together with a given morphism $\operatorname{Spec} R \to S$ making it an S-scheme, show that the restriction map $\mathfrak{Quot}_{E/X/S}(\operatorname{Spec} R) \to \mathfrak{Quot}_{E/X/S}(\operatorname{Spec} K)$ is bijective, where K is the quotient field of R and $\operatorname{Spec} K$ is regarded as an S-scheme in the obvious way.

(8) Counterexample of Hironaka. Hironaka constructed a 3-dimensional smooth proper scheme X over complex numbers $\mathbb{C}$, together with a free action of the group $G = \mathbb{Z}/(2)$, for which the quotient X/G does not exist as a scheme. (See [**Har77**, Appendix B, Example 3.4.1] for construction of X. We leave the definition of the G action and the proof that X/G does not exist to the reader.) In particular, this means the Hilbert functor $\mathfrak{Hilb}_{X/\mathbb{C}}$ is not representable by a scheme.

5.1.6. Construction of Grassmannian. The following explicit construction of the Grassmannian scheme $\mathrm{Grass}(r, d)$ over $\mathbb{Z}$ is best understood as the construction of a quotient $GL_{d,\mathbb{Z}} \backslash V$, where V is the scheme of all $d \times r$ matrices of rank d, and the group-scheme $GL_{d,\mathbb{Z}}$ acts on V on the left by matrix multiplication. However, we will not use the language of group-scheme actions here, instead, we give a direct elementary construction of the Grassmannian scheme.

The reader can take $d = 1$ in what follows, in a first reading, to get the special case $\mathrm{Grass}(r,1) = \mathbb{P}_{\mathbb{Z}}^{r-1}$, which has another construction as $\mathrm{Proj}\,\mathbb{Z}[x_1, \ldots, x_r]$.

Construction by gluing together affine patches. For any integers $r \geq d \geq 1$, the Grassmannian scheme $\mathrm{Grass}(r,d)$ over $\mathbb{Z}$, together with the tautological quotient $u : \oplus^r \mathcal{O}_{\mathrm{Grass}(r,d)} \to \mathcal{U}$ where $\mathcal{U}$ is a rank d locally free sheaf on $\mathrm{Grass}(r,d)$, can be explicitly constructed as follows.

If M is a $d \times r$-matrix, and $I \subset \{1, \ldots, r\}$ with cardinality $\#(I)$ equal to d, the I-th minor M_I of M will mean the $d \times d$ minor of M whose columns are indexed by I.

For any subset $I \subset \{1, \ldots, r\}$ with $\#(I) = d$, consider the $d \times r$ matrix X^I whose I-th minor X_I^I is the $d \times d$ identity matrix $1_{d \times d}$, while the remaining entries of X^I are independent variables $x_{p,q}^I$ over $\mathbb{Z}$. Let $\mathbb{Z}[X^I]$ denote the polynomial ring in the variables $x_{p,q}^I$, and let $U^I = \mathrm{Spec}\,\mathbb{Z}[X^I]$, which is non-canonically isomorphic to the affine space $\mathbf{A}_{\mathbb{Z}}^{d(r-d)}$.

For any $J \subset \{1, \ldots, r\}$ with $\#(J) = d$, let $P_J^I = \det(X_J^I) \in \mathbb{Z}[X^I]$ where X_J^I is the J-th minor of X^I. Let $U_J^I = \mathrm{Spec}\,\mathbb{Z}[X^I, 1/P_J^I]$ the open subscheme of U^I where P_J^I is invertible. This means the $d \times d$-matrix X_J^I admits an inverse $(X_J^I)^{-1}$ on U_J^I.

For any I and J, a ring homomorphism $\theta_{I,J} : \mathbb{Z}[X^J, 1/P_I^J] \to \mathbb{Z}[X^I, 1/P_J^I]$ is defined as follows. The images of the variables $x_{p,q}^J$ are given by the entries of the matrix formula $\theta_{I,J}(X^J) = (X_J^I)^{-1} X^I$. In particular, we have $\theta_{I,J}(P_I^J) = 1/P_J^I$, so the map extends to $\mathbb{Z}[X^J, 1/P_I^J]$.

Note that $\theta_{I,I}$ is identity on $U_I^I = U^I$, and we leave it to the reader to verify that for any three subsets I, J and K of $\{1, \ldots, r\}$ of cardinality d, the co-cycle condition $\theta_{I,K} = \theta_{I,J}\theta_{J,K}$ is satisfied. Therefore the schemes U^I, as I varies over all the $\binom{r}{d}$ different subsets of $\{1, \ldots, r\}$ of cardinality d, can be glued together by the co-cycle $(\theta_{I,J})$ to form a finite-type scheme $\mathrm{Grass}(r,d)$ over $\mathbb{Z}$. As each U^I is isomorphic to $\mathbf{A}_{\mathbb{Z}}^{d(r-d)}$, it follows that $\mathrm{Grass}(r,d) \to \mathrm{Spec}\,\mathbb{Z}$ is smooth of relative dimension $d(r-d)$.

Separatedness. The intersection of the diagonal of $\mathrm{Grass}(r,d)$ with $U^I \times U^J$ can be seen to be the closed subscheme $\Delta_{I,J} \subset U^I \times U^J$ defined by entries of the matrix formula $X_I^J\, X^I - X^J = 0$, and so $\mathrm{Grass}(r,d)$ is a separated scheme.

Properness. We now show that $\pi : \mathrm{Grass}(r,d) \to \mathrm{Spec}\,\mathbb{Z}$ is proper. It is enough to verify the valuative criterion of properness for discrete valuation rings. Let $\mathcal{R}$ be a dvr, $\mathcal{K}$ its quotient field, and let $\varphi : \mathrm{Spec}\,\mathcal{K} \to \mathrm{Grass}(r,d)$ be a morphism. This is given by a ring homomorphism $f : \mathbb{Z}[X^I] \to \mathcal{K}$ for some I. Having fixed one such I, next choose J such that $\nu(f(P_J^I))$ is minimum, where $\nu : \mathcal{K} \to \mathbb{Z} \cup \{\infty\}$ denotes the discrete valuation. As $P_I^I = 1$, note that $\nu(f(P_J^I)) \leq 0$, therefore $f(P_J^I) \neq 0$ in $\mathcal{K}$ and so the matrix $f(X_J^I)$ lies in $GL_d(\mathcal{K})$.

Now consider the homomorphism $g : \mathbb{Z}[X^J] \to \mathcal{K}$ defined by entries of the matrix formula

$$g(X^J) = f((X_J^I)^{-1} X^I).$$

Then g defines the same morphism $\varphi : \mathrm{Spec}\,\mathcal{K} \to \mathrm{Grass}(r,d)$, and moreover all $d \times d$ minors X_K^J satisfy $\nu(g(P_K^J)) \geq 0$. As the minor X_J^J is identity, it follows from the above that in fact $\nu(g(x_{p,q}^J)) \geq 0$ for all entries of X^J. Therefore, the map $g : \mathbb{Z}[X^J] \to \mathcal{K}$ factors uniquely via $\mathcal{R} \subset \mathcal{K}$. The resulting morphism of schemes

$\operatorname{Spec}\mathcal{R} \to U^J \hookrightarrow \operatorname{Grass}(r,d)$ prolongs $\varphi : \operatorname{Spec}\mathcal{K} \to \operatorname{Grass}(r,d)$. We have already checked separatedness of $\operatorname{Grass}(r,d)$, so now we see that $\operatorname{Grass}(r,d) \to \operatorname{Spec}\mathbb{Z}$ is proper.

Universal quotient. We next define a rank d locally free sheaf $\mathcal{U}$ on $\operatorname{Grass}(r,d)$ together with a surjective homomorphism $\oplus^r \mathcal{O}_{\operatorname{Grass}(r,d)} \to \mathcal{U}$. On each U^I we define a surjective homomorphism $u^I : \oplus^r \mathcal{O}_{U^I} \to \oplus^d \mathcal{O}_{U^I}$ by the matrix X^I. Compatible with the co-cycle $(\theta_{I,J})$ for gluing the affine pieces U^I, we give gluing data $(g_{I,J})$ for gluing together the trivial bundles $\oplus^d \mathcal{O}_{U^I}$ by putting

$$g_{I,J} = (X^I_J)^{-1} \in GL_d(U^I_J).$$

This is compatible with the homomorphisms u^I, so we get a surjective homomorphism $u : \oplus^r \mathcal{O}_{\operatorname{Grass}(r,d)} \to \mathcal{U}$.

Projective embedding. As $\mathcal{U}$ is given by the transition functions $g_{I,J}$ described above, the determinant line bundle $\det(\mathcal{U})$ is given by the transition functions $\det(g_{I,J}) = 1/P^I_J \in GL_1(U^I_J)$. For each I, we define a global section

$$\sigma_I \in \Gamma(\operatorname{Grass}(r,d), \det(\mathcal{U}))$$

by putting $\sigma_I|_{U^J} = P^J_I \in \Gamma(U^J, \mathcal{O}_{U^J})$ in terms of the trivialization over the open cover (U^J). We leave it to the reader to verify that the sections σ_I form a linear system which is base point free and separates points relative to $\operatorname{Spec}\mathbb{Z}$, and so gives an embedding of $\operatorname{Grass}(r,d)$ into $\mathbb{P}^m_{\mathbb{Z}}$ where $m = \binom{r}{d} - 1$. This is a closed embedding by the properness of $\pi : \operatorname{Grass}(r,d) \to \operatorname{Spec}\mathbb{Z}$. In particular, $\det(\mathcal{U})$ is a relatively very ample line bundle on $\operatorname{Grass}(r,d)$ over $\mathbb{Z}$.

Note. The σ_I are known as the Plücker coordinates, and these satisfy certain quadratic polynomials known as the Plücker relations, which define the projective image of the Grassmannian. We will not need these facts.

5.2. Castelnuovo–Mumford regularity

Mumford's deployment of m-regularity led to a simplification in the construction of Quot schemes. The original construction of Grothendieck had instead relied on Chow coordinates.

Let k be a field and let $\mathcal{F}$ be a coherent sheaf on the projective space $\mathbb{P}^n$ over k. Let m be an integer. The sheaf $\mathcal{F}$ is said to be *m-regular* if we have

$$H^i(\mathbb{P}^n, \mathcal{F}(m-i)) = 0 \text{ for each } i \geq 1.$$

The definition, which may look strange at first sight, is suitable for making inductive arguments on $n = \dim(\mathbb{P}^n)$ by restriction to a suitable hyperplane. If $H \subset \mathbb{P}^n$ is a hyperplane which does not contain any associated point of $\mathcal{F}$, then we have a short exact sheaf sequence

$$0 \to \mathcal{F}(m-i-1) \xrightarrow{\alpha} \mathcal{F}(m-i) \to \mathcal{F}_H(m-i) \to 0$$

where the map α is locally given by multiplication with a defining equation of H, hence is injective. The resulting long exact cohomology sequence

$$\ldots \to H^i(\mathbb{P}^n, \mathcal{F}(m-i)) \to H^i(\mathbb{P}^n, \mathcal{F}_H(m-i)) \to H^{i+1}(\mathbb{P}^n, \mathcal{F}(m-i-1)) \to \ldots$$

shows that if $\mathcal{F}$ is m-regular, then so is its restriction $\mathcal{F}_H$ (with the *same value* for m) to a hyperplane $H \simeq \mathbb{P}^{n-1}$ which does not contain any associated point of

$\mathcal{F}$. Note that whenever $\mathcal{F}$ is coherent, the set of associated points of $\mathcal{F}$ is finite, so there will exist at least one such hyperplane H when the field k is infinite.

The following lemma is due to Castelnuovo, according to Mumford [**Mum66**].

LEMMA 5.1. If $\mathcal{F}$ is an m-regular sheaf on $\mathbb{P}^n$ then the following statements hold:

(a) The canonical map $H^0(\mathbb{P}^n, \mathcal{O}_{\mathbb{P}^n}(1)) \otimes H^0(\mathbb{P}^n, \mathcal{F}(r)) \to H^0(\mathbb{P}^n, \mathcal{F}(r+1))$ is surjective whenever $r \geq m$.

(b) We have $H^i(\mathbb{P}^n, \mathcal{F}(r)) = 0$ whenever $i \geq 1$ and $r \geq m - i$. In other words, if $\mathcal{F}$ is m-regular, then it is m'-regular for all $m' \geq m$.

(c) The sheaf $\mathcal{F}(r)$ is generated by its global sections, and all its higher cohomologies vanish, whenever $r \geq m$.

PROOF. As the cohomologies base-change correctly under a field extension, we can assume that the field k is infinite. We argue by induction on n. The statements **(a)**, **(b)** and **(c)** clearly hold when $n = 0$, so next let $n \geq 1$. As k is infinite, there exists a hyperplane H which does not contain any associated point of $\mathcal{F}$, so that the restriction $\mathcal{F}_H$ is again m-regular as explained above. As H is isomorphic to $\mathbb{P}^{n-1}_k$, by the inductive hypothesis the assertions of the lemma hold for the sheaf $\mathcal{F}_H$.

When $r = m - i$, the equality $H^i(\mathbb{P}^n, \mathcal{F}(r)) = 0$ in statement **(b)** follows for all $n \geq 0$ by definition of m-regularity. To prove **(b)**, we now proceed by induction on r where $r \geq m - i + 1$. Consider the exact sequence

$$H^i(\mathbb{P}^n, \mathcal{F}(r-1)) \to H^i(\mathbb{P}^n, \mathcal{F}(r)) \to H^i(H, \mathcal{F}_H(r)).$$

By inductive hypothesis for $r-1$ the first term is zero, while by inductive hypothesis for $n-1$ the last term is zero, which shows that the middle term is zero, completing the proof of **(b)**.

Now consider the commutative diagram

$$\begin{array}{ccccc} & & H^0(\mathbb{P}^n, \mathcal{F}(r)) \otimes H^0(\mathbb{P}^n, \mathcal{O}_{\mathbb{P}^n}(1)) & \xrightarrow{\sigma} & H^0(H, \mathcal{F}_H(r)) \otimes H^0(H, \mathcal{O}_H(1)) \\ & & \downarrow{\scriptstyle\mu} & & \downarrow{\scriptstyle\tau} \\ H^0(\mathbb{P}^n, \mathcal{F}(r)) & \xrightarrow{\alpha} & H^0(\mathbb{P}^n, \mathcal{F}(r+1))arr & \xrightarrow{\nu_{r+1}} & H^0(H, \mathcal{F}_H(r+1)) \end{array}$$

The top map σ is surjective, for the following reason: By m-regularity of $\mathcal{F}$ and using the statement **(b)** already proved, we see that $H^1(\mathbb{P}^n, \mathcal{F}(r-1)) = 0$ for $r \geq m$, and so the restriction map $\nu_r : H^0(\mathbb{P}^n, \mathcal{F}(r)) \to H^0(H, \mathcal{F}_H(r))$ is surjective. Also, the restriction map $\rho : H^0(\mathbb{P}^n, \mathcal{O}_{\mathbb{P}^n}(1)) \to H^0(H, \mathcal{O}_H(1))$ is surjective. Therefore the tensor product $\sigma = \nu_r \otimes \rho$ of these two maps is surjective.

The second vertical map τ is surjective by inductive hypothesis for $n - 1 = \dim(H)$.

Therefore, the composite $\tau \circ \sigma$ is surjective, so the composite $\nu_{r+1} \circ \mu$ is surjective, hence $H^0(\mathbb{P}^n, \mathcal{F}(r+1)) = \mathrm{im}(\mu) + \ker(\nu_{r+1})$. As the bottom row is exact, we get $H^0(\mathbb{P}^n, \mathcal{F}(r+1)) = \mathrm{im}(\mu) + \mathrm{im}(\alpha)$. However, we have $\mathrm{im}(\alpha) \subset \mathrm{im}(\mu)$, as the map α is given by tensoring with a certain section of $\mathcal{O}_{\mathbb{P}^n}(1)$ (which has divisor H). Therefore, $H^0(\mathbb{P}^n, \mathcal{F}(r+1)) = \mathrm{im}(\mu)$. This completes the proof of **(a)** for all n.

To prove **(c)**, consider the map $H^0(\mathbb{P}^n, \mathcal{F}(r)) \otimes H^0(\mathbb{P}^n, \mathcal{O}_{\mathbb{P}^n}(p)) \to H^0(\mathbb{P}^n, \mathcal{F}(r+p))$, which is surjective for $r \geq m$ and $p \geq 0$ as follows from a repeated use of **(a)**. For $p \gg 0$, we know that $H^0(\mathbb{P}^n, \mathcal{F}(r+p))$ is generated by its global sections. It follows that $H^0(\mathbb{P}^n, \mathcal{F}(r))$ is also generated by its global sections for $r \geq m$. We already know from **(b)** that $H^i(\mathbb{P}^n, \mathcal{F}(r)) = 0$ for $i \geq 1$ when $r \geq m$. This proves **(c)**, completing the proof of the lemma. □

REMARK 5.2. The following fact, based on the diagram used in the course of the above proof, will be useful later: With notation as above, let the restriction map $\nu_r : H^0(\mathbb{P}^n, \mathcal{F}(r)) \to H^0(H, \mathcal{F}_H(r))$ be surjective. Also, let $\mathcal{F}_H$ be r-regular, so that by Lemma 5.1.**(a)** the map $H^0(H, \mathcal{O}_H(1)) \otimes H^0(H, \mathcal{F}_H(r)) \to H^0(H, \mathcal{F}_H(r+1))$ is surjective. Then the restriction map $\nu_{r+1} : H^0(\mathbb{P}^n, \mathcal{F}(r+1)) \to H^0(H, \mathcal{F}_H(r+1))$ is again surjective. As a consequence, if $\mathcal{F}_H$ is m regular and if for some $r \geq m$ the restriction map $\nu_r : H^0(\mathbb{P}^n, \mathcal{F}(r)) \to H^0(H, \mathcal{F}_H(r))$ is surjective, then the restriction map $\nu_p : H^0(\mathbb{P}^n, \mathcal{F}(p)) \to H^0(H, \mathcal{F}_H(p))$ is surjective for all $p \geq r$.

Exercise. Find all the values of m for which the invertible sheaf $\mathcal{O}_{\mathbb{P}^n}(r)$ is m-regular.

Exercise. Suppose $0 \to \mathcal{F}' \to \mathcal{F} \to \mathcal{F}'' \to 0$ is an exact sequence of coherent sheaves on $\mathbb{P}^n$. Show that if $\mathcal{F}'$ and $\mathcal{F}''$ are m-regular, then $\mathcal{F}$ is also m-regular, if $\mathcal{F}'$ is $(m+1)$-regular and $\mathcal{F}$ is m-regular, then $\mathcal{F}''$ is m-regular, and if $\mathcal{F}$ is m-regular and $\mathcal{F}''$ is $(m-1)$-regular, then $\mathcal{F}'$ is m-regular.

The use of m-regularity for making Quot schemes is via the following theorem.

THEOREM 5.3. (Mumford) For any non-negative integers p and n, there exists a polynomial $F_{p,n}$ in $n+1$ variables with integral coefficients, which has the following property:

Let k be any field, and let $\mathbb{P}^n$ denote the n-dimensional projective space over k. Let $\mathcal{F}$ be any coherent sheaf on $\mathbb{P}^n$, which is isomorphic to a subsheaf of $\oplus^p \mathcal{O}_{\mathbb{P}^n}$. Let the Hilbert polynomial of $\mathcal{F}$ be written in terms of binomial coefficients as

$$\chi(\mathcal{F}(r)) = \sum_{i=0}^{n} a_i \binom{r}{i}$$

where $a_0, \ldots, a_n \in \mathbb{Z}$.

Then $\mathcal{F}$ is m-regular, where $m = F_{p,n}(a_0, \ldots, a_n)$.

PROOF. (From Mumford [**Mum66**]) As before, we can assume that k is infinite. We argue by induction on n. When $n = 0$, clearly we can take $F_{p,0}$ to be any polynomial. Next, let $n \geq 1$. Let $H \subset \mathbb{P}^n$ be a hyperplane which does not contain any of the finitely many associated points of $\oplus^p \mathcal{O}_{\mathbb{P}^n}/\mathcal{F}$ (such an H exists as k is infinite). Then the following torsion sheaf vanishes:

$$\underline{Tor}_1^{\mathcal{O}_{\mathbb{P}^n}}(\mathcal{O}_H, \oplus^p \mathcal{O}_{\mathbb{P}^n}/\mathcal{F}) = 0.$$

Therefore the sequence $0 \to \mathcal{F} \to \oplus^p \mathcal{O}_{\mathbb{P}^n} \to \oplus^p \mathcal{O}_{\mathbb{P}^n}/\mathcal{F} \to 0$ restricts to H to give a short exact sequence $0 \to \mathcal{F}_H \to \oplus^p \mathcal{O}_H \to \oplus^p \mathcal{O}_H/\mathcal{F}_H \to 0$. This shows that $\mathcal{F}_H$ is isomorphic to a subsheaf of $\oplus^p \mathcal{O}_{\mathbb{P}_k^{n-1}}$ (under an identification of H with $\mathbb{P}_k^{n-1}$), which is a basic step needed for our inductive argument.

Note that $\mathcal{F}$ is torsion free if non-zero, and so we have a short exact sequence $0 \to \mathcal{F}(-1) \to \mathcal{F} \to \mathcal{F}_H \to 0$. ¿From the associated cohomology sequence we get $\chi(\mathcal{F}_H(r)) = \chi(\mathcal{F}(r)) - \chi(\mathcal{F}(r-1)) = \sum_{i=0}^n a_i \binom{r}{i} - \sum_{i=0}^n a_i \binom{r-1}{i} = \sum_{i=0}^n a_i \binom{r-1}{i-1} = \sum_{j=0}^{n-1} b_j \binom{r}{j}$ where the coefficients $b_0, \ldots, b_{n-1}$ have expressions $b_j = g_j(a_0, \ldots, a_n)$

where the g_j are polynomials with integral coefficients independent of the field k and the sheaf $\mathcal{F}$. (**Exercise**: Write down the g_j explicitly.)

By inductive hypothesis on $n-1$ there exists a polynomial $F_{p,n-1}(x_0,\ldots,x_{n-1})$ such that $\mathcal{F}_H$ is m_0-regular where $m_0 = F_{p,n-1}(b_0,\ldots,b_{n-1})$. Substituting $b_j = g_j(a_0,\ldots,a_n)$, we get $m_0 = G(a_0,\ldots,a_n)$, where G is a polynomial with integral coefficients independent of the field k and the sheaf $\mathcal{F}$.

For $m \geq m_0 - 1$, we therefore get a long exact cohomology sequence

$$0 \to H^0(\mathcal{F}(m-1)) \to H^0(\mathcal{F}(m)) \stackrel{\nu_m}{\to} H^0(\mathcal{F}_H(m)) \to H^1(\mathcal{F}(m-1)) \to H^1(\mathcal{F}(m)) \to 0 \to \ldots$$

which for $i \geq 2$ gives isomorphisms $H^i(\mathcal{F}(m-1)) \xrightarrow{\sim} H^i(\mathcal{F}(m))$. As we have $H^i(\mathcal{F}(m)) = 0$ for $m \gg 0$, these equalities show that

$$H^i(\mathcal{F}(m)) = 0 \text{ for all } i \geq 2 \text{ and } m \geq m_0 - 2.$$

The surjections $H^1(\mathcal{F}(m-1)) \to H^1(\mathcal{F}(m))$ show that the function $h^1(\mathcal{F}(m))$ is a monotonically decreasing function of the variable m for $m \geq m_0 - 2$. We will in fact show that for $m \geq m_0$, the function $h^1(\mathcal{F}(m))$ is strictly decreasing until its value reaches zero, which would imply that

$$H^1(\mathcal{F}(m)) = 0 \text{ for } m \geq m_0 + h^1(\mathcal{F}(m_0)).$$

Next we will put a suitable upper bound on $h^1(\mathcal{F}(m_0))$ to complete the proof of the theorem. Note that $h^1(\mathcal{F}(m-1)) \geq h^1(\mathcal{F}(m))$ for $m \geq m_0$, and moreover equality holds for some $m \geq m_0$ if and only if the restriction map $\nu_m : H^0(\mathcal{F}(m)) \to H^0(\mathcal{F}_H(m))$ is surjective. As $\mathcal{F}_H$ is m-regular, it follows from Remark 5.2 that the restriction map $\nu_j : H^0(\mathcal{F}(j)) \to H^0(\mathcal{F}_H(j))$ is surjective for all $j \geq m$, so $h^1(\mathcal{F}(j-1)) = h^1(\mathcal{F}(j))$ for all $j \geq m$. As $h^1(\mathcal{F}(j)) = 0$ for $j \gg 0$, this establishes our claim that $h^1(\mathcal{F}(m))$ is strictly decreasing for $m \geq m_0$ until its value reaches zero.

To put a bound on $h^1(\mathcal{F}(m_0))$, we use the fact that as $\mathcal{F} \subset \oplus^p \mathcal{O}_{\mathbb{P}^n}$ we must have $h^0(\mathcal{F}(r)) \leq p h^0(\mathcal{O}_{\mathbb{P}^n}(r)) = p\binom{n+r}{n}$. ¿From the already established fact that $h^i(\mathcal{F}(m)) = 0$ for all $i \geq 2$ and $m \geq m_0 - 2$, we now get

$$\begin{aligned} h^1(\mathcal{F}(m_0)) &= h^0(\mathcal{F}(m_0)) - \chi(\mathcal{F}(m_0)) \\ &\leq p\binom{n+m_0}{n} - \sum_{i=0}^{n} a_i \binom{m_0}{i} \\ &= P(a_0, \ldots a_n) \end{aligned}$$

where $P(a_0,\ldots,a_n)$ is a polynomial expression in $a_0,\ldots,a_n$, obtained by substituting $m_0 = G(a_0,\ldots,a_n)$ in the second line of the above (in)equalities. Therefore, the coefficients of the corresponding polynomial $P(x_0,\ldots,x_n)$ are again independent of the field k and the sheaf $\mathcal{F}$. Note moreover that as $h^1(\mathcal{F}(m_0)) \geq 0$, we must have $P(a_0,\ldots,a_n) \geq 0$.

Substituting in an earlier expression, we get

$$H^1(\mathcal{F}(m)) = 0 \text{ for } m \geq G(a_0,\ldots,a_n) + P(a_0,\ldots,a_n).$$

Taking $F_{p,n}(x_0,\ldots,x_n)$ to be $G(x_0,\ldots,x_n) + P(x_0,\ldots,x_n)$, and noting the fact that $P(a_0,\ldots,a_n) \geq 0$, we see that $\mathcal{F}$ is $F_{p,n}(a_0,\ldots,a_n)$-regular. This completes the proof of the theorem. □

Exercise. Write down such polynomials $F_{p,n}$.

5.3. Semi-continuity and base-change

5.3.1. Base-change without flatness. The following lemma on base-change does not need any flatness hypothesis. The price paid is that the integer r_0 may depend on ϕ.

LEMMA 5.4. Let $\phi : T \to S$ be a morphism of noetherian schemes, let $\mathcal{F}$ be a coherent sheaf on $\mathbb{P}^n_S$, and let $\mathcal{F}_T$ denote the pull-back of $\mathcal{F}$ under the induced morphism $\mathbb{P}^n_T \to \mathbb{P}^n_S$. Let $\pi_S : \mathbb{P}^n_S \to S$ and $\pi_T : \mathbb{P}^n_T \to T$ denote the projections. Then there exists an integer r_0 such that the base-change homomorphism

$$\phi^*\pi_{S*}\mathcal{F}(r) \to \pi_{T*}\mathcal{F}_T(r)$$

is an isomorphism for all $r \geq r_0$.

PROOF. As base-change holds for open embeddings, using a finite affine open cover U_i of S and a finite affine open cover $V_{i,j}$ of each $\phi^{-1}(U_i)$ (which is possible by noetherian hypothesis), it is enough to consider the case where S and T are affine.

Note that for all integers i, the base-change homomorphism

$$\phi^*\pi_{S*}\mathcal{O}_{\mathbb{P}^n_S}(i) \to \pi_{T*}\mathcal{O}_{\mathbb{P}^n_T}(i)$$

is an isomorphism. Moreover, if a and b are any integers and if $f : \mathcal{O}_{\mathbb{P}^n_S}(a) \to \mathcal{O}_{\mathbb{P}^n_S}(b)$ is any homomorphism and $f_T : \mathcal{O}_{\mathbb{P}^n_T}(a) \to \mathcal{O}_{\mathbb{P}^n_T}(b)$ denotes its pull-back to $\mathbb{P}^n_T$, then for all i we have the following commutative diagram where the vertical maps are base-change isomorphisms.

$$\begin{array}{ccc}
\phi^*\pi_{S*}\mathcal{O}_{\mathbb{P}^n_S}(a+i) & \xrightarrow{\phi^*\pi_{S*}f(i)} & \phi^*\pi_{S*}\mathcal{O}_{\mathbb{P}^n_S}(b+i) \\
\downarrow & & \downarrow \\
\pi_{T*}\mathcal{O}_{\mathbb{P}^n_T}(a+i) & \xrightarrow{\pi_{T*}f_T(i)} & \pi_{T*}\mathcal{O}_{\mathbb{P}^n_T}(b+i)
\end{array}$$

As S is noetherian and affine, there exists an exact sequence

$$\bigoplus^p \mathcal{O}_{\mathbb{P}^n_S}(a) \xrightarrow{u} \bigoplus^q \mathcal{O}_{\mathbb{P}^n_S}(b) \xrightarrow{v} \mathcal{F} \to 0$$

for some integers a, b, $p \geq 0$, $q \geq 0$. Its pull-back to $\mathbb{P}^n_T$ is an exact sequence

$$\bigoplus^p \mathcal{O}_{\mathbb{P}^n_T}(a) \xrightarrow{u_T} \bigoplus^q \mathcal{O}_{\mathbb{P}^n_S}(b) \xrightarrow{v_T} \mathcal{F}_T \to 0.$$

Let $\mathcal{G} = \ker(v)$ and let $\mathcal{H} = \ker(v_T)$. For any integer r, we get exact sequences

$$\pi_{S*}\bigoplus^p \mathcal{O}_{\mathbb{P}^n_S}(a+r) \to \pi_{S*}\bigoplus^q \mathcal{O}_{\mathbb{P}^n_S}(b+r) \to \pi_{S*}\mathcal{F}(r) \to R^1\pi_{S*}\mathcal{G}(r)$$

and

$$\pi_{T*}\bigoplus^p \mathcal{O}_{\mathbb{P}^n_T}(a+r) \to \pi_{T*}\bigoplus^q \mathcal{O}_{\mathbb{P}^n_T}(b+r) \to \pi_{T*}\mathcal{F}_T(r) \to R^1\pi_{T*}\mathcal{H}(r).$$

There exists an integer r_0 such that $R^1\pi_{S*}\mathcal{G}(r) = 0$ and $R^1\pi_{T*}\mathcal{H}(r) = 0$ for all $r \geq r_0$. Hence for all $r \geq r_0$, we have exact sequences

$$\pi_{S*}\bigoplus^p \mathcal{O}_{\mathbb{P}^n_S}(a+r) \xrightarrow{\pi_{S*}u(r)} \bigoplus^q \mathcal{O}_{\mathbb{P}^n_S}(b+r) \xrightarrow{\pi_{S*}v(r)} \pi_{S*}\mathcal{F}(r) \to 0$$

and

$$\pi_{T*}\bigoplus^p \mathcal{O}_{\mathbb{P}^n_T}(a+r) \xrightarrow{\pi_{T*}u_T(r)} \pi_{T*}\bigoplus^q \mathcal{O}_{\mathbb{P}^n_T}(b+r) \xrightarrow{\pi_{T*}v_T(r)} \pi_{T*}\mathcal{F}_T(r) \to 0.$$

Pulling back the second-last exact sequence under $\phi : T \to S$, we get the commutative diagram with exact rows

$$\begin{array}{ccccccc}
\phi^*\pi_{S*}\bigoplus^p \mathcal{O}_{\mathbb{P}^n_S}(a+r) & \xrightarrow{\phi^*\pi_{S*}u} & \phi^*\bigoplus^q \mathcal{O}_{\mathbb{P}^n_S}(b+r) & \xrightarrow{\phi^*\pi_{S*}v} & \phi^*\pi_{S*}\mathcal{F}(r) & \to & 0 \\
\downarrow & & \downarrow & & \downarrow & & \\
\pi_{T*}\bigoplus^p \mathcal{O}_{\mathbb{P}^n_T}(a+r) & \xrightarrow{\pi_{T*}u_T(r)} & \pi_{T*}\bigoplus^q \mathcal{O}_{\mathbb{P}^n_T}(b+r) & \xrightarrow{\pi_{T*}v_T(r)} & \pi_{T*}\mathcal{F}_T(r) & \to & 0
\end{array}$$

in which the first row is exact by the right-exactness of tensor product. The vertical maps are base-change homomorphisms, the first two of which are isomorphisms for all r. Therefore by the five lemma, $\phi^*\pi_{S*}\mathcal{F}(r) \to \pi_{T*}\mathcal{F}_T(r)$ is an isomorphism for all $r \geq r_0$. □

The following elementary proof of the above result is taken from Mumford [**Mum66**]: Let M be the graded $\mathcal{O}_S$-module $\bigoplus_{m\in\mathbb{Z}} \pi_{S*}\mathcal{F}(m)$, so that $\mathcal{F} = M^\sim$. Let ϕ^*M be the graded $\mathcal{O}_T$-module which is the pull-back of M. Then we have $\mathcal{F}_T = (\phi^*M)^\sim$. On the other hand, let $N = \bigoplus_{m\in\mathbb{Z}} \pi_{T*}\mathcal{F}_T(m)$, so that we have $\mathcal{F}_T = N^\sim$. Therefore, in the category of graded $\mathcal{O}_T[x_0,\ldots,x_n]$-modules, we get an induced equivalence between ϕ^*M and N, which means the natural homomorphisms of graded pieces $(\phi^*M)_m \to N_m$ are isomorphisms for all $m \gg 0$. □

5.3.2. Flatness of $\mathcal{F}$ from local freeness of $\pi_*\mathcal{F}(r)$.

LEMMA 5.5. *Let S be a noetherian scheme and let $\mathcal{F}$ be a coherent sheaf on $\mathbb{P}^n_S$. Suppose that there exists some integer N such that for all $r \geq N$ the direct image $\pi_*\mathcal{F}(r)$ is locally free. Then $\mathcal{F}$ is flat over S.*

PROOF. Consider the graded module $M = \bigoplus_{r\geq N} M_r$ over $\mathcal{O}_S$, where $M_r = \pi_*\mathcal{F}(r)$. The sheaf $\mathcal{F}$ is isomorphic to the sheaf $M^\sim$ on $\mathbb{P}^n_S = \mathbf{Proj}_S\, \mathcal{O}_S[x_0,\ldots,x_n]$ made from the graded sheaf M of $\mathcal{O}_S$-modules. As each M_r is flat over $\mathcal{O}_S$, so is M. Therefore for any x_i the localisation M_{x_i} is flat over $\mathcal{O}_S$. There is a grading on M_{x_i}, indexed by $\mathbb{Z}$, defined by putting $\deg(v_p/x_i^q) = p - q$ for $v_p \in M_p$ (this is well-defined). Hence the component $(M_{x_i})_0$ of degree zero, being a direct summand of M_{x_i}, is again flat over $\mathcal{O}_S$. But by definition of $M^\sim$, this is just $\Gamma(U_i,\mathcal{F})$, where $U_i = \mathbf{Spec}_S\, \mathcal{O}_S[x_0/x_i,\ldots,x_n/x_i] \subset \mathbb{P}^n_S$. As the U_i form an open cover of $\mathbb{P}^n_S$, it follows that $\mathcal{F}$ is flat over $\mathcal{O}_S$. □

Exercise. Show that the converse of the above lemma holds: if $\mathcal{F}$ is flat over S then $\pi_*\mathcal{F}(r)$ is locally free for all sufficiently large r.

5.3.3. Grothendieck complex for semi-continuity. The following is a very important basic result of Grothendieck, and the complex $K^\cdot$ occurring in it is called the Grothendieck complex.

THEOREM 5.6. *Let $\pi : X \to S$ be a proper morphism of noetherian schemes where $S = \operatorname{Spec} A$ is affine, and let $\mathcal{F}$ be a coherent $\mathcal{O}_X$-module which is flat over $\mathcal{O}_S$. Then there exists a finite complex*

$$0 \to K^0 \to K^1 \to \ldots \to K^n \to 0$$

of finitely generated projective A-modules, together with a functorial A-linear isomorphism

$$H^p(X, \mathcal{F} \otimes_A M) \xrightarrow{\sim} H^p(K^\cdot \otimes_A M)$$

on the category of all A-modules M.

The above theorem is the foundation for all results about direct images and base-change for flat families of sheaves, such as Theorem 5.10.

As another consequence of the above theorem, we have the following.

THEOREM 5.7. ([EGA] III 7.7.6) Let S be a noetherian scheme and $\pi : X \to S$ a proper morphism. Let $\mathcal{F}$ be a coherent sheaf on X which is flat over S. Then there exists a coherent sheaf $\mathcal{Q}$ on S together with a functorial $\mathcal{O}_S$-linear isomorphism

$$\theta_{\mathcal{G}} : \pi_*(\mathcal{F} \otimes_{\mathcal{O}_X} \pi^*\mathcal{G}) \to \underline{Hom}_{\mathcal{O}_S}(\mathcal{Q}, \mathcal{G})$$

on the category of all quasi-coherent sheaves $\mathcal{G}$ on S. By its universal property, the pair $(\mathcal{Q}, \theta)$ is unique up to a unique isomorphism.

PROOF. If $S = \operatorname{Spec} A$, then we can take $\mathcal{Q}$ to be the coherent sheaf associated to the A-module Q which is the cokernel of the transpose $\partial^\vee : (K^1)^\vee \to (K^0)^\vee$ where $\partial : K^0 \to K^1$ is the differential of any chosen Grothendieck complex of A-modules $0 \to K^0 \to K^1 \to \ldots \to K^n \to 0$ for the sheaf $\mathcal{F}$, whose existence is given by Theorem 5.6. For any A-module M, the right-exact sequence $(K^1)^\vee \to (K^0)^\vee \to Q \to 0$ with M gives on applying $Hom_A(-, M)$ a left-exact sequence

$$0 \to Hom_A(Q, M) \to K^0 \otimes_A M \to K^1 \otimes_A M.$$

Therefore by Theorem 5.6, we have an isomorphism

$$\theta^A_M : H^0(X_A, \mathcal{F}_A \otimes_A M) \to Hom_A(Q, M).$$

Thus, the pair (Q, θ^A) satisfies the theorem when $S = \operatorname{Spec} A$. More generally, we can cover S by affine open subschemes. Then on their overlaps, the resulting pairs $(\mathcal{Q}, \theta)$ glue together by their uniqueness. □

A *linear scheme* $\mathbf{V} \to S$ over a noetherian base scheme S is a scheme of the form $\mathbf{Spec}\,\mathrm{Sym}_{\mathcal{O}_S}\,\mathcal{Q}$ where $\mathcal{Q}$ is a coherent sheaf on S. This is naturally a group scheme. Linear schemes generalise the notion of (geometric) vector bundles, which are the special case where $\mathcal{Q}$ is locally free of constant rank.

The *zero section* $\mathbf{V}_0 \subset \mathbf{V}$ of a linear scheme $\mathbf{V} = \mathbf{Spec}\,\mathrm{Sym}_{\mathcal{O}_S}\,\mathcal{Q}$ is the closed subscheme defined by the ideal generated by $\mathcal{Q}$. Note that the projection $\mathbf{V}_0 \to S$ is an isomorphism, and $\mathbf{V}_0$ is just the image of the zero section $0 : S \to \mathbf{V}$ of the group-scheme.

THEOREM 5.8. ([EGA] III 7.7.8, 7.7.9) Let S be a noetherian scheme and $\pi : X \to S$ a projective morphism. Let $\mathcal{E}$ and $\mathcal{F}$ be coherent sheaves on X. Consider the set-valued contravariant functor $\mathfrak{Hom}(\mathcal{E}, \mathcal{F})$ on S-schemes, which associates to any $T \to S$ the set of all $\mathcal{O}_{X_T}$-linear homomorphisms $Hom_{X_T}(\mathcal{E}_T, \mathcal{F}_T)$ where $\mathcal{E}_T$ and $\mathcal{F}_T$ denote the pull-backs of $\mathcal{E}$ and $\mathcal{F}$ under the projection $X_T \to X$. If $\mathcal{F}$ is flat over S, then the above functor is representable by a linear scheme $\mathbf{V}$ over S.

PROOF. First note that if $\mathcal{E}$ is a locally free $\mathcal{O}_X$-module, then $\mathfrak{Hom}(\mathcal{E}, \mathcal{F})$ is the functor $T \mapsto H^0(X_T, (\mathcal{F} \otimes_{\mathcal{O}_X} \mathcal{E}^\vee)_T)$. The sheaf $\mathcal{F} \otimes_{\mathcal{O}_X} \mathcal{E}^\vee$ is again flat over S, so we can apply Theorem 5.7 to get a coherent sheaf $\mathcal{Q}$, such that we have

$\pi_*(\mathcal{F} \otimes_{\mathcal{O}_X} \mathcal{E}^\vee \otimes_{\mathcal{O}_X} \pi^*\mathcal{G}) = \underline{Hom}_{\mathcal{O}_S}(\mathcal{Q}, \mathcal{G})$ for all quasi-coherent sheaves $\mathcal{G}$ on S. In particular, if $f : \operatorname{Spec} R \to S$ is any morphism then taking $\mathcal{G} = f_*\mathcal{O}_R$ we get

$$\begin{aligned} Mor_S(\operatorname{Spec} R, \mathbf{Spec}\,\mathrm{Sym}_{\mathcal{O}_S} \mathcal{Q}) &= Hom_{\mathcal{O}_S\text{-mod}}(\mathcal{Q}, f_*\mathcal{O}_R) \\ &= H^0(X, \mathcal{F} \otimes_{\mathcal{O}_X} \mathcal{E}^\vee \otimes_{\mathcal{O}_X} \pi^* f_*\mathcal{O}_R) \\ &= H^0(X_R, (\mathcal{F} \otimes_{\mathcal{O}_X} \mathcal{E}^\vee)_R) \\ &= Hom_{X_R}(\mathcal{E}_R, \mathcal{F}_R). \end{aligned}$$

This shows that $\mathbf{V} = \mathbf{Spec}\,\mathrm{Sym}_{\mathcal{O}_S} \mathcal{Q}$ is the required linear scheme when $\mathcal{E}$ is locally free on X. More generally for an arbitrary coherent $\mathcal{E}$, over any affine open $U \subset S$ there exist vector bundles E_1 and E_0 on X_U and a right exact sequence $E_1 \to E_0 \to \mathcal{E} \to 0$. (This is where we need projectivity of $X \to S$. Instead, we could have assumed just properness together with the condition that locally over S we have such a resolution of $\mathcal{E}$.) Then applying the above argument to the functors $\mathfrak{Hom}(E_1, \mathcal{F})$ and $\mathfrak{Hom}(E_0, \mathcal{F})$, we get coherent sheaves $\mathcal{Q}_1$ and $\mathcal{Q}_0$ on U, and from the natural transformation $\mathfrak{Hom}(E_0, \mathcal{F}) \to \mathfrak{Hom}(E_1, \mathcal{F})$ induced by the homomorphism $E_1 \to E_0$, we get a homomorphism $\mathcal{Q}_1 \to \mathcal{Q}_0$. Let $\mathcal{Q}_U$ be its cokernel, and put $\mathbf{V}_U = \mathbf{Spec}\,\mathrm{Sym}_{\mathcal{O}_U} \mathcal{Q}_U$. It follows from its definition (and the left exactness of Hom) that the scheme $\mathbf{V}_U$ has the desired universal property over U. Therefore all such $\mathbf{V}_U$, as U varies over an affine open cover of S, patch together to give the desired linear scheme $\mathbf{V}$. (In sheaf terms, the sheaves $\mathcal{Q}_U$ will patch together to give a coherent sheaf $\mathcal{Q}$ on S with $\mathbf{V} = \mathbf{Spec}\,\mathrm{Sym}_{\mathcal{O}_S} \mathcal{Q}$.) □

REMARK 5.9. In particular, note that the zero section $\mathbf{V}_0 \subset \mathbf{V}$ is where the universal homomorphism vanishes. If $f \in Hom_{X_T}(\mathcal{E}_T, \mathcal{F}_T)$ defines a morphism $\varphi_f : T \to \mathbf{V}$, then the inverse image $f^{-1}\mathbf{V}_0$ is a closed subscheme T' of T with the universal property that if $U \to T$ is any morphism of schemes such that the pull-back of f is zero, then $U \to T$ factors via T'.

5.3.4. Base-change for flat sheaves. The following is the main result of Grothendieck on base change for flat families of sheaves, which is a consequence of Theorem 5.6.

THEOREM 5.10. Let $\pi : X \to S$ be a proper morphism of noetherian schemes, and let $\mathcal{F}$ be a coherent $\mathcal{O}_X$-module which is flat over $\mathcal{O}_S$. Then the following statements hold:

(1) For any integer i the function $s \mapsto \dim_{\kappa(s)} H^i(X_s, \mathcal{F}_s)$ is upper semi-continuous on S,

(2) The function $s \mapsto \sum_i (-1)^i \dim_{\kappa(s)} H^i(X_s, \mathcal{F}_s)$ is locally constant on S.

(3) If for some integer i, there is some integer $d \geq 0$ such that for all $s \in S$ we have $\dim_{\kappa(s)} H^i(X_s, \mathcal{F}_s) = d$, then $R^i\pi_*\mathcal{F}$ is locally free of rank d, and $(R^{i-1}\pi_*\mathcal{F})_s \to H^{i-1}(X_s, \mathcal{F}_s)$ is an isomorphism for all $s \in S$.

(4) If for some integer i and point $s \in S$ the map $(R^i\pi_*\mathcal{F})_s \to H^i(X_s, \mathcal{F}_s)$ is surjective, then there exists an open subscheme $U \subset S$ containing s such that for any quasi-coherent $\mathcal{O}_U$-module $\mathcal{G}$ the natural homomorphism

$$(R^i\pi_{U*}\mathcal{F}_{X_U}) \otimes_{\mathcal{O}_U} \mathcal{G} \to R^i\pi_{U*}(\mathcal{F}_{X_U} \otimes_{\mathcal{O}_{X_U}} \pi_U{}^*\mathcal{G})$$

is an isomorphism, where $X_U = \pi^{-1}(U)$ and $\pi_U : X_U \to U$ is induced by π. In particular, $(R^i\pi_*\mathcal{F})_{s'} \to H^i(X_{s'}, \mathcal{F}_{s'})$ is an isomorphism for all s' in U.

(5) If for some integer i and point $s \in S$ the map $(R^i\pi_*\mathcal{F})_s \to H^i(X_s, \mathcal{F}_s)$ is surjective, then the following conditions (a) and (b) are equivalent:

(a) The map $(R^{i-1}\pi_*\mathcal{F})_s \to H^{i-1}(X_s, \mathcal{F}_s)$ is surjective.
(b) The sheaf $R^i\pi_*\mathcal{F}$ is locally free in a neighbourhood of s in S.

See for example [**Har77**, Chapter III, Section 12] for a proof. It is possible to replace the use of the formal function theorem in [**Har77**] (or the original argument in [EGA] based on completions) in proving the statement (4) above, with an elementary argument based on applying Nakayama lemma to the Grothendieck complex.

5.4. Generic flatness and flattening stratification

5.4.1. Lemma on generic flatness.

LEMMA 5.11. Let A be a noetherian domain, and B a finite type A-algebra. Let M be a finite B-module. Then there exists an $f \in A$, $f \neq 0$, such that the localisation M_f is a free module over A_f.

PROOF. Over the quotient field K of A, the K-algebra $B_K = K \otimes_A B$ is of finite type, and $M_K = K \otimes_A M$ is a finite module over B_K. Let n be the dimension of the support of M_K over $\operatorname{Spec} B_K$. We argue by induction on n, starting with $n = -1$ which is the case when $M_K = 0$. In this case, as $K \otimes_A M = S^{-1}M$ where $S = A - \{0\}$, each $v \in M$ is annihilated by some non-zero element of A. Taking a finite generating set, and a common multiple of corresponding annihilating elements, we see there exists an $f \neq 0$ in A with $fM = 0$. Hence $M_f = 0$, proving the lemma when $n = -1$.

Now let $n \geq 0$, and let the lemma be proved for smaller values. As B is noetherian and M is assumed to be a finite B-module, there exists a finite filtration

$$0 = M_0 \subset \ldots \subset M_r = M$$

where each M_i is a B-submodule of M such that for each $i \geq 1$ the quotient module M_i/M_{i-1} is isomorphic to $B/\mathfrak{p}_i$ for some prime ideal $\mathfrak{p}_i$ in B.

Note that if $0 \to M' \to M \to M'' \to 0$ is a short exact sequence of B-modules, and if f' and f'' are non-zero elements of A such that $M'_{f'}$ and $M''_{f''}$ are free over respectively $A_{f'}$ and $A_{f''}$, then M_f is a free module over A_f where $f = f'f''$. We will use this fact repeatedly. Therefore it is enough to prove the result when M is of the form $B/\mathfrak{p}$ for a prime ideal $\mathfrak{p}$ in B. This reduces us to the case where B is a domain and $M = B$.

As by assumption $K \otimes_A B$ has dimension $n \geq 0$ (that is, $K \otimes_A B$ is non-zero), the map $A \to B$ must be injective. By Noether normalisation lemma, there exist elements $b_1, \ldots, b_n \in B$, such that $K \otimes_A B$ is finite over its subalgebra $K[b_1, \ldots, b_n]$ and the elements $b_1, \ldots, b_n$ are algebraically independent over K. (For simplicity of notation, we write $1 \otimes b$ simply as b.) If $g \neq 0$ in A is chosen to be a 'common denominator' for coefficients of equations of integral dependence satisfied by a finite set of algebra generators for $K \otimes_A B$ over $K[b_1, \ldots, b_n]$, we see that B_g is finite over $A_g[b_1, \ldots, b_n]$.

Let m be the generic rank of the finite module B_g over the domain $A_g[b_1, \ldots, b_n]$. Then we have a short exact sequence of $A_g[b_1, \ldots, b_n]$-modules of the form

$$0 \to A_g[b_1, \ldots, b_n]^{\oplus m} \to B_g \to T \to 0$$

where T is a finite torsion module over $A_g[b_1, \ldots, b_n]$. Therefore, the dimension of the support of $K \otimes_{A_g} T$ as a $K \otimes_{A_g} (B_g)$-module is strictly less than n. Hence by induction on n (applied to the data A_g, B_g, T), there exists some $h \neq 0$ in A with T_h free over A_{gh}. Taking $f = gh$, the lemma follows from the above short exact sequence. □

The above theorem has the following consequence, which follows by restricting attention to a non-empty affine open subscheme of S.

THEOREM 5.12. Let S be a noetherian and integral scheme. Let $p : X \to S$ be a finite type morphism, and let $\mathcal{F}$ be a coherent sheaf of $\mathcal{O}_X$-modules. Then there exists a non-empty open subscheme $U \subset S$ such that the restriction of $\mathcal{F}$ to $X_U = p^{-1}(U)$ is flat over $\mathcal{O}_U$.

5.4.2. Existence of flattening stratification.

THEOREM 5.13. Let S be a noetherian scheme, and let $\mathcal{F}$ be a coherent sheaf on the projective space $\mathbb{P}^n_S$ over S. Then the set I of Hilbert polynomials of restrictions of $\mathcal{F}$ to fibers of $\mathbb{P}^n_S \to S$ is a finite set. Moreover, for each $f \in I$ there exist a locally closed subscheme S_f of S, such that the following conditions are satisfied.

(i) Point-set: The underlying set $|S_f|$ of S_f consists of all points $s \in S$ where the Hilbert polynomial of the restriction of $\mathcal{F}$ to $\mathbb{P}^n_s$ is f. In particular, the subsets $|S_f| \subset |S|$ are disjoint, and their set-theoretic union is $|S|$.

(ii) Universal property: Let $S' = \coprod S_f$ be the coproduct of the S_f, and let $i : S' \to S$ be the morphism induced by the inclusions $S_f \hookrightarrow S$. Then the sheaf $i^*(\mathcal{F})$ on $\mathbb{P}^n_{S'}$ is flat over S'. Moreover, $i : S' \to S$ has the universal property that for any morphism $\varphi : T \to S$ the pullback $\varphi^*(\mathcal{F})$ on $\mathbb{P}^n_T$ is flat over T if and only if φ factors through $i : S' \to S$. The subscheme S_f is uniquely determined by the polynomial f.

(iii) Closure of strata: Let the set I of Hilbert polynomials be given a total ordering, defined by putting $f < g$ whenever $f(n) < g(n)$ for all $n \gg 0$. Then the closure in S of the subset $|S_f|$ is contained in the union of all $|S_g|$ where $f \leq g$.

PROOF. It is enough to prove the theorem for open subschemes of S which cover S, as the resulting strata will then glue together by their universal property.

Special case: Let $n = 0$, so that $\mathbb{P}^n_S = S$. ¿For any $s \in S$, the *fiber* $\mathcal{F}|_s$ of $\mathcal{F}$ over s will mean the pull-back of $\mathcal{F}$ to the subscheme $\operatorname{Spec} \kappa(s)$, where $\kappa(s)$ is the residue field at s. (This is obtained by tensoring the stalk of $\mathcal{F}$ at s with the residue field at s, both regarded as $\mathcal{O}_{S,s}$-modules.) The Hilbert polynomial of the restriction of $\mathcal{F}$ to the fiber over s is the degree 0 polynomial $e \in \mathbb{Q}[\lambda]$, where $e = \dim_{\kappa(s)} \mathcal{F}|_s$.

By Nakayama lemma, any basis of $\mathcal{F}|_s$ prolongs to a neighbourhood U of s to give a set of generators for $F|_U$. Repeating this argument, we see that there exists a smaller neighbourhood V of s in which there is a right-exact sequence

$$\mathcal{O}_V^{\oplus m} \xrightarrow{\psi} \mathcal{O}_V^{\oplus e} \xrightarrow{\phi} \mathcal{F} \to 0.$$

Let $I_e \subset \mathcal{O}_V$ be the ideal sheaf formed by the entries of the $e \times m$ matrix $(\psi_{i,j})$ of the homomorphism $\mathcal{O}_V^{\oplus m} \xrightarrow{\psi} \mathcal{O}_V^{\oplus e}$. Let V_e be the closed subscheme of V defined by I_e. For any morphism of schemes $f : T \to V$, the pull-back sequence

$$\mathcal{O}_T^{\oplus m} \xrightarrow{f^*\psi} \mathcal{O}_T^{\oplus e} \xrightarrow{f^*\phi} f^*\mathcal{F} \to 0$$

is exact, by right-exactness of tensor products. Hence the pull-back $f^*\mathcal{F}$ is a locally free $\mathcal{O}_T$-module of rank e if and only if $f^*\psi = 0$, that is, f factors via the subscheme

$V_e \hookrightarrow V$ defined by the vanishing of all entries $\psi_{i,j}$. Thus we have proved assertions **(i)** and **(ii)** of the theorem.

As the rank of the matrix $(\psi_{i,j})$ is lower semi-continuous, it follows that the function e is upper semi-continuous, which proves the assertion **(iii)** of the theorem, completing its proof when $n = 0$.

General case: We now allow the integer n to be arbitrary. The idea of the proof is as follows: We show the existence of a stratification of S which is a 'g.c.d.' of the flattening stratifications for direct images $\pi_*\mathcal{F}(i)$ for all $i \geq N$ for some integer N (where the flattening stratifications for $\pi_*\mathcal{F}(i)$ exist by case $n = 0$ which we have treated above). This is the desired flattening stratification of $\mathcal{F}$ over S, as follows from Lemma 5.5.

As S is noetherian, it is a finite union of irreducible components, and these are closed in S. Let Y be an irreducible component of S, and let U be the non-empty open subset of Y which consists of all points which do not lie on any other irreducible component of S. Let U be given the reduced subscheme structure. Note that this makes U an integral scheme, which is a locally closed subscheme of S. By Theorem 5.12 on generic flatness, U has a non-empty open subscheme V such that the restriction of $\mathcal{F}$ to $\mathbb{P}^n_V$ is flat over $\mathcal{O}_V$. Now repeating the argument with S replaced by its reduced closed subscheme $S - V$, it follows by noetherian induction on S that there exist finitely many reduced, locally closed, mutually disjoint subschemes V_i of S such that set-theoretically $|S|$ is the union of the $|V_i|$ and the restriction of $\mathcal{F}$ to $\mathbb{P}^n_{V_i}$ is flat over $\mathcal{O}_{V_i}$. As each V_i is a noetherian scheme, and as the Hilbert polynomials are locally constant for a flat family of sheaves, it follows that only finitely many polynomials occur in V_i in the family of Hilbert polynomials $P_s(m) = \chi(\mathbb{P}^n_s, \mathcal{F}_s(m))$ as s varies over points of V_i. This allows us to conclude the following:

(A) Only finitely many distinct Hilbert polynomials $P_s(m) = \chi(\mathbb{P}^n_s, \mathcal{F}_s(m))$ occur, as s varies over all of S.

By the semi-continuity theorem applied to the flat families $\mathcal{F}_{V_i} = \mathcal{F}|_{\mathbb{P}^n_{V_i}}$ parametrised by the finitely many noetherian schemes V_i, we get the following:

(B) There exists an integer N_1 such that $R^r\pi_*\mathcal{F}(m) = 0$ for all $r \geq 1$ and $m \geq N_1$, and moreover $H^r(\mathbb{P}^n_s, \mathcal{F}_s(m)) = 0$ for all $s \in S$.

For each V_i, by Lemma 5.4 there exists an integer $r_i \geq N_1$ with the property that for any $m \geq r_i$ the base change homomorphism

$$(\pi_*\mathcal{F}(m))|_{V_i} \to \pi_{i*}\mathcal{F}_{V_i}(m)$$

is an isomorphism, where $\mathcal{F}_{V_i}$ denotes the restriction of $\mathcal{F}$ to $\mathbb{P}^n_{V_i}$, and $\pi_i : \mathbb{P}^n_{V_i} \to V_i$ the projection. As the higher cohomologies of all fibers (in particular, the first cohomology) vanish by **(B)**, it follows by semi-continuity theory for the flat family $\mathcal{F}_{V_i}$ over V_i that for any $s \in V_i$ the base change homomorphism

$$(\pi_{i*}\mathcal{F}_{V_i}(m))|_s \to H^0(\mathbb{P}^n_s, \mathcal{F}_s(m))$$

is an isomorphism for $m \geq r_i$. Taking N to be the maximum of all r_i over the finitely many non-empty V_i, and composing the above two base change isomorphisms, we get the following.

(C) There exists an integer $N \geq N_1$ such that the base change homomorphism

$$(\pi_*\mathcal{F}(m))|_s \to H^0(\mathbb{P}^n_s, \mathcal{F}_s(m))$$

is an isomorphism for all $m \geq N$ and $s \in S$.

Note. We now forget the subschemes V_i but retain the facts **(A)**, **(B)**, **(C)** which were proved using the $\boldsymbol{V_i}$.

Let $\pi : \mathbb{P}^n_S \to S$ denote the projection. Consider the coherent sheaves $E_0, \dots, E_n$ on S, defined by

$$E_i = \pi_* \mathcal{F}(N+i) \text{ for } i = 0, \dots, n.$$

By applying the special case of the theorem (where the relative dimension n of $\mathbb{P}^n_S$ is 0) to the sheaf E_0 on $\mathbb{P}^0_S = S$, we get a stratification (W_{e_0}) of S indexed by integers e_0, such that for any morphism $f : T \to S$ the pull-back f^*E_0 is a locally free $\mathcal{O}_T$-module of rank e_0 if and only if f factors via $W_{e_0} \hookrightarrow S$. Next, for each stratum W_{e_0}, we take the flattening stratification (W_{e_0,e_1}) for $E_1|_{W_{e_0}}$, and so on. Thus in $n+1$ steps, we obtain finitely many locally closed subschemes

$$W_{e_0,\dots,e_n} \subset S$$

such that for any morphism $f : T \to S$ the pull-back f^*E_i for $i = 0, \dots, n$ is a locally free $\mathcal{O}_T$-module of constant rank e_i if and only if f factors via $W_{e_0,\dots,e_n} \hookrightarrow S$.

For any integer N and n where $n \geq 0$, there is a bijection from the set of numerical polynomials $f \in \mathbb{Q}[\lambda]$ of degree $\leq n$ to the set $\mathbb{Z}^{n+1}$, given by

$$f \mapsto (e_0, \dots, e_n) \text{ where } e_i = f(N+i).$$

Thus, each tuple $(e_0, \dots, e_n) \in \mathbb{Z}^{n+1}$ can be uniquely replaced by a numerical polynomial $f \in \mathbb{Q}[\lambda]$ of degree $\leq n$, allowing us to re-designate $W_{e_0,\dots,e_n} \subset S$ as $W_f \subset S$.

Note that at any point $s \in S$, by **(B)** we have $H^r(\mathbb{P}^n_s, \mathcal{F}_s(m)) = 0$ for all $r \geq 1$ and $m \geq N$. The polynomial $P_s(m) = \chi(\mathbb{P}^n_s, \mathcal{F}_s(m))$ has degree $\leq n$, so it is determined by its $n+1$ values $P_s(N), \dots, P_s(N+n)$. This shows that at any point $s \in W_f$, the Hilbert polynomial $P_s(m)$ equals f. The desired locally closed subscheme $S_f \subset S$, whose existence is asserted by the theorem, will turn out to be a certain closed subscheme $S_f \subset W_f$ whose underlying subset is all of $|W_f|$. The scheme structure of S_f (which may in general differ from that of W_f) is defined as follows.

For any $i \geq 0$ and $s \in S$, the base change homomorphism

$$(\pi_* \mathcal{F}(N+i))|_s \to H^0(\mathbb{P}^n_s, \mathcal{F}_s(N+i))$$

is an isomorphism by statement **(C)**. Hence each $\pi_*\mathcal{F}(N+i)$ has fibers of constant rank $f(N+i)$ on the subscheme W_f. However, this does not mean $\pi_*\mathcal{F}(N+i)$ restricts to a locally constant sheaf of rank $f(N+i)$. But it means that W_f has a closed subscheme $\boldsymbol{W_f^i}$, whose underlying set is all of $|W_f|$, such that $\pi_*\mathcal{F}(N+i)$ is locally free of rank $f(N+i)$ when restricted to $W_f^{(i)}$, and moreover has the property that any base-change $T \to S$ under which $\pi_*\mathcal{F}(N+i)$ pulls back to a locally free sheaf of rank $f(N+i)$ factors via $\boldsymbol{W_f^i}$. The scheme structure of $\boldsymbol{W_f^{(i)}}$ is defined by a coherent ideal sheaf $I_i \subset \mathcal{O}_{W_f}$. Let $I \subset \mathcal{O}_{W_f}$ be the sum of the I_i over $i \geq 0$. By noetherian condition, the increasing sequence

$$I_0 \subset I_0 + I_1 \subset I_0 + I_1 + I_2 \subset \dots$$

terminates in finitely many steps, showing I is again a coherent ideal sheaf. Let $S_f \subset W_f$ be the closed subscheme defined by the ideal sheaf I. Note therefore that

$|S_f| = |W_f|$ and for all $i \geq 0$, the sheaf $\pi_* \mathcal{F}(N+i)$ is locally free of rank $f(N+i)$ when restricted to S_f.

It follows that from their definition that the S_f satisfy property (i) of the theorem.

We now show the morphism $\coprod_f S_f \to S$ indeed has the property (ii) of the theorem. By Lemma 5.4, there exists some $N' \geq N$ such that for all $i \geq N'$, the base-change $(\pi_* \mathcal{F}(i))|_{S_f} \to (\pi_{S_f})_* \mathcal{F}_{S_f}(i)$ is an isomorphism for each S_f. Therefore $\mathcal{F}_{S_f}$ is flat over S_f by Lemma 5.5, as the direct images $\pi_* \mathcal{F}(i)$ for all $i \geq N'$ are locally free over S_f. Conversely, if $\phi : T \to S$ is a morphism such that $\mathcal{F}_T$ is flat, then the Hilbert polynomial is locally constant over T. Let T_f be the open and closed subscheme of T where the Hilbert polynomial is f. Clearly, the set map $|T_f| \to |S|$ factors via $|S_f|$. But as the direct images $\pi_{T*} \mathcal{F}_T(i)$ are locally free of rank $f(i)$ on T_f, it follows in fact that the schematic morphism $T_f \to S$ factors via S_f, proving the property (ii) of the theorem.

As by **(A)** only finitely many polynomials f occur, there exists some $p \geq N$ such that for any two polynomials f and g that occur, we have $f < g$ if and only if $f(p) < g(p)$. As S_f is the flattening stratification for $\pi_* \mathcal{F}(p)$, the property (iii) of the theorem follows from the corresponding property in the case $n = 0$, applied to the sheaf $\pi_* \mathcal{F}(p)$ on S.

This completes the proof of the theorem. □

Exercise. What is the flattening stratification of S for the coherent sheaf $\mathcal{O}_{S^{red}}$ on S, where S^{red} is the underlying reduced scheme of S?

5.5. Construction of Quot schemes

5.5.1. Notions of projectivity. Let S be a noetherian scheme. Recall that as defined by Grothendieck, a morphism $X \to S$ is called a *projective morphism* if there exists a coherent sheaf E on S, together with a closed embedding of X into $\mathbb{P}(E) = \mathbf{Proj}\, \mathrm{Sym}_{\mathcal{O}_S} E$ over S. Equivalently, $X \to S$ is projective when it is proper and there exists a relatively very ample line bundle L on X over S. These conditions are related by taking L to be the restriction of $\mathcal{O}_{\mathbb{P}(E)}(1)$ to X, or in the reverse direction, taking E to be the direct image of L on S. A morphism $X \to S$ is called *quasi-projective* if it factors as an open embedding $X \hookrightarrow Y$ followed by a projective morphism $Y \to S$.

A stronger version of projectivity was introduced by Altman and Kleiman: a morphism $X \to S$ of noetherian schemes is called *strongly projective* (respectively, *strongly quasi-projective*) if there exists a vector bundle E on S together with a closed embedding (respectively, a locally closed embedding) $X \subset \mathbb{P}(E)$ over S.

Finally, the strongest version of (quasi-)projectivity is as follows (used for example in the textbook [**Har77**] by Hartshorne): a morphism $X \to S$ of noetherian schemes is projective in the strongest sense if X admits a (locally-)closed embedding into $\mathbb{P}^n_S$ for some n.

Note that none of the three versions of projectivity is local over the base S.

Exercises. (i) Give examples to show that the above three notions of projectivity are in general distinct.

(ii) Show that if $X \to S$ is projective and flat, where S is noetherian, then $X \to S$ is strongly projective.

(iii) Note that if every coherent sheaf of $\mathcal{O}_S$-modules is the quotient of a vector bundle, then projectivity over the base S is equivalent to strong projectivity. If S admits an ample line bundle (for example, if S is quasi-projective over an affine base), then all three notions of projectivity over S are equivalent to each other.

5.5.2. Main existence theorems. Grothendieck's original theorem on Quot schemes, whose proof is outlined in [FGA] TDTE-IV, is the following.

THEOREM 5.14. (Grothendieck) Let S be a noetherian scheme, $\pi : X \to S$ a projective morphism, and L a relatively very ample line bundle on X. Then for any coherent $\mathcal{O}_X$-module E and any polynomial $\Phi \in \mathbb{Q}[\lambda]$, the functor $\mathfrak{Quot}^{\Phi,L}_{E/X/S}$ is representable by a projective S-scheme $\mathrm{Quot}^{\Phi,L}_{E/X/S}$.

Altman and Kleiman gave a complete and detailed proof of the existence of Quot schemes in [**AK80**]. They could remove the noetherian hypothesis, by instead assuming strong (quasi-)projectivity of $X \to S$ together with an assumption about the nature of the coherent sheaf E, and deduce that the scheme $\mathrm{Quot}^{\Phi,L}_{E/X/S}$ is then strongly (quasi-)projective over S.

For simplicity, in these lecture notes we state and prove the result in [**AK80**] in the noetherian context.

THEOREM 5.15. (Altman-Kleiman) Let S be a noetherian scheme, X a closed subscheme of $\mathbb{P}(V)$ for some vector bundle V on S, $L = \mathcal{O}_{\mathbb{P}(V)}(1)|_X$, E a coherent quotient sheaf of $\pi^*(W)(\nu)$ where W is a vector bundle on S and ν is an integer, and $\Phi \in \mathbb{Q}[\lambda]$. Then the functor $\mathfrak{Quot}^{\Phi,L}_{E/X/S}$ is representable by a scheme $\mathrm{Quot}^{\Phi,L}_{E/X/S}$ which can be embedded over S as a closed subscheme of $\mathbb{P}(F)$ for some vector bundle F on S.

The vector bundle F can be chosen to be an exterior power of the tensor product of W with a symmetric power of V.

Taking both V and W to be trivial in the above, we get the following.

THEOREM 5.16. If S is a noetherian scheme, X is a closed subscheme of $\mathbb{P}^n_S$ for some $n \geq 0$, $L = \mathcal{O}_{\mathbb{P}^n_S}(1)|_X$, E is a coherent quotient sheaf of $\bigoplus^p \mathcal{O}_X(\nu)$ for some integers $p \geq 0$ and ν, and $\Phi \in \mathbb{Q}[\lambda]$, then the functor $\mathfrak{Quot}^{\Phi,L}_{E/X/S}$ is representable by a scheme $\mathrm{Quot}^{\Phi,L}_{E/X/S}$ which can be embedded over S as a closed subscheme of $\mathbb{P}^r_S$ for some $r \geq 0$.

The rest of this section is devoted to proving Theorem 5.15, with an extra noetherian hypothesis. At the end, we will remark on how the proof also gives us the original version of Grothendieck.

5.5.3. Reduction to the case of $\mathfrak{Quot}^{\Phi,L}_{\pi^*W/\mathbb{P}(V)/S}$. It is enough to prove Theorem 5.15 in the special case that $X = \mathbb{P}(V)$ and $E = \pi^*(W)$ where V and W are vector bundles on S, as a consequence of the next lemma.

LEMMA 5.17. (i) Let ν be any integer. Then tensoring by L^ν gives an isomorphism of functors from $\mathfrak{Quot}^{\Phi,L}_{E/X/S}$ to $\mathfrak{Quot}^{\Psi,L}_{E(\nu)/X/S}$ where the polynomial $\Psi \in \mathbb{Q}[\lambda]$ is defined by $\Psi(\lambda) = \Phi(\lambda + \nu)$.

(ii) Let $\phi : E \to G$ be a surjective homomorphism of coherent sheaves on X. Then the corresponding natural transformation $\mathfrak{Quot}^{\Phi,L}_{G/X/S} \to \mathfrak{Quot}^{\Phi,L}_{E/X/S}$ is a closed embedding.

PROOF. The statement (i) is obvious. The statement (ii) just says that given any locally noetherian scheme T and a family $\langle \mathcal{F}, q\rangle \in \mathfrak{Quot}^{\Phi,L}_{E/X/S}(T)$, there exists a closed subscheme $T' \subset T$ with the following universal property: for any locally noetherian scheme U and a morphism $f : U \to T$, the pulled back homomorphism of $\mathcal{O}_{X_U}$-modules $q_U : E_U \to \mathcal{F}_U$ factors via the pulled back homomorphism $\phi_U : E_U \to G_U$ if and only if $U \to T$ factors via $T' \hookrightarrow T$. This is satisfied by taking T' to be the vanishing scheme for the composite homomorphism $\ker(\phi) \hookrightarrow E$ $xrightarrowq \to \mathcal{F}$ of coherent sheaves on X_T (see Remark 5.9), which makes sense here as both $\ker(\phi)$ and $\mathcal{F}$ are coherent on X_T and $\mathcal{F}$ is flat over T. $\square$

Therefore if $\mathfrak{Quot}^{\Phi,L}_{\pi^* W/\mathbb{P}(V)/S}$ is representable, then for any coherent quotient E of $\pi^* W(\nu)|_X$, we can take $\mathrm{Quot}^{\Phi,L}_{E/X/S}$ to be a closed subscheme of $\mathrm{Quot}^{\Phi,L}_{\pi^* W/\mathbb{P}(V)/S}$.

5.5.4. Use of m-regularity. We consider the sheaf $E = \pi^*(W)$ on $X = \mathbb{P}(V)$ where V is a vector bundle on S, and take $L = \mathcal{O}_{\mathbb{P}(V)}(1)$. For any field k and a k-valued point s of S, we have an isomorphism $\mathbb{P}(V)_s \simeq \mathbb{P}^n_k$ where $n = \mathrm{rank}(V) - 1$, and the restricted sheaf E_s on $\mathbb{P}(V)_s$ is isomorphic to $\bigoplus^p \mathcal{O}_{\mathbb{P}(V)_s}$ where $p = \mathrm{rank}(W)$. It follows from Theorem 5.3 that given any $\Phi \in \mathbb{Q}[\lambda]$, there exists an integer m which depends only on $\mathrm{rank}(V)$, $\mathrm{rank}(W)$ and Φ, such that for any field k and a k-valued point s of S, the sheaf E_s on $\mathbb{P}(V)_s$ is m-regular, and for any coherent quotient $q : E_s \to \mathcal{F}$ on $\mathbb{P}(V)_s$ with Hilbert polynomial Φ, the sheaf $\mathcal{F}$ and the kernel sheaf $\mathcal{G} \subset E_s$ of q are both m-regular. In particular, it follows from the Castelnuovo Lemma 5.1 that for $r \geq m$, all cohomologies $H^i(X_s, E_s(r))$, $H^i(X_s, \mathcal{F}(r))$, and $H^i(X_s, \mathcal{G}(r))$ are zero for $i \geq 1$, and $H^0(X_s, E_s(r))$, $H^0(X_s, \mathcal{F}(r))$, and $H^0(X_s, \mathcal{G}(r))$ are generated by their global sections.

¿From the above it follows by Theorem 5.10 that if T is an S-scheme and $q : E_T \to \mathcal{F}$ is a T-flat coherent quotient with Hilbert polynomial Φ, then we have the following, where $\mathcal{G} \subset E_T$ is the kernel of q.

(*) The sheaves $\pi_{T*}\mathcal{G}(r)$, $\pi_{T*}E_T(r)$, $\pi_{T*}\mathcal{F}(r)$ are locally free of fixed ranks determined by the data n, p, r, and Φ, the homomorphisms $\pi_T{}^*\pi_{T*}(\mathcal{G}(r)) \to \mathcal{G}(r)$, $\pi_T{}^*\pi_{T*}(E_T(r)) \to E_T(r)$, $\pi_T{}^*\pi_{T*}(\mathcal{F}(r)) \to \mathcal{F}(r)$ are surjective, and the higher direct images $R^i\pi_{T*}\mathcal{G}(r)$, $R^i\pi_{T*}E_T(r)$, $R^i\pi_{T*}\mathcal{F}(r)$ are zero, for all $r \geq m$ and $i \geq 1$.

(**) In particular we have the following commutative diagram of locally free sheaves on X_T, in which both rows are exact, and all three vertical maps are surjective.

$$\begin{array}{ccccccccc}
0 & \to & \pi_T{}^*\pi_{T*}(\mathcal{G}(r)) & \to & \pi_T{}^*\pi_{T*}(E_T(r)) & \to & \pi_T{}^*\pi_{T*}(\mathcal{F}(r)) & \to & 0 \\
\downarrow & & \downarrow & & \downarrow & & & & \\
0 & \longrightarrow & \mathcal{G}(r) & \longrightarrow & E(r) & \longrightarrow & \mathcal{F}(r) & \longrightarrow & 0
\end{array}$$

5.5.5. Embedding Quot into Grassmannian. We now fix a positive integer r such that $r \geq m$. Note that the rank of $\pi_{T*}\mathcal{F}(r)$ is $\Phi(r)$ and $\pi_* E(r) = W \otimes_{\mathcal{O}_S} \mathrm{Sym}^r V$. Therefore the surjective homomorphism $\pi_{T*}E_T(r) \to \pi_{T*}\mathcal{F}(r)$ defines an element of the set $\mathfrak{Grass}(W \otimes_{\mathcal{O}_S} \mathrm{Sym}^r V, \Phi(r))(T)$. We thus get a morphism of functors

$$\alpha : \mathfrak{Quot}^{\Phi,L}_{E/X/S} \to \mathfrak{Grass}(W \otimes_{\mathcal{O}_S} \mathrm{Sym}^r V, \Phi(r)).$$

It associates to $q : E_T \to \mathcal{F}$ the quotient $\pi_{T*}(q(r)) : \pi_{T*}E_T(r) \to \pi_{T*}\mathcal{F}(r)$.

The above morphism α is injective because the quotient $q : E_T \to \mathcal{F}$ can be recovered from $\pi_{T*}(q(r)) : \pi_{T*}E_T(r) \to \pi_{T*}\mathcal{F}(r)$ as follows.

If $G = \mathrm{Grass}(W \otimes_{\mathcal{O}_S} \mathrm{Sym}^r V, \Phi(r))$ with projection $p_G : G \to S$, and $u : p_G{}^*E \to \mathcal{U}$ denotes the universal quotient on G with kernel $v : \mathcal{K} \to p_G{}^*E$, then the homomorphism $\pi_T{}^*\pi_{T*}(\mathcal{G}(r)) \to \pi_T{}^*\pi_{T*}E_T(r)$ can be recovered from the morphism $T \to G$ as the pull-back of $v : \mathcal{K} \to p_G{}^*E$. Let h be the composite $\pi_T{}^*\pi_{T*}(\mathcal{G}(r)) \to \pi_T{}^*\pi_{T*}(E_T(r)) \to E_T(r)$. As a consequence of the properties of the diagram **(**)**, the following is a right exact sequence on X_T

$$\pi_T{}^*\pi_{T*}(\mathcal{G}(r)) \xrightarrow{h} E_T(r) \xrightarrow{q(r)} \mathcal{F} \to 0$$

and so $q(r) : E_T(r) \to \mathcal{F}(r)$ can be recovered as the cokernel of h. Finally, twisting by $-r$, we recover q, proving the desired injectivity of the morphism of functors $\alpha : \mathfrak{Quot}^{\Phi,L}_{E/X/S} \to \mathfrak{Grass}(W \otimes_{\mathcal{O}_S} \mathrm{Sym}^r V, \Phi(r))$.

5.5.6. Use of flattening stratification. We will next prove that

$$\alpha : \mathfrak{Quot}^{\Phi,L}_{E/X/S} \to \mathfrak{Grass}(W \otimes_{\mathcal{O}_S} \mathrm{Sym}^r V, \Phi(r))$$

is relatively representable. In fact, we will show that given any locally noetherian S-scheme T and a surjective homomorphism $f : W_T \otimes_{\mathcal{O}_T} \mathrm{Sym}^r V_T \to \mathcal{J}$ where $\mathcal{J}$ is a locally free $\mathcal{O}_T$-module of rank $\Phi(r)$, there exists a locally closed subscheme T' of T with the following universal property **(F)**:

(F) Given any locally noetherian S-scheme Y and an S-morphism $\phi : Y \to T$, let f_Y be the pull-back of f, and let $\mathcal{K}_Y = \ker(f_Y) = \phi^* \ker(f)$. Let $\pi_Y : X_Y \to Y$ be the projection, and let $h : \pi_Y{}^*\mathcal{K}_Y \to E_Y$ be the composite map

$$\pi_Y{}^*\mathcal{K}_Y \to \pi_Y{}^*(W \otimes_{\mathcal{O}_S} \mathrm{Sym}^r V) = \pi_Y{}^*\pi_{Y*}E_Y \to E_Y.$$

Let $q : E_Y \to \mathcal{F}$ be the cokernel of h. Then $\mathcal{F}$ is flat over Y with its Hilbert polynomial on all fibers equal to Φ if and only if $\phi : Y \to T$ factors via $Y' \hookrightarrow Y$.

The existence of such a locally closed subscheme T' of T is given by Theorem 5.13, which shows that T' is the stratum corresponding to Hilbert polynomial Φ for the flattening stratification over T for the sheaf $\mathcal{F}$ on X_T.

When we take T to be $\mathrm{Grass}(W \otimes_{\mathcal{O}_S} \mathrm{Sym}^r V, \Phi(r))$ with universal quotient $u : p_G{}^*E \to \mathcal{U}$, the corresponding locally closed subscheme T' represents the functor $\mathfrak{Quot}^{\Phi,L}_{E/X/S}$ by its construction.

Hence we have shown that $\mathfrak{Quot}^{\Phi,L}_{E/X/S}$ is represented by a locally closed subscheme of $\mathrm{Grass}(W \otimes_{\mathcal{O}_S} \mathrm{Sym}^r V, \Phi(r))$. As $\mathrm{Grass}(W \otimes_{\mathcal{O}_S} \mathrm{Sym}^r V, \Phi(r))$ embeds as a closed subscheme of $\mathbb{P}(\bigwedge^{\Phi(r)} W \otimes_{\mathcal{O}_S} \mathrm{Sym}^r V)$, we get a locally closed embedding of S-schemes

$$\mathrm{Quot}^{\Phi,L}_{E/X/S} \subset \mathbb{P}(\bigwedge^{\Phi(r)} (W \otimes_{\mathcal{O}_S} \mathrm{Sym}^r V)).$$

In particular, the morphism $\mathrm{Quot}^{\Phi,L}_{E/X/S} \to S$ is separated and of finite type.

5.5.7. Valuative criterion for properness. The original reference for the following argument is EGA IV (2) 2.8.1.

The functor $\mathfrak{Quot}^{\Phi,L}_{E/X/S}$ satisfies the following valuative criterion for properness over S: given any discrete valuation ring R over S with quotient field K, the restriction map

$$\mathfrak{Quot}^{\Phi,L}_{E/X/S}(\operatorname{Spec} R) \to \mathfrak{Quot}^{\Phi,L}_{E/X/S}(\operatorname{Spec} K)$$

is bijective. This can be seen as follows. Given any coherent quotient $q : E_K \to \mathcal{F}$ on X_R which defines an element $\langle \mathcal{F}, q\rangle$ of $\mathfrak{Quot}^{\Phi,L}_{E/X/S}(\operatorname{Spec} K)$, let $\overline{\mathcal{F}}$ be the image of the composite homomorphism $E_R \to j_*(E_K) \to j_*\mathcal{F}$ where $j : X_K \hookrightarrow X_R$ is the open inclusion. Let $\overline{q} : E_R \to \overline{\mathcal{F}}$ be the induced surjection. Then we leave it to the reader to verify that $\langle \mathcal{F}, q\rangle$ is an element of $\mathfrak{Quot}^{\Phi,L}_{E/X/S}(\operatorname{Spec} R)$ which maps to $\langle \mathcal{F}, q\rangle$, and is the unique such element. (Use the basic fact that being flat over a dvr is the same as being torsion-free.)

As S is noetherian and as we have already shown that $\operatorname{Quot}^{\Phi,L}_{E/X/S} \to S$ is of finite type, it follows that $\operatorname{Quot}^{\Phi,L}_{E/X/S} \to S$ is a proper morphism. Therefore the embedding of $\operatorname{Quot}^{\Phi,L}_{E/X/S}$ into $\mathbb{P}(\bigwedge^{\Phi(r)}(W \otimes_{\mathcal{O}_S} \operatorname{Sym}^r V))$ is a closed embedding.

This completes the proof of Theorem 5.15. □

5.5.8. The version of Grothendieck. We now describe how to get Theorem 5.14 from the above proof. As S is noetherian, we can find a common m such that given any field-valued point $s : \operatorname{Spec} k \to S$ and a coherent quotient $q : E_s \to \mathcal{F}$ on X_s with Hilbert polynomial Φ, the sheaves $E_s(r)$, $\mathcal{F}(r)$, $\mathcal{G}(r)$ (where $\mathcal{G} = \ker(q)$) are generated by global sections and all their higher cohomologies vanish, whenever $r \geq m$. This follows from the theory of m-regularity, and semi-continuity.

Because we have such a common m, we get as before an injective morphism from the functor $\mathfrak{Quot}^{\Phi,L}_{E/X/S}$ into the Grassmannian functor $\mathfrak{Grass}(\pi_*E(r), \Phi(r))$. The sheaf $\pi_*E(r)$ is coherent, but need not be the quotient of a vector bundle on S. Consequently, the scheme $\operatorname{Grass}(\pi_*E(r), \Phi(r))$ is projective over the base, but not necessarily strongly projective.

Finally, the use of flattening stratification, which can be made over an affine open cover of S, gives a locally closed subscheme of $\operatorname{Grass}(\pi_*E(r), \Phi(r))$ which represents $\mathfrak{Quot}^{\Phi,L}_{E/X/S}$, which is in fact a closed subscheme by the valuative criterion. Thus, we get $\operatorname{Quot}^{\Phi,L}_{E/X/S}$ as a projective scheme over S.

5.6. Some variants and applications

5.6.1. Quot scheme in quasi-projective case.

Exercise. Let $\pi : Z \to S$ be a proper morphism of noetherian schemes. Let $Y \subset Z$ be a closed subscheme, and let $\mathcal{F}$ be a coherent sheaf on Z. Then there exists an open subscheme $S' \subset S$ with the universal property that a morphism $T \to S$ factors through S' if and only if the support of the pull-back $\mathcal{F}_T$ on $Z_T = Z \times_S T$ is disjoint from $Y_T = Y \times_S T$.

Exercise. As a consequence of the above, show the following: If $\pi : Z \to S$ is a proper morphism with S noetherian, if $X \subset Z$ is an open subscheme, and if E is a coherent sheaf on Z, then $\mathfrak{Quot}_{E|_X/X/S}$ is an open subfunctor of $\mathfrak{Quot}_{E/Z/S}$.

With the above preparation, the construction of a quot scheme extends to the strongly quasi-projective case, to give the following.

THEOREM 5.18. (Altman and Kleiman) Let S be a noetherian scheme, X a locally closed subscheme of $\mathbb{P}(V)$ for some vector bundle V on S, $L = \mathcal{O}_{\mathbb{P}(V)}(1)|_X$, E a coherent quotient sheaf of $\pi^*(W)(\nu)|_X$ where W is a vector bundle on S and ν is an integer, and $\Phi \in \mathbb{Q}[\lambda]$. Then the functor $\mathfrak{Quot}^{\Phi,L}_{E/X/S}$ is representable by a scheme $\mathrm{Quot}^{\Phi,L}_{E/X/S}$ which can be embedded over S as a locally closed subscheme of $\mathbb{P}(F)$ for some vector bundle F on S. Moreover, the vector bundle F can be chosen to be an exterior power of the tensor product of W with a symmetric power of V.

PROOF. Let $\overline{X} \subset \mathbb{P}(V)$ be the schematic closure of $X \subset \mathbb{P}(V)$, and let $\overline{E}$ be the coherent sheaf on $\overline{X}$ defined as the image of the composite homomorphism

$$\pi^*(W)(\nu)|_{\overline{X}} \to j_*(\pi^*(W)(\nu)|_X) \to j_*E.$$

Then we get a quotient $\pi^*(W)(\nu)|_{\overline{X}} \to \overline{E}$ which restricts on $X \subset \overline{X}$ to the given quotient $\pi^*(W)(\nu)|_X \to E$. Therefore by the above exercise, $\mathfrak{Quot}_{E/X/S}$ is an open subfunctor of $\mathfrak{Quot}_{\overline{E}/\overline{X}/S}$. Now the result follows from the Theorem 5.15. □

In order to extend Grothendieck's construction of a quot scheme to the quasi-projective case, one first needs the following lemma which is of independent interest.

LEMMA 5.19. Any coherent sheaf on an open subscheme of a noetherian scheme S can be prolonged to a coherent sheaf on all of S.

PROOF. First consider the case where $S = \operatorname{Spec} A$ is affine, and let $j : U \hookrightarrow S$ denote the inclusion. The quasi-coherent sheaf $j_*(\mathcal{F})$ corresponds to the A-module $M = H^0(S, j_*(\mathcal{F}))$, in the sense that $j_*(\mathcal{F}) = M^\sim$. Given any $u \in U$, there exist finitely many elements $e_1, \ldots, e_n \in M$ which generate the fiber $\mathcal{F}_u$ regarded as a vector space over the residue field $\kappa(u)$. By Nakayama these elements will generate the stalks of $\mathcal{F}$ in an open neighbourhood of u in U. Therefore by the noetherian hypothesis, there exist finitely many elements $e_1, \ldots, e_r \in M$ which generate the stalk of $\mathcal{F}$ at each point of U. If $N \subset M$ is the submodule generated by these elements, then $\mathcal{G} = N^\sim$ is a coherent prolongation of $\mathcal{F}$ to $S = \operatorname{Spec} A$, proving the result in the affine case.

In the general case, by the noetherian condition there exists a maximal coherent prolongation $(U', \mathcal{F}')$ of $\mathcal{F}$. Then unless $U' = S$, we can obtain a further prolongation of $\mathcal{F}'$ by using the affine case. For, if $u \in S - U'$, we can take an affine open subscheme V containing u, and a coherent prolongation $\mathcal{G}'$ of $\mathcal{F}'|_{U' \cap V}$ to all of V, and then glue together $\mathcal{G}'$ and $\mathcal{F}'$ along $U' \cap V$ to further prolong $\mathcal{F}'$ to $U' \cup V$, contradicting the maximality of $(U', \mathcal{F}')$. □

THEOREM 5.20. (Grothendieck) Let S be a noetherian scheme, X a quasi-projective scheme over S, L a line bundle on X which is relatively very ample over S, E a quotient sheaf on X, and $\Phi \in \mathbb{Q}[\lambda]$. Then the functor $\mathfrak{Quot}^{\Phi,L}_{E/X/S}$ is representable by a scheme $\mathrm{Quot}^{\Phi,L}_{E/X/S}$ which is quasi-projective over S.

PROOF. By definition of quasi-projectivity of $X \to S$, note that X can be embedded over S as a locally closed subscheme of $\mathbb{P}(V)$ for some coherent sheaf V on S, such that L is isomorphic to $\mathcal{O}_{\mathbb{P}(V)}(1)|_X$. Let $\overline{X} \subset \mathbb{P}(V)$ be the schematic closure of X in $\mathbb{P}(V)$. This is a projective scheme over S, and X is embedded as an open subscheme in it. By Lemma 5.19 the coherent sheaf E has a coherent prolongation $\overline{E}$ to $\overline{X}$. For any such prolongation $\overline{E}$, the functor $\mathfrak{Quot}_{E/X/S}$ is an open subfunctor of $\mathfrak{Quot}_{\overline{E}/\overline{X}/S}$. Therefore the desired result now follows from Theorem 5.14. □

5.6.2. Scheme of morphisms. We recall the following basic facts about flatness.

LEMMA 5.21. **(1)** Any finite-type flat morphism between noetherian schemes is open.

(2) Let $\pi : Y \to X$ be a finite-type morphism of noetherian schemes. Then all $y \in Y$ such that π is flat at y (that is, $\mathcal{O}_{Y,y}$ is a flat $\mathcal{O}_{X,\pi(y)}$-module) form an open subset of Y.

(3) Let S be a noetherian scheme, and let $f : X \to S$ and $g : Y \to S$ be finite type flat morphisms. Let $\pi : Y \to X$ be any morphism such that $g = f \circ \pi$. Let $y \in Y$, let $x = \pi(y)$, and let $s = g(y) = f(x)$. If the restricted morphism $\pi_s : Y_s \to X_s$ between the fibers over s is flat at $y \in Y_s$, then π is flat at $y \in Y$.

PROOF. See for example Altman and Kleiman [**AK70**] Chapter V. The statement **(3)** is a consequence of what is known as the **local criterion for flatness.** □

THEOREM 5.22. Let S be a noetherian scheme, and let $f : X \to S$ and $g : Y \to S$ be proper flat morphisms. Let $\pi : Y \to X$ be any projective morphism with $g = f \circ \pi$. Then S has open subschemes $S_2 \subset S_1 \subset S$ with the following universal properties:

(a) For any locally noetherian S-scheme T, the base change $\pi_T : Y_T \to X_T$ is a flat morphism if and only if the structure morphism $T \to S$ factors via S_1. (This does not need π to be projective.)

(b) For any locally noetherian S-scheme T, the base change $\pi_T : Y_T \to X_T$ is an isomorphism if and only if the structure morphism $T \to S$ factors via S_2.

PROOF. **(a)** By Lemma 5.21.(2), all $y \in Y$ such that π is flat at y form an open subset $Y' \subset Y$. Then $S_1 = S - g(Y - Y')$ is an open subset of S as g is proper. We give S_1 the open subscheme structure induced from S. It follows from the local criterion of flatness (Lemma 5.21.(3)) that S_1 exactly consists of all $s \in S$ such that the restricted morphism $\pi_s : Y_s \to X_s$ between the fibers over s is flat. Therefore again by the local criterion of flatness, S_1 has the desired universal property.

(b) Let $\pi_1 : Y_1 \to X_1$ be the pull-back of π under the inclusion $S_1 \hookrightarrow S$. Let L be a relatively very ample line bundle for the projective morphism $\pi_1 : Y_1 \to X_1$. Then by noetherianness there exists an integer $m \geq 1$ such that $\pi_{1*}L^m$ is generated by its global sections and $R^i\pi_{1*}L^m = 0$ for all $i \geq 1$. By flatness of π_1, it follows that $\pi_{1*}L^m$ is a locally free sheaf. Let $U \subset X_1$ be the open subschemes such that $\pi_{1*}L^m$ is of rank 1 on U. Finally, let $S_2 = S_1 - f(X_1 - U)$, which is open as f is proper. We give S_2 the induced open subscheme structure, and leave it to the reader to verify that it indeed has the required universal property **(b)**. □

If X and Y are schemes over a base S, then for any S-scheme T, an **S-morphism from X to Y parametrised by T** will mean a T-morphism from $X \times_S T$ to $Y \times_S T$. The set of all such will be denoted by $\mathfrak{Mor}_S(X,Y)(T)$. The association $T \mapsto \mathfrak{Mor}_S(X,Y)(T)$ defines a contravariant functor $\mathfrak{Mor}_S(X,Y)$ from S-schemes to *Sets*.

Exercise. Let k be a field, let $S = \operatorname{Spec} k[[t]]$, $X = \operatorname{Spec} k = \operatorname{Spec}(k[[t]]/(t))$, and let $Y = \mathbb{P}^1_S$. Is $\mathfrak{Mor}_S(X,Y)$ representable?

THEOREM 5.23. Let S be a noetherian scheme, let X be a projective scheme over S, and let Y be quasi-projective scheme over S. Assume moreover that X is flat over S. Then the functor $\mathfrak{Mor}_S(X,Y)$ is representable by an open subscheme $\operatorname{Mor}_S(X,Y)$ of $\operatorname{Hilb}_{X\times_S Y/S}$.

PROOF. We can associate to each morphism $f : X_T \to Y_T$ (where T is a scheme over S) its graph $\Gamma_T(f) \subset (X \times_S Y)_T$, which is closed in $(X \times_S Y)_T$ by separatedness of $Y \to S$. We regard $\Gamma_T(f)$ as a closed subscheme of $(X \times_S Y)_T$ which is isomorphic to X under the graph morphism $(\mathrm{id}_X, f) : X \to \Gamma_T(f)$ and the projection $\Gamma_T(f) \to X$, which are inverses to each other. As X is proper and flat over S, so is $\Gamma_T(f)$, therefore this defines a set-map $\Gamma_T : \mathfrak{Mor}_S(X,Y)(T) \to \mathfrak{Hilb}_{X\times_S Y/S}(T)$ which is functorial in T, so we obtain a morphism of functors

$$\Gamma : \mathfrak{Mor}_S(X,Y) \to \mathfrak{Hilb}_{X\times_S Y/S}\,.$$

Given any element of $\mathfrak{Hilb}_{X\times_S Y/S}(T)$, represented by a family $Z \subset (X \times_S Y)_T$, it follows by applying Theorem 5.22.(b) to the projection $Z \to X$ that T has an open subscheme T' with the following universal property: for any base-change $U \to T$, the pull-back $Z_U \subset (X \times_S Y)_U$ maps isomorphically on to X_U under the projection $p : (X \times_S Y)_U \to X_U$ if and only if $U \to T$ factors via T'. Note therefore that over T', the scheme $Z_{T'}$ will be the graph of a uniquely determined morphism $X_{T'} \to Y_{T'}$.

This shows that the morphism of functors $\Gamma : \mathfrak{Mor}_S(X,Y) \to \mathfrak{Hilb}_{X\times_S Y/S}$ is a representable morphism which is an open embedding. Therefore a representing scheme $\operatorname{Mor}_S(X,Y)$ for $\mathfrak{Mor}_S(X,Y)$ exists as an open subscheme of $\operatorname{Hilb}_{X\times_S Y/S}$. $\square$

Exercise. Let S be a noetherian scheme and $X \to S$ a flat projective morphism. Consider the set-valued contravariant functor $\mathfrak{Aut}_{X/S}$ on locally noetherian S-schemes, which associates to any T the set of all automorphisms of X_T over T. Show that this functor is representable by an open subscheme of $\operatorname{Mor}_S(X,X)$.

Exercise. Let S be a noetherian scheme and $\pi : Z \to X$ a morphism of S-schemes, where X is proper over S and Z is quasi-projective over S. Consider the set-valued contravariant functor $\Pi_{Z/X/S}$ on locally noetherian S-schemes, which associates to any T the set of all sections of $\pi_T : Z_T \to X_T$. Show that this functor is representable by an open subscheme of $\operatorname{Hilb}_{Z/S}$.

5.6.3. Quotient by a flat projective equivalence relation. Let X be a scheme over a base S. A *schematic equivalence relation* on X over S will mean an S-scheme R together with a morphism $f : R \to X \times_S X$ over S such that for any S-scheme T the set map $f(T) : R(T) \to X(T) \times X(T)$ is injective and its image is the graph of an equivalence relation on $X(T)$. (Here, we denote by $Z(T)$ the set $Mor_S(T,Z) = h_Z(T)$ of all T-valued points of Z, where Z and T are S-schemes.)

We will say that a morphism $q : X \to Q$ of S-schemes is a *quotient* for a schematic equivalence relation $f : R \to X \times_S X$ over S if q is a *co-equaliser* for the component morphisms $f_1, f_2 : R \rightrightarrows X$ of $f : R \to X \times_S X$. This means $q \circ f_1 = q \circ f_2$, and given any S-scheme Z and an S-morphism $g : X \to Z$ such that $g \circ f_1 = g \circ f_2$, there exists a unique S-morphism $h : Q \to Z$ such that $g = h \circ q$. A schematic quotient $q : X \to Q$, when it exists, is unique up to a unique isomorphism. **Exercise:** A schematic quotient, when it exists, is necessarily an epimorphism in the category of S-schemes.

Caution. Even if $q : X \to Q$ is a schematic quotient for R, for a given T the map $q(T) : X(T) \to Q(T)$ may not be a quotient for $R(T)$ in the category of sets. The map $q(T)$ may fail to be surjective, and moreover it may identify two distinct equivalence classes. **Exercise:** Give examples where such phenomena occur.

We will say that the quotient $q : X \to Q$ is *effective* if the induced morphism $(f_1, f_2) : R \to X \times_Q X$ is an isomorphism of S-schemes. In particular, it will ensure that distinct equivalence classes do not get identified under $q(T) : X(T) \to Q(T)$. But $q(T)$ can still fail to be surjective, as in the following example.

Exercise. Let $S = \operatorname{Spec} \mathbb{Z}$, and let $X \subset \mathbf{A}^n_{\mathbb{Z}}$ be the complement of the zero section of $\mathbf{A}^n_{\mathbb{Z}}$. Note that for any ring B, an element of $X(\operatorname{Spec} B)$ is a vector $u \in B^n$ such that at least one component of u is invertible in B. Show that $X \times_S X$ has a closed subscheme R whose B-valued points for any ring B are all pairs $(u, v) \in X(\operatorname{Spec} B) \times X(\operatorname{Spec} B)$ such that there exists an invertible element $\lambda \in B^\times$ with $\lambda u = v$. Show that an effective quotient $q : X \to Q$ exists, where $Q = \mathbb{P}^{n-1}_{\mathbb{Z}}$. However, show that q does not admit a global section, and so $q(Q) : X(Q) \to Q(Q)$ is not surjective.

The famous example by Hironaka (see in [**Har77**, Appendix B, Example 3.4.1]) of a non-projective smooth complete variety X over $\mathbb{C}$ together with a schematic equivalence relation R (for which the morphisms $f_i : R \to X$ are finite flat, in fact, étale of degree 2) shows that schematic quotients do not always exist. But under the powerful assumption of projectivity, Grothendieck proved an existence result for quotients, to which we devote the rest of this section.

We will need the following elementary lemma from Grothendieck's theory of faithfully flat descent (this is a special case of [**SGA1**, Exposé VIII, Corollary 1.9]). The reader can consult Part 1 for an exposition of descent.

LEMMA 5.24. (1) Any faithfully flat quasi-compact morphism of schemes $f : X \to Y$ is an effective epimorphism, that is, f is a co-equaliser for the projections $p_1, p_2 : X \times_Y X \rightrightarrows X$.

(2) Let $p : D \to H$ be a faithfully flat quasi-compact morphism. Let $Z \subset D$ be a closed subscheme such that

$$p_1^{-1} Z = p_2^{-1} Z \subset D \times_H D$$

where $p_1, p_2 : D \times_H D \rightrightarrows D$ are the projections, and $p_i^{-1} Z$ is the schematic inverse image of Z under p_i. Then there exists a unique closed subscheme Q of H such that $Z = p^{-1} Q \subset D$. By base-change from $p : D \to H$, it follows that the induced morphism $p|_Z : Z \to Q$ is faithfully flat and quasi-compact. $\square$

The idea of using Hilbert schemes to make quotients of flat projective equivalence relations is due to Grothendieck, who used it in his construction of a relative

Picard scheme. In set-theoretic terms, the idea is actually very simple: Let X be a set, and $R \subset X \times X$ an equivalence relation on X. Let H be the power set of X (means the set of all subsets of X), and let $\varphi : X \to H$ be the map which sends $x \in X$ to its equivalence class $[x] \in H$. If $Q \subset H$ is the image of φ, then the induced map $q : X \to Q$ is the quotient of X modulo R in the category of sets. The scheme-theoretic analogue of the above is the following theorem of Grothendieck, where the Hilbert scheme of X plays the role of power set. The first detailed proof appeared in Altman and Kleiman [**AK80**].

THEOREM 5.25. *Let S be a noetherian scheme, and let $X \to S$ be a quasi-projective morphism. Let $f : R \to X \times_S X$ be a schematic equivalence relation on X over S, such that the projections $f_1, f_2 : R \rightrightarrows X$ are proper and flat. Then a schematic quotient $X \to Q$ exists over S. Moreover, Q is quasi-projective over S, the morphism $X \to Q$ is faithfully flat and projective, and the induced morphism $(f_1, f_2) : R \to X \times_Q X$ is an isomorphism (the quotient is effective).*

PROOF. (Following Altman and Kleiman [**AK80**]) The properness of f_i together with separatedness of $X \to S$ implies properness of $f : R \to X \times_S X$. Also, f is functorially injective by definition of a schematic equivalence relation. It follows that f is a closed embedding, which allows us to regard R as a closed subscheme of $X \times_S X$ (**Exercise:** *Any proper morphism of noetherian schemes, which is injective at the level of functor of points, is a closed embedding*). This defines an element (R) of $\mathfrak{H}ilb_{X/S}(X)$, as the projection $p_2|_R = f_2$ is proper and flat.

By Theorem 5.20, there exists a scheme $\mathrm{Hilb}_{X/S}$ which represents the functor $\mathfrak{H}ilb_{X/S}$. As the parameter scheme X is noetherian and as the Hilbert polynomial is locally constant, only finitely many polynomials Φ occur as Hilbert polynomials of fibers of $f_2 : R \to X$ with respect to a chosen relatively very ample line bundle L on X over S. Let H be the finite disjoint union of the corresponding open subschemes $\mathrm{Hilb}^{\Phi,L}_{X/S}$ of $\mathrm{Hilb}_{X/S}$. Then H is a quasi-projective scheme over S as each $\mathrm{Hilb}^{\Phi,L}_{X/S}$ is so by Theorem 5.20. The family $(R) \in \mathfrak{H}ilb_{X/S}(X)$ therefore defines a classifying morphism $\varphi : X \to H$, with the property that $\varphi^* D = R$ where $D \subset X \times_S H$ denotes the restriction to H of the universal family over $\mathrm{Hilb}_{X/S}$, and $\varphi^* D$ denotes $(\mathrm{id}_X \times \varphi)^{-1} D$. Also, note that the projection $p : D \to H$ is proper and flat. If X is non-empty then each fiber of $f_2 : R \to X$ is also non-empty as the diagonal Δ_X is contained in R, and therefore the Hilbert polynomial of each fiber is non-zero. Hence $p : D \to H$ is surjective, and so p is faithfully flat.

For any S-scheme T, it follows from its definition that a T-valued point of D is a pair (x, V) with $x \in X(T)$ and $V \in H(T)$ such that

$$x \in V$$

where the notation "$x \in V$" more precisely means that the graph morphism $(x, \mathrm{id}_T) : T \to X \times_S T$ factors via $V \subset X \times_S T$. With this notation, we will establish the following crucial property:

(***) *For any S-scheme T, and T-valued points $x, y \in X(T)$, the following equivalences hold: $(x, y) \in R(T) \Leftrightarrow x \in \varphi(y) \Leftrightarrow \varphi(x) = \varphi(y) \in H(T)$.*

For this, note that for any $x, y \in X(T)$, the morphism $(x, y) : T \to X \times_S X$ factors as the composite

$$T \xrightarrow{(x,\mathrm{id}_T)} X \times_S T \xrightarrow{\mathrm{id}_X \times y} X \times_S X.$$

As $\varphi^*D = R$ and $(\varphi \circ y)^*D = \varphi(y)$, it follows that $y^*R = \varphi(y)$, in other words, the schematic inverse image of $R \subset X \times_S X$ under $\mathrm{id}_X \times y : X \times_S T \to X \times_S X$ is $\varphi(y) \subset X \times_S T$. Hence the above factorisation of $(x, y) : T \to X \times_S X$ shows that

$$\begin{aligned}(x, y) \in R(T) &\Leftrightarrow \text{the morphism } (x, \mathrm{id}_T) : T \to X \times_S T \text{ factors via } \varphi(y) \subset X \times_S T \\ &\Leftrightarrow x \in \varphi(y).\end{aligned}$$

Moreover, $x \in \varphi(x)$ as $\Delta_X \subset R$. Therefore, if $\varphi(x) = \varphi(y)$ then $x \in \varphi(y)$.

It now only remains to prove that if $(x, y) \in R(T)$ then $\varphi(x) = \varphi(y)$, that is, the subschemes $\varphi(x)$ and $\varphi(y)$ of $X \times_S T$ are identical. Note that $\varphi(x) = (\varphi \circ x)^*D = x^*\varphi^*D = x^*R$ and similarly $\varphi(y) = y^*R$, therefore we wish to show that $x^*R = y^*R$. To show this in terms of functor of points, for any T-scheme $u : U \to T$ we just have to show that $(x^*R)(U) = (y^*R)(U)$ as subsets of $(X \times_S T)(U)$. As x^*R is the inverse image of R under $\mathrm{id}_X \times x : X \times_S T \to X \times_S X$, it follows that a U-valued point of x^*R is the same as an element $z \in X(U)$ such that $(\mathrm{id}_X \times x) \circ (z, u) \in R(U)$. But as $(\mathrm{id}_X \times x) \circ (z, u) = (z, x \circ u)$, it follows that

$$z \in (x^*R)(U) \Leftrightarrow (z, x \circ u) \in R(U).$$

As $R(U)$ is an equivalence relation on the set $X(U)$, and as by assumption $(x, y) \in R(T)$, we have $(x \circ u, y \circ u) \in R(U)$, and so by transitivity we have

$$z \in (x^*R)(U) \Leftrightarrow (z, x \circ u) \in R(U) \Leftrightarrow (z, y \circ u) \in R(U) \Leftrightarrow z \in (y^*R)(U).$$

Hence the subschemes x^*R and y^*R of $X \times_S T$ have the same U-valued points for any T-scheme U, and therefore $x^*R = y^*R$, as was to be shown. This completes the proof of the assertion **(***)**.

The graph morphism $(\mathrm{id}_X, \varphi) : X \to X \times_S H$ is a closed embedding as H is separated over S. As $\Delta_X \subset R$ and as $\varphi^*D = R$, it follows that (id_X, φ) factors through $D \subset X \times_S H$. Thus, we get a closed subscheme $\Gamma_\varphi \subset D$, which is the isomorphic image of X under (id_X, φ). We wish to apply the Lemma 5.24.(2) to the faithfully flat quasi-compact morphism $p : D \to H$ and the closed subscheme $Z = \Gamma_\varphi \subset D$.

Any T-valued point of Γ_φ is a pair $(x, \varphi(x)) \in D(T)$ where $x \in X(T)$. Any T-valued point of $D \times_H D$ is a triple (x, y, V) where $x, y \in X(T)$ and $V \in H(T)$ such that $x, y \in V$. Under the projections $p_1, p_2 : D \times_H D \rightrightarrows D$, we have $p_1(x, y, V) = (x, V)$ and $p_2(x, y, V) = (y, V)$. We now have

$$\begin{aligned} p_1(x, y, V) \in \Gamma_\varphi(T) &\Leftrightarrow (x, V) \in \Gamma_\varphi(T) \text{ and } y \in V \\ &\Leftrightarrow V = \varphi(x) \text{ and } y \in V \\ &\Leftrightarrow y \in \varphi(x) = V \\ &\Leftrightarrow \varphi(y) = \varphi(x) = V \quad \text{(by the property } (\ast\ast\ast)).\end{aligned}$$

Similarly, we have $p_2(x, y, V) \in \Gamma_\varphi(T)$ if and only if $\varphi(y) = \varphi(x) = V$. Therefore, $p_1(x, y, V) \in \Gamma_\varphi(T)$ if and only if $p_2(x, y, V) \in \Gamma_\varphi(T)$. This holds for all T-valued points for all S-schemes T, and so $p_1^{-1}\Gamma_\varphi = p_2^{-1}\Gamma_\varphi \subset D \times_H D$. Therefore by Lemma 5.24(2) there exists a unique closed subscheme $Q \subset H$ such that Γ_φ is the pull-back of Q under $D \to H$. Let $p : \Gamma_\varphi \to Q$ be the morphism induced by the restriction to Γ_φ of $p : D \to H$. Let $q : X \to Q$ be defined as the composite

$$X \xrightarrow{(\mathrm{id}_X, \varphi)} \Gamma_\varphi \xrightarrow{p} Q.$$

Then note that the composite $X \xrightarrow{q} Q \hookrightarrow H$ equals φ.

We will now show that $q : X \to Q$ as defined above is the desired quotient of X by R, with the required properties.

(i) Quasi-projectivity of $Q \to S$: This is satisfied as Q is closed in H and H is quasi-projective over S.

(ii) Faithful flatness and projectivity of q: This follows by base change from the faithfully flat projective morphism $p : D \to H$, as the following square is Cartesian.

$$\begin{array}{ccc} X & \xrightarrow{(\mathrm{id}_X, \varphi)} & D \\ {\scriptstyle q}\downarrow & & \downarrow{\scriptstyle p} \\ Q & \longrightarrow & H \end{array}$$

(iii) Exactness of $R \rightrightarrows X \to Q$ and the isomorphism $R \to X \times_Q X$: By **(***)**, for any T-valued points $x, y \in X(T)$, we have $(x, y) \in R(T)$ if and only if $\varphi(x) = \varphi(y)$. This shows that the composite $R \xrightarrow{f_1} X \xrightarrow{q} Q$ equals the composite $R \xrightarrow{f_2} X \xrightarrow{q} Q$, and the induced morphism $(f_1, f_2) : R \to X \times_Q X$ is an isomorphism, by showing these statements hold at the level of functor of points. Under the isomorphism $R \to X \times_Q X$, the morphisms $f_1, f_2 : R \rightrightarrows X$ become the projection morphisms $p_1, p_2 : X \times_Q X \rightrightarrows X$. By Lemma 5.24, the morphism $q : X \to Q$ is a co-equaliser for p_1, p_2, and so q is a co-equaliser for f_1, f_2.

This completes the proof of Theorem 5.25. □

What Altman and Kleiman actually prove in [**AK80**] is a strongly projective form of the above theorem (without a noetherian assumption), using the hypothesis of strong quasi-projectivity in the following places in the above proof: if $X \to S$ is strongly quasi-projective then $H \to S$ will again be so by Theorem 5.18, and therefore Q will be strongly quasi-projective over S. Moreover, $D \to H$ will be strongly projective, and therefore by base-change $X \to Q$ will be strongly projective. In the noetherian case, this gives us the following result.

THEOREM 5.26. Let S be a noetherian scheme, and let $X \to S$ be a strongly quasi-projective morphism. Let $f : R \to X \times_S X$ be a schematic equivalence relation on X over S, such that the projections $f_1, f_2 : R \rightrightarrows X$ are proper and flat. Then a schematic quotient $X \to Q$ exists over S. Moreover, the quotient is effective, the morphism $X \to Q$ is faithfully flat and strongly projective, and Q is strongly quasi-projective over S.

Part 3

Local properties and Hilbert schemes of points

Barbara Fantechi and Lothar Göttsche

Introduction

In Chapter 6 we give an elementary introduction to deformation theory and then use this to study local properties of the Hilbert and Quot scheme, like tangent space, dimension and nonsingularity. The idea is that much of the information about the local properties of a scheme X at a point p can be obtained by studying the maps from the spectrum of local Artin algebras to X, sending the closed point to p. This is particularly useful if the scheme X in a natural way represents a functor, like for instance the Hilbert scheme. Then one has to study the infinitesimal deformations of an object Z parametrized by a point $[Z] \in X$, i.e. families over the spectrum of a local Artin algebra, whose restriction to the closed point is Z. It turns out that much information can be captured by a tangent-obstruction theory, i.e. one has to look at the tangent space of the functor (i.e. the families over $\operatorname{Spec} k[\varepsilon]/(\varepsilon)^2$) and the obstructions to extending a deformation from a local Artin to a slightly larger one. From the tangent-obstruction theory one can then at least obtain a description of a formal neighborhood of the point $[Z]$ in X.

In Chapter 7 we will study the Hilbert schemes $X^{[n]}$ of points which parametrize finite subschemes of length n on a smooth quasiprojective variety X. In the case that X is a surface, the Hilbert schemes of points have recently received a lot of attention because of their nice properties and their relations to many other fields of mathematics, including moduli of sheaves, vertex algebras, and also theoretical physics.

We will construct the Hilbert–Chow morphism from the Hilbert scheme of n points to the n-th symmetric power that sends a scheme to its support with multiplicities. We study some simple cases for small n and we show that the Hilbert schemes of points on a smooth curve are isomorphic to the corresponding symmetric powers. We then study a natural stratification of the Hilbert scheme of points $S^{[n]}$ on a surface and use it to sketch the computation of the Betti numbers. It turns out that the Betti numbers organize themselves in a natural way in a generating function over all n. This implies that the cohomology groups of the $S^{[n]}$ for different n should be intimately tied together. This is for instance provided by the action of the Heisenberg algebra on their direct sum over all n. In the final section of these notes we give a brief outline of this result, which leads to the connection of the Hilbert schemes with vertex algebras.

We would like to thank Leila Khatami and Angelo Vistoli for some help with the commutative algebra.

CHAPTER 6

Elementary Deformation Theory

In this chapter we will give an elementary introduction to the beginnings of deformation theory. Deformation theory is also treated in Part 4 from a more advanced point of view.

CONVENTION 6.0.1. Let k be a field. In this chapter we will assume all schemes to be schemes locally of finite type over k, and by point we will mean a k-valued point, unless otherwise specified.

Let Y be a scheme and $p \in Y$ a point. We want to study the local properties of Y near p like tangent space, (non)singularity and dimension. It turns out that this can be done via studying morphisms $f : \operatorname{Spec} A \to Y$, where A is a local Artin algebra and f maps the closed point to p.

In particular we are interested in the case of the Hilbert scheme. Let X be a projective scheme over a field k and let $Z \subset X$ be a closed subscheme (or more generally X quasiprojective and $Z \subset X$ closed and proper), and we want to study the local properties of $Hilb(X)$ near the point $[Z]$ given by Z. Then the above means that we have to study infinitesimal deformations of Z, i.e. flat families over $\operatorname{Spec} A$ with the closed fibre equal to Z.

6.1. Infinitesimal study of schemes

CONVENTION 6.1.1. For a local ring A we denote by m_A its maximal ideal.

DEFINITION 6.1.2. Let (Art/k) be the category of local Artinian k-algebras with residue field k.

REMARK 6.1.3. Let A be a local k–algebra with residue field k. Then the following are equivalent:

(1) A is Artinian;
(2) A is finite dimensional as a k-vector space;
(3) the maximal ideal of A is finitely generated and nilpotent, i.e. there exists an $N > 0$ such that $m_A^N = 0$.

Geometrically, a k-algebra A is in (Art/k) if $S := \operatorname{Spec} A$ is a k-scheme of finite type such that $S_{red} = \operatorname{Spec} k$.

DEFINITION 6.1.4. A *deformation functor* D will be a covariant functor

$$(Art/k) \to (Sets)$$

such that $D(k)$ is a single point. If $\varphi : B \to A$ is a homomorphism in (Art/k) and $\alpha \in D(A)$, we call $\beta \in D(B)$ a *lifting* or *lift* of α if $\varphi_*(\beta) = \alpha$. A *morphism* $D \to D'$ of deformation functors is a natural transformation; it is an *isomorphism* if $D(A) \to D'(A)$ is bijective for every $A \in (Art/k)$.

The idea is that $D(k)$ is the object that we want to deform, and $D(A)$ is the set of (isomorphism classes of) deformations over $S := \operatorname{Spec} A$.

EXAMPLE 6.1.5. Let X be a scheme and Z a closed subscheme; define $H_{Z,X}(A)$ to be the set of deformations of Z in X over $S := \operatorname{Spec} A$, i.e. of S-flat closed subschemes Z_S of $X \times S$ whose fibre over S_{red} is Z. Clearly $H_{Z,X}(k)$ is a set with one element (namely Z itself), and functoriality is due to the fact that if Z_S is S-flat and $T \to S$ is a morphism, then $Z_T := Z_S \times_S T \subset X \times T$ is T-flat.

EXAMPLE 6.1.6. Many interesting deformation functors arise as infinitesimal local versions of a moduli functor, as follows. Let $F : (schemes) \to (sets)$ be a contravariant functor. Let $p \in F(\operatorname{Spec} k)$ be a point of F. Then we can associate to (F, p) a deformation functor $D_{F,p}$ by letting

$$D_{F,p}(A) := \{\alpha \in F(\operatorname{Spec} A) \mid \alpha|_{\operatorname{Spec}(A/m_A)} = p\}.$$

For instance the functor $H_{Z,X}$ defined above is associated to the point $[Z]$ of the Hilbert functor H_X of X, defined by letting $H_X(S)$ be the set of S-flat closed subschemes W of $X \times S$.

EXAMPLE 6.1.7. Let X be a scheme, and $\mathcal{F}$ a coherent sheaf on X. The Quot functor $Q_{\mathcal{F}}$ associated to $\mathcal{F}$ is defined as follows. For any scheme S, let $p : X \times S \to S$ be the projection; define $Q_{\mathcal{F}}(S)$ to be the set of coherent subsheaves $\mathcal{S}$ of $p^*\mathcal{F}$ such that the quotient $p^*\mathcal{F}/\mathcal{S}$ is flat over S; pullback is defined in the obvious way. See Part 2 for a proof that this functor is representable if X is projective. Note that the Hilbert scheme is the special case where $\mathcal{F} = \mathcal{O}_X$. We will study later the induced deformation functor.

REMARK 6.1.8. In particular if X is a scheme and $p \in X$, we can view p as a point in $h_X(\operatorname{Spec} k)$, where h_X is the functor of points of X defined by $h_X(S) = \operatorname{Hom}(S, X)$. We write $D_{X,p}$ for $D_{h_X,p}$.

To understand a deformation functor D it is crucial to understand what happens for surjective morphisms $\sigma : B \to A$ in (Art/k). In particular we will focus on answering the following questions.

(1) What deformations $F \in D(A)$ lift to elements in $D(B)$, i.e. what is the image of $D(\sigma) : D(B) \to D(A)$?
(2) How unique is such a lift, i.e. how big is a fibre of $D(\sigma)$ when it is nonempty?

DEFINITION 6.1.9. A *small extension*

$$0 \to M \to B \to A \to 0$$

is a surjection $B \to A$ in (Art/k) with kernel M such that $M \cdot m_B = 0$. Note that this is equivalent to saying that the structure of M as a B–module is induced via the natural map $B \to k$ by its structure of a k–vector space.

REMARK 6.1.10. It is easy to show that every surjection in (Art/k) can be written as a finite composition of small extensions; thus, it is enough to restrict oneself to the case where σ is a small extension in answering the questions above.

REMARK 6.1.11. Let D be a deformation functor, and $B \in (Art/k)$. Then $D(B) \to D(k)$ is surjective (i.e. $D(B)$ is nonempty). In fact $D(B)$ contains a distinguished element 0. As B is a k-algebra we have a canonical homomorphism

$k \to B$, such that $k \to B \to k$ is the identity: let 0 be the image of the unique element of $D(k)$ under $D(k) \to D(B)$.

The fundamental example of a deformation functor $D : (Art/k) \to Sets$ is the one defined by a local ring.

DEFINITION 6.1.12. Let (Loc/k) be the category of local k-algebras R with residue field k such that the *tangent space* $T_R = (m_R/m_R^2)^\vee$ is a finite dimensional k–vector space. To every object R of (Loc/k) we can associate a deformation functor h_R defined by $h_R(A) = Hom_k(R, A)$. It is easy to check that this defines a covariant functor from (Loc/k) to the category of deformation functors. For any object R in (Loc/k) we can consider its *completion* $\hat{R} = \varprojlim R/m_R^N$. There is a natural homomorphism $\varphi_R : R \to \hat{\boldsymbol{R}}$, and if it is an isomorphism we say that R is *complete*. We denote by $(CLoc/k)$ the full subcategory of complete rings.

EXERCISE 6.1.13.
(1) Prove that any homomorphism $\psi : R \to S$ in (Loc/k) is local, i.e. $\psi(m_R) \subset m_S$.
(2) Let A be an object in (Art/k); prove that A is complete. Prove that every homomorphism $\psi : R \to A$ in (Loc/k) factors via R/m_R^N for some $N > 0$ (hint: use Remark 6.1.3(3)).
(3) Prove that any homomorphism $\psi : R \to S$ in (Loc/k) induces a homomorphism $\hat{\psi} : \hat{R} \to \hat{S}$ such that $\varphi_S \circ \psi = \hat{\psi} \circ \varphi_R$.
(4) Prove that any homomorphism $\varphi : R \to S$ in (Loc/k) induces a morphism $\varphi^* : h_S \to h_R$. In fact, $R \mapsto h_R$ defines a contravariant functor h from (Loc/k) to the category of deformation functors.
(5) Prove that the restriction of h to $(CLoc/k)$ is fully faithful, that is, induces a bijection on the sets of morphisms.
(6) Prove that the natural map $h_{\hat{R}} \to h_R$ is an isomorphism for every $R \in (Loc/k)$.
(7) Let $\psi : R \to S$ be a morphism in (Loc/k); prove that $\psi^* : h_S \to h_R$ is an isomorphism if and only if $\hat{\psi} : \hat{R} \to \hat{S}$ is an isomorphism.
(8) Let $S = k[[t_1, \ldots, t_n]]$ be the formal power series ring; prove that S is complete, and that for every $R \in (CLoc/k)$ the map $Mor(S, R) \to m_R \times \ldots \times m_R$ defined by $\psi \mapsto (\psi(t_1), \ldots, \psi(t_n))$ is a bijection.

REMARK 6.1.14. Note in particular the case where $R = \mathcal{O}_{X,p}$, where X is a scheme and p is a point of X. Let A be a local Artin ring. Then $h_R(A)$ is the same as the set of morphisms $\operatorname{Spec} A \to X$ sending the closed point to p. In other words, $h_R = D_{X,p}$.

REMARK 6.1.15. If $R = \mathcal{O}_{X,p}$, it is easy to see that $d = \dim T_R$ is the minimal dimension of a smooth scheme containing as closed subscheme an open neighborhood of p in X.

DEFINITION 6.1.16. The dimension of the tangent space of R is called the *embedding dimension* of R.

LEMMA 6.1.17. A homomorphism $\varphi : R \to S$ in $(CLoc/k)$ is surjective if and only if the induced linear map $T_S \to T_R$ is injective.

PROOF. By duality, $T_S \to T_R$ injective is equivalent to $m_R/m_R^2 \to m_S/m_S^2$ surjective; therefore the only if part is obvious. So assume that $T_S \to T_R$ is injective.

It is enough to prove that for every $N > 0$ the induced map $R/m_R^N \to S/m_S^N$ is surjective (this can be either checked directly, or one can apply Prop. II.9.1 in [**Har77**]). The case $N = 2$ is clear by assumption. Since $\mathrm{Sym}^{N-1}(m_S/m_S^2) \to m_S^{N-1}/m_S^N$ is surjective, we deduce that $m_R^{N-1}/m_R^N \to m_S^{N-1}/m_S^N$ is surjective for every $N > 0$; this completes the proof by induction on N. □

PROPOSITION 6.1.18. (1) Let $R \in (Loc/k)$, and $b := (b_1, \dots, b_d)$ be a basis of m_R/m_R^2; write $S := k[[t_1, \dots, t_d]]$. Then the homomorphism $\psi_b : S \to \hat{R}$ such that $\psi(b)(t_i) = b_i$ is surjective, and its kernel J is contained in m_S^2.
(2) R is regular if for some (and hence any) basis b of m_R/m_R^2 the morphism φ_b is an isomorphism, and this is equivalent to $\hat{R}$ being isomorphic to a free power series ring.

PROOF. (1) is a consequence of Lemma 6.1.17 and Exercise 6.1.13(8).

(2) R is regular if and only if $\hat{R}$ is by [**AM69**] Corollary 11.24. Since S is regular and $\dim T_S = \dim T_R$, one has R regular iff $\dim \hat{R} = \dim S$ iff ψ_b is an isomorphism. □

For simplicity of notation we will often in the future also write n instead if m_S for the maximal ideal of S. We now want to see that the behavior of h_R under small extensions can be nicely described in terms of T and of J/nJ. This also means that we can find out a lot about R by looking at h_R.

THEOREM 6.1.19. Let $R \in (Loc/k)$ have embedding dimension d, and assume we have fixed an isomorphism of R with S/J, where S is $k[[t_1, \dots, t_d]]$ and $J \subset m_S^2$. Write $n := m_S$. For every small extension $0 \to M \to B \to A \to 0$ there is a natural exact sequence of sets (this is defined in Remark 6.1.20(1) below)

$$0 \to T \otimes_k M \to h_R(B) \to h_R(A) \xrightarrow{ob} (J/nJ)^\vee \otimes_k M.$$

Furthermore this sequence is functorial in small extensions.

REMARK 6.1.20. (1) The exactness of the sequence at $h_R(A)$ means that an element $a \in h_R(A)$ lifts to B if and only if $ob(a) = 0$, i.e. $ob(a)$ is an obstruction to the lifting. The exactness at $h_R(B)$ means that, if a lifting exists, $T \otimes_k M$ acts transitively on the liftings. Finally that the sequence starts with 0 means that the liftings form an affine space under $T \otimes_k M$.
(2) Functoriality of the sequence means the following: For any morphism of small extensions, i.e. a commutative diagram

$$\begin{array}{ccccccccc} 0 & \longrightarrow & M & \longrightarrow & B & \longrightarrow & A & \longrightarrow & 0 \\ & & \downarrow{\scriptstyle \varphi_M} & & \downarrow{\scriptstyle \varphi_B} & & \downarrow{\scriptstyle \varphi_A} & & \\ 0 & \longrightarrow & M' & \longrightarrow & B' & \longrightarrow & A' & \longrightarrow & 0, \end{array}$$

we get a commutative diagram

$$\begin{array}{ccccccccc} 0 & \longrightarrow & T \otimes_k M & \longrightarrow & h_R(B) & \longrightarrow & h_R(A) & \longrightarrow & (J/nJ)^\vee \otimes_k M \\ & & \downarrow{\scriptstyle id\otimes\varphi_M} & & \downarrow{\scriptstyle \varphi_{B*}} & & \downarrow{\scriptstyle \varphi_{A*}} & & \downarrow{\scriptstyle id\otimes\varphi_M} \\ 0 & \longrightarrow & T \otimes_k M' & \longrightarrow & h_R(B') & \longrightarrow & h_R(A') & \longrightarrow & (J/nJ)^\vee \otimes_k M' \end{array}$$

Note that we require also that φ_{B*} be $T \otimes_k M$-equivariant, where the action of $T \otimes B$ on $h_R(B')$ is induced by that of $T \otimes_k M'$ via $id \otimes \varphi_M$.

PROOF. Let $0 \to M \to B \to A \to 0$ be a small extension. Assume given a k-algebra homomorphism $\varphi : R \to A$. This induces a homomorphism $\varphi' : S \to A$ (by composing with the natural map $S \to S/J = R$). As S is a power series ring, this lifts to a homomorphism $\widetilde{\varphi} : S \to B$. The lifts $\psi : R \to B$ of φ correspond bijectively to the lifts $\widetilde{\varphi} : S \to B$ with $\widetilde{\varphi}|_J = 0$.

Fix a lifting $\alpha : S \to B$ of φ and let $\beta : S \to B$ be another lifting. Then $h := (\beta - \alpha)$ is a linear map $S \to M$. As $M \cdot m_B = 0$, we see that $\alpha(f)x = \beta(f)x = f(0)x$ for any $f \in S$, $x \in M$, where $f(0)$ is the residue class of f in k. Let $f, g \in S$. Then

$$\begin{aligned} h(fg) &= \alpha(f)\alpha(g) - \beta(f)\alpha(g) + \beta(f)\alpha(g) - \beta(f)\beta(g) \\ &= h(f)g(0) + f(0)h(g). \end{aligned}$$

Thus h is a derivation from S to M, and we get that the set of liftings $\widetilde{\varphi} : S \to B$ of φ is an affine space over the space of derivations from S to M i.e. under $(n/n^2)^\vee \otimes_k M$.

As h is a derivation and $J \subset n^2$, it follows that $h|_J = 0$. Thus $\widetilde{\varphi}|_J$ does not depend on the lifting $\widetilde{\varphi}$, and by $m_B \cdot M = 0$ it has nJ in its kernel. Let $ob(\varphi) : J/nJ \to M$ be the induced map. Then the liftings $\widetilde{\varphi} : S \to B$ of φ give homomorphisms $R \to B$ if and only if $ob(\varphi) = 0$.

The functoriality of the exact sequence is an exercise. □

For many functors occurring in deformation theory, a property similar to Theorem 6.1.19 can be proven (we will give some examples later in this chapter).

DEFINITION 6.1.21. A deformation functor D is said to have a *tangent-obstruction theory* if there exist finite dimensional k-vector spaces T_1 (also called tangent space), T_2 (also called obstruction space) such that the following holds.

(1) For all small extensions $0 \to M \to B \to A \to 0$ there exists an exact sequence of sets

$$T_1 \otimes_k M \to D(B) \to D(A) \stackrel{ob}{\to} T_2 \otimes_k M.$$

(2) In case $A = k$, the sequence becomes

$$0 \to T_1 \otimes_k M \to D(B) \to D(A) \stackrel{ob}{\to} T_2 \otimes_k M.$$

(3) The exact sequences in (1) and (2) are functorial in small extensions, in the sense defined above.

We say that D has a *generalized tangent-obstruction theory* if we remove the finite dimensionality condition in the previous definition.

REMARK 6.1.22. If D is a deformation functor with tangent-obstruction theory as above and $i : T_2 \to T_2'$ is a linear injective map, we can define a tangent-obstruction theory with $T_1' = T_1$ and $ob' := (i \otimes \mathrm{id}_M) \circ ob$. So the obstruction space is not uniquely determined by D. On the other hand, T_1 is uniquely determined by D.

PROPOSITION 6.1.23. Let D be a functor with a generalized tangent-obstruction theory T_1, T_2. Then the vector space T_1 is determined by D up to canonical isomorphism.

PROOF. Consider the small extension $0 \to (\varepsilon) \to k[\varepsilon]/\varepsilon^2 \to k \to 0$. Write for brevity $A_1 := k[\varepsilon]/\varepsilon^2$. By point (2) in the definition, we see that $D(A_1)$ is an affine space over T_1 (since (ε) is canonically isomorphic to k). Remark 6.1.11 tells us that $D(A_1)$ has a distinguished element 0. Hence there is a canonical bijection $\iota : D(A_1) \to T_1$. If $\lambda \in K$, we can define multiplication by λ on $D(A_1)$ by $D(\varphi_\lambda)$ where $\varphi_\lambda : A_1 \to A_1$ is defined as $\varphi_\lambda(\varepsilon) = \lambda\varepsilon$. The reader is invited to check that functoriality implies that $\iota(\lambda\alpha) = \lambda\iota(\alpha)$ for every $\alpha \in D(A_1)$.

Let $A_2 = k[\varepsilon_1, \varepsilon_2]/(\varepsilon_1^2, \varepsilon_2^2, \varepsilon_1\varepsilon_2)$ and define maps $p_1, p_2 : A_2 \to A_1$ $p_i(\varepsilon_j) = \delta_{ij}\varepsilon$. It follows from Definition 6.1.21 and functoriality that the natural map $(D(p_1), D(p_2)) : D(A_2) \to D(A_1) \times D(A_1)$ is a bijection. Define a sum operation on $D(A_1)$ by composing the inverse of this bijection with the map $D(f) : D(A_2) \to D(A_1)$, where $f : A_2 \to A_1$ is defined by $f(\varepsilon_i) = \varepsilon$. Again, the reader is invited to check that $\iota(\alpha + \alpha') = \iota(\alpha) + \iota(\alpha')$ for every $\alpha, \alpha' \in D(A_1)$. □

In the next two sections we will see what we can learn about D from knowing its tangent and obstruction space, first in the case where D is pro-representable and then in the general case. We will also discuss to what extent T_2 in Definition 6.1.21 is unique.

6.2. Pro-representable functors

DEFINITION 6.2.1. A deformation functor D is called *pro-representable* if there exists an algebra $R \in (Loc/k)$ such that D is isomorphic to h_R. We say then that R *pro-represents* D.

Note that if D is the deformation functor associated to a point p in a contravariant functor $F : (schemes) \to (sets)$, then D is pro-representable if F is representable; the converse, however, is not true.

REMARK 6.2.2. In the usual language of category theory, F is representable if it is isomorphic to h_A for some $A \in (Art/k)$. Pro-representable usually means that it is isomorphic to h_R where R is a pro-object, i.e. an inverse limit of objects of (Art/k)). This would mean something more general than what we use (namely, R any complete local k-algebra with residue field k, with or without finite-dimensional tangent space). The slight abuse of terminology has been introduced in [**Sch68**] and is now standard in deformation theory.

REMARK 6.2.3. Proposition 6.1.18 implies that, if D is a pro-representable deformation functor, then D is isomorphic to h_R where $R = S/J$, $S = k[[x_1, \ldots, x_d]]$, $J \subset m_S^2$. Moreover, by Theorem 6.1.19 the functor h_R has a natural tangent-obstruction theory $T_1(R)$, $T_2(R)$ with $T_1(R) = T_R = T_S$ and $T_2(R) = (J/m_SJ)^\vee$.

THEOREM 6.2.4. Consider a functor h_R as in Remark 6.2.3. Let T_1, T_2 be a tangent-obstruction theory for h_R. Then there is a canonical isomorphism $T_1 \to T_1(R)$ and there is a canonical *injective* linear map $T_2(R) \to T_2$, commuting with the obstruction maps.

PROOF. The first statement was already proven in Proposition 6.1.23. So we only need to construct the injective linear map $T_2(R) \to T_2$. As before we write for brevity n instead of m_S.

Step 1. To make the idea of the proof clear, we first make the (wrong) assumption that T_2 is also an obstruction space for small extensions of complete local rings. In the second step we indicate the necessary changes.

Let $M := J/nJ$, $B := S/nJ$. Then we have a small extension $0 \to M \to B \to R \to 0$. The obstruction to lifting $id : R \to R$ to a map $\varphi : R \to B$ is a canonical element

$$o = ob(id) \in (J/nJ) \otimes_k T_2 = Hom((J/nJ)^\vee, T_2).$$

Assume o is not injective. Then there exists a sub-vector space $V \subset J/nJ$ of codimension 1, such that $\pi(o) = 0$ under the projection $\pi : M \to M/V$. Thus there exists a lifting $R \to B/V$ of $id : R \to R$. On the other hand in the diagram

$$\begin{array}{ccccccccc} 0 & \longrightarrow & J & \longrightarrow & S & \longrightarrow & R & \longrightarrow & 0 \\ & & \downarrow & & \downarrow & & \downarrow{=} & & \\ 0 & \longrightarrow & J/nJ = M & \longrightarrow & B & \longrightarrow & R & \longrightarrow & 0 \\ & & \downarrow & & \downarrow & & \downarrow{=} & & \\ 0 & \longrightarrow & M/V & \longrightarrow & B/V & \longrightarrow & R & \longrightarrow & 0 \end{array}$$

we see that the map $J \to M \to M/V$ is just the canonical quotient map. In particular it is nonzero. Thus by Theorem 6.1.19 no lift exists. Thus $(J/nJ)^\vee \to T_2$ is injective.

Step 2. We have instead to work with small extensions of Artin algebras. We can still use basically the same argument with some minor changes.

By the theorem of Artin Rees ([**AM69**] Proposition 10.9), there exists an $i > 0$, such that $n^i \cap J \subset nJ$. Let $M = (J+n^i)/(nJ+n^i) = J/nJ$, $B := S/(nJ+n^i)$. Thus we have a small extension $0 \to M \to B \to R/n^i \to 0$. The obstruction to lifting the quotient map $p : R \to R/n^i$ to $\varphi : R \to B$ is $o := ob(p) \in Hom((J/nJ)^\vee, T_2)$. If o is not injective, there exists again a subvector space $V \subset M$ of codimension 1 such that there exists a lifting $R \to B/V$ of p. On the other hand the induced map $J \to M/V$ is again as above just the canonical quotient map, and thus nonzero. Thus by Theorem 6.1.19 no lift exists and o is injective. □

Now we want to show that the tangent-obstruction theory can be used to get information about the structure of R.

A local ring A in (Loc/k) is called *complete intersection* if $A = R/I$ for R a regular local ring and I an ideal generated by $codim(A, R)$ elements. A scheme X is called a *local complete intersection* at a point p if $\mathcal{O}_{X,p}$ is a complete intersection: equivalently, if there is an open neighborhood U of p in X and a closed embedding $i : U \to Y$ such that Y is smooth, U is pure-dimensional and $\mathcal{I}_{U/Y}/\mathcal{I}^2_{U/Y}$ is locally free of rank $codim_p(X, Y)$.

COROLLARY 6.2.5. Under the assumptions of Theorem 6.2.4, let $d := dim(T_1)$ and $r := dim(T_2)$.

Then $d \geq dim(R) \geq d - r$. Furthermore if $dim(R) = d - r$, then R is a complete intersection. If $r = 0$, then $R \simeq k[[x_1, \dots, x_d]]$.

PROOF. We have $R = S/J$ with $S = k[[x_1, \dots, x_d]]$. By Theorem 6.2.4 we get $dim(J/nJ) \leq r$. Thus, by Nakayama's Lemma, J has at most r generators. Thus by the principal ideal theorem we have $dim(R) \geq d - r$ and if $dim(R) = d - r$, then R is a complete intersection in S.

Finally if $r = 0$, then $J = 0$ and thus $R = S = k[[x_1, \ldots, x_d]]$. □

COROLLARY 6.2.6. If R is the completion $\widehat{\mathcal{O}}_{V,p}$ of the local ring of a scheme at a point, then we get under the assumptions of Theorem 6.2.4, that $d \geq dim_p(V) \geq d - r$. Furthermore, if $dim_p(V) = d - r$, then V is a local complete intersection at p and if $r = 0$, then V is nonsingular at p.

PROOF. We know that $dim(R) = dim(\mathcal{O}_{V,p}) = dim_p(V)$ ([**AM69**] Corollary 11.19). Furthermore $\mathcal{O}_{V,p}$ is regular if and only if $\widehat{\mathcal{O}}_{V,p}$ is regular ([**AM69**] Corollary 11.24). Finally assume that $\mathcal{O}_{V,p} = A/I$ where A is a regular local ring and I is an ideal. Then $R = \widehat{A}/\widehat{I}$ with $\widehat{I}$ the m_A-adic completion of I. Then $I/m_AI = \widehat{I}/m_A\widehat{I}$. By Nakayama's Lemma, if $\widehat{I}$ is generated by s elements, so is I. □

6.3. Non-pro-representable functors

In the previous section we studied the case where the functor $D : (Art/k) \to (Sets)$ is pro-representable, but many functors occurring in deformation theory are not (usually because the objects that we want to deform might have infinitesimal automorphisms, i.e. the tangent space to the space of automorphisms is nonzero). If D has a tangent-obstruction theory it is however close to being pro-representable, in the following sense.

DEFINITION 6.3.1. Let $\alpha : F \to G$ be a morphism of deformation functors. The morphism α is called *smooth* if for all small extensions $0 \to M \to B \to A \to 0$ the canonical map $F(B) \to F(A) \times_{G(A)} G(B)$ is surjective. This implies that $a \in F(A)$ can be lifted to $F(B)$ if and only if $\alpha(a) \in G(A)$ can be lifted to $G(B)$.

Let F be a deformation functor. If $R \in (Loc/k)$, a smooth morphism $h_R \to F$ makes h_R into a *pro-representable hull* for F if $h_R(k[\epsilon]/\epsilon^2) \to F(k[\epsilon]/\epsilon^2)$ is bijective.

REMARK 6.3.2.
(1) Note that by Proposition 6.1.23 if F has a tangent-obstruction theory then $F(k[\epsilon]/\epsilon^2)$ can be viewed as the tangent space to the functor F. Thus $\alpha : h_R \to F$ is a hull if and only if it is smooth and an isomorphism of tangent spaces.
(2) The formal criterion for smoothness says that a morphism $f : X \to Y$ of schemes is smooth at a point $p \in X$ such that $f(p)$ has residue field k, if and only if the corresponding morphism of functors $D_{X,p} \to D_{Y,f(p)}$ is smooth; a smooth morphism is étale if and only if it is also an isomorphism on tangent spaces.

LEMMA 6.3.3. If F is a functor with tangent-obstruction theory T_1, T_2, and $\alpha : h_R \to F$ is a pro-representable hull, then T_1, T_2 are also a tangent-obstruction theory for h_R. In particular T_1 is canonically isomorphic to T_R and we can use the estimates of Corollary 6.2.5 on the dimension and singularities of R.

PROOF. The fact that T_1 is naturally isomorphic to T_R follows from the definition and the fact that the vector space structure on $F(k[\varepsilon]/\varepsilon^2)$ is determined by the functor F itself if F has a tangent-obstruction theory (as in the proof of Proposition 6.1.23).

To prove that T_2 is an obstruction theory for h_R, fix a small extension $B \to A$ with kernel M, and define $o : h_R(A) \to T_2 \otimes M$ as $ob_F \circ \alpha_{A*}$. Let $a \in h_R(A)$; then by smoothness a lifts to $h_R(B)$ if and only if $\alpha_{A*}(a)$ lifts to $F(B)$ if and only if $o(a) = 0$. $\square$

The main result is now that functors $D : (Art/k) \to (Sets)$ that have a tangent-obstruction theory have a pro-representable hull.

THEOREM 6.3.4. (Schlessinger) Let $D : (Art/k) \to (Sets)$ have a tangent-obstruction theory. Then there is a pro-representable functor h_R which is a hull for D.

We will not give a proof of this theorem: the proof was originally given in Schlessinger's thesis, and we refer the reader to [**Sch68**] where the author also gives necessary and sufficient conditions for a deformation functor to admit a pro-representable hull. These conditions are weaker than having a tangent-obstruction theory: for an example, see [**FM98**]. For an expository proof of Schlessinger's Theorem, see also [**Art74a**].

In practice it is often easier to verify (or find proofs in the literature) that a deformation functor has a tangent-obstruction theory than that it satisfies Schlessinger's axioms. The main reason why this is true is that functors which come from the local study of an algebraic stack in the sense of Artin all have a tangent-obstruction theory: the connection between algebraic stacks and deformation theory goes back to Artin's original paper [**Art74b**].

The existence of a hull for D is the infinitesimal counterpart to the Kuranishi family or a miniversal deformation in complex-analytic deformation theory. An important corollary to Schlessinger's theorem is the following. Note that Schlessinger in [**Sch68**] gives a different set of necessary and sufficient conditions for pro-representability.

COROLLARY 6.3.5. A deformation functor is pro-representable if and only if it has a tangent-obstruction theory such that the exact sequence (2) in Definition 6.1.21 is exact for every small extension.

PROOF. One implication is Theorem 6.2.4. To prove the other one, assume that $\alpha : h_R \to F$ is a pro-representable hull for F. We will prove that $h_R(B) \to F(B)$ is bijective for every $B \in (Art/k)$ by induction on $\dim_k B$. The case $\dim_k B = 1$ is clear since this implies $B = k$.

In general, let $M := \{b \in B \,|\, b \cdot m_B = 0\}$, and $A = B/M$. Then $B \to A$ is a small extension and by assumption we can assume that $h_R(A) \to F(A)$ is bijective. Fix $a \in h_R(A)$; by smoothness, a lifts to $h_R(B)$ if and only if $\alpha(a) \in F(A)$ lifts to $F(B)$.

Assume that this is the case, and let $h_R(B)_a := \{b \in h_R(B) \,|\, b|_A = a\}$, and similarly for $F(B)_{\alpha(a)}$. Since the map $\alpha : h_R(B)_a \to F(B)_{\alpha(a)}$ is $T_1 \otimes M$ equivariant and source and target are affine spaces over $T_1 \otimes M$, it is necessarily bijective. $\square$

REMARK 6.3.6. It is easy to see that a hull is unique up to isomorphism in the following sense: if $\alpha : h_R \to F$ and $\beta : h'_R \to F$ are two pro-representable hulls, then there is an isomorphism $\gamma : h_R \to h'_R$ such that $\alpha = \beta \circ \gamma$. In general this isomorphism is not unique.

As said before, most functors appearing in deformation theory are not pro-representable. We give without proof some examples of not necessarily pro-representable functors with a tangent-obstruction theory. The statements are proved in a slightly different language in Part 4.

EXAMPLE 6.3.7. (1) Let X be a smooth variety. The functor of deformations of X is given by letting $Def_X(A)$ be the set of isomorphism classes of pairs (W, α) where W is a flat scheme over $\operatorname{Spec} A$ and α $W \times_{\operatorname{Spec} A} \operatorname{Spec} k \to X$ is an isomorphism; pullback is defined by fibre product.

In this case $T_1 = H^1(X, T_X)$ and $T_2 = H^2(X, T_X)$ (see Part 4, Section 5.B). The functor Def_X is pro-representable if $H^0(X, T_X) = 0$ (but not conversely).

(2) Let $\mathcal{E}$ be a coherent sheaf on a variety X. The functor of deformations of $\mathcal{E}$ is given by letting $Def_{\mathcal{E}}(A)$ be the set of isomorphism classes of pairs $(\mathcal{F}, \alpha)$ where $\mathcal{F}$ is a coherent sheaf on $X \times \operatorname{Spec} A$, flat over $\operatorname{Spec} A$, and α is an isomorphism of the restriction of $\mathcal{F}$ to $X \times \operatorname{Spec} k$ with $\mathcal{E}$.

In this case $T_1 = \operatorname{Ext}^1(\mathcal{E}, \mathcal{E})$ and $T_2 = \operatorname{Ext}^2(\mathcal{E}, \mathcal{E})$ and in case $\mathcal{E}$ is locally free $T_1 = H^1(X, End(\mathcal{E}))$ and $T_2 = H^2(X, End(\mathcal{E}))$ (see Part 4, Section 8.5.A).

This functor is pro-representable if (but not only if) $\mathcal{E}$ is simple, i.e., $\operatorname{Hom}(\mathcal{E}, \mathcal{E}) = k \cdot \operatorname{id}_{\mathcal{E}}$.

As one can see in these examples, usually the T_i are cohomology groups or Ext groups. It usually happens that if T_1 is the group with index i, then T_2 is the group with index $i+1$. In this case usually the group with index $i-1$ is the group of infinitesimal automorphisms, e.g. $H^0(X, T_X)$ is the tangent space to the automorphisms of X: its vanishing usually implies pro-representability of the functor.

6.4. Examples of tangent–obstruction theories

Let X be a scheme over k and $Y \subset X$ a proper closed subscheme; let $\mathcal{I}_Y$ be the ideal sheaf of Y in X. We will prove that the deformation functor $H_{Y,X}$ of Y in X of Example 6.1.5 admits a tangent-obstruction theory with $T_1 = \operatorname{Hom}_{\mathcal{O}_X}(\mathcal{I}_Y, \mathcal{O}_Y)$ and $T_2 = \operatorname{Ext}^1_{\mathcal{O}_X}(\mathcal{I}_Y, \mathcal{O}_Y)$. When X is quasiprojective, the functor $H_{Y,X}$ is pro-representable since it is the deformation functor associated to the point $[Y]$ in the Hilbert functor of X (see Example 6.1.6), since by Part 2 the Hilbert functor is representable. Therefore Corollary 6.2.6 yields information about the local structure of the Hilbert scheme.

This result will be obtained as a corollary from the computation of a tangent-obstruction theory for the deformation functor associated to the Quot scheme.

SET-UP 6.4.1. Let X be a scheme over k and F a coherent sheaf on X. Let

$$0 \to M \to B \to A \to 0$$

be a short exact sequence in (Art/k) such that $M^2 = 0$: in particular M is naturally an A-module (this is a generalization of a small extension since $M \subset m_B$). Write $\otimes$ instead of $\otimes_k$ and $\times$ instead of $\times_{\operatorname{Spec} k}$; write $X_A := X \times \operatorname{Spec} A$ and $X_B = X \times \operatorname{Spec} B$.

DEFINITION 6.4.2. Assume given a short exact sequence

$$0 \to S \to F \otimes A \to Q \to 0 \qquad (e)$$

of coherent sheaves on X_A, flat over $\operatorname{Spec} A$. We say that an exact sequence

$$0 \to S' \to F \otimes B \to Q' \to 0 \qquad (e')$$

of coherent sheaves on X_B, flat over $\operatorname{Spec} B$ is an *extension* of (e) if $S' \otimes_B A$ coincides with S as a subsheaf of $F \otimes A$.

LEMMA 6.4.3. Let Q' be a coherent sheaf on X_B. Then Q' is flat over B if and only if the map $Q' \otimes_B M \to Q'$ is injective and $Q' \otimes_B A$ is flat over A.

PROOF. We may assume that $B \to A$ is a small extension (if not, factor it as a finite number of small extensions and proceed by induction). Recall that Q' is flat over B if and only if for all ideals $I \subset B$ the map $Q' \otimes_B I \to Q'$ is injective ([**Har77**] III,9.1A(a)).

"$\Rightarrow$" The first statement is the special case $I = M$. The second is true since flatness is invariant under base extension. ([**Har77**], III, 9.1A(b)).

"$\Leftarrow$" Let $I \subset B$ be an ideal. Define an ideal $I' \subset A$ by requiring that there be an exact sequence

$$0 \to M \cap I \to I \to I' \to 0.$$

We apply $Q' \otimes_B \cdot$ and get a commutative diagram with exact rows

$$\begin{array}{ccccccccc} & & Q' \otimes_B (I \cap M) & \longrightarrow & Q' \otimes_B I & \longrightarrow & (Q' \otimes_B A) \otimes_A I' & \longrightarrow & 0 \\ & & \downarrow{\scriptstyle\alpha} & & \downarrow{\scriptstyle\beta} & & \downarrow{\scriptstyle\gamma} & & \\ 0 & \longrightarrow & Q' \otimes_B M & \longrightarrow & Q' & \longrightarrow & Q' \otimes_B A & \longrightarrow & 0 \end{array}$$

γ is injective, because $Q' \otimes_B A$ is A-flat. Thus α injective implies β injective. Because $m_B \cdot M = 0$, as a B-module M is a k-vector space and therefore so is $I \cap M$. Choose a sub-vector space (and hence a B-submodule) C of M so that $M = (I \cap M) \oplus C$. Hence $Q' \otimes_B (I \cap M) \to Q' \otimes_B M$ is injective and $Q' \otimes_B M \to Q'$ is injective. □

We recall a standard lemma about extensions (see e.g. [**Har77**] Ex. III.6.1).

LEMMA 6.4.4. Let $0 \to \mathcal{E} \to \mathcal{F} \xrightarrow{\pi} \mathcal{G} \to 0$ be an exact sequence of coherent sheaves. Then π has a section if and only if the extension class of the sequence in $\operatorname{Ext}^1(\mathcal{G}, \mathcal{F})$ vanishes. In this case the sections of π are an affine space over $\operatorname{Hom}(\mathcal{G}, \mathcal{F})$.

THEOREM 6.4.5. (a) To the exact sequence (e) we can associate an element $\operatorname{ob}(e) \in \operatorname{Ext}^1_{X_A}(S, Q \otimes_A M)$.

(b) An extension (e') of (e) exists if and only if $ob(e) = 0$. If $\operatorname{ob}(e) = 0$ then the possible extensions are naturally a torsor under $\operatorname{Hom}_{X_A}(S, Q \otimes_A M)$.

PROOF. Step (a): construction of $\operatorname{ob}(e)$. Since Q is A-flat the following diagram of coherent sheaves on $\mathcal{O}_{X_B}$ has exact row and columns:

$$\begin{array}{ccccccccc}
 & & 0 & & & & 0 & & \\
 & & \downarrow & & & & \downarrow & & \\
 & & S\otimes_A M & & & & S & & \\
 & & \downarrow & & & & \downarrow & & \\
0 & \to & F\otimes M & \to & F\otimes B & \to & F\otimes A & \to & 0 \\
 & & \downarrow & & & & \downarrow & & \\
 & & Q\otimes_A M & & & & Q & & \\
 & & \downarrow & & & & \downarrow & & \\
 & & 0 & & & & 0 & &
\end{array}$$

Consider the induced maps $\alpha : S\otimes_A M \to F\otimes B$ and $\beta : F\otimes B \to Q$; clearly one has $\operatorname{im}\alpha \subset \operatorname{Ker}\beta$. Define $\tilde{F} := \operatorname{Ker}(\beta)/\operatorname{im}(\alpha)$; it is easy to see that there is an exact sequence of coherent sheaves on X_B

$$0 \to Q\otimes_A M \to \tilde{F} \to S \to 0. \tag{6.4.1}$$

To conclude Step (a) it is enough to prove that (e) is in fact a sequence of $\mathcal{O}_{X_A}$ modules, i.e., verify that $M\cdot\tilde{F} = 0$; then we can define $\operatorname{ob}(e)$ to be the class of the extension (6.4.1).

Fix $m\in M$ and $f = \sum f_i\otimes b_i \in F\otimes B$. Then $m\cdot\sum f_i\otimes b_i = \sum f_i\otimes mb_i \in F\otimes M$. Let $a_i \in A$ be the image of b_i. Since $M^2 = 0$, we obtain that $m\cdot\sum f_i\otimes b_i = \sum f_i\otimes ma_i$, where we now view M as an A-module. Therefore $f\otimes m\in (F\otimes B)\otimes_B M$ is equal to $\bar{f}\otimes m \in (F\otimes A)\otimes_A M$, where $\bar{f}$ is the image of f in $F\otimes A$.

If $f\in\operatorname{Ker}\beta$ then $\bar{f}$ is a section of S; in this case $m\cdot f = m\cdot\bar{f}$ is contained in $S\otimes_A M \subset F\otimes M$; therefore its image in $F\otimes B$ is contained in $\operatorname{im}\alpha$. So multiplication by m is the zero map on $\tilde{F}$, concluding the proof.

Step (b). It is enough to prove that extensions are in a natural bijection with splittings of the exact sequence (6.4.1). Assume we are given a splitting, i.e. a morphism $\xi : S\to\tilde{F}$ which is a right inverse to the surjection $\tilde{F}\to S$. Let S' be the inverse image of $\xi(S)$ in $\operatorname{Ker}\beta$. We can complete the H-shaped diagram above to a commutative diagram with exact rows and columns

$$\begin{array}{ccccccccc}
 & & 0 & & 0 & & 0 & & \\
 & & \downarrow & & \downarrow & & \downarrow & & \\
0 & \to & S\otimes_A M & \to & S' & \to & S & \to & 0 \\
 & & \downarrow & & \downarrow & & \downarrow & & \\
0 & \to & F\otimes M & \to & F\otimes B & \to & F\otimes A & \to & 0 \\
 & & \downarrow & & \downarrow & & \downarrow & & \\
0 & \to & Q\otimes_A M & \to & Q' & \to & Q & \to & 0 \\
 & & \downarrow & & \downarrow & & \downarrow & & \\
 & & 0 & & 0 & & 0 & &
\end{array}$$

and Q' is the requested extension (to prove that it is flat over B, we use Lemma 6.4.3).

Conversely, given an extension Q' of Q, we can produce the diagram above just by exactness, and the subsheaf $S'/\operatorname{im}(\alpha)$ of $\tilde{F}$ maps isomorphically to S, therefore defining a splitting of $\tilde{F}\to S$. $\square$

REMARK 6.4.6. In statement (b) above we can replace $\mathrm{Hom}_{X_A}(S, Q\otimes_A M)$ with $\mathrm{Hom}_X(S\otimes_A k, Q\otimes_A M)$, since these two vector spaces are naturally isomorphic.

PROPOSITION 6.4.7. *If $m_A \cdot M = 0$, then we can replace $\mathrm{Ext}^1_{X_A}(S, Q\otimes_A M)$ with $\mathrm{Ext}^1_X(S\otimes_A k, Q\otimes_A M)$ in statement (a) of Theorem 6.4.5.*

PROOF. Since the sequence (6.4.1) is exact and S is flat over A, the sequence stays exact if we apply $\otimes_A k$. This yields a commutative diagram with exact rows and column:

$$\begin{array}{ccccccccc}
 & & & & & & 0 & & \\
 & & & & & & \downarrow & & \\
 & & & & & & S\otimes_A m_A & & \\
 & & & & & & \downarrow & & \\
0 & \to & Q\otimes_A M & \to & \tilde{F} & \to & S & \to & 0 \\
 & & \| & & \downarrow & & \downarrow & & \\
0 & \to & Q\otimes_A M & \to & \tilde{F}\otimes_A k & \to & S\otimes_A k & \to & 0 \\
 & & & & & & \downarrow & & \\
 & & & & & & 0 & &
\end{array}$$

where the right column is exact (again, since S is A-flat), and $Q\otimes_A M\otimes_A k = Q\otimes_A M$ (since $M\cdot m_A = 0$ implies that $M\otimes_A k = M$). An elementary diagram chasing argument proves that we can complete the diagram above to a commutative diagram with exact rows and columns

$$\begin{array}{ccccccccc}
 & & & & 0 & & 0 & & \\
 & & & & \downarrow & & \downarrow & & \\
 & & & & S\otimes_A m_A & = & S\otimes_A m_A & & \\
 & & & & \downarrow & & \downarrow & & \\
0 & \to & Q\otimes_A M & \to & \tilde{F} & \to & S & \to & 0 \\
 & & \| & & \downarrow & & \downarrow & & \\
0 & \to & Q\otimes_A M & \to & \tilde{F}\otimes_A k & \to & S\otimes_A k & \to & 0 \\
 & & & & \downarrow & & \downarrow & & \\
 & & & & 0 & & 0 & &
\end{array}$$

and this in turn shows that the middle row splits if and only if the bottom row splits. Note that the bottom row is (by definition) a sequence of sheaves of $\mathcal{O}_X$-modules, therefore whether it splits is determined by its class in

$$\mathrm{Ext}^1_X(S\otimes_A k, Q\otimes_A M) = \mathrm{Ext}^1_X(S|_X, Q|_X)\otimes M.$$

□

DEFINITION 6.4.8. Let X be a scheme over k, F_0 a coherent sheaf on X, S_0 a coherent subsheaf of F_0 and $Q_0 = F_0/S_0$ the quotient. Define a deformation functor by letting $D_{F_0,Q_0}(A)$ be the set of coherent subsheaves S of $F := F_0\otimes A$ such that $Q = F/S$ is flat over A and $S\otimes_A k = S_0$ as subsheaves of F_0. If X is projective, then D_{F_0,Q_0} is the deformation functor associated to the Quot scheme of F_0 at the point Q_0.

THEOREM 6.4.9. *The deformation functor D_{F_0,Q_0} just defined has a generalized tangent-obstruction theory with $T_1 = \mathrm{Hom}_{\mathcal{O}_X}(S_0, Q_0)$ and $T_2 = \mathrm{Ext}^1_{\mathcal{O}_X}(S_0, Q_0)$. The exact sequence (2) in Definition 6.1.21 is exact for any A.*

PROOF. The only thing left to prove is functoriality of the obstruction. We leave this to the reader. □

COROLLARY 6.4.10. The deformation functor $H_{Y,X}$ has a generalized tangent-obstruction theory with $T_1 = \mathrm{Hom}_{\mathcal{O}_X}(\mathcal{I}_Y, \mathcal{O}_Y)$ and $T_2 = \mathrm{Ext}^1_{\mathcal{O}_X}(I_Y, \mathcal{O}_Y)$.

PROOF. This is just the special case $F_0 = \mathcal{O}_X$ and $S_0 = I_Y$ of the theorem. □

Applying Corollary 6.4.10 and Corollary 6.2.6 now gives information on the local structure of the Hilbert scheme.

COROLLARY 6.4.11. Let X be quasiprojective and $[Z] \in Hilb(X)$ (i.e., assume that Z is a proper closed subscheme of X). Let $d := dim(\mathrm{Hom}_{\mathcal{O}_X}(\mathcal{I}_Z, \mathcal{O}_Z))$ and $r := dim(\mathrm{Ext}^1_{\mathcal{O}_X}(\mathcal{I}_Z, \mathcal{O}_Z))$.

Then $d \geq dim_{[Z]} Hilb(X) \geq d - r$. Furthermore if $dim_{[Z]} Hilb(X) = d$, then $Hilb(X)$ is nonsingular at $[Z]$. If $dim_{[Z]} Hilb(X) = d - r$, then $Hilb(X)$ is a local complete intersection at $[Z]$.

We now want to sketch another example of functor having tangent-obstruction theory. Let X, Y be schemes and assume Y is nonsingular. Let $f : X \to Y$ be a morphism, and let D_f be its deformation functor: that is,

$$D_f(A) := \{f_A : X \times \mathrm{Spec}\, A \to Y \mid f_A|_X = f\}.$$

If X is projective, we can consider the open subset $Mor(X,Y) \subset Hilb(X \times Y)$ parametrizing graphs of morphisms from X to Y; in this case D_f is the deformation functor associated to the point $[f] \in Mor(X,Y)$.

THEOREM 6.4.12. Assume that X is separated. The functor D_f has a generalized tangent-obstruction theory with $T_i = H^{i-1}(X, f^*T_Y)$. The exact sequence (2) in Definition 6.1.21 is exact for any A.

PROOF. Assume first that X and Y are affine; the theorem in this case is Exercise II.8.6 in [**Har77**]. For the general case, choose affine open covers Y_i of Y and X_i of X such that $f(X_i) \subset Y_i$ (exercise: show that such a cover exists). Let $X_{ij} = X_i \cap X_j$.

Assume we are given $g : X \times \mathrm{Spec}\, A \to Y \in D_f(A)$, and let $B \to A$ be a small extension with kernel M. Choose arbitrarily morphisms $h_i : X_i \times \mathrm{Spec}\, B \to Y_i \subset Y$ such that $h_i|_{X_i \times \mathrm{Spec}\, A} = g|_{X_i \times \mathrm{Spec}\, A}$. By the affine case, there exist unique elements $\eta_{ij} \in \Gamma(X_{ij}, f^*T_Y) \otimes M$ such that $\eta_{ij}(h_i|_{X_{ij} \times \mathrm{Spec}\, B}) = h_j|_{X_{ij} \times \mathrm{Spec}\, B}$. It is easy to check that the η_{ij} so defined are a Čech cocycle, and that a global extension $h : X \times \mathrm{Spec}\, B \to Y$ of g exists if and only if this cocycle is a coboundary. Moreover, the class of this cocycle does not depend on the particular choice of the h_i's, nor indeed of the open covers X_i and Y_i. We leave it to the reader to complete this argument. □

6.5. More tangent-obstruction theories

In this section we briefly sketch further examples of deformation functors with tangent-obstruction theories. We will not give complete proofs in these cases, since we are not going to need them in the sequel.

PROPOSITION 6.5.1. Let X be a projective scheme, $\mathcal{E}$ a coherent sheaf on X. Then the deformation functor $Def_{\mathcal{E}}$ has a tangent-obstruction theory defined by $T_i = \text{Ext}^i(\mathcal{E},\mathcal{E})$.

PROOF. The proof goes in two steps. The tangent space can be constructed directly as follows. Let

$$0 \to \mathcal{E} \xrightarrow{f} \mathcal{E}' \xrightarrow{p} \mathcal{E} \to 0$$

be an extension of $\mathcal{E}$ by $\mathcal{E}$. We want to give $\mathcal{E}'$ the structure of a coherent sheaf on $X_\varepsilon := X \times \operatorname{Spec} k[\varepsilon]/\varepsilon^2$, flat over X_ε, and such that the sequence above is obtained by tensoring $\mathcal{E}'$ with the exact sequence $0 \to k \xrightarrow{\cdot\varepsilon} k[\varepsilon]/\varepsilon^2 \to k \to 0$. This is easy; it is enough to define, for e' a section of $\mathcal{E}'$, $\varepsilon \cdot e' := f(p(e'))$. We leave it to the reader to check that the map $\text{Ext}^1(\mathcal{E},\mathcal{E}) \to Def_{\mathcal{E}}(k[\varepsilon]/\varepsilon^2)$ so defined induces an isomorphism between $\text{Ext}^1(\mathcal{E},\mathcal{E})$ and $T_1(Def_{\mathcal{E}})$ (where the structure of a vector space on the latter is defined as in Proposition 6.1.23). So if $Def_{\mathcal{E}}$ has a tangent-obstruction theory, then its tangent space must be $\text{Ext}^1(\mathcal{E},\mathcal{E})$.

For the obstruction space, fix $\mathcal{O}_X(1)$ a very ample line bundle on X. Then there exists $N \gg 0$ such that $\mathcal{E}(N)$ is generated by global sections and has no higher cohomology. The first statement can be rephrased by saying that we can obtain $\mathcal{E}$ as a quotient of $\mathcal{F} := \mathcal{O}_X(-N) \otimes V$, where V is the vector space $H^0(X,\mathcal{E}(N))$. Let $\mathcal{S}$ be the kernel of the surjection $\mathcal{F} \to \mathcal{E}$. Using Cohomology and Base Change ([**Har77**] Theorem III.12.11), one can prove that the natural morphism $D_{\mathcal{F},\mathcal{E}} \to Def_{\mathcal{E}}$ is smooth; hence $\text{Ext}^1(\mathcal{S},\mathcal{E})$ is an obstruction space for $Def_{\mathcal{E}}$. It is now easy to check that the natural map $\text{Ext}^1(\mathcal{S},\mathcal{E}) \to \text{Ext}^2(\mathcal{E},\mathcal{E})$ is an isomorphism.

To complete the proof, one still has to show that T_1 satisfies the claimed properties; this can also be achieved by studying the morphism of deformation functors $D_{\mathcal{F},\mathcal{E}} \to Def_{\mathcal{E}}$ but we will not do this. □

A subscheme Z of a scheme X is called a *local complete intersection in* X, if at all points $p \in Z$, $\mathcal{I}_{Z/X}$ is locally generated by $codim_p(Z,X)$ elements. In the case of local complete intersections in X we can replace the Ext groups by cohomology groups. Let $Z \subset X$ be a local complete intersection in X. Let $N_{Z/X} = (\mathcal{I}_Z/\mathcal{I}_Z^2)^\vee$ be the normal bundle of Z in X, which is locally free.

PROPOSITION 6.5.2. If Z is a local complete intersection in X, then $H_{Z,X}$ has a generalized tangent-obstruction theory with $T_i = H^{i-1}(Z, N_{Z/X})$.

PROOF. For the tangent space there is nothing to prove, since for any closed subscheme Y one has $\text{Hom}_{\mathcal{O}_X}(\mathcal{I}_Y,\mathcal{O}_Y) = \text{Hom}_{\mathcal{O}_Y}(\mathcal{I}_Y/\mathcal{I}_Y^2,\mathcal{O}_Y)$. The proof for the obstruction space can be carried out similarly to that of Theorem 6.4.12 (of which this is indeed a generalization). The first step is to prove the result in case X, and hence Y, are affine. We do not prove this, but refer to [**Art74a**] §I.3 for an argument.

For the general case, assume we are given $Z_S \subset X \times S$ flat over S and having Z as closed fibre. Since Z is a local complete intersection, so is Z_S, and locally (on every affine open subset of X) the deformation functor is unobstructed (see [**Art74a**] §I.3 for an argument). Therefore locally on X liftings always exist. Cover X by open affines X^i, and choose liftings Z_T^i of $Z_S^i = Z_S \cap X^i$. On $X^{ij} := X^i \cap X^j$ the two possible liftings $Z_T^i \cap X^{ij}$ and $Z_T^j \cap X^{ij}$ differ by an element $\eta_{ij} \in N_{Z/X}(Z^{ij}) \otimes M$; it is easy to check that the η_{ij} are a Čech cocyle and that the associated cohomology class $\eta \in H^1(Z, N_{Z/X}) \otimes M$ is well-defined (i.e, changing the choice of the liftings Z_T^i changes the cocycle by a coboundary) and is an obstruction to the existence of a global lifting. We invite the reader unfamiliar with this kind of argument to work out the details. □

REMARK 6.5.3. A similar Čech cohomology argument can be used to show that if X is a separated smooth scheme, then Def_X has a generalized tangent-obstruction space with $T_i = H^i(X, T_X)$; the key step here is to prove the affine case (that is, every infinitesimal deformation of a smooth affine variety is trivial). A proof of this can be found in [**Art74a**]. Once this is established, any deformation of an arbitrary smooth separated X can be trivialized on an affine open cover, and the gluing maps are deformations of the identity; then one only needs to apply Theorem 6.4.12 to conclude the proof.

To study deformation theory in general one has to take into account some version of the cotangent complex, defined by Illusie in [**Ill71**], [**Ill72a**]; the same work also contains a very comprehensive study of tangents and obstructions, in the more general, and in a sense more natural, context of square zero (instead of small) extensions. A more elementary approach can be used if one restricts one's attention to lci morphisms, as in [**Vis97**].

CHAPTER 7

Hilbert Schemes of Points

Introduction

In this chapter let k be an algebraically closed field of characteristic 0. Let X be a quasiprojective scheme over k with an ample line bundle $\mathcal{O}(1)$. The Hilbert scheme $Hilb(X)$ of X parametrizes all closed, proper subschemes of X. We know that $Hilb(X)$ can be written as a disjoint union

$$Hilb(X) = \coprod_P Hilb^P(X)$$

of quasiprojective schemes where $Hilb^P(X)$ parametrizes the subschemes with Hilbert polynomial P (i.e. it represents the contravariant functor sending a scheme T to the set of all closed subschemes $Z \subset X \times T$, which are flat over T and the Hilbert polynomial of the fibres is P). See Part 2.

We want to deal with the simplest case that P is the constant polynomial n. As the degree of the Hilbert polynomial is the dimension of the subscheme, we see that $Hilb^n(X)$ parametrizes 0-dimensional subschemes of length n of X. In other words this means that

$$\dim H^0(Z, \mathcal{O}_Z) = \sum_{p \in supp(Z)} \dim_k(\mathcal{O}_{Z,p}) = n.$$

Here $len(Z) := \dim H^0(Z, \mathcal{O}_Z)$ is the length of Z as module over itself. In the future we will also write $X^{[n]}$ for $Hilb^n(X)$. The simplest example of an element in $X^{[n]}$ is just a set $\{p_1, \ldots, p_n\}$ of n distinct points on X. It is easy to see that these form an open subset of $X^{[n]}$. $X^{[n]}$ parametrizes sets of n not necessarily distinct points on X, with additional non-reduced structure when some of these points come together. Another space that parametrizes in a different way sets of n points on X is the symmetric power $X^{(n)}$ of X, the quotient of X^n by the action of the symmetric group $\mathcal{S}_n$ in n-letters by permuting the factors. $X^{(n)}$ parametrizes effective 0-cycles of degree n on X, i.e. formal sums $\sum_i n_i[p_i]$ with $p_i \in X$, $n_i \in \mathbb{Z}_{>0}$ and $\sum n_i = n$. There is an obvious set-theoretic map

$$\rho : X^{[n]} \to X^{(n)}, Z \mapsto \sum_p len(\mathcal{O}_{Z,p})[p],$$

which sends Z to its support with multiplicities. We shall see below that this is indeed a morphism of schemes.

In case X is a nonsingular curve, we will show that ρ is an isomorphism for all n. If X is a nonsingular surface, then $X^{[n]}$ is nonsingular and irreducible and ρ is a birational resolution of singularities of $X^{(n)}$. This is not true for X of dimension at least 3.

Hilbert schemes of points on a surface have received a lot of interest both in mathematics and in theoretical physics. Partially this is because they are canonical resolutions of singularities of the symmetric powers, but also because of their relations to moduli of vector bundles and to infinite dimensional Lie algebras. See [**Göt02**] for an overview of some of these relations. A nice and very readable introduction to Hilbert schemes of points and some of the newer results is [**Nak99**].

7.1. The symmetric power and the Hilbert–Chow morphism

As said in the introduction, the Hilbert scheme $X^{[n]}$ of n points on X is closely related to the symmetric power $X^{(n)} = X^n/\mathcal{S}_n$. We first need to know that the symmetric power exists as an algebraic variety.

DEFINITION 7.1.1. Let X be a quasiprojective variety over k, and let G be a group acting (by automorphisms) on X. A variety Y together with a surjective morphism $\pi : X \to Y$ is called a *quotient* of X by G if the following holds.

(1) The fibres of π are the orbits of G.
(2) Any G-invariant morphism $\varphi : X \to Z$ to a scheme Z factors through π.

It follows that the quotient is unique up to isomorphism, if it exists. We denote it by X/G.

In general it is a difficult question whether a quotient exists. However if G is finite and X quasiprojective, the problem is easy.

THEOREM 7.1.2. Let X be a quasiprojective variety with an action of a finite group G. Then the quotient X/G exists as a variety.

PROOF. (Sketch) First assume that X is affine. Let $k[X]$ be the affine coordinate ring. Condition (2) in the definition of the quotient implies that $k[X/G]$ should be the ring of invariants $k[X]^G \subset k[X]$. It is easy to see that $k[X]^G$ is a finitely generated k-algebra, so we define $X/G := \mathrm{Spec}(k[X]^G)$, which is an affine variety. Let $\pi : X \to X/G$ be the morphism induced by the inclusion $k[X]^G \subset k[X]$. It is not difficult to show that π is surjective and the fibres are the G-orbits.

If X is not affine, it has an open cover (U_i) by affines, and as G is finite, we can choose the affine sets in such a way that each orbit is contained in one of the U_i. Replacing the U_i with $W_i = \bigcap_{g\in G} g(U_i)$ we get an open cover of X by G-invariant affine open subsets (as X is a quasiprojective variety, hence separated, the intersection of affine open sets is affine). Then it is not difficult to show that the W_i/G glue to give the quotient X/G. □

With some additional care (see e.g. [**Har92**] Lecture 10) one shows that if G is finite and X quasiprojective, then X/G is quasiprojective, and projective if X is projective.

In particular if X is a quasiprojective variety, then the symmetric power $X^{(n)} := X^n/\mathcal{S}_n$ (where the symmetric group $\mathcal{S}_n$ acts by permutation of the factors) exists as a quasiprojective variety.

EXAMPLE 7.1.3.

(1) By the fundamental theorem on symmetric functions $k[x_1, \dots, x_n]^{\mathcal{S}_n} = k[s_1, \dots, s_n]$ where the s_i are the elementary symmetric functions in the x_i. Thus $(\mathbb{A}^1)^{(n)} = \mathbb{A}^n$. Similarly one shows $(\mathbb{P}^1)^{(n)} = \mathbb{P}^n$.

(2) We want to see that $(\mathbb{A}^2)^{(2)}$ is isomorphic to the product of $\mathbb{A}^2$ with a quadric cone $\mathrm{Spec}(k[u,v,w]/(uw-v^2))$. Let x_1, y_1, and x_2, y_2 be the coordinates on the two factors. We put $x := x_1 - x_2$, $y = y_1 - y_2$. Then also $x, y, x' := x_1+x_2, y' := y_1+y_2$ are coordinates on $(\mathbb{A}^2)^2$, on which the transposition τ of $\mathcal{S}_2$ acts by $\tau(x) = -x$, $\tau(y) = -y$, $\tau(x') = x'$, $\tau(y') = y'$. Thus $k[x,y,x',y']^\tau = (k[x,y]^\tau)[x'y']$. Note that every monomial in x, y is an eigenvector for τ, by $\tau(x^iy^j) = (-1)^{i+j}x^iy^j$, i.e. $k[x,y]$ has a basis of eigenvectors for τ. Therefore $k[x,y]^\tau$ is the linear subspace generated by the eigenvectors x^iy^j with eigenvalue 1, i.e. with $i+j$ even. Putting $u = x^2$, $v = xy$, $w = y^2$, this shows that $k[x,y,x',y']^{\mathcal{S}_2}$ is the subalgebra generated by u, v, w, x', y', which is isomorphic to $k[u,v,w,x',y']/(uw - v^2)$. In particular we see that the singular locus of $(\mathbb{A}^2)^{(2)}$ is the image of the diagonal in $(\mathbb{A}^2)^2$.

More generally if S is a nonsingular surface, then locally in the étale topology S is isomorphic to $\mathbb{A}^2$, thus the same argument shows that the singular locus of $S^{(2)}$ is the image of the diagonal. If $k = \mathbb{C}$, then the same is true in the analytic topology.

By definition the points of the symmetric power $X^{(n)}$ are the orbits of the n-tuples of points on X under permutation, i.e. they are the effective 0-cycles $\sum n_i[x_i]$ with $x_i \in X$, $n_i > 0$ and $\sum n_i = n$. This allows us to give a different description of $X^{(n)}$ as a Chow variety of 0-cycles.

Let $X \subset \mathbb{P}^d$ be locally closed. We see that $X^{(n)}$ is a locally closed subvariety of $(\mathbb{P}^d)^{(n)}$. Let $\check{\mathbb{P}}^d$ be the dual projective space of hyperplanes in $\mathbb{P}^d$. Let $Div^n(\check{\mathbb{P}}^d) \simeq \mathbb{P}^{\binom{n+d}{d}-1}$be the space of effective divisors of degree n on $\check{\mathbb{P}}^d$. For any $p \in \mathbb{P}^d$ let

$$H_p := \{l \in \check{\mathbb{P}}^d \mid p \in l\}.$$

Then $p \mapsto H_p$ defines an isomorphism $\mathbb{P}^d \simeq Div^1(\check{\mathbb{P}}^d)$. For $(x_1, \ldots, x_n) \in (\mathbb{P}^d)^n$ let

$$ch(x_1, \ldots, x_n) := \sum_i H_{x_i} \in Div^n(\check{\mathbb{P}}^d).$$

Then $ch : (\mathbb{P}^d)^n \longrightarrow Div^n(\check{\mathbb{P}}^d)$ is a $\mathcal{S}_n$-invariant morphism, and thus gives a morphism $ch : (\mathbb{P}^d)^{(n)} \longrightarrow Div^n(\check{\mathbb{P}}^d)$. As $(\mathbb{P}^d)^n$ is projective , we see that the image is closed. One checks that ch is an isomorphism onto its image. In particular we can also identify $X^{(n)}$ with its image in $Div^n(\check{\mathbb{P}}^d)$.

Now we want to define the Hilbert–Chow morphism $\rho : X^{[n]}_{red} \to X^{(n)}$. We will give it by defining a morphism $X^{[n]} \longrightarrow Div^n(\check{\mathbb{P}}^d)$ whose image is supported in $X^{(n)}$. $Div^n(\check{\mathbb{P}}^d)$ represents the contravariant functor associating to each scheme T the effective relative Cartier divisors $D \subset \check{\mathbb{P}}^d \times T$ of degree n. Relative means equivalently, either that D is flat over T or that the restriction to each fibre over a point $t \in T$ is a Cartier divisor. Thus in order to construct ρ we need a way to obtain effective Cartier divisors.

For this we first review a construction of Mumford which associates under suitable conditions to a coherent sheaf $\mathcal{F}$ on a scheme Y an effective Cartier divisor $div(\mathcal{F})$ on Y. We will not carry out the construction in full detail or in full generality but only give a sketch in the case of interest to us. The general construction can be found in [**MFK94**] Chap. 5 Sec. 3.

LEMMA 7.1.4. Let R be a local ring and M an R-module that admits a resolution

$$0 \to R^n \xrightarrow{\varphi} R^n \to M \to 0.$$

Then the class of $\det \varphi \in R/R^*$ depends only on M and not on the resolution chosen (that is, $\det \varphi$ is well defined up to multiplication by a unit in R); moreover, $\det \varphi$ is not a zero divisor.

PROOF. Fixing the map $R^n \xrightarrow{\psi} M \to 0$, the possible $\varphi' : R^n \to R^n$ are obtained from φ by composing with an automorphism α of R^n. Thus $det(\varphi') = det(\alpha)det(\varphi)$ and $det(\alpha) \in R^*$.

The map $\psi : R^n \to M$ corresponds to the choice of a set of generators of M as R-module. We obtain any other choice by successively adding and removing generators. Thus let $\psi : R^n \to M$ be given by $m_1, \ldots, m_n \in M$ and $\psi' : R^{n+1} \to M$ given by $m_1, \ldots, m_n, x \in M$. Then in M we have a relation

$$\sum_{i=1}^{n} a_i m_i + x = 0, \qquad a_i \in R,$$

and given a resolution $0 \to R^n \xrightarrow{\varphi} R^n \xrightarrow{\psi} M \to 0$ we get a resolution

$$0 \to R^{n+1} \xrightarrow{\varphi'} R^{n+1} \xrightarrow{\psi'} M \to 0, \qquad \varphi' = \begin{pmatrix} & \varphi & & 0 \\ a_1 & \ldots & a_n & 1 \end{pmatrix}$$

and $det(\varphi') = det(\varphi)$. □

REMARK 7.1.5. Let X be a smooth connected variety over k and let $\mathcal{F}$ be a coherent sheaf on X with $supp(\mathcal{F}) \neq X$. For an irreducible hypersurface $V \subset X$ let $[V]$ be its generic point, and put $R := \mathcal{O}_{X,[V]}$. Let M be the stalk of $\mathcal{F}$ at $[V]$. Then R is a discrete valuation ring, and by [**Har77**] III.6.11A, III.6.12A, M has homological dimension 1, i.e. there exists a free resolution

$$0 \to R^n \xrightarrow{\varphi} R^n \to M \to 0.$$

Note that the two modules on the left have the same rank, because $supp(\mathcal{F}) \neq X$. Let $m_V \in \mathbb{Z}_{\geq 0}$ be the valuation of $\det(\varphi) \in R$.

More geometrically we can describe this as follows: There is an open subset $U \subset X$ whose intersection with V is open, on which we have a resolution

$$0 \to \mathcal{O}^n \xrightarrow{\varphi} \mathcal{O}^n \to \mathcal{F} \to 0.$$

Then let m_V be the order of vanishing of $det(\varphi)$ on an open subset of V.

By the previous Lemma, m_V is independent of the choice of the resolution.

DEFINITION 7.1.6. Under the assumptions of Remark 7.1.5, let

$$div(\mathcal{F}) := \sum_V m_V V.$$

This is by definition an effective Cartier divisor on X. The sum is finite because m_V can only be nonzero if $V \subset supp(\mathcal{F})$.

We want to construct such a divisor $div(\mathcal{F})$ in a relative situation. Before stating and proving the main theorem to this effect, we collect a few useful results.

LEMMA 7.1.7. Let X be a smooth projective variety of dimension n and $\mathcal{F}$ a coherent sheaf on $X \times S$, flat over S. Then $\mathcal{F}$ admits a locally free resolution of length n.

PROOF. Step 1: prove that $\mathcal{F}$ is a quotient of a locally free sheaf.

For every $s \in S$, write $\mathcal{F}_s$ for $\mathcal{F}|_{X\times\{s\}}$. Since X is projective and S is of finite type, there exists $N > 0$ such that $\mathcal{F}_s(N)$ is generated by global sections and $H^i(\mathcal{F}_s(N)) = 0$ for all $i > 0$ and all $s \in S$. By flatness $\pi_*\mathcal{F}(N)$ is locally free, where $\pi : X \times S \to S$ is the projection. Therefore we obtain a surjection $\pi^*\pi_*\mathcal{F}(N) \to \mathcal{F}(N)$; tensoring by $\mathcal{O}(-N)$ yields a surjection $\mathcal{E}_0 \to \mathcal{F}$ where $\mathcal{E}_0 := (\pi^*\pi_*\mathcal{F}(N)) \otimes \mathcal{O}(-N)$ is locally free.

Step 2: there exists an exact sequence of coherent sheaves on $X \times S$

$$0 \to \mathcal{E}_n \to \mathcal{E}_{n-1} \to \ldots \to \mathcal{E}_0 \to \mathcal{F} \to 0$$

where each $\mathcal{E}_i$ with $i = 0, \ldots, n-1$ is locally free, and $\mathcal{E}_n$ is S-flat.

Let $\mathcal{E}_0 \to \mathcal{F}$ be the surjection constructed in step 1; note that its kernel is S-flat. Repeat step 1 on this kernel, and go on by induction.

Step 3 and final: prove that $\mathcal{E}_n$ is locally free.

The restriction of $\mathcal{E}_n$ to $X \times \{s\}$ is locally free for every $s \in S$ (by [**Har77**], Proposition III.6.12A). The statement is local: to conclude the proof it is therefore enough to use the following Lemma. □

LEMMA 7.1.8. Let R be a regular local ring and A any local ring in (Loc/k) (see Definition 6.1.12). Let M be a finitely generated $S := A \otimes_k R$-module, flat over A, and assume that $M \otimes_S R$ is free as an R-module. Then M is free as an S-module.

PROOF. Note that $M \otimes_S R = M \otimes_A k$. Let $\bar{m}_1, \ldots, \bar{m}_r$ be free generators of $M \otimes_S R$ as an R-module. Lift them to elements $m_1, \ldots, m_r$ in M, and consider the homomorphism $\alpha : S^{\oplus r} \to M$ defined by $\alpha(e_i) = m_i$; this is surjective by Nakayama's lemma. Consider the induced exact sequence

$$0 \to K \to S^{\oplus r} \xrightarrow{\alpha} M \to 0;$$

since M is A-flat, it stays exact after $\otimes_A k$, which implies that $K \otimes_A k = K \otimes_S R$ is zero. Again by Nakayama, this implies that K is zero. □

In the rest of this section we will consider also nonclosed points.

DEFINITION 7.1.9. Let Y be a scheme. A (not necessarily closed) point p of Y has depth d if *depth* $\mathcal{O}_{Y,p}$ is d. In particular it has depth 1 if $m_p \subset \mathcal{O}_{Y,p}$ contains a non zero divisor and for every non zero divisor $t \in m_p$ every element in the maximal ideal of the quotient ring $\mathcal{O}_{Y,p}/(t)$ is a zero divisor.

REMARK 7.1.10. If Y is a smooth variety, then points of depth 1 are generic points of prime divisors, i.e., closed irreducible hypersurfaces.

The proof of the following Lemma can be found in [**EGAIV4**], Prop. 21.1.8.

LEMMA 7.1.11. Let Y be a scheme and f, g two Cartier divisors on Y (i.e., global sections of $\mathcal{K}^*/\mathcal{O}^*$). Then $f = g$ if and only if their stalk at every depth 1 point is the same.

LEMMA 7.1.12. (1) Let $\mathcal{E}$ be a free coherent sheaf of rank r on a scheme Y. Then $\det \mathcal{E} := \Lambda^r \mathcal{E}$ is isomorphic to $\mathcal{O}_Y$; the isomorphism is unique up to a section of $\mathcal{O}_Y^*$.

(2) Let

$$0 \to \mathcal{E}_n \to \mathcal{E}_{n-1} \to \ldots \to \mathcal{E}_0 \to 0$$

be an exact sequence of locally free coherent sheaves on a scheme Y. Then there is a canonical isomorphism

$$\bigotimes_{i=0}^{n} (\det \mathcal{E}_i)^{(-1)^i} \to \mathcal{O}_Y.$$

PROOF. (1) is trivial. (2) will be proven by induction. Do first the case $n = 2$.

If $0 \to \mathcal{E}_2 \to \mathcal{E}_1 \to \mathcal{E}_0 \to 0$ is a short exact sequence of locally free sheaves, then there exists a *canonical* isomorphism

$$\det \mathcal{E}_2 \otimes \det \mathcal{E}_0 \simeq \det \mathcal{E}_1.$$

In fact, locally on an open set choose a basis $f_1, \ldots, f_{r_2}, g_1, \ldots, g_{r_0}$ of $\mathcal{E}_1$ such that the f_i are the image of a basis of $\mathcal{E}_2$ and the g_i map to a basis of $\mathcal{E}_0$ (both denoted by the same letters). Then the isomorphism is given by

$$(f_1 \wedge \ldots \wedge f_{r_2}) \otimes (g_1 \wedge \ldots \wedge g_{r_0}) \mapsto f_1 \wedge \ldots \wedge f_{r_2} \wedge g_1 \wedge \ldots \wedge g_{r_0}.$$

It is easy to verify that this is independent of the choice of the g_i and thus glues to a global isomorphism.

For the induction step, note that we can split the given exact sequence up into two exact sequences of locally free sheaves

$$0 \to \widetilde{\mathcal{E}} \to \mathcal{E}_1 \to \mathcal{E}_0 \to 0, \quad 0 \to \mathcal{E}_n \to \ldots \to \mathcal{E}_2 \to \widetilde{\mathcal{E}} \to 0$$

and obtain by induction a canonical isomorphism $\bigotimes_{i=0}^{n} \big(\det \mathcal{E}_i\big)^{(-1)^i} \simeq \mathcal{O}_Y$. □

THEOREM 7.1.13. Let X be a smooth irreducible projective variety. Let S be a scheme and let $\mathcal{F}$ be a coherent sheaf on $X \times S$, flat over S. Assume that $supp(\mathcal{F}_s) \neq X$ for all $s \in S$. Then there exists an effective Cartier divisor $div(\mathcal{F})$ on $X \times S$ such that

(1) The formation of $div(\mathcal{F})$ commutes with base change.
(2) If S is a point, then $div(\mathcal{F})$ is the same as in Definition 7.1.6.

PROOF. Let

$$0 \to \mathcal{E}_n \to \ldots \to \mathcal{E}_0 \to \mathcal{F} \to 0 \tag{7.1.1}$$

be a finite locally free resolution of $\mathcal{F}$ on $X \times S$, which exists by Lemma 7.1.7. Let $U \subset X \times S$ be an open subset such that the $\mathcal{E}_i$ are free on U. Let $V := U \setminus supp(\mathcal{F})$. Then by Lemma 7.1.12(2) we have on V a canonical isomorphism

$$\mathcal{O}_{X\times S}|_V \simeq \bigotimes_{i=0}^{n} \big(\det \mathcal{E}_i\big)^{(-1)^i}|_V.$$

On the other hand, as the $\mathcal{E}_i$ are free, by Lemma 7.1.12(1) there is an isomorphism

$$\bigotimes_{i=0}^{n} \big(\det \mathcal{E}_i\big)^{(-1)^i} \simeq \mathcal{O}_{X\times S} \qquad \text{on } U,$$

unique up to a unit. The composition of these isomorphisms defines a nonzero section $f \in \mathcal{O}_X^*(V)$, unique up to a section of $\mathcal{O}_X^*(U)$. Since V is an open dense subset of U, any section of $\mathcal{O}_X^*(V)$ defines a section of $\mathcal{K}^*(U)$; therefore we can associate to our choice of the resolution $\mathcal{E}_i$ and of the open subset U a Cartier divisor on U. We now want to prove that the Cartier divisors thus locally constructed depend only on $\mathcal{F}$ and not on the choice of resolution; this will imply that they glue to yield a global Cartier divisor.

To check that the Cartier divisor is well-defined, let $v \in X \times S$ be a point of depth 1, and let V be the closed subset of $X \times S$ whose generic point is v. Let $\widetilde{\mathcal{E}}$ be the kernel of $\mathcal{E}_0 \to \mathcal{F} \to 0$, so that we have an exact sequence of S-flat sheaves

$$0 \to \widetilde{\mathcal{E}} \xrightarrow{\varphi} \mathcal{E}_0 \to \mathcal{F} \to 0.$$

This sequence stays exact when we pass to the stalk at v; it becomes a sequence of finitely generated $\mathcal{O}_{X\times S,v}$-modules with $\mathcal{E}_0$ free. Since the local ring $\mathcal{O}_{X\times S,v}$ has depth 1, the module $\widetilde{\mathcal{E}}_v$ must also be free; therefore we can find an open neighborhood W of v in $X \times S$ (i.e., an open subset not disjoint from V) such that both $\mathcal{E}_i|_W$ and $\widetilde{\mathcal{E}}|_W$ are free.

On W we have a canonical isomorphism

$$\det \widetilde{\mathcal{E}} \simeq \bigotimes_{i=1}^{n} \left(\det \mathcal{E}_i\right)^{(-1)^{i-1}},$$

and $div(\mathcal{F})$ coincides on W with $\det \varphi$ (as in Lemma 7.1.4).

As two Cartier divisors are equal if they coincide at points of depth 1 by Lemma 7.1.11, this shows that (f) is independent of the choice of the resolution and thus glues to give an effective Cartier divisor $div(\mathcal{F})$ on $X \times S$.

Moreover, as the points of depth 1 on a smooth variety are precisely the generic points of prime divisors, it also shows that in case S is a point we get the same definition as in Definition 7.1.6.

Let $h : T \to S$ be a morphism. For a sheaf $\mathcal{E}$ on $X \times S$ we denote by $\mathcal{E}_T$ its pullback via $id_X \times h$. Using flatness of $\mathcal{F}$, one checks that if $0 \to \mathcal{E}_n \to \ldots \to \mathcal{E}_0 \to \mathcal{F} \to 0$ is a resolution of $\mathcal{F}$ on $X \times S$, then

$$0 \to (\mathcal{E}_n)_T \to \ldots \to (\mathcal{E}_0)_T \to (\mathcal{F})_T \to 0$$

is a resolution of $(\mathcal{F})_T$. It follows that the pullback of $div(\mathcal{F})$ is $div(\mathcal{F}_T)$, thus $div(\mathcal{F})$ is compatible with base change. □

Finally we can construct the Hilbert–Chow morphism. Let $H \subset \mathbb{P}^d \times \check{\mathbb{P}}^d$ be the incidence correspondence. H is a fibre bundle over $\mathbb{P}^d$ with fibre $\mathbb{P}^{d-1}$. We denote by $p, \check{p}$ the projections to $\mathbb{P}^d$ and $\check{\mathbb{P}}^d$ respectively. Let S be a scheme, and let $Z \subset \mathbb{P}^d \times S$ be a closed subscheme, flat of degree n over S. Let $p_S := p \times id_S$, $\check{p}_S := \check{p} \times id_S$. Let $Z^* := p_S^{-1}(Z) \subset H \times S$. Let $\mathcal{F} := (\check{p}_S)_*(\mathcal{O}_{Z^*})$. Then $\mathcal{F}$ is a coherent sheaf on $\check{\mathbb{P}}^d \times S$, flat over S. Let Z_s be the fibre of Z over $s \in S$. Then

$$supp(\mathcal{F}_s) = \left\{ l \in \check{\mathbb{P}}^d \mid l \cap Z_s \neq 0 \right\},$$

in particular $supp(\mathcal{F}_s) \neq \check{\mathbb{P}}^d$. Thus $div(\mathcal{F})$ is a relative Cartier divisor on $\check{\mathbb{P}}^d \times S$. Thus we have constructed a morphism $\rho : X^{[n]} \to Div^n(\check{\mathbb{P}}^d)$. Finally we can check from the definitions that $div(\mathcal{F}_s) = \sum_{p \in supp(Z_s)} len(\mathcal{O}_{Z,p}) H_p$. Thus we see that the support of the image of $X^{[n]}$ is $X^{(n)}$, so if we give $X^{[n]}$ the reduced structure, the morphism will factor through $X^{(n)}$. Thus we have shown the following theorem.

THEOREM 7.1.14. Let X be a smooth quasiprojective variety. There is a surjective morphism $\rho : X^{[n]}_{red} \to X^{(n)}$, given on the level of points by $Z \mapsto \sum_{p \in supp(Z)} len(\mathcal{O}_{Z,p})[p]$.

7.2. Irreducibility and nonsingularity

We will show that, if X is a nonsingular quasiprojective curve or surface over k, then $X^{[n]}$ is irreducible and nonsingular. The proof of the following Lemma and Theorem is in part inspired by [**Leh04**].

LEMMA 7.2.1. Let X be a connected variety over k. Then $X^{[n]}$ is connected for all $n \geq 0$.

PROOF. First we recall that the Quot scheme allows us to define the projectivization of any coherent sheaf on X.

For a coherent sheaf $\mathcal{F}$ on X let $\mathbb{P}(\mathcal{F}) := Quot^1(\mathcal{F})$, thus $\mathbb{P}(\mathcal{F})$ parametrizes 1-dimensional quotients of the fibres of $\mathcal{F}$. $\mathbb{P}(\mathcal{F})$ is a quasiprojective scheme with a morphism to X, and the fibre of $\mathbb{P}(\mathcal{F})$ over $x \in X$ is

$$\mathbb{P}(\{\lambda : \mathcal{F}(x) \to k(x) \text{ surjection}\}) \simeq \mathbb{P}(\mathcal{F}(x)).$$

Now we want to show the claim by induction on n. $X^{[0]}$ is one point corresponding to the empty set. Assume we have shown that $X^{[n]}$ is connected. Then $X \times X^{[n]}$ is connected. Let $Z_n(X) \subset X \times X^{[n]}$ be the universal family with ideal sheaf $\mathcal{I}_{Z_n(X)}$. Let $\mathbb{P} := \mathbb{P}(\mathcal{I}_{Z_n(X)})$, with the projection $\pi : \mathbb{P} \to X \times X^{[n]}$. The fibre of $\mathbb{P}$ over (x, Z) is a projective space and thus connected. Thus $\mathbb{P}$ is connected.

On $X \times \mathbb{P}$ we have a universal exact sequence

$$0 \to \mathcal{I} \to (\mathcal{I}_{\pi^{-1}(Z_n(X))}) \to Q \to 0.$$

Here Q is the universal quotient line bundle, and the kernel $\mathcal{I}$ is an ideal sheaf on $X \times \mathbb{P}$ defining a subscheme $\mathcal{Z} \subset X \times \mathbb{P}$. The above exact sequence gives rise to an exact sequence

$$0 \to Q \to \mathcal{O}_{\mathcal{Z}} \to \mathcal{O}_{\pi^{-1}Z_n(X)} \to 0.$$

As $\mathcal{O}_{\pi^{-1}Z_n(X)}$ is flat of degree n over $\mathbb{P}$ and Q is flat of degree 1, we see that $\mathcal{Z}$ is flat of degree $(n+1)$ over $\mathbb{P}$. Thus we have a morphism $\psi : \mathbb{P} \to X^{[n+1]}$, which on points is given by sending $(\lambda : \mathcal{I}_Z \to k(x))$ to the subscheme of X with ideal $ker(\lambda)$. We want to see that ψ is surjective. Let $W \in X^{[n+1]}$. Let $p \in supp(W)$. Choose $f \in \mathcal{O}_W$ an element in the kernel of the multiplication by the maximal ideal m at p. Let $Z \subset W$ be the subscheme with ideal (f). Then $Z \in X^{[n]}$. Let $f = g_0, g_1, \ldots, g_k$ be a basis of $\mathcal{I}_Z/m\mathcal{I}_Z$ and define $\lambda : \mathcal{I}_Z \to k(p)$ by $\sum a_i g_i \mapsto a_0$. Then $\mathcal{I}_W = ker(\lambda)$. Thus ψ is surjective, and thus $X^{[n+1]}$ is connected. □

REMARK 7.2.2. If X is a smooth irreducible surface S, one can show that $\mathbb{P} = \mathbb{P}(I_{Z_n(S)})$ is just the blowup of $S \times S^{[n]}$ along the universal family $Z_n(S)$ [**Ell**],[**ES98**]. There it is also shown that $\mathbb{P}(I_{Z(S)})$ is the incidence correspondence

$$S^{[n,n+1]} := \Big\{(Z, W) \in S^{[n]} \times S^{[n+1]} \mid Z \subset W\Big\}.$$

Here $Z \subset W$ means that Z is a subscheme of W. One often uses $S^{[n,n+1]}$ to understand properties of the Hilbert schemes of points inductively.

Let X be a nonsingular quasiprojective variety of dimension d. Let $X_0^n \subset X^n$ be the dense open set of $(p_1, \ldots, p_n)$ with the p_i distinct. Let $X_0^{(n)}$ be its image in $X^{(n)}$, which parametrizes effective zero cycles $\sum_i [p_i]$ with the p_i distinct. This is also open and dense. As $\mathcal{S}_n$ acts freely on $(X^n)_0$ we see that $X_0^{(n)}$ is nonsingular of dimension nd. Let $X_0^{[n]}$ be the preimage in $X^{[n]}$. One checks that at any point of $X_0^{[n]}$ the dimension of the tangent space is nd and that $\rho|_{X_0^{[n]}}$ is an isomorphism. Thus $X^{[n]}$ contains a nonsingular open subset which is isomorphic to an open subset of $X^{(n)}$. In the case that X is a curve or a surface one can use this to show that $X^{[n]}$ is nonsingular and irreducible.

THEOREM 7.2.3.

(1) Let C be an irreducible nonsingular quasiprojective curve and $n \geq 0$. Then $C^{[n]}$ is nonsingular and irreducible of dimension n.
(2) (Fogarty [**Fog68**]) Let S be an irreducible nonsingular quasiprojective surface and $n \geq 0$. Then $S^{[n]}$ is nonsingular and irreducible of dimension $2n$.

PROOF. Let $X = C$ or $X = S$ and let $d = dim(X)$. As $X^{[n]}$ is connected and contains a nonsingular open subset of dimension nd, it is enough to show that the dimension of the tangent space $T_{[Z]}X^{[n]}$ is nd for all $[Z] \in X^{[n]}$. This will first show the nonsingularity of $X^{[n]}$ at any point in the closure $\overline{X_0^{[n]}}$. If $X^{[n]}$ was reducible, then by connectedness there would be another irreducible component intersecting $\overline{X_0^{[n]}}$, and any intersection point would be a singular point of $X^{[n]}$.

We know $T_{[Z]}X^{[n]} = \mathrm{Hom}_{\mathcal{O}_X}(\mathcal{I}_Z, \mathcal{O}_Z)$. Applying $\mathrm{Hom}(\bullet, \mathcal{O}_Z)$ to $0 \to \mathcal{I}_Z \to \mathcal{O}_X \to \mathcal{O}_Z \to 0$ we obtain

$$\begin{aligned} \mathrm{Hom}_{\mathcal{O}_X}(\mathcal{O}_Z, \mathcal{O}_Z) \to & \mathrm{Hom}_{\mathcal{O}_X}(\mathcal{O}_X, \mathcal{O}_Z) \to \mathrm{Hom}_{\mathcal{O}_X}(\mathcal{I}_Z, \mathcal{O}_Z) \\ & \to \mathrm{Ext}^1_{\mathcal{O}_X}(\mathcal{O}_Z, \mathcal{O}_Z). \end{aligned}$$

The first map is an isomorphism $k^n \to k^n$. Thus $\mathrm{Hom}(\mathcal{I}_Z, \mathcal{O}_Z) \subset \mathrm{Ext}^1(\mathcal{O}_Z, \mathcal{O}_Z)$, and it is enough to show that $ext^1(\mathcal{O}_Z, \mathcal{O}_Z) \leq nd$.

In the case of a curve C we have $\mathrm{Hom}(\mathcal{O}_Z, \mathcal{O}_Z) = H^0(\mathcal{O}_Z) = k^n$, and by Serre duality $\mathrm{Ext}^1(\mathcal{O}_Z, \mathcal{O}_Z) = H^0(\mathcal{O}_Z \otimes K_C)^\vee = k^n$.

In the case of a surface S we have $\mathrm{Hom}(\mathcal{O}_Z, \mathcal{O}_Z) = H^0(\mathcal{O}_Z) = k^n$ and by Serre duality $\mathrm{Ext}^2(\mathcal{O}_Z, \mathcal{O}_Z) = H^0(\mathcal{O}_Z \otimes K_S)^\vee = k^n$. Thus it suffices to show that

$$\chi(\mathcal{O}_Z, \mathcal{O}_Z) = \sum_{i=0}^{2} ext^i(\mathcal{O}_Z, \mathcal{O}_Z) = 0.$$

Let $0 \to \mathcal{E}_l \to \ldots \to \mathcal{E}_0 \to \mathcal{O}_Z \to 0$ be a locally free resolution of $\mathcal{O}_Z$ on S. Then $\sum_i (-1)^i rk(\mathcal{E}_i) = 0$ and

$$\chi(\mathcal{O}_Z, \mathcal{O}_Z) = \sum_{i=0}^{l} (-1)^i \chi(\mathcal{E}_i, \mathcal{O}_Z) = \sum_{i=0}^{l} (-1)^i n \cdot rk(\mathcal{E}_i) = 0.$$

□

REMARK 7.2.4. Note that in this proof we do not show that the obstruction space $\mathrm{Ext}^1_{\mathcal{O}_X}(\mathcal{I}_Z, \mathcal{O}_Z)$ vanishes; in fact it usually will not.

REMARK 7.2.5. Let X be a nonsingular variety. Then $X^{[n]}$ is nonsingular for $n \le 3$.

PROOF. Let $d = dim(X)$. It is enough to show that $hom_{\mathcal{O}_X}(\mathcal{I}_Z, \mathcal{O}_Z) \le dn$ for all $[Z] \in X^{[n]}$. Obviously

$$\mathrm{Hom}_{\mathcal{O}_X}(\mathcal{I}_Z, \mathcal{O}_Z) = \bigoplus_{p \in supp(Z)} \mathrm{Hom}_{\mathcal{O}_{X,p}}(\mathcal{I}_{Z,p}, \mathcal{O}_{Z,p}).$$

Thus we can reduce to the case that $supp(Z)$ is a point p. Let m be the maximal ideal at p. Is is easy to show that there are local parameters $x_1, \dots, x_d$ at p such that $\mathcal{O}_{Z,p}$ is of the form

$$\begin{cases} k[x_1, \dots, x_d]/m & n = 1, \\ k[x_1, \dots, x_d]/(m^2 + (x_2, \dots, x_d)) & n = 2, \\ \\ k[x_1, \dots, x_d]/(m^3 + (x_2, \dots, x_d)) & n = 3. \\ \text{or } k[x_1, \dots, x_d]/(m^2 + (x_3, \dots, x_d)) \end{cases}$$

In all cases one easily checks that $Hom_{\mathcal{O}_{X,p}}(\mathcal{I}_{Z,p}, \mathcal{O}_{Z,p}) = d\,len(\mathcal{O}_{Z,p})$. □

REMARK 7.2.6. Let X be nonsingular of dimension 3. Then $X^{[4]}$ is singular.

PROOF. Let $[Z] \in X^{[4]}$ be the point $\mathcal{O}_Z = \mathcal{O}_p/m^2$. Then

$$\mathrm{Hom}_{\mathcal{O}_X}(\mathcal{I}_Z, \mathcal{O}_Z) = \mathrm{Hom}_k(m^2/m^3, m/m^2) = k^{18}.$$

So the dimension is bigger than $dn = 12$. □

REMARK 7.2.7. Let X be a nonsingular variety. Note that a 0-dimensional subscheme $Z \subset X$ of length at most 3 has embedding dimension 2, i.e. is locally a subscheme of a nonsingular surface. In fact it is not difficult to see that if $Z \subset X^{[n]}$ has embedding dimension at most 2, then $[Z]$ is a nonsingular point of X.

EXAMPLE 7.2.8. An example of Iarrobino [**Iar72**] shows that if X is nonsingular of dimension $d \ge 3$ and n is sufficiently large, then $X^{[n]}$ is reducible.

The idea is very simple: Iarrobino shows that for large n the dimension of the closed subset $\rho^{-1}(n[p])$ of subschemes of $X^{[n]}$ concentrated in a point $p \in X$ is larger or equal to nd. On the other hand we know that the locus $X_0^{[n]}$ consisting of n distinct points is open in $X^{[n]}$ and has dimension nd. Thus its closure is an irreducible component of dimension nd, which cannot contain $\rho^{-1}(n[p])$.

Now we check the result about the dimension of $\rho^{-1}(n[p])$. Let m be the maximal ideal in $\mathcal{O}_{X,p}$. We consider subvector spaces of the form $I = q^{-1}(V) \subset \mathcal{O}_{X,p}$ where V is a subvector space of codimension l of m^k/m^{k+1} and $q := \mathcal{O}_{X,p} \to \mathcal{O}_{X,p}/m^{k+1}$ is the quotient map. Note that, for any I as above, in I/m^{k+1} the multiplication by any element of m is the zero map. Thus I is an ideal in $\mathcal{O}_{X,p}$ defining a subscheme $Z \in \rho^{-1}(n[p])$, with $n := \binom{k+d-1}{d} + l$. Thus we get a closed subset of $\rho^{-1}(n[p])$ which is isomorphic to the Grassmannian of l-dimensional quotients of the $s := \binom{k+d-1}{d-1}$-dimensional vector space m^k/m^{k+1}, thus it has dimension $l(s-l)$.

If $d \geq 3$ and k is sufficiently large and l is near $s/2$ then it is easy to see that $l(s-l) \geq nd$. For instance for $d=3, k=7$ we get $s := 36$ and $\binom{k+d-1}{d} := 84$. Thus e.g. for $l=12$ we get $n=96$ and $l(s-l) = 288 = 3n$. Thus $X^{[96]}$ is reducible for X a smooth variety of dimension 3.

REMARK 7.2.9. One can show that (see [**Iar85**])

(1) $(\mathbb{P}^d)^{[n]}$ is irreducible for $n \leq 7$ and all d,
(2) $(\mathbb{P}^d)^{[8]}$ is irreducible for $d=3$ and reducible for $d \geq 4$.

It is not difficult to show that these results also hold for X any smooth quasiprojective variety of dimension d instead of $\mathbb{P}^d$.

7.3. Examples of Hilbert schemes

Let X be a nonsingular projective variety of dimension d. We give some examples of $X^{[n]}$ for small values of n.

EXAMPLE 7.3.1.

(1) $X^{[0]}$ is one reduced point, corresponding to the empty subscheme of X.
(2) Subschemes of length 1 of X are just points of X and $X^{[1]} = X$. The universal family is just the diagonal $\Delta \subset X \times X$.
(3) Points in $X^{[2]}$ are either a set $\{p_1, p_2\}$ of distinct points on X or a subscheme Z of length 2 concentrated in one point p. Let m be the ideal of p. Then $m \supset \mathcal{I}_Z \supset m^2$. Thus $\mathcal{I}_Z$ is given by a one-codimensional subspace of m/m^2, i.e. by a point in $\mathbf{P}(T_{X,p}) = \mathbb{P}^{n-1}$. In other words a point in $X^{[2]}$ is either a set of two points in X or a point p and a tangent direction at p.

    This allows us to describe $X^{[2]}$ globally as follows: Let $\widehat{X^2}$ be the blowup of $X \times X$ along the diagonal. Let E be the exceptional divisor. The action of $\mathcal{S}_2$ on X^2 extends to $\widehat{X^2}$. Let Y be the quotient. E is the fixed locus of the nontrivial element of $\mathcal{S}_2$. As this is a divisor, we see that Y is nonsingular. It is easy to see that $\pi : \widehat{X^2} \to Y$ is flat of degree 2. This gives a morphism $Y \to X^{[2]}$, which is birational and bijective. Thus by Zariski's Main Theorem it is an isomorphism.

EXERCISE 7.3.2. Verify that $Hilb^2\mathbb{A}^2$ is the blow-up of the second symmetric power of $\mathbb{A}^2$ along the diagonal. Hint: construct explicitly a universal family. In fact, this is true if you replace $\mathbb{A}^2$ by any smooth algebraic variety.

For general n we can say something about $X^{[n]}$ if X has dimension 1 or 2. In the case of a nonsingular curve the Hilbert schemes of points are just the symmetric powers.

PROPOSITION 7.3.3. Let C be a nonsingular quasiprojective curve. Then $\rho : C^{[n]} \to C^{(n)}$ is an isomorphism.

PROOF. As the local ring of C at a point p is a discrete valuation ring, all ideals in $\mathcal{O}_{C,p}$ are powers of the maximal ideal m_p. Thus for all $[Z] \in C^{[n]}$ we have

$$\mathcal{O}_Z = \bigoplus_i \mathcal{O}_{C,p_i}/m_{p_i}^{n_i}, \qquad \sum_i n_i = n.$$

Then ρ sends Z to $\sum_i n_i[p_i]$. Thus ρ is bijective. As it is also birational, it is an isomorphism by Zariski's Main Theorem.

Alternatively one can see that

$$\pi : C \times C^{(n-1)} \to C^{(n)}, \left(p, \sum n_i[p_i]\right) \mapsto \sum n_i[p_i] + [p]$$

is flat of degree n over $C^{(n)}$, defining an inverse to ρ. □

Now let S be a nonsingular surface. We have seen that the open subset $S_0^{(n)}$ of $S^{(n)}$ consisting of sums of n distinct points is nonsingular. Now we want to see that $S^{(n)} \setminus S_0^{(n)}$ is the singular locus of $S^{(n)}$. By Example 7.1.3 this is true for $n = 2$, because $S^{(2)} \setminus S_0^{(2)}$ is the image of the diagonal in S^2 under the quotient map $S^2 \to S^{(2)}$. Let $\Delta = \bigcup_{1 \le i < j \le n} \Delta_{i,j}$ be the big diagonal in S^n, where $\Delta_{i,j} = \{(x_1, \ldots, x_n) \in S^n \mid x_i = x_j\}$, and let Δ_0 be the open subset where precisely 2 of the x_i coincide. Let $p \in \Delta_0$; we can assume that $p = (z, z, z_3, \ldots, z_n) \in \Delta_{1,2}$. Then the stabilizer of p in S_n is $\{1, \tau\}$ where τ is the transposition of the first and second factor. Thus by Example 7.1.3 we see that a formal neighborhood of the image of p in $S^{(n)}$ (i.e. the completed local ring of $S^{(n)}$ at p) is isomorphic to $k[[u, v, w, x', y', x_3, y_3, \ldots, x_n, y_n]]/(uw - v^2)$. Therefore all the points in the image of Δ_0 are singular points of $S^{(n)}$. The closure of the image of Δ_0 is $S^{(n)} \setminus S_0^{(n)}$. As the singular locus of $S^{(n)}$ is closed, it follows that the singular locus of $S^{(n)}$ is precisely $S^{(n)} \setminus S_0^{(n)}$.

THEOREM 7.3.4. (Fogarty [**Fog68**]) *Let S be a nonsingular quasiprojective surface. Then $\rho : S^{[n]} \to S^{(n)}$ is a resolution of singularities.*

PROOF. $S^{[n]}$ is nonsingular and irreducible, and ρ is an isomorphism over the open subset $S_0^{(n)}$, which is the nonsingular locus of $S^{(n)}$. Thus it is a resolution of singularities. □

REMARK 7.3.5.

(1) One of the reasons for the interest in the Hilbert schemes of points on a surface is that they give canonical resolutions of the singularities of the symmetric powers.
(2) In fact it turns out [**Bea83**] that $\rho : S^{[n]} \to S^{(n)}$ has a particular nice property: $S^{(n)}$ is Gorenstein and $K_{S^{[n]}} = \rho^*(K_{S^{(n)}})$. One says that ρ is a *crepant* resolution of singularities.
(3) In case S is a K3-surface or an Abelian variety it is also shown in [**Bea83**] that $S^{[n]}$ is homomorphic symplectic, i.e. it carries an everywhere nondegenerate holomorphic two-form. This is an additional reason for interest in Hilbert schemes of points because holomorphic symplectic varieties are very rare. In fact almost all known holomorphic symplectic projective varieties can be related to Hilbert schemes of points.

7.4. A stratification of the Hilbert schemes

For the rest of Chapter 7 let S be a smooth projective surface over the complex numbers.

We want to study a natural stratification of $S^{(n)}$ and $S^{[n]}$.

DEFINITION 7.4.1. For any partition $\nu = (n_1, \dots, n_r)$ of n (i.e. $n_1 \geq n_2 \geq \dots \geq n_r > 0$ and $\sum n_i = n$), we define a locally closed subset

$$S_\nu^{(n)} := \{\sum n_i[x_i] \in S^{(n)} \mid x_i \in S \text{ distinct points}\}$$

of $S^{(n)}$. We have thus a stratification

$$S^{(n)} = \coprod_\nu S_\nu^{(n)}$$

into locally closed strata. Putting $S_\nu^{[n]} := \rho^{-1}(S_\nu^{(n)})$ we also obtain a stratification of $S^{[n]}$ into locally closed strata.

Now we want to study the strata $S_\nu^{(n)}$. First we note that they are nonsingular. Let $\nu = (n_1, \dots, n_r)$. Write $\nu = (1^{\alpha_1}, 2^{\alpha_2}, \dots, s^{\alpha_s})$ where α_i is the number of times i occurs in $(n_1, \dots, n_r)$. Let $(S^{\alpha_1} \times \dots \times S^{\alpha_r})_* \subset S^{\alpha_1} \times \dots \times S^{\alpha_r}$ be the open subset where all components $p_i \in S$ are distinct. Then $S_\nu^{(n)}$ is the quotient of $(S^{\alpha_1} \times \dots \times S^{\alpha_s})_*$ by the action of $\mathcal{S}_{\alpha_1} \times \dots \times \mathcal{S}_{\alpha_s}$, where each $\mathcal{S}_{\alpha_i}$ permutes the factors of S^{α_i}. As this action is free, $S_\nu^{(n)}$ is nonsingular, and it has dimension $\alpha_1 + \dots + \alpha_s = r$.

EXERCISE 7.4.2. Let $\overline{S}_\nu^{(n)}$ be the closure of $S_\nu^{(n)}$ in $S^{(n)}$. Let

$$\pi_\nu : S^{(\alpha_1)} \times \dots \times S^{(\alpha_s)} \to S^{(n)}; \ (\xi_1, \dots, \xi_s) \mapsto \sum_{i=1}^s i\xi_i.$$

Let $(S^{(\alpha_1)} \times \dots \times S^{(\alpha_s)})_*$ be the quotient of $(S^{\alpha_1} \times \dots \times S^{\alpha_s})_*$ by $\mathcal{S}_{\alpha_1} \times \dots \times \mathcal{S}_{\alpha_s}$).

Check that the argument above shows that the restriction of π_ν to $(S^{(\alpha_1)} \times \dots \times S^{(\alpha_s)})_*$ is an isomorphism with $S_\nu^{(n)}$. Thus the image of π_ν is the closure $\overline{S}_\nu^{(n)}$. Show that $\pi_\nu : S^{(\alpha_1)} \times \dots \times S^{(\alpha_s)} \to \overline{S}_\nu^{(n)}$ is the normalization.

DEFINITION 7.4.3. Let $H_n := Hilb^n(\mathrm{Spec}(k[[x,y]]/(x,y)^n)$, parametrizing the ideals of colength n in $k[[x,y]]$ (note that every ideal of colength n in $k[[x,y]]$ contains $(x,y)^n$). The H_n are called the *punctual* Hilbert schemes.

PROPOSITION 7.4.4. Let $\nu := (n_1, \dots, n_r)$ be a partition of n. Then via the Hilbert–Chow morphism $S_\nu^{[n]}$ is a locally trivial fibre bundle (in the analytic topology and in the étale topology) with fibre $H_{n_1} \times \dots \times H_{n_r}$ over $S_\nu^{(n)}$.

PROOF. First we look at the worst stratum $\rho : S_{(n)}^{[n]} \to S_{(n)}^{(n)} \simeq S$. For a point p in S the fibre $\rho^{-1}(n[p])$ is the set of subschemes Z of length n in S with support p. The ideal of any such scheme is contained in m_p^n. Then the choice of holomorphic local coordinates x, y in a neighborhood U of p determines an isomorphism

$$\rho^{-1}(U) \simeq U \times H_n.$$

Thus $\rho : S_{(n)}^{(n)} \to S$ is a locally trivial fibre bundle with fibre H_n in the analytic topology. Being a bit more careful, we can replace in this argument local coordinates by local parameters in a Zariski open neighborhood of p (i.e. x, y, such that dx, dy span the cotangent space at every point of U). Thus this is even a locally trivial fibre bundle in the Zariski topology.

Now let $\nu = (n_1, \dots, n_r) = (1^{\alpha_1}, 2^{\alpha_2}, \dots, s^{\alpha_s})$ be a partition of n. Let $\xi := n_1[p_1] + \dots + n_r[p_r] \in S_\nu^{(n)}$. Then the fibre $\rho^{-1}(\xi)$ is isomorphic to $H_{n_1} \times \dots \times H_{n_r}$.

In fact we can choose (in the analytic topology) disjoint open neighborhoods U_i of the p_i in S, and these give rise to an open neighborhood U of ξ in $S_\nu^{(n)}$ such that

$$\rho^{-1}(U) \simeq U \times H_{n_1} \times \ldots \times H_{n_r}.$$

Thus $\rho : S_\nu^{[n]} \to S_\nu^{(n)}$ is a locally trivial fibre bundle in the analytic topology with fibre $H_{n_1} \times \ldots \times H_{n_r}$.

Again being more careful we can prove more: the bundle is locally trivial in the étale topology: We have

$$(S^{\alpha_1+\ldots+\alpha_s})_* \times_{S_\nu^{(n)}} S_\nu^{[n]} = (S_{(n_1)}^{[n_1]} \times \ldots \times S_{(n_r)}^{[n_r]})_*,$$

where the $*$ on the right hand side indicates that the supports of the subschemes are distinct. $(S^{\alpha_1+\ldots+\alpha_s})_* \to S_\nu^{(n)}$ is étale, and $(S_{(n_1)}^{[n_1]} \times \ldots \times S_{(n_r)}^{[n_r]})_* \to S^r$ is a locally trivial fibre bundle in the Zariski topology; thus $S_\nu^{[n]} \to S_\nu^{(n)}$ is étale locally trivial. □

Now we want to have a look at the fibres of ρ. By the above, we only need to look at the punctual Hilbert schemes H_n. They have been studied quite extensively (see for instance [**Iar77**], [**Bri77**]). For an overview see also [**Iar87**].

It is clear that H_1 is one point corresponding to the maximal ideal (x, y) and $H_2 = \mathbb{P}_1$ corresponding to the ideals given by the 1-codimensional linear subspaces of $(x, y)/(x, y)^2$.

For $n \geq 3$ we have to distinguish two cases depending on the embedding dimension of the scheme Z. A scheme $Z \in H_n$ is called *curvilinear* if its embedding dimension is 1. This means that it locally lies on a smooth curve. We denote $H_n^c \subset H_n$, the open subscheme of curvilinear subschemes.

REMARK 7.4.5. H_n^c is a locally trivial $\mathbb{A}^{n-2}$ bundle over $\mathbb{P}^1$. In particular H_n^c is irreducible of dimension $n-1$.

PROOF. We have a morphism $H_n^c \to \mathbb{P}^1$, sending Z to its tangent space at $(0,0)$. Let $U \subset H_n^c$ be the inverse image of $\mathbb{A}^1 = \mathbb{P}^1 \setminus (x = 0)$. Then $Z \in U$ lies on a curve C which is given as the graph of a power series $y = \sum_{i \geq 1} a_i x^i$. Thus we can write

$$I_Z = (y - a_1 x - \ldots - a_{n-1} x^{n-1}, x^n).$$

Thus U is isomorphic to $\mathbb{A}^{n-1}$ with coordinates $a_1, \ldots, a_{n-1}$, and the map to $\mathbb{A}^1$ is given by $(a_1, \ldots, a_{n-1}) \mapsto a_1$. Similarly one argues over the inverse image of $\mathbb{P}^1 \setminus (y = 0)$. □

In the case $n = 3$ we see that the only subscheme which is not curvilinear is the scheme with ideal $(x, y)^2$ and H_3^c is dense in H_3. In fact this is true in general.

THEOREM 7.4.6. [**Bri77**] H_n is irreducible and H_n^c is open and dense in H_n.

Another proof of this irreducibility in [**Ell**], [**ES98**] makes use of the fact that one has an inductive description of the $S^{[n]}$ via the incidence varieties $S^{[n,n+1]} = \{(Z, W) \in S^{[n]} \times S^{[n+1]} \mid Z \subset W\} = P(I_{Z_n(S)})$ (see Remark 7.2.2).

7.5. The Betti numbers of the Hilbert schemes of points

Now I want to summarize what we have shown so far and put it into context. Then we will use it to compute the Betti numbers of the Hilbert schemes of points on a surface. In this section we assume that S is a smooth projective surface over $\mathbb{C}$.

DEFINITION 7.5.1. Let $f : X \to Y$ be a projective morphism of varieties over $\mathbb{C}$. Suppose that Y has a stratification

$$Y = \coprod_{\alpha} Y_\alpha$$

into locally closed subvarieties, such that the Y_α are all nonsingular. Write $X_\alpha := f^{-1}(Y_\alpha)$. Assume that for all α the restriction $f : X_\alpha \to Y_\alpha$ is a locally trivial fibre bundle with fibre F_α in the analytic topology.

Then f is called *strictly semismall* (with respect to the stratification) if for all α

$$2\, dim(F_\alpha) = codim(Y_\alpha).$$

The results of the last section can be translated into the following.

PROPOSITION 7.5.2. $\rho : S^{[n]} \to S^{(n)}$ is strictly semismall with respect to the stratification by the $S^{(n)}_\nu$. Furthermore the fibres of ρ are irreducible.

PROOF. We know that $S^{[n]}$ is nonsingular and irreducible of dimension $2n$, which is also the dimension of $S^{(n)}$. For $\nu = (n_1, \ldots, n_r)$ a partition of n, we have seen that $S^{(n)}_\nu$ is nonsingular of dimension $2r$, i.e. its codimension is $\sum_{i=1}^r 2(n_i - 1)$. We have shown that $S^{[n]}_\nu \to S^{(n)}_\nu$ is a locally trivial fibre bundle in the analytic topology with fibre $H_{n_1} \times \ldots \times H_{n_r}$ and the H_{n_i} are irreducible of dimension $n_i - 1$. This shows the claim. □

This result can be used to compute the Betti numbers of the Hilbert schemes. One can use the properties of intersection cohomology and perverse sheaves to get a short argument, almost entirely without computations. The disadvantage of this approach is that one has to use advanced tools and results. We will here describe these tools very briefly. One can then use them as a black box, i.e. one can determine the Betti numbers of the Hilbert schemes quite easily without having to really understand what intersection cohomology and perverse sheaves are.

Let Y be an algebraic variety over $\mathbb{C}$. We will only use the analytic topology over $\mathbb{C}$ and all cohomology considered is with $\mathbb{Q}$ coefficients. We write $H^i(Y) := H^i(Y, \mathbb{Q})$.

Let IC_Y be the intersection cohomology complex on Y. This is a complex of sheaves on Y (strictly speaking an element in the derived category), with the property that its cohomology $IH^i(Y) := H^i(Y, IC_Y)$ is the intersection cohomology of Y. The intersection cohomology groups $IH^i(Y)$ are defined for any algebraic variety Y and fulfill Poincaré duality between $IH^{-i}(Y)$ and $IH^i(Y)$. If Y is smooth and projective of complex dimension n, then $IC_Y = \mathbb{Q}_Y[n]$ is just the constant sheaf $\mathbb{Q}$ put in degree n. Thus in this case $IH^{i-n}(Y) = H^i(Y)$. More generally the same is true if $Y = X/G$ is the quotient of a smooth projective variety by a finite group. One of the most powerful results about intersection cohomology and perverse sheaves is the decomposition theorem of [**BBD82**] for proper morphisms of algebraic varieties. This result becomes much simpler for semismall morphisms and in this case computes the intersection cohomology of X in terms of the intersection cohomologies of the closures $\overline{Y}_\alpha$ of the strata of Y.

LEMMA 7.5.3. [**GS93**] Let $f : X \to Y$ be a proper morphism, which is strictly semismall with irreducible fibres with respect to a stratification $Y = \coprod_\alpha Y_\alpha$ into locally closed nonsingular strata. Then $Rf_*(IC_X) = \sum_\alpha IC_{\overline{Y}_\alpha}$.

Here Rf_* is the pushforward in the derived category, and $\overline{Y}_\alpha$ is the closure of Y_α.

LEMMA 7.5.4. [**GS93**] Let $\pi : X \to Y$ be a finite birational morphism of irreducible algebraic varieties. Then $R\pi_*(IC_X) = IC_Y$.

Now we want to use these two results to compute the Betti numbers of the Hilbert schemes. Write $b_i(Y) := \dim H^i(Y)$ for the i-th Betti number of Y and $p(Y) := \sum_i b_i(Y)z^i$ for the Poincaré polynomial.

THEOREM 7.5.5. Let S be a smooth projective surface over $\mathbb{C}$. Then

$$\sum_{n=0}^{\infty} p(S^{[n]})t^n = \prod_{k>0} \frac{(1+z^{2k-1}t^k)^{b_1(S)}(1+z^{2k+1}t^k)^{b_3(S)}}{(1-z^{2k-2}t^k)^{b_0(S)}(1-z^{2k}t^k)^{b_2(S)}(1-z^{2k+2}t^k)^{b_4(S)}}.$$

PROOF. Write as above $\nu = (n_1, \ldots, n_r) = (1^{\alpha_1}, 2^{\alpha_2}, \ldots, s^{\alpha_s})$. As the Hilbert–Chow morphism ρ is semismall with irreducible fibres for the stratification of $S^{(n)}$ by the $S^{(n)}_\nu$ and $S^{[n]}$ is nonsingular projective of dimension $2n$, we get by Lemma 7.5.3 that

$$R\rho_*(\mathbb{Q}_{X^{[n]}}[2n]) = R\rho_*(IC_{X^{[n]}}) = \sum_\nu IC_{\overline{X}^{(n)}_\nu}.$$

In Exercise 7.4.2 we saw that $\pi_\nu : S^{(\alpha_1)} \times \ldots \times S^{(\alpha_r)} \to \overline{S}^{(n)}_\nu$ is the normalization of $\overline{S}^{(n)}_\nu$. Thus Lemma 7.5.4 gives

$$IC_{\overline{S}^{(n)}_\nu)} = R(\pi_\nu)_*(\mathbb{Q}_{S^{(\alpha_1)}\times\ldots\times S^{(\alpha_s)}}[2(\alpha_1 + \ldots + \alpha_2)]).$$

Thus we obtain

$$R\rho_*(\mathbb{Q}_{S^{[n]}}) = \sum_\nu R(\pi_\nu)_*(\mathbb{Q}_{S^{(\alpha_1)}\times\ldots\times S^{(\alpha_s)}}[2(\alpha_1 + \ldots + \alpha_2)]),$$

where again $\nu = (1^{\alpha_1}, \ldots, s^{\alpha_s})$ runs through all the partitions of n. Now we take the cohomology of this relation. Recall that the total cohomology $H^*(Y)$ is just the pushforward $Rp_*(\mathbb{Q})$, where $p : Y \to pt$ is the map to a point. Therefore taking cohomology commutes with push-forward. Thus we get

$$H^{i+2n}(S^{[n]}) = \sum_\nu H^{i+2(\alpha_1+\ldots+\alpha_s)}(S^{(\alpha_1)} \times \ldots \times S^{(\alpha_s)}).$$

Thus we have expressed the Betti numbers of the Hilbert schemes in terms of those of the symmetric powers of S. Therefore we can use Macdonald's formula [**Mac62**] for the Betti numbers of the $S^{(n)}$:

$$\sum_{n>0} p(S^{(n)})t^n = \frac{(1+zt)^{b_1(S)}(1+z^3t)^{b_3(S)}}{(1-t)^{b_0(S)}(1-z^2t)^{b_2(S)}(1-z^4t)^{b_4(S)}}.$$

It is now easy to put everything together:

$$\begin{aligned}\sum_{n>0} p(S^{[n]})t^n &= \sum_{n>0} \sum_{\alpha_1+\ldots s\alpha_s=n} p(S^{(\alpha_1)})\ldots p(S^{(\alpha_s)})z^{2(n-\alpha_1-\ldots-\alpha_s)}t^n \\ &= \prod_{k>0}\Big(\sum_{l\geq 0} p(S^{(l)})z^{2l(k-1)}t^{lk}\Big) \\ &= \prod_{k>0} \frac{(1+z^{2k-1}t^k)^{b_1(S)}(1+z^{2k+1}t^k)^{b_3(S)}}{(1-z^{2k-2}t^k)^{b_0(S)}(1-z^{2k}t^k)^{b_2(S)}(1-z^{2k+2}t^k)^{b_4(S)}}.\end{aligned}$$

□

This result can be proven in a number of different ways: In the case of $\mathbb{P}^2$ Ellingsrud and Strømme [**ES87**] give a cell-decomposition of $(\mathbb{P}^2)^{[n]}$ (i.e. $(\mathbb{P}^2)^{[n]}$ can be written as union of disjoint locally closed subvarieties isomorphic to affine spaces). This implies that they can compute $H^{2i}((\mathbb{P}^2)^{[n]})$ as the number of i-dimensional cells. To obtain this cell-decomposition they use a natural $\mathbb{C}^*$-action on $(\mathbb{P}^2)^{[n]}$ with finitely many fixed points and then apply the results of Białynicki-Birula [**BB73**]. The cell-decomposition also induces a cell-decomposition of H_n. This result is used in [**Göt90**] together with the fact that $S^{[n]}_{(n)} \to S$ is Zariski locally trivial and $S^{[n]}_{(\nu)} \to S^{(n)}_{(\nu)}$ is étale locally trivial to compute the numbers of points of $S^{[n]}$ over finite fields and then compute the Betti numbers of the $S^{[n]}$ using the Weil conjectures. In [**Che96**] so-called virtual Hodge polynomials are used to determine the Hodge numbers of $S^{[n]}$. These are polynomials $h(Z,x,y)$ which are defined for any variety over $\mathbb{C}$ and additive for unions of disjoint varieties and multiplicative in products, and which coincide with the usual Hodge polynomial $\sum_{p,q} h^{p,q}(Z)x^py^q$ if Z is smooth and projective. The argument we gave above can be refined by using mixed Hodge modules (basically a mixed Hodge structure on the intersection cohomology) instead of perverse sheaves to also give the Hodge numbers of the $S^{[n]}$, see [**GS93**]. The result was extended to the Douady spaces (the analogue of the Hilbert schemes of points) of a complex surface in [**dCM00**]. In [**Göt01**] and [**dCM02**] the result is refined to the Grothendieck ring of varieties and to motives.

7.6. The Heisenberg algebra

In this section all the cohomology that we consider is with $\mathbb{Q}$-coefficients. The generating formula of Theorem 7.5.5 suggests that somehow all the cohomology groups of $S^{[n]}$ for different n are tied together, and that one should try to look at all of them at the same time. So we denote $\mathbb{H}_n := H^*(S^{[n]})$ and consider the direct sum of all these cohomologies

$$\mathbb{H} := \bigoplus_{n\geq 0} \mathbb{H}_n.$$

We want to see that this has an additional structure: $\mathbb{H}$ carries an irreducible representation of the Heisenberg algebra modelled on the cohomology of $S^{[n]}$. This was conjectured by Vafa and Witten [**VW94**] and proven by Nakajima and Grojnowski [**Nak97**], [**Gro96**]. The result and its proof are explained in [**Nak99**] and also in [**EG00**]. Here I will just try to briefly explain the result. For simplicity of exposition we will assume that $H^1(S) = H^3(S) = 0$.

We want to relate the Hilbert schemes $S^{[n]}$ for different n. Thus we need to find a way to go from $S^{[n]}$ to $S^{[n+m]}$. To relate $S^{[n]}$ and $S^{[n+1]}$ the obvious thing is to add to any subscheme $Z \in S^{[n]}$ a point in S. As we have seen above this can be done by looking at the incidence correspondence,

$$S^{[n,n+1]} := \{(Z,W) \in S^{[n]} \times S^{[n+1]} \mid Z \subset W\},$$

where we mean by $Z \subset W$ that Z is a subscheme of W. Then $S^{[n,n+1]}$ parametrizes all ways to obtain a subscheme of length $n+1$ by adding a point to a subscheme of length n.

Thus to relate $S^{[n]}$ to $S^{[n+m]}$ we also want to use an incidence correspondence. The obvious generalization would be to use just the incidence variety of $S^{[n]}$ and $S^{[n+m]}$, which parametrizes all ways to obtain a subscheme of length $n+m$ by adding a subscheme of length m to a subscheme of length n, but it turns out that we want the difference to be supported at a point of S. Thus we put

$$Z_{n,m} := \{(Z,p,W) \in S^{[n]} \times S \times S^{[n+m]} \mid \rho(W) - \rho(Z) = m[p]\},$$

with the projections pr_1, pr_2, pr_3 to $S^{[n]}$, S, $S^{[n+m]}$ respectively. Note that in the above $\rho(W)$ and $\rho(Z)$ are effective 0-cycles of degree $n+m$ and n respectively, thus their difference is a zero cycle of degree m, which is required to be effective and supported in the point p.

This correspondence defines for each $\alpha \in H^*(S)$ a map

$$p_m(\alpha) : H^*(S^{[n]}) \to H^*(S^{[n+m]});\ y \mapsto PD(pr_{3*}(pr_2^*(\alpha) \cup pr_1^*(y) \cap [Z_{n,m}])).$$

where $[Z_{n,m}]$ is the fundamental class of $Z_{n,m}$ and PD denotes Poincaré duality. We call the $p_m(\alpha)$ the *creation operators*. Intuitively this map can be described as follows: Assume that α and y can be represented as the cohomology classes Poincaré dual to the fundamental class of submanifolds $A \subset S$ and $Y \subset S^{[n]}$.

Then $p_m(\alpha)y$ will be the class of the closure of

$$\{Z \sqcup P \mid [Z] \in Y, P \in S^{[n]}_{(n)} \text{ with } supp(P) \in A\}.$$

Thus $p_m(\alpha)$ is the operation of adding a fat point in A. Thus for all $m > 0$ and all $\alpha \in H^*(S)$ we obtain operators $p_m(\alpha) : \mathbb{H} \to \mathbb{H}$, sending $\mathbb{H}_n$ to $\mathbb{H}_{n+m}$. Note that $S^{[0]}$ is a point, and thus $\mathbb{H}_0 = \mathbb{Q}$. Let $\mathbf{1}$ be its unit element.

A weak version of the result of Nakajima and Grojnowski says that all the cohomology of the Hilbert schemes can be obtained by just applying the creation operators to $\mathbf{1}$. In fact given a basis of $H^*(S)$ this gives a canonical basis of $H^*(S^{[n]})$.

THEOREM 7.6.1. The $p_m(\alpha)$, with $m > 0$ and $\alpha \in H^*(S)$ commute.
Let $\{a_i\}_{i \in L}$ be a basis of $H^*(S)$. Then the set all monomials

$$p_{n_1}(\alpha_{i_1}) \dots p_{n_k}(\alpha_{i_k})\mathbf{1}, \quad ,k \geq 0,\ i_j \in L,\ \sum n_j = n$$

is a basis of $\mathbb{H}_n$.

This means that, given the intuitive description of the $p_m(\alpha)$ above, we get at least intuitively a very explicit description of the cohomologies of the Hilbert schemes: Assume that the α_i are represented by submanifolds $A_i \subset S$. Then a basis of $H^*(S^{[n]})$ is given by the closures of the classes of subsets of $S^{[n]}$ of the form

$$\{P_1 \sqcup P_2 \dots \sqcup P_l \mid P_j \in S^{[n_j]}_{(n_j)} \text{ with support in } A_{i_j}\}.$$

In order to get the Heisenberg algebra, we also need to consider *annihilation operators*, $p_{-m}(\alpha) : \mathbb{H}_{n+m} \longrightarrow \mathbb{H}_n$. We define $p_0(\alpha) = 0$ and let $p_{-m}(\alpha)$ be the adjoint operator of $p_m(\alpha)$ with respect to the intersection pairing on the cohomology of $S^{[n]}$ and $S^{[n+m]}$.

Again one can get an analogous intuitive interpretation. If y is the class of $Y \subset S^{[n+m]}$, and a the class of a submanifold $A \subset S$, then $p_{-m}(a)y$ should be the class of the closure of

$$\{Z \in S^{[n]} \mid Z \sqcup P \in Y \text{ for some } P \in S^{[n]}_{(n)} \text{ with support in } A\}.$$

Thus $p_{-m}(a)$ is obtained by subtracting a fat point in A.

We denote by $\langle\ ,\ \rangle$ the intersection pairing on S. Denote by $[p_n(\alpha), p_m(\beta)] := p_n(\alpha)p_m(\beta) - p_m(\beta)p_n(\alpha)$ the commutator. Then the main result of [**Nak97**], [**Gro96**] is:

THEOREM 7.6.2. $[p_n(\alpha), p_m(\beta)] = n\delta_{n,-m}\langle\alpha,\beta\rangle id_{\mathbb{H}}$

Thus all the creation operators $p_n(\alpha)$ commute with each other, and also all annihilation operators $p_{-m}(\beta)$ commute. Furthermore $p_n(\alpha)$ and $p_{-m}(\beta)$ commute unless $n = m$, when we just get a multiple of the identity. From our intuitive description this is quite plausible: Adding a fat point of length n in A and a fat point of length m in B should commute and similarly for subtracting fat points. Also adding a fat point of length n in A and subtracting a fat point of length m in B should commute unless $n = m$. However if $n = m$ this will no longer be true, because we get an extra term from subtracting the point that we have just added, and this extra term will just be a multiple of what we started with. Obviously this is not a proof, but still it gives the basic idea. Also this argument will not show what multiple of the identity to take. This multiple has been determined in [**ES98**] using again the inductive description of the $S^{[n]}$ via the $S^{[n,n+1]} = \mathbb{P}(\mathcal{I}_{Z_n(S)})$ as in Remark 7.2.2, and in [**Nak97**] using an argument with vertex operators.

How does Theorem 7.6.2 imply Theorem 7.6.1, and what does it have to do with the Heisenberg algebra? Let V be a $\mathbb{Q}$-vector space with a nondegenerate bilinear form $\langle\ ,\ \rangle$. Let T be the tensor algebra on $V[t,t^{-1}]$. Elements of T are of the form

$$v_1t^{i_1} \otimes \ldots \otimes v_kt^{i_k}, \quad v_j \in V,\ i_j \in \mathbb{Z}, \quad j \geq 0.$$

Let $\mathbf{e}$ be the neutral element of the tensor algebra corresponding to the empty tensor product. We have $T = \bigoplus_{n\in\mathbb{Z}} T^i$, where the grading is determined by giving t^i the degree i. The *Heisenberg algebra* $H(V)$ modelled on V is obtained from T by imposing the relations

$$[ut^i, vt^j] = i\delta_{i,-j}\langle u,v\rangle\mathbf{e}.$$

The *Fock space* $F(V)$ is the subalgebra of $H(V)$ obtained by replacing $V[t,t^{-1}]$ by $tV[t]$. $F(V)$ becomes an $H(V)$-module, by putting $ut^0 \cdot w := 0$ for all $w \in F(V)$ and $ut^{-i} \cdot \mathbf{e} := 0$ for all $i > 0$. Then one can show that $F(V)$ is an irreducible module for $H(V)$. Let $F(V)^d$ be the part of degree d, where t has degree 1. Then it is not difficult to see that

$$\sum_{n\geq 0} \dim\,(F(V)^d)z^d = \prod_{k\geq 1} \frac{1}{(1-z^k)^{\dim(V)}}.$$

Now let $V = H^*(S)$ with the intersection pairing. Then Theorem 7.6.2 says that there is an $H(V)$-module homomorphism

$$F(V) \to \mathbb{H}, ut^i \mapsto p_i(u)\mathbf{1}.$$

As $F(V)$ is irreducible and by Theorem 7.5.5 both have the same Poincaré series, this is an isomorphism. This implies in particular that $H^*(S^{[n]})$ has the basis given in Theorem 7.6.1. Note that the fact that $p_i(u)$ and $p_{-i}(u)$ are adjoint operators for the intersection pairings on the Hilbert schemes, also makes it easy to determine the intersection pairing in this basis.

This Heisenberg algebra action has been further used to study the ring structure of the $H^*(S^{[n]})$, first in [**Leh99**] then building on this in [**LQW02**], [**LS01**]. The main point is to understand the relation of the ring structure and the multiplicative structure: one has to understand the commutation relations between the Heisenberg operators and the operators of multiplying by certain canonical (or tautological) cohomology classes. Again $S^{[n,n+1]}$ plays an important rôle in the argument.

Part 4

Grothendieck's existence theorem in formal geometry

with a letter of Jean-Pierre Serre

Luc Illusie

CHAPTER 8

Grothendieck's existence theorem in formal geometry

Introduction

The main theme of these notes is Grothendieck's exposé 182 at the Séminaire Bourbaki [**Gro95a**]. Most of the material in (loc. cit.) has been treated at length in [**EGAIII1**], [**EGAIII2**] and [**SGA1**]. Our purpose here is to provide an introduction, explaining the proofs of the key theorems, discussing typical applications, and updating when necessary. The central results are the comparison theorem between formal and algebraic cohomology for proper morphisms and the existence theorem for formal sheaves. We give the highlights of the proofs in §8.2, 8.4 after recalling some basic terminology on formal schemes in §8.1 (sticking to the locally noetherian context, which suffices here). In §8.3 we revisit some points of [**EGAIII2**, 7]: base change formula and cohomological flatness. We believe that the use of derived categories and, especially, perfect complexes, simplifies the exposition. This section, however, is not essential for the sequel. In §8.5 we discuss several applications to the existence of formal or algebraic liftings, by combining Grothendieck's theorems of the preceding sections with basic results of deformation theory, mostly in the smooth case. Finally, in §8.6 we discuss Serre's examples [**Ser61**] of projective smooth schemes in characteristic $p > 0$ which cannot be lifted to characteristic zero.

I am very grateful to Serre for a conversation about his examples in [**Ser61**], for sending me a copy of Mumford's unpublished letter to him [**Mum**], and for permitting me to include a letter to me in which he solves a question left open in [**Ser61**]. Esnault, Messing, Raynaud and Serre read a preliminary version of these notes and suggested several corrections and modifications. I thank them heartily, as well as the students of the school for numerous questions and comments.

Convention. We use the symbol lim (resp. colim) to denote a projective (resp. inductive) limit.

8.1. Locally noetherian formal schemes

8.1.1. An *adic noetherian ring* is a noetherian ring A equipped with a topology having the following property: there exists a fundamental system of neighborhoods of 0 in A consisting of the powers I^n $(n > 0)$ of an ideal I and A is separated and complete for this topology; i.e., A is the projective limit of the discrete rings $A_n = A/I^{n+1}$ $(n \geq 0)$. An *ideal of definition* of A is an ideal I which has this property, or, equivalently, which is open and whose powers tend to 0. If I is an ideal of definition, one says that A is *I-adic*, the topology is called the *I-adic topology*, and the filtration of A by the powers of I the *I-adic filtration*. If I is an ideal of definition, for an ideal J to be an ideal of definition, it is necessary and sufficient that there exist positive integers p, q such that $J \supset I^p \supset J^q$.

To an I-adic noetherian ring A is associated a topologically ringed space

$$\mathcal{X} = (\operatorname{Spf} A, \mathcal{O}_{\mathcal{X}}), \tag{8.1.1.1}$$

defined as follows. For $n \in \mathbb{N}$, let $X_n = \operatorname{Spec} A_n$ ($A_n = A/I^{n+1}$). These schemes form an increasing sequence of closed subschemes of $\operatorname{Spec} A$, (with closed, nilpotent immersions as transition maps)

$$X_0 = \operatorname{Spec} A_0 \to X_1 \to \cdots \to X_n \to \cdots .$$

They all have the same underlying space $\mathcal{X}$, called the *formal spectrum* of A. Note that I is contained in the radical of A [**EGAI**, Ch. 0, 7.1.10]; i.e., $1 - x$ is invertible for all $x \in I$, which means that $\mathcal{X}$, as a closed subspace of $\operatorname{Spec} A$, contains all its closed points. Every open subset of $\operatorname{Spec} A$ containing $\mathcal{X}$ is equal to $\operatorname{Spec} A$. The sheaf of rings $\mathcal{O}_{\mathcal{X}}$ is defined to be the inverse limit of the sheaves $\mathcal{O}_{X_n}$ on $\mathcal{X}$, equipped with the natural topology such that on any open subset U of $\mathcal{X}$, $\Gamma(U, \mathcal{O}_{\mathcal{X}}) = \lim_n \Gamma(U, \mathcal{O}_{X_n})$, where $\Gamma(U, \mathcal{O}_{X_n})$ has the discrete topology. In particular, $\Gamma(\mathcal{X}, \mathcal{O}_{\mathcal{X}}) = A$, and for $f \in A$, and $\mathcal{D}(f) = \mathcal{X}_f$ the open subset of $\mathcal{X}$ where the image of f in A_0 is invertible, $\Gamma(\mathcal{X}_f, \mathcal{O}_{\mathcal{X}}) = A_{\{f\}}$, the completed fraction ring $\lim_n S_f^{-1} A / S_f^{-1} I^{n+1}$. The stalks $\mathcal{O}_{\mathcal{X},x} = \operatorname{colim}_{f \in A, f(x) \neq 0} A_{\{f\}}$ are local (noncomplete) rings.

The topologically ringed space (8.1.1.1) depends only on A as a *topological* ring. It doesn't change if one replaces I by another ideal of definition. The space $\mathcal{X}$ is the subspace of $\operatorname{Spec} A$ consisting of *open* prime ideals, and $\mathcal{O}_{\mathcal{X}}$ is the inverse limit of the sheaves $(A/J)\tilde{}$ where J runs through the ideals of definition of A.

An *affine noetherian formal scheme* is a topologically ringed space isomorphic to one of the form (8.1.1.1). A *locally noetherian formal scheme* is a topologically ringed space such that any point has an open neighborhood which is an affine noetherian formal scheme. It is called *noetherian* if its underlying space is noetherian. A *morphism* $f : \mathcal{X} \to \mathcal{Y}$ between locally noetherian formal schemes is a morphism of ringed spaces which is *local* (i.e., such that for each point $x \in X$, the map $\mathcal{O}_{\mathcal{Y},f(x)} \to \mathcal{O}_{\mathcal{X},x}$ is local) and *continuous* (i. e. for every open affine $V \subset \mathcal{Y}$, the map $\Gamma(V, \mathcal{O}_V) \to \Gamma(f^{-1}(V), \mathcal{O}_{\mathcal{X}})$ is continuous). Locally noetherian formal schemes form a category in an obvious way.

As in the case of usual schemes, one checks that if $\mathcal{Y} = \operatorname{Spf}(A)$ is a noetherian affine formal scheme (in the sequel we will usually omit the sheaf of rings from the notation) and $\mathcal{X}$ is any locally noetherian formal scheme, then we have

$$\operatorname{Hom}(\mathcal{X}, \mathcal{Y}) = \operatorname{Hom}_{cont}(A, \Gamma(\mathcal{X}, \mathcal{O}_{\mathcal{X}})), \tag{8.1.1.2}$$

where $\operatorname{Hom}_{cont}$ means the set of *continuous* ring homomorphisms. In particular, if $\mathcal{X}$ is affine, of ring B, then

$$\operatorname{Hom}(\mathcal{X}, \mathcal{Y}) = \operatorname{Hom}_{cont}(A, B).$$

8.1.2. Let $\mathcal{X} = \operatorname{Spf} A$ be an affine noetherian formal scheme, and let I be an ideal of definition of A. Let M be an A-module *of finite type*. With M is associated a coherent module $\tilde{M}$ on $X = \operatorname{Spec} A$. In an analogous way, one associates with M a module M^{Δ} on $\mathcal{X}$, defined as follows. For $n \in \mathbb{N}$, let $X_n = \operatorname{Spec} A/I^{n+1}$ as in 8.1.1. We put

$$M^{\Delta} = \lim_n \tilde{M}_n,$$

where $M_n = M/I^{n+1}M$. It is easily checked that M^{Δ} does not depend on the choice of I, that the functor $M \mapsto M^{\Delta}$ is exact, that

$$\Gamma(\mathcal{X}, M^{\Delta}) = M,$$

and that the formation of M^{Δ} commutes with tensor products and internal *Hom*. The main point is that, if

$$i : \mathcal{X} \to X$$

is the natural morphism, defined by the inclusion on the underlying topological spaces and the canonical map $\mathcal{O}_X \to \mathcal{O}_{\mathcal{X}}$ on the sheaves of rings, then, since M is of finite type, Krull's theorem implies that

$$M^{\Delta} = i^* \tilde{M}.$$

Since, for any $f \in A$, $A_{\{f\}}$ is adic noetherian, it follows that $\mathcal{O}_{\mathcal{X}}$ is a *coherent* sheaf of rings, M^{Δ} is *coherent*, and the coherent modules on $\mathcal{X}$ are exactly those of the form M^{Δ} for M of finite type over A.

8.1.3. Locally noetherian formal schemes are more conveniently described – and in practice usually appear – as colimits of increasing chains of nilpotent thickenings. By a *thickening* we mean a closed immersion of schemes $X \to X'$ whose ideal I is a nilideal; the schemes X and X' then have the same underlying space. If X' is noetherian, so is X, and I is nilpotent; conversely, if X is noetherian and I/I^2 coherent (as an $\mathcal{O}_X$-module), X' is noetherian [**EGAI**, Ch. 0, 7.2.6, 10.6.4]. If X' is noetherian, X' is affine if and only if X is [**EGAI**, 6.1.7]. We say that a thickening is *of order n* if $I^{n+1} = 0$.

Let $\mathcal{X}$ be a locally noetherian formal scheme. It follows from the discussion in the affine case that $\mathcal{O}_{\mathcal{X}}$ is a *coherent* sheaf of rings, and that the coherent modules on $\mathcal{X}$ are exactly the modules which are of finite presentation, or equivalently, which on any affine open $U = \operatorname{Spf} A$ are of the form M^{Δ} for an A-module M of finite type.

An *ideal of definition* of $\mathcal{X}$ is a coherent ideal $\mathcal{I}$ of $\mathcal{O}_{\mathcal{X}}$ such that, for any point $x \in \mathcal{X}$, there exists an affine neighborhood $U = \operatorname{Spf} A$ of x such that $\mathcal{I}|U$ is of the form I^{Δ} for an ideal of definition I of A. A coherent ideal $\mathcal{I}$ is an ideal of definition if and only if the ringed space $(\mathcal{X}, \mathcal{O}_{\mathcal{X}}/\mathcal{I})$ is a scheme having $\mathcal{X}$ as an underlying space. Ideals of definition of $\mathcal{X}$ exist. In fact, there is a *largest* one,

$$\mathcal{T} = \mathcal{T}_{\mathcal{X}}, \tag{8.1.3.1}$$

which is the unique ideal of definition $\mathcal{I}$ such that $(\mathcal{X}, \mathcal{O}_{\mathcal{X}}/\mathcal{I})$ is *reduced.* If $U = \operatorname{Spf} A$ is an affine open subset, then $\mathcal{T}|U = T^{\Delta}$, where T is the ideal of elements a of A which are *topologically nilpotent*, i.e., whose image in A/I is nilpotent. If $\mathcal{I}$ is an ideal of definition of $\mathcal{X}$, so is any power $\mathcal{I}^n$ for $n \geq 1$. If $\mathcal{X}$ is noetherian, then, as in the affine case, if $\mathcal{I}$ is an ideal of definition of $\mathcal{X}$ and $\mathcal{J}$ is a coherent ideal, then for $\mathcal{J}$ to be an ideal of definition it is necessary and sufficient that there exist positive integers p, q such that $\mathcal{J} \supset \mathcal{I}^p \supset \mathcal{J}^q$.

Fix an ideal of definition $\mathcal{I}$ of $\mathcal{X}$. For $n \in \mathbb{N}$, the ringed space $(\mathcal{X}, \mathcal{O}_{\mathcal{X}}/\mathcal{I}^{n+1})$ is a *locally noetherian scheme* X_n, and we have an increasing chain of thickenings

$$X. = (X_0 \to X_1 \to \cdots \to X_n \to \cdots), \tag{8.1.3.2}$$

whose colimit (in the category of locally noetherian formal schemes) is $\mathcal{X}$: the thickenings induce the identity on the underlying spaces, which are all equal to the underlying space of $\mathcal{X}$, and we have

$$\mathcal{O}_{\mathcal{X}} = \lim_n \mathcal{O}_{X_n},$$

as topological rings ($\Gamma(U, \mathcal{O}_{X_n})$ having the discrete topology on any affine open U). Let J_n be the ideal of X_0 in X_n; i.e., $J_n = \operatorname{Ker} \mathcal{O}_{X_n} \to \mathcal{O}_{X_0}$. Then, for $m \leq n$, the ideal of X_m in X_n is J_n^{m+1} (in particular, $J_n^{n+1} = 0$), J_1 is a coherent module on X_0, and $J_n = \mathcal{I}/\mathcal{I}^{n+1}$.

Conversely, consider a sequence (8.1.3.2) of ringed spaces satisfying:

(i) X_0 is a locally noetherian scheme,

(ii) the underlying maps of topological spaces are homeomorphisms and, using them to identify the underlying spaces, the maps of rings $\mathcal{O}_{X_{n+1}} \to \mathcal{O}_{X_n}$ are surjective,

(iii) if $J_n = \operatorname{Ker} \mathcal{O}_{X_n} \to \mathcal{O}_{X_0}$, then for $m \leq n$, $\operatorname{Ker} \mathcal{O}_{X_n} \to \mathcal{O}_{X_m} = J_n^{m+1}$

(iv) J_1 (as an $\mathcal{O}_{X_0}$-module) is coherent,

Then the topologically ringed space $\mathcal{X} = (X_0, \lim_n \mathcal{O}_{X_n})$ is a locally noetherian formal scheme, and if $\mathcal{I} = \operatorname{Ker} \mathcal{O}_{\mathcal{X}} \to \mathcal{O}_{X_0} = \lim J_n$, $\mathcal{I}$ is an ideal of definition of $\mathcal{X}$, and $\mathcal{I}^{n+1} = \operatorname{Ker} \mathcal{O}_{\mathcal{X}} \to \mathcal{O}_{X_n}$.

The verification is straightforward [**EGAI**, 10.6.3 – 10.6.5], by reduction to the case where X_0 is *affine*, of ring A_0, in which case every X_n is automatically affine noetherian, of ring A_n, and $\mathcal{X} = \operatorname{Spf} A$, where $A = \lim_n A_n$.

8.1.4. Let $\mathcal{X}$ be a locally noetherian formal scheme, $\mathcal{I}$ an ideal of definition of $\mathcal{X}$, and consider the corresponding chain of thickenings (8.1.3.2). For $m \leq n$ denote by

$$u_{mn} : X_m \to X_n \quad , \quad u_n : X_n \to \mathcal{X}$$

the canonical morphisms. If E is a coherent module on $\mathcal{X}$, then $E_n := u_n^* E$ is a coherent module on X_n, these modules form an inverse system, with $\mathcal{O}_{X_n}$-linear transition maps $E_n \to E_m$ inducing isomorphism $u_{mn}^* E_n \xrightarrow{\sim} E_m$, and $E = \lim_n E_n$. Conversely, let $F_. = (F_n, f_{mn} : F_n \to F_m)$ be an inverse system of $\mathcal{O}_{X_n}$-modules, with transition maps f_{mn} which are $\mathcal{O}_{X_n}$-linear. We will say that $F_.$ is *coherent* if each F_n is coherent and the transition maps f_{mn} induce isomorphisms $u_{mn}^* F_n \xrightarrow{\sim} F_m$. If $F_.$ is coherent, and $F := \lim_n F_n$ is the corresponding $\mathcal{O}_{\mathcal{X}}$-module, then F is coherent and $F_.$ is canonically isomorphic to the inverse system $(u_n^* F)$. The functor

$$\text{(8.1.4.1)} \qquad \operatorname{Coh}(\mathcal{X}) \to \operatorname{Coh}(X_.), \quad E \mapsto (u_n^* E)$$

from the category of coherent sheaves on $\mathcal{X}$ to the category $\operatorname{Coh}(X_.)$ of coherent inverse systems (F_n) is an equivalence. For $E = \lim_n E_n \in \operatorname{Coh}(\mathcal{X})$ as above, the *support* of E is *closed* (as E is coherent) and coincides with that of E_0. By (a special case of) the flatness criterion [**Bou61**, III, §5, th. 1], E is flat (equivalently, locally free of finite type) if and only if E_n is locally free of finite type for all n.

8.1.5. Let $f : \mathcal{X} \to \mathcal{Y}$ be a morphism of locally noetherian formal schemes, and let $\mathcal{J}$ be an ideal of definition of $\mathcal{Y}$. Since $\mathcal{J} \subset \mathcal{T}_{\mathcal{Y}}$, the continuity of f implies that the ideal $f^*(\mathcal{J})\mathcal{O}_{\mathcal{X}}$ is contained in $\mathcal{T}_{\mathcal{X}}$ (8.1.3.1). Fix an ideal of definition $\mathcal{I}$ such that $f^*(\mathcal{J})\mathcal{O}_{\mathcal{X}} \subset \mathcal{I}$ (e. g. $\mathcal{I} = \mathcal{T}_{\mathcal{X}}$), and consider the inductive systems $X_.$, $Y_.$ defined by $\mathcal{I}$ and $\mathcal{J}$ respectively, as in (8.1.3.2). Then, since $f^*(\mathcal{J}^{n+1})\mathcal{O}_{\mathcal{X}} \subset \mathcal{I}^{n+1}$, f induces a morphism of inductive systems

$$\text{(8.1.5.1)} \qquad f_. : X_. \to Y_.,$$

i.e., morphisms of schemes $f_n : X_n \to Y_n$ such that the squares

(8.1.5.2)
$$\begin{array}{ccc} X_m & \longrightarrow & X_n \\ {\scriptstyle f_m}\downarrow & & \downarrow{\scriptstyle f_n} \\ Y_m & \longrightarrow & Y_n \end{array}$$

are commutative, and f is the colimit of the morphisms f_n, characterized by making the squares

(8.1.5.3)
$$\begin{array}{ccc} X_n & \xrightarrow{u_n} & \mathcal{X} \\ {\scriptstyle f_n}\downarrow & & \downarrow{\scriptstyle f} \\ Y_n & \xrightarrow{u_n} & \mathcal{Y} \end{array}$$

commutative. It is easily checked [**EGAI**, 10.6.8] that $f \mapsto f_{\cdot}$ defines a bijection from the set of morphisms from $\mathcal{X}$ to $\mathcal{Y}$ such that $f^*(\mathcal{J})\mathcal{O}_{\mathcal{X}} \subset \mathcal{I}$ to the set of morphisms of inductive systems of the type (8.1.5.1).

In general, $f^*(\mathcal{J})\mathcal{O}_{\mathcal{X}}$ is not an ideal of definition of $\mathcal{X}$. When this is the case, f is called an *adic morphism* (and $\mathcal{X}$ a $\mathcal{Y}$-adic formal scheme). One can then take $\mathcal{I} = f^*(\mathcal{J})\mathcal{O}_{\mathcal{X}}$, and the squares (8.1.5.2) are *cartesian*. Conversely any morphism of inductive systems (8.1.5.1) such that the squares (8.1.5.2) are cartesian define an adic morphism from $\mathcal{X}$ to $\mathcal{Y}$.

Let $f : \mathcal{X} \to \mathcal{Y}$ be an adic morphism, and let E be a coherent sheaf on $\mathcal{X}$. Then the following conditions are equivalent:

(i) E is *flat over* $\mathcal{Y}$ (or $\mathcal{Y}$-flat); i.e., for every point x of X, the stalk E_x is flat over $\mathcal{O}_{\mathcal{Y},f(x)}$;

(ii) with the notations of (8.1.5.3), $E_n = u_n^* E$ is Y_n-flat for all $n \geq 0$;

(iii) E_0 is Y_0-flat and the natural (surjective) map

$$\mathrm{gr}^n \mathcal{O}_{\mathcal{Y}} \otimes_{\mathrm{gr}^0 \mathcal{O}_{\mathcal{Y}}} \mathrm{gr}^0 E \to \mathrm{gr}^n E,$$

where the associated graded gr is taken with respect to the $\mathcal{J}$-adic filtration, is an isomorphism for all $n \geq 0$.

This is a consequence of the flatness criterion [**Bou61**, III, §5, th. 2, prop. 2]. When $E = \mathcal{O}_{\mathcal{X}}$ satisfies the above equivalent conditions, we say that f is *flat*.

8.1.6. Let X be a locally noetherian *scheme*, and let X' be a closed subset of (the underlying space of) X. Choose a coherent ideal I of $\mathcal{O}_X$ such that the closed subscheme of X defined by I has X' as an underlying space. Such ideals exist, there is in fact a largest one, consisting of local sections of $\mathcal{O}_X$ vanishing on X'; for this one, X' has the reduced scheme structure. Consider the inductive system of (locally noetherian) schemes, all having X' as underlying space,

$$X_0 \to X_1 \to \cdots \to X_n \to \cdots,$$

where X_n is the closed subscheme of X defined by I^{n+1}. It satisfies the conditions (i) - (iv) of 8.1.3 and therefore the colimit

(8.1.6.1)
$$X_{/X'} := \mathrm{colim}_n X_n$$

is a locally noetherian formal scheme, having X' as underlying space, called the *formal completion of* X *along* X'. It is sometimes denoted simply $\hat{X}$, when no confusion can arise. It is easily checked that $X_{/X'}$ does not depend on the choice of the ideal I. In fact, $\mathcal{O}_{\hat{X}}$ is the inverse limit of the rings $\mathcal{O}_X/J$, where J runs through all the coherent ideals of $\mathcal{O}_X$ such that the support of $\mathcal{O}_X/J$ is X' (on any noetherian open subset of X, the powers of I form a cofinal system). If X is affine, $X = \operatorname{Spec} A$, and $I = \tilde{J}$, then $\hat{X} = \operatorname{Spf} \hat{A}$, where $\hat{A}$ is the completion of A with respect to the J-adic topology.

The closed immersions $i_n : X_n \to X$ define a morphism of ringed spaces

$$i_X : \hat{X} \to X \tag{8.1.6.2}$$

(or i), which is *flat*, and for any *coherent* sheaf F on X, the natural map

$$i^*F \to F_{/X'} := \lim_n i_n^* F. \tag{8.1.6.3}$$

is an isomorphism. When $X = \operatorname{Spec} A$ and $F = \tilde{M}$, with M an A-module of finite type, then $F_{/\ X} = \hat{M}^{\Delta}$ (8.1.2), where $\hat{M}$ is the module $\lim_n(M/I^{n+1}M)$ on the I-adic completion $\hat{A}$ of A. The above assertion follows from Krull's theorem: if A is noetherian, and J is an ideal of A, then the J-adic completion $\hat{A}$ is flat over A, and for any A-module M of finite type, $\hat{M} = M \otimes \hat{A}$. One sometimes writes $\hat{F}$ for $F_{/\ X'}$ when no confusion can arise. Note that if F is not coherent, (8.1.6.3) is not in general an isomorphism. One checks similarly that the kernel of the adjunction map

$$F \to i_* i^* F \tag{8.1.6.4}$$

consists of sections of F which are zero in a neighborhood of X'.

Let $f : X \to Y$ be a morphism of locally noetherian schemes, X' (resp. Y') a closed subset of X (resp. Y) such that $f(X') \subset Y'$. Choose coherent ideals $J \subset \mathcal{O}_X$, $K \subset \mathcal{O}_Y$ defining closed subschemes with underlying spaces X' and Y' respectively and such that $f^*(K)\mathcal{O}_X \subset J$ (one can take for example for K any ideal defining a closed subscheme with underlying space Y' and for J the ideal of sections of $\mathcal{O}_X$ vanishing on X'). Then f induces a morphism of inductive systems

$$f_. : X_. \to Y_.,$$

where X_n (resp. Y_n) is defined as above. By the correspondence explained in 8.1.5 we get from $f_.$ a morphism

$$\hat{f} : X_{/X'} \to Y_{/\ Y'}, \tag{8.1.6.5}$$

which does not depend on the choices of J, K, and is called the *extension* of f to the completions $X_{/X'}$ and $Y_{/\ Y}$. This morphism sits in a commutative square

$$\begin{array}{ccc} X_{/X'} & \xrightarrow{i} & X \\ {\scriptstyle \hat{f}}\downarrow & & \downarrow{\scriptstyle f} \\ Y_{/\ Y} & \xrightarrow{i} & Y \end{array} \tag{8.1.6.6}$$

where the horizontal maps are the canonical morphisms (8.1.6.2). When $X' = f^{-1}(Y')$, one can take $J = f^*(K)\mathcal{O}_X$, all the squares

$$\begin{array}{ccc} X_n & \longrightarrow & X \\ {\scriptstyle f_n}\downarrow & & \downarrow{\scriptstyle f} \\ Y_n & \longrightarrow & Y \end{array}$$

are cartesian, hence the same holds for the square (8.1.5.2), and therefore $\hat{f}$ is an *adic* morphism.

8.2. The comparison theorem

8.2.1. Let $f : X \to Y$ be a morphism of locally noetherian schemes, let Y' be a closed subset of Y, $X' = f^{-1}(Y')$. Write $\hat{X} = X_{/X'}$, $\hat{Y} = Y_{/Y'}$. If F is an $\mathcal{O}_X$-module, the square (8.1.6.6) defines base change maps (see 8.2.19)

$$i^* R^q f_* F \to R^q \hat{f}_*(i^* F) \tag{8.2.1.1}$$

(for all $q \in \mathbb{Z}$), which are maps of $\mathcal{O}_{\hat{Y}}$-modules. As the functor i_* is exact, the general definition given in 8.2.19 boils down here to the following: the map

$$R^q f_* F \to i_* R^q \hat{f}_*(i^* F) = R^q f_*(i_* i^* F)$$

adjoint to (8.2.1.1) is the map obtained by applying $R^q f_*$ to the adjunction map $F \to i_* i^* F$. If F is *coherent*, then $i^* F$ can be identified with $\hat{F} = F_{/X'}$ by (8.1.6.3), and similarly $i^* R^q f_* F$ can be identified with $(R^q f_* F)_{/Y'}$ if $R^q f_* F$ is coherent: this is the case when F is coherent and f is *proper* (or f is of finite type and *the support of F is proper over Y*, which means ([**EGAII**, 5.4.10]) that there exists a closed subscheme Z of X which is proper over Y and whose underlying space is the support of F), by the finiteness theorem for proper morphisms [**EGAIII1**, 3.2.1, 3.2.4]. In this case, (8.2.1.1) can be rewritten

$$(R^q f_* F)^{\wedge} \to R^q \hat{f}_* \hat{F}. \tag{8.2.1.2}$$

On the other hand, if F is coherent, the squares (8.1.5.3), with $\mathcal{X} = \hat{X}$, $\mathcal{Y} = \hat{Y}$ define $\mathcal{O}_{Y_n}$-linear base change maps

$$u_n^* R^q \hat{f}_* \hat{F} \to R^q (f_n)_* F_n,$$

where $F_n = u_n^* \hat{F} = i_n^* F$ (in the notation of (8.1.6.2)). By adjunction, these maps can be viewed as $\mathcal{O}_{\hat{Y}}$-linear maps

$$R^q \hat{f}_* \hat{F} \to R^q (f_n)_* F_n$$

(obtained by applying $R^q \hat{f}_*$ to the adjunction map $\hat{F} \to (u_n)_* F_n$), hence define $\mathcal{O}_{\hat{Y}}$-linear maps

$$R^q \hat{f}_* \hat{F} \to \lim_n R^q (f_n)_* F_n. \tag{8.2.1.3}$$

Note the base change map (8.2.1.1) is defined more generally for $F \in D^+(X, \mathcal{O}_X)$, as induced on the sheaves $\mathcal{H}^q$ from the base change map in $D^+(\hat{Y}, \mathcal{O}_{\hat{Y}})$

$$i^* R f_* F \to R \hat{f}_* i^* F. \tag{8.2.1.4}$$

THEOREM 8.2.2. *Let $f : X \to Y$ be a finite type morphism of noetherian schemes, Y' a closed subset of Y, $X' = f^{-1}(Y')$, $\hat{f} : \hat{X} \to \hat{Y}$ the extension of f to the formal completions of X and Y along X' and Y'. Let F be a coherent sheaf on X whose support is proper over Y. Then, for all q, the canonical maps (8.2.1.2), (8.2.1.3) are topological isomorphisms.*

REMARKS 8.2.3.

(a) Under the assumptions of 8.2.2 on f, it follows that for any $F \in D^+(X, \mathcal{O}_X)$ such that, for all i, $\mathcal{H}^i F$ is coherent and properly supported over Y, (8.2.1.4) is an isomorphism. Using that the natural functor from the bounded derived category $D^b(\mathrm{Coh}(X))$ of coherent sheaves on X to the full subcategory $D^b(X)_{coh}$ of $D^b(X) := D^b(X, \mathcal{O}_X)$ consisting of complexes with coherent cohomology is an equivalence [**SGA6**, II 2.2.2.1], one can extend the isomorphism (8.2.1.3) of 8.2.2 to the case $F \in D^b(X)_{coh}$. We omit the details.

(b) By considering a closed subscheme Z of X whose underlying space is the support of F, 8.2.2 is reduced to the case where f is *proper.*

(c) Grothendieck's original proof has not been published. From [**Gro95a**, p. 05], one can guess that it consisted of two steps: (i) proof in the case where f is projective, using *descending* induction on q (see [**Har77**, III 11.1] for the case where Y' is a point); (ii) proof in the general case by reducing to the projective case via Chow's lemma and noetherian induction. The proof given in [**EGAIII1**, 4.1.7, 4.1.8] follows an argument due to Serre.

(d) It is easily seen that 8.2.2 is actually equivalent to the following special case:

COROLLARY 8.2.4. *Under the assumptions of 8.2.2, suppose that* $Y = \operatorname{Spec} A$, *with A a noetherian ring, let I be an ideal of A such that* $\operatorname{Supp}(\mathcal{O}_Y/\mathcal{I}) = Y'$, *where* $\mathcal{I} = \tilde{I}$. *Let* $Y_n = \operatorname{Spec}(A/I^{n+1})$, $X_n = Y_n \times_Y X$, $F_n = i_n^* F = F/\mathcal{I}^{n+1}F$. *Then, for all q, the natural maps*

$$\varphi_q : H^q(X, F)^{\wedge} \to \lim_n H^q(X, F_n), \tag{8.2.4.1}$$

defined by the composition of (8.2.1.2) and (8.2.1.3), and

$$\psi_q : H^q(\hat{X}, \hat{F}) \to \lim_n H^q(X, F_n), \tag{8.2.4.2}$$

defined by (8.2.1.3), are topological isomorphisms.

The proof of 8.2.4 [**EGAIII1**, 4.1.7] uses two ingredients, the first one is standard, elementary commutative and homological algebra, the second is much deeper: (a) the Artin–Rees lemma and the Mittag-Leffler conditions; (b) the finiteness theorem for proper morphisms [**EGAIII1**, 3.2], especially a *graded* variant [**EGAIII1**, 3.3.2]

We will briefly review (a) and (b) and then give the highlights of the proof of 8.2.4.

8.2.5. Artin–Rees and Mittag-Leffler.

8.2.5.1. Let A be a noetherian ring, I an ideal of A, M a finitely generated A-module endowed with a decreasing filtration by submodules $(M_n)_{n \in \mathbb{Z}}$. The filtration (M_n) is called *I-good* if it is *exhaustive* (i.e., there exists n_1 such that $M_{n_1} = M$) and it satisfies the following two conditions:

(i) $IM_n \subset M_{n+1}$ for all $n \in \mathbb{Z}$ (which means that M, filtered by (M_n) is a filtered module over the ring A filtered by the I-adic filtration);

(ii) there exists an integer n_0 such that $M_{n+1} = IM_n$ for all $n \geq n_0$.

For example, the *I-adic filtration* of M, defined by $M_n = M$ for $n \leq 0$ and $M_n = I^n M$ for $n \geq 0$ is I-good. All I-good filtrations define on M the same topology, namely the I-adic topology.

Assume that condition (i) holds. Consider the graded ring

$$A' := \bigoplus_{n \in \mathbb{N}} I^n,$$

sometimes written $\oplus I^n t^n$, where t is an indeterminate, to make clear that $I^n = I^n t^n$ is the n-th component of A', and the graded module over A',

$$M' = \bigoplus_{n \in \mathbb{N}} M_n,$$

also sometimes written $\bigoplus_{n \in \mathbb{N}} M_n t^n$. A basic observation [**Bou61**, III, §3, th. 1] – whose proof is straightforward – is that condition (ii) is equivalent to

(ii′) M' is a finitely generated A'-module.

Since A' is noetherian, this immediately implies the classical *Artin–Rees theorem*: if N is a submodule of M, then the filtration induced on N by the I-adic filtration of M is I-good, in other words, there exists $n_0 \geq 0$ such that, for all $n \geq n_0$,

$$I^n M \cap N = I^{n-n_0}(I^{n_0} M \cap N).$$

That (ii′) implies (ii) is a particular case of the following (equally straightforward) property [**EGAII**, 2.1.6]:

(iii) Let A be a commutative ring, $S = \bigoplus_{n \in \mathbb{N}} S_n$ a graded A-algebra, of finite type over S_0 and generated by S_1, and $M = \bigoplus_{n \in \mathbb{Z}} M_n$ a graded S-module of finite type. Then there exists $n_0 \in \mathbb{N}$ such that for all $n \geq n_0$, $M_{n+1} = S_1 M_n$

8.2.5.2. Let A be a commutative ring. Let $M. = (M_n, u_{mn} : M_n \to M_m)$ be a projective system of A-modules, indexed by $\mathbb{N}$. We say that:

(i) $M.$ is *strict* if the transition maps u_{mn} are surjective,

(ii) $M.$ is *essentially zero* if for each m there exists $n \geq m$ such that $u_{mn} = 0$, in other words the *pro-object* defined by $M.$ is zero,

(iii) $M.$ satisfies the *Mittag-Leffler condition* (ML for short) if for each m there exists $n \geq m$ such that, for all $n' \geq n$, $\operatorname{Im} u_{mn'} = \operatorname{Im} u_{mn}$ in M_m.

It is sometimes useful to consider the following stronger conditions: we say that:

(ii′) $M.$ is *Artin–Rees zero* (AR zero for short) if there exist an integer $r \geq 0$ such that for all n, $u_{n,n+r} = 0$,

(iii′) $M.$ satisfies the *Artin–Rees-Mittag-Leffler condition* (ARML for short) if there exists an integer $r \geq 0$ such that, for all m and all $n' \geq m + r$, $\operatorname{Im} u_{mn'} = \operatorname{Im} u_{m,m+r}$.

We refer to [**EGAIII1**, Ch. 0, 13] for a discussion of the Mittag-Leffler condition. Let us just recall two basic (easy) points:

(a) If $M.$ is essentially zero, then $\lim_n M_n = 0$.

(b) The functor $M. \mapsto \lim_n M_n$ is left exact. Moreover, let

$$0 \to L. \to M. \to N. \to 0$$

be an exact sequence of inverse systems of A-modules; if $L.$ satisfies ML then the sequence

$$0 \to \lim_n L_n \to \lim_n M_n \to \lim_n N_n \to 0$$

is exact.

The stronger condition (iii′) (sometimes called *uniform* Mittag-Leffler condition) has a close relationship with the Artin–Rees theorem. See [**SGA5**, V] for a discussion of this. The terminology *AR zero*, *ARML* is taken from there.

We will need a (very) particular case of a general result [**EGAIII1**, Ch. 0, 13.3.1] of commutation of $H^q(X, -)$ with inverse limits, whose proof is elementary:

PROPOSITION 8.2.5.3. *Let X be a scheme, and let $(F_n)_{n \in \mathbb{N}}$ be an inverse system of quasi-coherent sheaves on X with* surjective *transition maps. Assume that, for all $i \in \mathbb{Z}$, the inverse system (of $\mathbb{Z}$-modules) $H^i(X, F_.)$ satisfies ML. Then, for all $i \in \mathbb{Z}$, the natural map*

$$H^i(X, \lim_n F_n) \to \lim_n H^i(X, F_n)$$

is an isomorphism.

8.2.6. *The finiteness theorem.*

The fundamental finiteness theorem for proper morphisms [**EGAIII1**, 3.2.1] asserts that if $f : X \to Y$ is a proper morphism, with Y locally noetherian, and F is a coherent sheaf on X, then, for all $q \in \mathbb{Z}$, the sheaves $R^q f_* F$ on Y are coherent. We will need the following variant [**EGAIII1**, 3.3.1]:

THEOREM 8.2.6.1. *Let $f : X \to Y$ is a proper morphism, with Y noetherian. Let $S = \bigoplus_{n \in \mathbb{N}} S_n$ be a quasi-coherent, graded $\mathcal{O}_Y$-algebra of finite type over S_0 and generated by S_1. Let $M = \bigoplus_{n \in \mathbb{Z}} M_n$ be a quasi-coherent, graded $f^*(S)$-module of finite type. Then, for all $q \in \mathbb{Z}$,*

$$R^q f_* M := \bigoplus_{n \in \mathbb{Z}} R^q f_* M_n$$

is a graded S-module of finite type, and there exists an integer n_0 such that, for any $n \geq n_0$,

$$R^q f_* M_n = S_{n-n_0} R^q f_* M_{n_0}.$$

Here the structure of graded S-module on $R^q f_* M$ comes from the multiplication maps, which are the composites

$$S_k \otimes R^q f_* M_n \to f_* f^* S_k \otimes R^q f_* M_n \to R^q f_*((f^* S_k) \otimes M_n) \to R^q f_* M_{n+k}.$$

The last assertion in 8.2.6.1 is a consequence of the first one (thanks to 8.2.5.1 (iii)), and the first one follows from the finiteness theorem applied to the (proper) morphism $\tilde{f} : \tilde{X} \to \tilde{Y}$ defined by the cartesian square

$$\begin{array}{ccc} X & \longleftarrow & \tilde{X} \\ {\scriptstyle f}\downarrow & & \downarrow{\scriptstyle \tilde{f}} \\ Y & \longleftarrow & \tilde{Y} \end{array},$$

where $\tilde{Y} = \operatorname{Spec} S$, $\tilde{X} = \operatorname{Spec} f^*(S)$, and to the coherent module $\tilde{M}$ on $\tilde{X}$.

COROLLARY 8.2.6.2. *Under the assumptions of 8.2.4, let $B := \bigoplus_{n \in \mathbb{N}} I^n$. Then, for all q, $\bigoplus_{n \in \mathbb{N}} H^q(X, I^n F)$ is a finitely generated graded B-module, and there exists $n_0 \geq 0$ such that, for all $n \geq n_0$, $H^q(X, I^n F) = I^{n-n_0} H^q(X, I^{n_0} F)$.*

8.2.7. *Proof of 8.2.4.*

In contrast with Grothendieck's original proof, the proof given in [**EGAIII1**, 4.1.7] does not go by descending induction on q. The integer q remains fixed in the whole proof, which consists of a careful analysis of the inverse system of maps

$$H^q(F) \to H^q(F_n), \tag{8.2.7.1}$$

where $H^q = H^q(X, -)$ for brevity. The map (8.2.7.1) sits in a portion of the long exact sequence of cohomology associated with the short exact sequence

$$0 \to I^{n+1}F \to F \to F_n \to 0,$$

namely

$$H^q(I^{n+1}F) \to H^q(F) \to H^q(F_n) \to H^{q+1}(I^{n+1}F) \to H^{q+1}(F). \tag{8.2.7.2}$$

We deduce from (8.2.7.2) an exact sequence

$$0 \to R_n \to H^q(F) \to H^q(F_n) \to Q_n \to 0, \tag{8.2.7.3}$$

where

$$R_n = \text{Im } H^q(I^{n+1}F) \to H^q(F),$$

and

$$Q_n = \text{Im } H^q(F_n) \to H^{q+1}(I^{n+1}F) = \text{Ker } H^{q+1}(I^{n+1}F) \to H^{q+1}(F).$$

The main points are the following:

(1) The filtration (R_n) on $H^q(F)$ is I-good (8.2.5.1); in particular, the topology defined by (R_n) on $H^q(F)$ is the I-adic topology.

(2) The inverse system $Q_. = (Q_n)$ is AR zero (8.2.5.2 (ii′)).

(3) The inverse system $H^q(F_.) = (H^q(F_n))$ satisfies ARML (8.2.5.2 (iii′)).

Let us first show that (1), (2), (3) imply 8.2.4. Consider the exact sequence of inverse systems defined by (8.2.7.3):

$$0 \to H^q(F)/R_n \to H^q(F_n) \to Q_n \to 0. \tag{$*$}$$

By (2) we have $\lim_n Q_n = 0$ (8.2.5.2 (a)). By the left exactness of the functor $\lim_n$ we thus get an isomorphism

$$\lim_n H^q(F)/R_n \xrightarrow{\sim} \lim_n H^q(F_n). \tag{$**$}$$

By (1) the map

$$H^q(F)^{\wedge} = \lim_n H^q(F)/I^{n+1}H^q(F) \to \lim_n H^q(F)/R_n \tag{$***$}$$

deduced from the surjections $H^q(F)/I^{n+1}H^q(F) \to H^q(F)/R_n$ is an isomorphism. Putting (**) and (***) together, we get that (8.2.4.1) is an isomorphism. By definition, we have

$$H^q(\hat{X}, \hat{F}) = H^q(\hat{X}, \lim F_n) = H^q(X, i_* \lim F_n) = H^q(X, \lim (i_n)_* F_n).$$

Thanks to (3), the assumptions of 8.2.5.3 are satisfied, therefore 8.2.4.2 is an isomorphism.

It remains to show (1), (2), (3).

Proof of (1). We have $R_{-1} = H^q(F)$. The inclusions

$$I^m R_n \subset R_{m+n}$$

follow from the fact that the natural map

$$\bigoplus_{n \in \mathbb{N}} H^q(I^{n+1}F)t^n \to \bigoplus_{n \in \mathbb{N}} H^q(F)t^n$$

is a map of graded B-modules, where $B = \bigoplus_{n \in \mathbb{N}} I^n t^n$ (8.2.6.2). By 8.2.6.2 (applied to IF), $\bigoplus_{n \in \mathbb{N}} H^q(I^{n+1}F)t^n$ is of finite type over B, and therefore so is its quotient $R := \oplus_n R_n$, which proves (1), thanks to the equivalence between conditions (ii) and (ii′) in 8.2.5.1.

Proof of (2). This is the most delicate point. By 8.2.6.2 again, the module $N := \oplus_n H^{q+1}(I^{n+1}F)$ is finitely generated over B. Since B is noetherian, $Q := \oplus_n Q_n$, which is a (graded) sub-B-module of N is also finitely generated, and therefore there exists $r \geq 0$ such that $Q_{n+1} = IQ_n$ for all $n \geq r$. Since Q_k, as a quotient of $H^q(F_k)$ is killed by I^{k+1} (as an A-module), each Q_n is therefore killed by I^{r+1} (as an A-module). Now, for $a \in I^p$, the composition of the multiplication by a from $H^{q+1}(I^{n+1}F)$ to $H^{q+1}(I^{p+n+1}F)$ with the transition map from $H^{q+1}(I^{p+n+1}F)$ to $H^{q+1}(I^{n+1}F)$ is the multiplication by a in $H^{q+1}(I^{n+1}F)$. Since $Q_{n+r+1} = I^{r+1}Q_n$ for $n \geq r$, it follows that, for all $n \geq r$, the transition map $Q_{n+r+1} \to Q_n$ is zero, and hence, if $s = 2r+1$, for all n the transition map $Q_{n+s} \to Q_n$ is zero.

Proof of (3). This is a formal consequence of (2). In the exact sequence (*), the left term has surjective transition maps (thus trivially satisfies ARML) and the right one is AR zero so they both trivially satisfy ARML. Therefore the middle one satisfies ARML in view of the second assertion of the following lemma [**SGA5**, V 2.1.2], whose proof is elementary:

LEMMA 8.2.7.4. *Let*

$$0 \to L'_{\cdot} \to L_{\cdot} \to L''_{\cdot} \to 0$$

be an exact sequence of inverse systems of A-modules. If $L_{\cdot}$ satisfies ARML, so does $L''_{\cdot}$, and if $L'_{\cdot}$ and $L''_{\cdot}$ satisfy ARML, so does L.

REMARKS 8.2.8.

(a) Property (2) in 8.2.7 is the key technical point in Deligne's construction of the $Rf_!$ functor from $\mathrm{pro}D^b(X)_{coh}$ to $\mathrm{pro}D^b(S)_{coh}$ for $f : X \to S$ a compactifiable morphism of noetherian schemes (i.e., of the form gi with g proper and i an open immersion) [**Har66**, Appendix, Prop. 5] (more precisely, if, in the situation of 8.2.4, f is assumed to induce an isomorphism from $X - X'$ to $Y - Y'$, then what is shown in (loc. cit.) is that the inverse systems $H^q(I^{n+1}F)$ are AR zero for $q > 0$, as follows from the fact that, for $q > 0$, $H^q(F)$ is killed by a power of I.

(b) The proof of 8.2.2 shows that if $f : X \to Y$, Y', X' are as in 8.2.2 and F is a coherent sheaf on X such that, for some integer q, the graded modules $\oplus_n R^q f_*(I^n F)$ and $\oplus_n R^{q+1} f_*(I^n F)$ over the graded $\mathcal{O}_Y$-algebra $\oplus_n I^n$ are finitely generated, then $R^q f_* F$ is coherent and (8.2.1.3) and (8.2.1.4) are isomorphisms. See [**SGA2**, IX] for details and examples. This refined comparison theorem is a key tool in Grothendieck's Lefschetz type theorems for the fundamental group and the Picard group [**SGA2**, X, XI].

The comparison theorem 8.2.2 has many corollaries and applications. We will mention only a few of them. The following one (for $r = 0, 1$) is the main ingredient in the proof of Grothendieck's existence theorem, which will be discussed in §8.3.

COROLLARY 8.2.9. [**EGAIII1**, 4.5.1]

Let A be a noetherian ring, I an ideal of A, $f : X \to Y$ a morphism of finite type, $\hat{f} : \hat{X} \to \hat{Y}$ its completion along $Y' = V(I)$ and $X' = f^{-1}(Y')$ as in 8.2.4. Let F, G be coherent sheaves on X whose supports have an intersection which is proper over Y. Then, for all $r \in \mathbb{Z}$, $\mathrm{Ext}^r(F,G)$ is an A-module of finite type, and the natural map $\mathrm{Ext}^r(F,G) \to \mathrm{Ext}^r(\hat{F},\hat{G})$ induces an isomorphism

$$\mathrm{Ext}^r(F,G)^{\wedge} \xrightarrow{\sim} \mathrm{Ext}^r(\hat{F},\hat{G}). \tag{8.2.9.1}$$

We have

$$\mathrm{Ext}^r(F,G) = H^r R\mathrm{Hom}(F,G) = H^r R\Gamma(X, R\mathcal{H}om(F,G)).$$

The hypotheses on F, G imply that the cohomology sheaves of $R\mathcal{H}om(F,G)$ are coherent and have proper support over Y. Therefore, by the finiteness theorem, the cohomology groups of $RHom(F,G) = R\Gamma(X, R\mathcal{H}om(F,G))$ are finitely generated over A, and by Remark 8.2.3 (a), the base change map

$$R\Gamma(X, R\mathcal{H}om(F,G))\hat{} \to R\Gamma(\hat{X}, R\mathcal{H}om(F,G)\hat{}),$$

where $(-)\hat{} = i^*$, is an isomorphism. But, since i is flat,

$$R\mathcal{H}om(F,G)\hat{} = R\mathcal{H}om(\hat{F},\hat{G}),$$

and the conclusion follows.

The next corollary is very useful in geometric applications:

COROLLARY 8.2.10 (theorem on formal functions, [**EGAIII1**, 4.2.1]). *Let $f : X \to Y$ be a proper morphism of locally noetherian schemes, y a point of Y, $X_y = X \times_Y \mathrm{Spec}\, k(y)$ the fiber of f at y, F a coherent sheaf on X. Let $F_n = F \otimes \mathcal{O}_y/\mathbf{m}_y^{n+1}$ on $X_n = X \times_Y \mathrm{Spec}\, \mathcal{O}_y/\mathbf{m}_y^{n+1}$. Then, for all $q \in \mathbb{Z}$, the stalk $R^q f_*(F)_y$ is an $\mathcal{O}_y$-module of finite type, and the natural map*

$$(R^q f_*(F)_y)\hat{} = \lim{}_n (R^q f_*(F)_y/\mathbf{m}_y^{n+1} R^q f_*(F)_y) \to \lim{}_n H^q(X_y, F_n) \tag{8.2.10.1}$$

is an isomorphism.

The map (8.2.10.1) is defined by the base change maps

$$(R^q f_*(F)_y/\mathbf{m}^{n+1} R^q f_*(F)_y) \to H^q(X_y, F_n),$$

where in the right hand side, X_y is viewed as the underlying space of the scheme X_n. When y is closed, 8.2.10 is a special case of 8.2.2. One reduces to this case by base changing by $\mathrm{Spec}\, \mathcal{O}_y \to Y$.

8.2.11. Let $f : X \to Y$ be a proper morphism of locally noetherian schemes. Then $f_*\mathcal{O}_X$ is a finite $\mathcal{O}_Y$-algebra. Its spectrum $Y' = \mathrm{Spec}\, f_*\mathcal{O}_X$ is a finite scheme over Y, and the identity map of $f_*\mathcal{O}_X$ defines a factorization of f into

$$X \xrightarrow{f'} Y' \xrightarrow{g} Y,$$

with f' proper and g finite, called the *Stein factorization* of f. Its main property is described in the following theorem:

THEOREM 8.2.12 (Zariski's connectedness theorem, [**EGAIII1**, 4.3.1]). *With the assumptions and notations of 8.2.11, $f'_*\mathcal{O}_X = \mathcal{O}_{Y'}$, and the fibers of f' are connected and nonempty.*

The first assertion follows trivially from the definitions. For the second one, one first reduces to the case where $Y' = Y$ and y is a closed point of Y. Then, if $\hat{X}$ is the completion of X along X_y, by 8.2.10,

$$\mathcal{O}_{Y,y}\hat{} = (f_*\mathcal{O}_X)_y\hat{} = H^0(X_y, \mathcal{O}_{\hat{X}}),$$

which cannot be the product of two nonzero rings.

In particular, if $Y' = Y$, i.e., $f_*\mathcal{O}_X = \mathcal{O}_Y$, the fibers of f are connected and nonempty. It is not hard to see, using the base change formula 8.3.3 below, that they are in fact *geometrically connected* (i.e., are connected and remain so after any field extension) [**EGAIII1**, 4.3.4].

The following corollaries are easy, see [**EGAIII1**, 4.3, 4.4] for details.

COROLLARY 8.2.13. *Under the assumptions of 8.2.12, for every point y of Y, the connected components of the fiber X_y correspond bijectively to the points of Y'_y, i.e., to the maximal ideals of the finite $\mathcal{O}_y$-algebra $f_*(\mathcal{O}_X)_y$.*

This is because the underlying space $g^{-1}(y)$ of Y'_y is finite and discrete.

COROLLARY 8.2.14. *Let $f : X \to Y$ be a proper and surjective morphism of integral noetherian schemes, with Y normal. Assume that the generic fiber of f is geometrically connected (which is the case, for example, if f is* birational*). Then all fibers of f are geometrically connected.*

Let ζ (resp. η) be the generic point of X (resp. Y) (so that $f(\zeta) = \eta$). The hypothesis on the generic fiber means that the algebraic closure K' of $K = k(\eta)$ in $k(\zeta)$ is a (finite) radicial extension of K [**EGAIV2**, 4.5.15]. Let $y \in Y$. Since $\mathcal{O}_y$ is normal et K' is radicial over K, the normalization A of $\mathcal{O}_y$ in K' is a local ring. and the residue field extension is radicial [**Bou64**, Ch. 5, §8.2, n°3, Lemme 4]. Since A contains $(f_*\mathcal{O}_X)_y$, the same holds for $(f_*\mathcal{O}_X)_y$. Therefore, by 8.2.12 (and the remark after it) the fiber X_y is geometrically connected.

COROLLARY 8.2.15. *Under the assumptions of 8.2.12, a point x of X is isolated in its fiber, i.e., is both open and closed in $f^{-1}(f(x))$, if and only if $f'^{-1}(f'(x)) = \{x\}$. The set U of such points is open in X, $U' = f'(U)$ is open in Y', and $f' : X \to Y'$ induces an isomorphism $f'_{U'} : U \xrightarrow{\sim} f'(U)$.*

Let $y = f(x)$, $y' = f'(x)$. Since $g^{-1}(y)$ is finite, discrete, x is isolated in $f^{-1}(y)$ if and only if it is in $f'^{-1}(y')$. So we may assume $Y' = Y$, i.e., $f_*\mathcal{O}_X = \mathcal{O}_Y$, and hence, by 8.2.13, x is isolated in $f^{-1}(y)$ if and only if $f^{-1}(y) = \{x\}$. For $x \in U$, choose open affine neighborhoods $V = \operatorname{Spec} B$, $W = \operatorname{Spec} A$ of x and y respectively, such that $f(V) \subset W$. Since f is closed, $f(X - V)$ is a closed subset of Y which does not contain y. Therefore, there exists an open affine neighborhood of y of the form $W_s = \operatorname{Spec} A_s$ for some $s \in A$ such that $f^{-1}(W_s) \subset V$. Then $f^{-1}(W_s) = V_s = \operatorname{Spec} B_s$. Since $f_*\mathcal{O}_X = \mathcal{O}_Y$, f induces on W_s an isomorphism $V_s \xrightarrow{\sim} W_s$.

COROLLARY 8.2.16. *Let $f : X \to Y$ be a proper morphism of locally noetherian schemes. If f is quasi-finite (i.e., has finite fibers), then f is finite.*

COROLLARY 8.2.17 (Zariski's Main Theorem). *Let $f : X \to Y$ be a compactifiable morphism of locally noetherian schemes (8.2. 8 (a)) (e. g. a quasi-projective morphism,* [**EGAII**, 5.3.2]*). If f is quasi-finite, then f can be factored as $f = gj$, where $j : X \to Z$ is an open immersion and $g : Z \to Y$ is a finite morphism.*

If Y is noetherian, one can remove the hypothesis that f should be compactifiable, provided that f is assumed to be separated and of finite type, see [**EGAIV3**, 8.12.6], whose proof makes no use of the comparison theorem 8.2.2 but relies on deeper commutative algebra.

Finally, we mention a useful application of 8.2.13. If X is a locally noetherian scheme, we denote by $\pi_0(X)$ the set of its connected components.

COROLLARY 8.2.18. *Let A be a henselian noetherian local ring, $S = \operatorname{Spec} A$, s its closed point, X a proper scheme over S. Then the natural map*

$$\pi_0(X_s) \to \pi_0(X)$$

is bijective.

Consider the Stein factorization

$$X \xrightarrow{f'} S' \longrightarrow S$$

of the structural morphism $f : X \to S$. We have $S' = \operatorname{Spec} A'$, where A' is a finite A-algebra. Since A is henselian, A' decomposes as a product of local A-algebras A_i, parametrized by the points i of S'_s. Let $S'_i = \operatorname{Spec} A_i$ and $X_i = S'_i \times_S X$, so that X is the disjoint union of the X_i's. By 8.2.13 the fiber $(X_i)_i = f'^{-1}(i)$ of X_i at i is connected. Since X_i is proper over S'_i and S'_i is local, no component of X_i can be disjoint from its special fiber, hence X_i is connected. Hence the X_i's are the connected components of X and they correspond bijectively to the connected components of X_s by associating to a component its special fiber.

8.2.19. Base change maps. Let

$$\begin{array}{ccc} X' & \xrightarrow{h} & X \\ \downarrow{\scriptstyle f'} & & \downarrow{\scriptstyle f} \\ Y' & \xrightarrow{g} & X \end{array} \tag{8.2.19.1}$$

be a commutative square of ringed spaces and let F be an $\mathcal{O}_X$-module. Then there is a canonical map of $\mathcal{O}_{Y'}$-modules

$$\gamma : g^*f_*F \to f'_*h^*F, \tag{8.2.19.2}$$

called the *base change map*, which is defined in the following two equivalent ways. Let $a = gf' = fh$.

(a) By adjunction between g^* and g_*, defining γ is equivalent to defining

$$\gamma_1 : f_*F \to g_*f'_*h^*F = a_*h^*F.$$

One has $a_*h^*F = f_*h_*h^*F$, and one defines γ_1 by applying f_* to the adjunction map $F \to h_*h^*F$.

(b) By adjunction between f'^* and f'_*, defining γ is equivalent to defining

$$\gamma_2 : f'^*g^*f_*F = a^*f_*F \to h^*F.$$

One has $a^*f_*F = h^*f^*f_*F$, and one defines γ_2 by applying h^* to the adjunction map $f^*f_*F \to F$.

That these two definitions are equivalent is a nontrivial fact, proved by Deligne [**SGA4**, XVII] in a much more general context.

Along the same lines, one defines, for all $q \in \mathbb{Z}$, a canonical map

$$\gamma : g^*R^qf_*F \to R^qf_*(h^*F), \tag{8.2.19.3}$$

also called *base change map*. Again, by adjunction between g^* and g_*, it is equivalent to define

$$\gamma_1 : R^qf_*F \to g_*R^qf_*(h^*F).$$

One defines γ_1 as the composition vu of the following two maps:

$$u : R^qf_*F \to R^qa_*(h^*F),$$

$$v : R^q a_*(h^*F) \to g_* R^q f_*(h^*F).$$

The map u is the classical *functoriality map* on cohomology. Namely, we have an adjunction map in $D^+(X)$:

$$\alpha : F \to Rh_* h^* F,$$

defined as the composition $F \to h_* h^* F \to h_* C(h^*F)$, where the first map is the classical adjunction map and the second one is given by the choice of a resolution $h^*F \to C(h^*F)$ of h^*F by modules acyclic for h_*. Applying Rf_* to α, we get a map

$$Rf_*(\alpha) : Rf_*F \to Rf_* Rh_* h^* F = Ra_* h^* F,$$

giving u by passing to cohomology sheaves. In other words, if V is an open subset of Y and $U = f^{-1}(V)$, $U' = a^{-1}(V) = h^{-1}(U)$, $R^q f_* F$ is the sheaf associated to the presheaf $V \mapsto H^q(U, F)$, $R^q a_*(h^*F)$ is the sheaf associated to the presheaf $V \mapsto H^q(U', h^*F)$, and u is associated to the functoriality map $H^q(U, F) \to H^q(U', h^*F)$.

The map v is an edge homomorphism $H^q \to E_\infty^{0q} \to E_2^{0q}$ for the spectral sequence

$$E_2^{ij} = R^i g_* R^j f'_*(h^*F) \Rightarrow R^{i+j} a_*(h^*F).$$

More explicitly, with the above notations and $V' = g^{-1}(V)$, v is associated to the map

$$H^q(U', h^*F) \to H^0(V', R^q f'_*(h^*F))$$

obtained by restricting an element of $H^q(U', h^*F)$ to open subsets $f'^{-1}(W)$ for W open in V'.

Under suitable assumptions of cohomological finiteness, it is possible to define a base change map in $D(Y')$,

$$Lg^* Rf_* F \to Rf'_* Lh^* F, \tag{8.2.19.4}$$

inducing (8.2.19.3) (cf. [**SGA4**, XVII 4.1.5]). However, when (8.2.19.1) is a cartesian square of schemes and F is a quasi-coherent sheaf, this map has no good properties in general (see 8.3.5).

8.3. Cohomological flatness

8.3.1. The results of this section will not be used in §8.4. They complement those of §8.2. More precisely, following [**EGAIII2**, 7] we address the following question: in the situation of 8.2.10, with F *flat* over Y, when can we assert that the individual base change maps

$$R^q f_*(F)_y / \mathbf{m}_y^{n+1} R^q f_*(F)_y \to H^q(X_y, F_n)$$

are isomorphisms? More generally, when can we assert that the formation of $R^q f_*(F)$ commutes with any base change, when is $R^q f_*(F)$ locally free of finite type? As was shown in [**SGA6**, III], the use of derived categories simplifies the presentation given in [**EGAIII2**, 7]. Other expositions are given in [**Har77**, III 12] and [**Mum70**, 5].

In what follows, if X is a scheme, we denote by $D(X)$ the derived category of the category of $\mathcal{O}_X$-modules. The main tool is the following *base change formula*:

THEOREM 8.3.2. *Let*

(8.3.2.1)
$$\begin{array}{ccc} X' & \xrightarrow{h} & X \\ \downarrow{\scriptstyle f'} & & \downarrow{\scriptstyle f} \\ Y' & \xrightarrow{g} & Y \end{array}$$

be a cartesian square of schemes, with X and Y quasi-compact and separated. Let F (resp. G) be a quasi-coherent sheaf on X (resp. Y'). Assume that F and G are tor-independent *on Y, i.e., that for all points $x \in X$, $y' \in Y'$ such that $g(y') = f(x)$ we have*

$$Tor_q^{\mathcal{O}_{Y,y}}(G_{y'}, F_x) = 0$$

for all $q > 0$ (this is the case for example if F or G is flat *over Y). Then there is a natural isomorphism in $D(Y')$:*

(8.3.2.2)
$$G \otimes_Y^L Rf_*F \xrightarrow{\sim} Rf'_*(G \otimes_Y F),$$

where $G \otimes_Y^L Rf_*F := G \otimes^L Lg^*Rf_*F$ and $G \otimes_Y F = f'^*G \otimes h^*F$.

When $Y = Y'$ (resp. $G = \mathcal{O}_{Y'}$), the isomorphism (8.3.2.2) is called the *projection isomorphism* (resp. the *base change isomorphism*). When $G = \mathcal{O}_{Y'}$, one deduces from (8.3.2.2) a canonical map, for $q \in \mathbb{Z}$,

(8.3.2.3)
$$g^*R^qf_*F \to R^qf'_*(h^*F).$$

This map is the composition of the canonical map $g^*R^qf_*F \to H^q(Lg^*Rf_*F)$ and the isomorphism $H^q(Lg^*Rf_*F) \xrightarrow{\sim} R^qf'_*(h^*F)$ deduced from (8.3.2.2) by applying H^q. It will follow from the construction of (8.3.2.2) that this map is the base change map defined in (8.2.19.3). It is *not* an isomorphism in general. This question is addressed in 8.3.10–8.3.11.

The following corollaries are the most useful particular cases:

COROLLARY 8.3.3. *If, in the cartesian square (8.3.2.1), g is flat, then (8.3.2.2) gives a base change isomorphism*

$$g^*Rf_*F \xrightarrow{\sim} Rf'_*h^*F,$$

and the induced base change maps (8.3.2.3) are isomorphisms.

COROLLARY 8.3.4. *Let $f : X \to Y$ be a morphism between quasi-compact and separated schemes. Let y be a point of Y, denote by X_y the fiber of f at y, i.e.,* $\operatorname{Spec} k(y) \times_Y X$, *and let F be a quasi-coherent sheaf on X, flat over Y. Then (8.3.2.2) gives a natural isomorphism (in the derived category of $k(y)$-vector spaces)*

$$k(y) \otimes_{\mathcal{O}_Y}^L Rf_*F \xrightarrow{\sim} R\Gamma(X_y, \mathcal{O}_{X_y} \otimes_{\mathcal{O}_X} F).$$

Let us prove 8.3.2. First, consider the case where X, Y, Y' are affine, with rings B, A, A' respectively, so that X' is affine of ring $B' = A' \otimes_A B$, and $F = \tilde{M}$, $G = \tilde{N}$ for some B-module M and A'-module N. Then Rf_*F is represented by the underlying A-module $M_{[A]}$ of M, and $Rf'_*(G \otimes_Y F)$ by the underlying A'-module $(N \otimes_A M)_{[A']}$ of $(N \otimes B') \otimes_{B'} (B' \otimes M)$. On the other hand, $G \otimes_Y^L Rf_*F$ is represented by $N \otimes_A^L M_{[A]} := N \otimes_{A'}^L (A' \otimes_A^L M_{[A]})$, which can be calculated as $N \otimes_A P$ where P is a flat resolution of $M_{[A]}$. The tor-independence hypothesis says that $Tor_q^A(N, M_{[A]}) = 0$ for $q > 0$; i.e., the natural map

(*)
$$N \otimes_A^L M_{[A]} \to N \otimes_A M_{[A]}$$

is an isomorphism (in $D(A')$). The isomorphism (8.3.2.2) is the composition of (*) and the (trivial) isomorphism

$$(**) \qquad N \otimes_A M_{[A]} \xrightarrow{\sim} (N \otimes_A M)_{[A']}.$$

Assume now that the morphism f (but not necessarily the scheme Y) is affine. Then $X = \operatorname{Spec} B$ for a quasi-coherent $\mathcal{O}_Y$-algebra B, and $F = \tilde{M}$ for a quasi-coherent B-module M. We have again $Rf_*F = f_*F$, which is represented by the underlying (quasi-coherent) $\mathcal{O}_Y$-module $M_{[A]}$ of M. The preceding discussion, applied to affine open subsets of Y' above affine open subsets of Y, shows that we have natural identifications

$$G \otimes_Y^L Rf_*F \xrightarrow{\sim} G \otimes g^* f_*F \xrightarrow{\sim} f'_*(G \otimes_Y F) \xrightarrow{\sim} Rf'_*(G \otimes_Y F).$$

Their composition defines the isomorphism (8.3.2.2).

In the general case, choose a finite open affine cover $\mathcal{U} = (U_i)_{i \in I}$ of X ($I = \{1, \cdots, r\}$). Since X and Y are separated, any finite intersection $U_{i_0 \cdots i_n} = U_{i_0} \cap \cdots \cap U_{i_n}$ ($i_0 < \cdots < i_n$) of the U_i's is affine over Y [**EGAII**, 1.6.2]. Therefore (by [**EGAIII1**, 1.4]) we have

$$Rf_*F = f_*\check{C}(\mathcal{U}, F),$$

where $\check{C}(\mathcal{U}, F)$ is the alternating Čech complex of $\mathcal{U}$ with values in F. By the discussion in the case f is affine, we get isomorphisms

$$(***) \quad G \otimes_Y^L Rf_*F \xrightarrow{\sim} G \otimes g^* f_*\check{C}(\mathcal{U}, F) \xrightarrow{\sim} f'_*\check{C}(\mathcal{U}', G \otimes_Y F) \xrightarrow{\sim} Rf'_*(G \otimes_Y F),$$

where $\mathcal{U}'$ is the cover of X' formed by the inverse images of the U_i's. It is easy to check that the composition (***) does not depend on the choice of $\mathcal{U}$. We take this composition as the definition of the isomorphism (8.3.2.2).

The compatibility between (8.3.2.3) and (8.2.19.3) is left to the reader.

REMARK 8.3.5. It is easy to generalize 8.3.2 to the case Y is quasi-compact and f is quasi-compact and quasi-separated (in the last part of the argument, the intersections $U_{i_0 \cdots i_n}$ are only quasi-compact, and one has to replace the Čech complex by a suitable "hyper Čech" variant).

It seems difficult, however, to get rid of the tor-independence assumption. For example, when (8.3.2.1) is a cartesian square of *affine* schemes, as in the beginning of the proof of 8.3.2, and $F = \mathcal{O}_X$, but no tor-independence assumption is made, we do have a base change map of the form (8.2.19.4), namely, the map corresponding to the map

$$A' \otimes_A^L B \to A' \otimes_A B$$

in $D(A')$, but this map is an isomorphism if and only if A' and B are tor-independent over A.

In order to obtain a satisfactory formalism one has to use some tools of homotopical algebra, such as derived tensor products of rings. No account has been written down as yet.

8.3.6. The main application of 8.3.2 is to the case f is a proper morphism of noetherian schemes and F is a coherent sheaf on X, which is flat over Y. In this case, the complex Rf_*F has nice properties, namely it's a *perfect* complex, and the base change formula 8.3.4 enables one to analyze the compatibility with base change of its cohomology sheaves $R^q f_*F$ around a point y of Y.

We first recall some basic finiteness conditions on objects of $D(X)$, where X is a locally noetherian scheme. These are discussed in much greater generality in [**SGA6**, I, II, III]. There are three main conditions: pseudo-coherence, finite tor-dimension, perfectness, the last one being a combination of the first two.

8.3.6.1. *Pseudo-coherence.* A complex $E \in D^b(X)$ is called *pseudo-coherent* if it has coherent cohomology (i.e., $H^q(E)$ is coherent for all q). One usually denotes by $D^b(X)_{coh}$ the full subcategory of $D^b(X)$ consisting of pseudo-coherent complexes. It is a triangulated subcategory. If X is affine and $E \in D^b(X)$ has $H^q(E) = 0$ for $q > a$, then E is pseudo-coherent if and only if E is isomorphic, in $D(X)$, to a complex $L \in D^b(X)$ such that $L^q = 0$ for $q > a$ and L^q is free of finite type for all q. In particular, on any locally noetherian scheme X, a pseudo-coherent complex is locally isomorphic, in the derived category, to a bounded above complex of $\mathcal{O}$-modules which are free of finite type, and for any point x of X, the stalk E_x, as a complex of $\mathcal{O}_{X,x}$-modules, is isomorphic (in the derived category $D(\mathcal{O}_{X,x})$) to a bounded above complex of $\mathcal{O}_{X,x}$-modules which are free of finite type.

When E has quasi-coherent components, the above assertion is proven by an easy step by step construction [**EGAIII1**, Ch. 0, 11.9.1]. In the general case, one has first to replace E by a bounded complex with quasi-coherent components, which is more delicate (see [**Har66**, II 7.19], [**SGA6**, II 2.2.1]).

8.3.6.2. *Finite tor-dimension.* Let $a, b \in \mathbb{Z}$ with $a \leq b$. A complex $E \in D(X)$ is said to be of *tor-amplitude* in $[a, b]$ if it satisfies the following equivalent conditions:

(i) E is isomorphic, in $D(X)$, to a complex L such that $L^q = 0$ for $q \notin [a, b]$ and L^q is *flat* for all q;

(ii) for any $\mathcal{O}_X$-module M, one has $H^q(M \otimes^L E) = 0$ for $q \notin [a, b]$.

The proof of the equivalence of (i) and (ii) is straightforward. A complex E is said to be of *finite tor-dimension* (or of *finite tor-amplitude*) if it is of tor-amplitude in $[a, b]$ for some interval $[a, b]$. The full subcategory of $D(X)$ consisting of complexes of finite tor-dimension is a triangulated subcategory.

For a complex E to be of tor-amplitude in $[a, b]$ it is necessary and sufficient that, for all $x \in X$, the stalk E_x, as a complex of $\mathcal{O}_{X,x}$-modules, be of tor-amplitude in $[a, b]$, i.e., isomorphic, in $D(\mathcal{O}_{X,x})$, to a complex L concentrated in degree in $[a, b]$ and flat in each degree, or, equivalently, such that, for any $\mathcal{O}_{X,x}$-module M, $H^q(M \otimes^L E_x) = 0$ for $q \notin [a, b]$.

8.3.6.3. *Perfectness.* A complex $E \in D^b(X)$ is called *perfect* if it is pseudo-coherent and locally of finite tor-dimension. It is said to be of *perfect amplitude* in $[a, b]$ (for $a, b \in \mathbb{Z}$ with $a \leq b$) if it is pseudo-coherent and of tor-amplitude in $[a, b]$. A *strictly perfect* complex is a bounded complex of locally free of finite type modules. The full subcategory of $D^b(X)$ consisting of perfect complexes is a triangulated subcategory.

Since an $\mathcal{O}_X$-module is locally free of finite type if and only if it is coherent and flat, it follows from 8.3.6.1 and 8.3.6.2 that a complex $E \in D^b(X)$ is perfect if and only if it is locally isomorphic, in the derived category, to a strictly perfect complex. In the same vein, we have the following useful criterion:

PROPOSITION 8.3.6.4. *Let x be a point of X and $E \in D^b(X)$ be a pseudo-coherent complex on X such that $H^q(E_x) = 0$ for $q \notin [a, b]$, for some interval $[a, b]$. Then the following conditions are equivalent:*

(i) in a neighborhood of x, E is of perfect amplitude in $[a, b]$;

(ii) $H^{a-1}(k(x) \otimes^L E) = 0$;

(iii) in a neighborhood of x, E is isomorphic, in the derived category, to a complex of free of finite type $\mathcal{O}$-modules concentrated in degree in $[a, b]$.

By 8.3.6.1 we may assume that E has coherent components and is concentrated in degree in $[a, b]$, with E^q free of finite type for $q > a$. We have to show that (ii) implies that E^a is locally free of finite type around x. From the exact sequence

$$0 \to E^{[a+1,b]} \to E \to E^a[-a] \to 0,$$

where $E^{[a+1,b]}$ is the naïve truncation of E in degree $\geq a+1$, we deduce that

$$Tor_1^{\mathcal{O}_{X,x}}(k(x), E_x^a) = 0.$$

By the standard flatness criterion [**Bou61**, III, §5, th. 3], this implies that E_x^a is free of finite type, hence that E^a is free of finite type in a neighborhood of x.

COROLLARY 8.3.6.5. *Let x be a point of X, $q \in \mathbb{Z}$, and $E \in D^b(X)$ be a pseudo-coherent complex on X such that $H^i E = 0$ for $i > b$ for some integer b. For $i \in \mathbb{Z}$, let*

$$\alpha^i(x) : k(x) \otimes H^i(E) \to H^i(k(x) \otimes^L E)$$

denote the canonical map.

(a) The following conditions are equivalent:

(i) $\alpha^q(x)$ is surjective;

(ii) $\tau_{>q}E$ is of perfect amplitude in $[q+1, b]$ in a neighborhood of x.

When these conditions are satisfied, there is an open neighborhood U of x such that $\alpha^q(y)$ is bijective for all $y \in U$, and such that for all quasi-coherent modules M on U, the natural map

$$\alpha^q(M) : M \otimes H^q(E) \to H^q(M \otimes^L E)$$

is bijective.

(b) Assume that (a) (i) holds. Then the following conditions are equivalent:

(i) $\alpha^{q-1}(x)$ is surjective;

(ii) $H^q(E)$ is locally free of finite type in a neighborhood of x.

Here, if L is a complex (in an abelian category) and $i \in \mathbb{Z}$, $\tau_{\geq i}L$ denotes the *canonical truncation* of L in degree $\geq i$, defined as $0 \to L^i/dL^{i-1} \to L^{i+1} \to \cdots$, and $\tau_{>i} = \tau_{\geq i+1}$.

Let us prove (a). The projection $E \to \tau_{\geq q}E$ induces an isomorphism

$$(*) \qquad H^q(k(x) \otimes^L E) \xrightarrow{\sim} H^q(k(x) \otimes^L \tau_{\geq q}E).$$

Consider the canonical distinguished triangle

$$(**) \qquad H^q(E)[-q] \to \tau_{\geq q}E \to \tau_{>q}E \to .$$

Taking (*) into account, we get from (**)

$$(***) \qquad \operatorname{Coker} \alpha^q(x) = H^q(k(x) \otimes^L \tau_{>q}E).$$

The equivalence between (i) and (ii) thus follows from 8.3.6.4. Assume that these conditions hold. It suffices to show the last assertion of (a). Let U be an open neighborhood of x such that $\tau_{>q}E|U$ is of perfect amplitude in $[q+1, b]$. Let M be a quasi-coherent sheaf on U. Applying $M \otimes^L -$ to (**), and taking into account that $H^i(M \otimes^L \tau_{>q}E) = 0$ for $i \leq q$, we get that $\alpha^q(M)$ is bijective. Let us prove (b). Since $\tau_{>q}E$ is of perfect amplitude in $[q+1, b]$ in a neighborhood of x, the triangle (**) shows that $H^q(E)$ is locally free of finite type in a neigborhood of x

if and only if $\tau_{\geq q}E$ is of perfect amplitude in $[q,b]$ in a neighborhood of x. But, by (i), this condition is equivalent to the surjectivity of $\alpha^{q-1}(x)$.

8.3.7. Let $f : X \to Y$ be a morphism of locally noetherian schemes and let F be a coherent sheaf of X. As said earlier, the main applications of 8.3.2 deal with the case where F is flat over Y. By definition, F is flat over Y if and only if for all $x \in X$, the $\mathcal{O}_{X,x}$-module F_x is flat over $O_{Y,y}$, where $y = f(x)$. It is often convenient to express this in the following way, given by the *flatness criterion* [**Bou61**, III, §5, th. 3]: F is flat over Y if and only if, for all $y \in Y$, the natural (surjective) map

$$\mathrm{gr}^n \mathcal{O}_{Y,y} \otimes_{k(y)} \mathrm{gr}^0 F_y \to \mathrm{gr}^n F_y \tag{8.3.7.1}$$

is an isomorphism for all $n \geq 0$, where gr means the associated graded for the $\mathbf{m}_y$-adic filtration on $\mathcal{O}_{Y,y}$ ($\mathbf{m}_y$ being the maximal ideal) and F_y, the inverse image of F on $\operatorname{Spec} \mathcal{O}_{Y,y} \times_Y X$. The bijectivity of (8.3.7.1) is also equivalent to the fact that $Tor_1^{\mathcal{O}_{Y,y}}(k(y), F_y) = 0$; i.e., the natural map $\tau_{\geq -1}(k(y) \otimes^L F) \to k(y) \otimes F$ is an isomorphism, or to the fact that, for each $n \geq 0$, $F_y/\mathbf{m}_y^{n+1}F$ is flat over $\operatorname{Spec} \mathcal{O}_{Y,y}/\mathbf{m}_y^{n+1}$.

THEOREM 8.3.8. *Let $f : X \to Y$ be a proper morphism of noetherian schemes, and let F be a coherent sheaf on X. Then Rf_*F is pseudo-coherent (8.3.6.1) on Y. If F is flat over Y, Rf_*F is perfect (8.3.6.3).*

The first assertion is just a rephrasing of Grothendieck's finiteness theorem [**EGAIII1**, 3.2.1], which says that the sheaves $R^q f_* F$ are coherent, together with [**EGAIII1**, 1.4.12], which implies that Rf_*F belongs to $D^b(Y)$. To prove the second assertion, we may assume that Y is affine. Let N be an integer such that $R^q f_* E = 0$ for all quasi-coherent sheaves E on X and $q > N$ (one can take N such that there is a covering of X by $N+1$ open affine subsets [**EGAIII1**, 1.4.12]). By (8.3.2.2), for any quasi-coherent $\mathcal{O}_Y$-module G, we have

$$G \otimes^L Rf_*F \xrightarrow{\sim} Rf_*(G \otimes_Y F),$$

and in particular,

$$H^q(G \otimes^L Rf_*F) = 0$$

for $q \neq [0,N]$. *A fortiori*, for any point y of Y and any $\mathcal{O}_{Y,y}$-module M, we have

$$H^q(M \otimes^L (Rf_*F)_y) = 0$$

for $q \neq [0,N]$. By 8.3.6.2, this means that Rf_*F is of perfect amplitude in $[0,N]$.

8.3.9. Under the assumptions of 8.3.8, with F flat over Y, assume Y affine, $Y = \operatorname{Spec} A$, and let Y' be a closed subscheme of Ydefined by an ideal I. As explained in [**EGAIII2**, 7.4.8], the pseudo-coherence of Rf_*F, together with the base change formula 8.3.2, gives another proof (in this particular case) of the fact that the maps φ_q (8.2.4.1) are isomorphisms.

By 8.3.2, we have, for $n \geq 0$,

$$A/I^{n+1} \otimes^L R\Gamma(X,F) \xrightarrow{\sim} R\Gamma(X,F_n). \tag{$*$}$$

The map φ_q is the inverse limit of the maps

$$\varphi_{q,n} : A/I^{n+1} \otimes H^q(X,F) \to H^q(X,F_n),$$

obtained by composing the natural map $A/I^{n+1} \otimes H^q(X,F) \to H^q(A/I^{n+1} \otimes^L R\Gamma(X,F))$ with the isomorphism $H^q(*)$. Since Rf_*F is pseudo-coherent, $R\Gamma(X,F)$

is isomorphic to a complex P of A-modules, which is bounded above and consists of free modules of finite type. The maps $\varphi_{q,n}$ can be rewritten

$$H^q(P)/I^{n+1}H^q(P) \to H^q(P/I^{n+1}P).$$

In general, none is an isomorphism, but it follows from Artin–Rees that the limit

$$\lim H^q(P)/I^{n+1}H^q(P) \to \lim H^q(P/I^{n+1}P).$$

is an isomorphism.

8.3.10. Let $f : X \to Y$ be a proper morphism of separated noetherian schemes, and let F be a coherent sheaf on X, flat over Y. Let $q \in \mathbb{Z}$. We say that F is *cohomologically flat over Y in degree q* if, for any morphism $g : Y' \to Y$, the base change map (8.3.2.3)

$$g^* R^q f_* F \to R^q f'_* F' \tag{8.3.10.1}$$

is an isomorphism, where, in the notations of (8.3.2.1), $F' = h^* F$. When $F = \mathcal{O}_X$ (i.e., f is flat), we just say that f is cohomologically flat in degree q.

If y is a point of Y and $X_y = \operatorname{Spec} k(y) \times_Y X$ is the fiber of f at y, the map (8.3.10.1) reads

$$k(y) \otimes R^q f_* F \to H^q(X_y, F/\mathbf{m}_y F). \tag{8.3.10.2}$$

We shall denote it by $\alpha^q(y)$ by analogy with the notation used in 8.3.6.5. The following criterion is a simple consequence of 8.3.6.5, applied to the pseudo-coherent complex $E = Rf_* F$ on Y (cf. [**EGAIII2**, 7.8.4], [**Har77**, III, 12.11]):

COROLLARY 8.3.11. *With $f : X \to Y$ and F as in 8.3.10, let $q \in \mathbb{Z}$ and let y be a point of Y. Let b be an integer such that $R^i f_* F = 0$ for $i > b$.*

(a) The following conditions are equivalent:

(i) the map $\alpha^q(y)$ (8.3.10.2) is surjective;

(ii) $\tau_{>q} Rf_* F$ is of perfect amplitude in $[q+1, b]$ in a neighborhood of y.

When these conditions are satisfied, there is an open neighborhood U of y such that $\alpha^q(z)$ is bijective for all $z \in U$ and such that $F|\, f^{-1}(U)$ is cohomologically flat over U in degree q.

(b) Assume that (a) (i) holds. Then the following conditions are equivalent:

(i) $\alpha^{q-1}(y)$ is surjective;

(ii) $R^q f_* F$ is locally free of finite type in a neighborhood of y.

REMARK 8.3.11.1. Since condition (a) for $q = -1$ is trivially satisfied for all y, we get that $Rf_* F$ is of perfect amplitude in $[0, b]$, as already observed at the end of the proof of 8.3.8.

On the other hand, condition (b) for $q = b$ is trivially satisfied for all y. Hence F is cohomologically flat in degree b.

REMARK 8.3.11.2. (cf. [**EGAIII1**, 4.6.1]) The following criterion is very useful: *if $H^{q+1}(X_y, F/\mathbf{m}_y F) = 0$, then $\alpha_q(y)$ is surjective; in particular, as follows from (b), if $H^1(X_y, F/\mathbf{m}_y F) = 0$, then, in a neighborhood of y, $f_* F$ is locally free of finite type and commutes with base change.*

Indeed, if $H^{q+1}(X_y, F/\mathbf{m}_y F) = 0$, then the theorem on formal functions 8.2.10 implies that

$$(R^q f_*(F)_y)\hat{} = 0,$$

hence $R^q f_*(F)_y = 0$, so that $\tau_{>q} Rf_*(F)_y = \tau_{>q+1} Rf_*(F)_y$. Since (trivially) $\alpha_{q+1}(y) = 0$, by (a) we have that $\tau_{>q} Rf_* F$ is of perfect amplitude in $[q+2, b]$, hence *a fortiori* in $[q+1, b]$, and therefore $\alpha_q(y)$ is surjective.

8.3.12. Assume that (a) (i) of 8.3.11 holds for all $y \in Y$, i.e., that $\tau_{>q} Rf_* F$ is of perfect amplitude in $[q+1, b]$. The dual

$$K = R\mathcal{H}om(\tau_{>q} Rf_* F, \mathcal{O}_Y)$$

is a perfect complex, of perfect amplitude in $[-b, -q-1]$. Let

$$Q := H^{-q-1}(K).$$

Then, for any quasi-coherent $\mathcal{O}_Y$-module M, there is a natural isomorphism

$$R^{q+1} f_*(M \otimes_Y F) \xrightarrow{\sim} \mathcal{H}om(Q, M). \tag{8.3.12.1}$$

Moreover, the formation of Q commutes with any base change. (This is the so-called *exchange property*, cf. [**EGAIII2**, 7.7.5, 7.7.6, 7.8.9].)

The proof is again a simple application of 8.3.2. By 8.3.2, we have

$$R^{q+1} f_*(M \otimes_Y F) = H^{q+1}(M \otimes^L Rf_* F). \tag{$*$}$$

The projection $Rf_* F \to \tau_{>q} Rf_* F$ induces an isomorphism

$$H^{q+1}(M \otimes^L Rf_* F) \xrightarrow{\sim} H^{q+1}(M \otimes^L \tau_{>q} Rf_* F). \tag{$**$}$$

Let $L := \tau_{>q} Rf_* F$, so that $K = R\mathcal{H}om(L, \mathcal{O}_Y)$. Since L is perfect, we have a natural biduality isomorphism

$$L \xrightarrow{\sim} R\mathcal{H}om(K, \mathcal{O}_Y),$$

which induces an isomorphism

$$M \otimes^L L \xrightarrow{\sim} R\mathcal{H}om(K, M). \tag{$***$}$$

Composing (*), (**) and $H^{q+1}(***)$, we get

$$R^{q+1} f_*(M \otimes_Y F) \xrightarrow{\sim} H^{q+1} R\mathcal{H}om(K, M) = \mathcal{E}xt^{q+1}(K, M).$$

But, since K is of perfect amplitude in $[-b, -q-1]$, i.e., locally isomorphic to a complex of free modules concentrated in degree in $[-b, -q-1]$, we have

$$\mathcal{E}xt^{q+1}(K, M) = \mathcal{H}om(H^{-q-1}K, M),$$

which gives (8.3.12.1). The proof of the compatibility of Q with base change is left to the reader.

8.3.13. Under the hypotheses of 8.3.8, the perfectness of $Rf_* F$ implies nice properties of the functions on Y:

$$y \mapsto \text{rk } H^q(X_y, F/\mathbf{m}_y F)$$

(for a fixed q), and

$$y \mapsto \sum (-1)^q \text{rk } H^q(X_y, F/\mathbf{m}_y F).$$

The first one is *upper semicontinuous*, while the second one is *locally constant*. This follows from 8.3.4. The verification is left to the reader. For a detailed discussion of these questions, see [**EGAIII2**, 7.7, 7.9] and [**SGA6**, III].

8.4. The existence theorem

8.4.1. Let A be an adic noetherian ring (8.1.1), I an ideal of definition of A, $Y = \operatorname{Spec} A$, $Y_n = \operatorname{Spec} A/I^{n+1}$, $\hat{Y} = \operatorname{colim}_n Y_n = \operatorname{Spf}(A)$. The problem which is addressed in this section is the following: given an adic noetherian $\hat{Y}$-formal scheme $\mathcal{Z} = \operatorname{colim}_n Z_n$ (8.1.5), when can we assert the existence (and uniqueness) of a (suitable) locally noetherian scheme X over Y whose I-adic completion $\hat{X} = \operatorname{colim}_n X_n$, where $X_n = X \times_Y Y_n$, is isomorphic to $\mathcal{Z}$? This is the so-called problem of *algebraization.* As for the analogous problem in complex analytic geometry (Serre's *GAGA*), Grothendieck's approach consists in first fixing X and comparing coherent sheaves on X and $\hat{X}$. The fundamental result is the following theorem:

THEOREM 8.4.2 ([**EGAIII1**, 5.1.4]). *Let X be a noetherian scheme, separated and of finite type over Y, and let $\hat{X}$ be its I-adic completion as in 8.4.1. Then the functor $F \mapsto \hat{F}$ (8.1.6) from the category of coherent sheaves on X whose support is proper over Y to the category of coherent sheaves on $\hat{X}$ whose support is proper over $\hat{Y}$ is an equivalence.*

Recall that the support of a coherent sheaf $\mathcal{E}$ on $\hat{X}$ is the support of $E_0 = \mathcal{E} \otimes \mathcal{O}_{X_0}$ on X_0 (8.1.4). It is called proper over $\hat{Y}$ if it is proper over Y_0 as a closed subset of X_0.

8.4.3. *Proof of 8.4.2.* Let F, G be coherent sheaves on X with proper supports over Y. By 8.2.9, $\operatorname{Hom}(F, G)$ is an A-module of finite type, hence separated and complete for the I-adic topology, and therefore the natural map

$$\operatorname{Hom}(F, G) \to \operatorname{Hom}(\hat{F}, \hat{G})$$

is an isomorphism. This proves that the $(-)^\wedge$ functor is fully faithful. It remains to prove that it is essentially surjective. This is done in several steps. We will outline the main points.

(a) *Projective case.* Assume $f : X \to Y$ to be projective. Let L be an ample line bundle on X. If M is an $\mathcal{O}_X$-module (resp. $\mathcal{O}_{\hat{X}}$-module) and $n \in \mathbb{Z}$, write, as usual, $M(n)$ for $M \otimes L^{\otimes n}$ (resp. $M \otimes \hat{L}^{\otimes n}$). The main point is the following result, which is a particular case of [**EGAIII1**, 5.2.4]:

LEMMA 8.4.3.1. *Let E be a coherent sheaf on $\hat{X}$. Then there exist nonnegative integers m, r and a surjective homomorphism*

$$\mathcal{O}_{\hat{X}}(-m)^r \to E.$$

Assuming 8.4.3.1, let us show how to prove the essential surjectivity in this case. Let E be a coherent sheaf on $\hat{X}$. By 8.4.3.1 we can find an exact sequence

$$\mathcal{O}_{\hat{X}}(-m_1)^{r_1} \xrightarrow{u} \mathcal{O}_{\hat{X}}(-m_0)^{r_0} \longrightarrow E \longrightarrow 0\,,$$

for some nonnegative integers m_0, m_1, r_0, r_1. By the full faithfulness of $(-)^\wedge$, there exists a unique morphism $v : \mathcal{O}_X(-m_1)^{r_1} \to \mathcal{O}_X(-m_0)^{r_0}$ such that $u = \hat{v}$. Let $F := \operatorname{Coker} v$. Then, by the exactness of $(-)^\wedge$ (on the category of coherent sheaves on X), $E = \hat{F}$.

Let us now prove 8.4.3.1. Since L is ample, so is $L_0 = \mathcal{O}_{X_0}(1)$ on X_0. Consider the graded $\mathcal{O}_{Y_0}$-algebra $S = \operatorname{gr}_I \mathcal{O}_Y = \bigoplus_{n \in \mathbb{N}} \tilde{I}^n/\tilde{I}^{n+1} = \bigoplus_{n \in \mathbb{N}} \mathcal{I}^n/\mathcal{I}^{n+1}$, and the graded $f_0^*(S)$-module $M = \operatorname{gr}_I E = \bigoplus_{n \in \mathbb{N}} \mathcal{I}^n E/\mathcal{I}^{n+1} E$, where $\mathcal{I} = I^\Delta$. Since $\operatorname{gr}_0 E$

is coherent on X_0 and the canonical map $\mathrm{gr}_I\,\mathcal{O}_Y \otimes_{\mathrm{gr}_0\mathcal{O}_Y} \mathrm{gr}_0\,E \to \mathrm{gr}_I\,E$ is surjective, M is of finite type over $f_0^*(S)$, hence corresponds to a coherent module $\tilde{M}$ on $X' := \mathrm{Spec}\,f_0^*(S)$. Since the inverse image $\mathcal{O}_{X'}(1)$ of $\mathcal{O}_X(1)$ on X' is ample, applying Serre's vanishing theorem [**EGAIII1**, 2.2.1] for $\tilde{M}$, $\mathcal{O}_{X'}(1)$ and the morphism $f' : X' \to Y' = \mathrm{Spec}\,S$ deduced from f_0 by base change by $Y' = \mathrm{Spec}\,S \to Y$, we find that there exists an integer n_0 such that, for all $n \geq n_0$, all $k \in \mathbb{N}$, and all $q > 0$,

$$H^q(X_0, \mathrm{gr}_k E(n)) = 0.$$

It follows that, for all $n \geq n_0$ and all $k \geq 0$, the transition map $H^0(X_0, E_{k+1}(n)) \to H^0(X_0, E_k(n))$ is surjective, and consequently the canonical map

$$H^0(\hat{X}, E(n)) = \lim_k H^0(X_0, E_k(n)) \to H^0(X_0, E_0(n))$$

is surjective. Since $\mathcal{O}_{X_0}(1)$ is ample, we may assume that n_0 has been chosen large enough for the existence of a finite number of global sections of $E_0(n_0)$ generating $E_0(n_0)$. Lifting these sections to $H^0(\hat{X}, E(n_0))$, we find a map

$$u : \mathcal{O}_{\hat{X}}(-n_0)^r \to E,$$

such that $u_0 = u \otimes \mathcal{O}_{X_0} : \mathcal{O}_{X_0}(-n_0)^r \to E_0$ is surjective. By Nakayama's lemma (since I is contained in the radical of A (8.1.1)), this implies that u is surjective.

REMARK 8.4.3.2. The above proof shows that the conclusion of 8.4.3.1 still holds if $\hat{X}$ is replaced by an adic $\hat{Y}$-formal scheme $\mathcal{X}$ such that $X_0 = \mathcal{X} \times_Y Y_0$ is proper and $\mathcal{O}_{\hat{X}}(1)$ by an invertible $\mathcal{O}_{\mathcal{X}}$-module L such that $L_0 = L \otimes \mathcal{O}_{X_0}$ is ample. It also shows that, under these hypotheses, there exists an integer n_0 such that $\Gamma(\mathcal{X}, E(n)) \to \Gamma(X_0, E_0(n))$ is surjective for all $n \geq n_0$, with the usual notation $E(n) = E \otimes L^{\otimes n}$.

(b) *Quasi-projective case.* Assume that we have an open immersion $j : X \to Z$, with Z projective over Y. Let E be a coherent sheaf on $\hat{X}$ whose support T_0 is proper over $\hat{Y}$. Then, the extension by zero $\hat{j}_!E$ is coherent on $\hat{Z}$, hence, by (a), of the form $\hat{F}$ for a coherent sheaf F on Z. The support T of F is contained in X (because $X \cap T$ is open in T and contains T_0 hence is equal to T), so that $F = j_!j^*F$, and $E = (j^*F)\hat{}$.

(c) *General case.* We proceed by noetherian induction on X. We assume that for every closed subscheme T of X whose underlying space is strictly contained in that of X, all coherent sheaves on $\hat{T}$ whose support is proper over $\hat{Y}$ are *algebraizable*, i.e., of the form $\hat{F}$ for some coherent sheaf F on T with proper support over Y, and we show that every coherent sheaf on $\hat{X}$ whose support is proper over $\hat{Y}$ is algebraizable. The main tool is Chow's lemma [**EGAIII1**, 5.6.1]: assuming X nonempty, one can find morphisms

$$Z \xrightarrow{g} X \xrightarrow{f} Y$$

such that g is projective and surjective, fg quasi-projective, and there exists an open immersion $j : U \to X$, with U nonempty, such that g induces an isomorphism over U. Let $T = X - U$ with the reduced scheme structure, and J be the ideal of T in X. Let E be a coherent sheaf on $\hat{X}$ whose support is proper over $\hat{Y}$. Consider the exact sequence

$$(*) \qquad 0 \to K \to E \to \hat{g}_*\hat{g}^*E \to C \to 0.$$

It suffices to show the following points:

(1) $\hat{g}_*\hat{g}^*E$ is algebraizable.

(2) K and C are killed by a positive power $\hat{J}^N$ of $\hat{J}$, hence can be viewed as coherent sheaves on $\hat{T}'$, where T' is the thickening of T defined by J^N.

(3) K and C, as coherent sheaves on $\hat{T}'$ are algebraizable.

(4) For a coherent sheaf on $\hat{X}$ whose support is proper over Y the property of being algebraizable is stable under kernel, cokernel and extension.

For (1), note that by case (b), $\hat{g}^*E$ (which has proper support over Y, g being proper) is algebraizable. The fact that $\hat{g}_*\hat{g}^*E$ is algebraizable then follows from the comparison theorem 8.2.2.

To prove (2), one may work locally on $\hat{X}$. One may replace X by $\operatorname{Spec} B$ with B adic noetherian such that IB is an ideal of definition of B. Then $E = \hat{F}$ for a coherent sheaf F on $\operatorname{Spec} B$, and by 8.2.9, (2) follows from the fact that the kernel and the cokernel of $F \to g_*g^*F$ are killed by a positive power of J.

In view of (2), (3) follows from the noetherian induction assumption.

In (4), the stability under kernel and cokernel is immediate, and the stability under extension follows from 8.2.9 for $r = 1$.

This completes the proof of 8.4.2.

8.4.4. We will first give applications of 8.4.2 to the algebraization of closed formal subschemes, finite formal schemes, and morphisms between formal schemes. We need some definitions.

(a) *Closed formal subschemes.* Let $\mathcal{X}$ be a locally noetherian formal scheme. If $\mathcal{A}$ is a coherent ideal of $\mathcal{O}_{\mathcal{X}}$, the topologically ringed space $\mathcal{Y}$ consisting of the support $\mathcal{Y}$ of $\mathcal{O}_{\mathcal{X}}/\mathcal{A}$, which is a closed subset of $\mathcal{X}$, and the sheaf of rings $\mathcal{O}_{\mathcal{X}}/\mathcal{A}$, restricted to $\mathcal{Y}$, is a locally noetherian formal scheme, adic over $\mathcal{X}$ (8.1.5), called the *closed formal subscheme* of $\mathcal{X}$ defined by $\mathcal{A}$. If $\mathcal{I}$ is an ideal of definition of $\mathcal{X}$ (8.1.3) and $X_n = (\mathcal{X}, \mathcal{O}_{\mathcal{X}}/\mathcal{I}^{n+1})$, so that $\mathcal{X} = \operatorname{colim}_n X_n$, then $\mathcal{Y} = \operatorname{colim}_n Y_n$, where Y_n is the closed subscheme of X_n such that $\mathcal{O}_{Y_n} = \mathcal{O}_{X_n} \otimes_{\mathcal{O}_{\mathcal{X}}} \mathcal{O}_{\mathcal{Y}}$. Conversely, any morphism of inductive systems $Y_. \to X_.$ such that $Y_n \to X_n$ is a closed subscheme and $Y_n = X_n \times_{X_{n+1}} Y_{n+1}$ (cf. 8.1.5) defines a closed formal subscheme $\mathcal{Y} = \operatorname{colim}_n Y_n$ of $\mathcal{X}$ such that $X_n \times_{\mathcal{X}} \mathcal{Y} = Y_n$. If $\mathcal{X}$ is affine, $X = \operatorname{Spf} A$, then $\mathcal{A} = \mathbf{a}^{\Delta}$ for an ideal $\mathbf{a}$ of A, and $\mathcal{Y} = \operatorname{Spf}(A/\mathbf{a})$. Finally, if X is a locally noetherian scheme and $\hat{X}$ is its completion along a closed subscheme X_0, then if Y is a closed subscheme of X and $\hat{Y}$ its completion along $Y_0 = X_0 \times_X Y$, $\hat{Y}$ is a closed formal subscheme of $\hat{X}$.

(b) *Finite morphisms.* Let $\mathcal{X} = \operatorname{colim} X_n$ be a locally noetherian formal scheme as in (a). A morphism $f : \mathcal{Z} \to \mathcal{X}$ of locally noetherian formal schemes is called *finite* [**EGAIII1**, 4.8.2] if f is an adic morphism (8.1.5) and $f_0 : Z_0 \to X_0$ is finite. By standard commutative algebra [**Bou61**, III §8.2, 11] (or [**EGAI**, Ch. 0, 7.2.9]), this is equivalent to saying that locally f is of the form Spf(B) $\to \operatorname{Spf}(A)$ with B finite over A and IB-adic, I being an ideal of definition of A, or that f is adic and each $f_n : Z_n = X_n \times_{\mathcal{X}} \mathcal{Z} \to X_n$ is finite. If f is finite, $f_*\mathcal{O}_{\mathcal{Z}}$ is a finite $\mathcal{O}_{\mathcal{Y}}$-algebra $\mathcal{B}$ such that $\mathcal{O}_{X_n} \otimes_{\mathcal{O}_{\mathcal{X}}} \mathcal{B} = f_*\mathcal{O}_{X_n}$ for all n. If $\mathcal{X} = \hat{X}$ with X as in (a), and Z is a finite scheme over X, then the completion $\hat{Z}$ of Z along $Z_0 = X_0 \times_X Z$ is finite over $\hat{X}$.

COROLLARY 8.4.5. *Let X/Y be as in 8.4.2. Then $Z \mapsto \hat{Z}$ is a bijection from the set of closed subschemes of X which are proper over Y to the set of closed formal subschemes of $\hat{X}$ which are proper over $\hat{Y}$ (8.4.4 (a)).*

The nontrivial point is the surjectivity. Let $\mathcal{Z} = \operatorname{colim} Z_n$ be a closed formal subscheme of $\hat{X}$ which is proper over $\hat{Y}$. It corresponds to a coherent quotient $\mathcal{O}_{\mathcal{Z}}$ of $\mathcal{O}_{\hat{X}}$ which has proper support over $\hat{Y}$. By 8.4.2 there exists a unique coherent $\mathcal{O}_X$-module F such that $\hat{F} = \mathcal{O}_{\mathcal{Z}}$. The problem is to algebraize the surjective map $u : \mathcal{O}_{\hat{X}} \to \mathcal{O}_{\mathcal{Z}}$. One cannot apply 8.4.2 because the support of $\mathcal{O}_X$ is not necessarily proper over Y. But the support of F, which is the intersection of the supports of $\mathcal{O}_X$ and F, is proper over Y. By 8.2.9, this is enough to ensure that the map $\operatorname{Hom}(\mathcal{O}_X, F) \to \operatorname{Hom}(\mathcal{O}_{\hat{X}}, \mathcal{O}_{\mathcal{Z}})$ is bijective. Therefore there exists a unique $v : \mathcal{O}_X \to F$ such that $\hat{v} = u$. Since $v_0 = u_0$ is surjective, so is v, hence $F = \mathcal{O}_Z$ for a closed subscheme Z of X which is proper over Y and such that $\hat{Z} = \mathcal{Z}$.

COROLLARY 8.4.6. *Let X/Y be as in 8.4.2. Then $Z \mapsto \hat{Z}$ is an equivalence from the category of finite X-schemes which are proper over Y to the category of finite $\hat{X}$-formal schemes which are proper over $\hat{Y}$ (8.4.4 (b)).*

By $Z \to g_*\mathcal{O}_Z$ (resp. $\mathcal{Z} \to g_*\mathcal{O}_{\mathcal{Z}}$), where g is the structural morphism, the first (resp. second) category is anti-equivalent to that of $\mathcal{O}_X$ (resp. $\mathcal{O}_{\hat{X}}$)-algebras which are finite and whose support is proper over Y (resp. $\hat{Y}$). If A and B are finite $\mathcal{O}_X$-algebras with proper supports over Y, and if $u : A \to B$ is a map of $\mathcal{O}_X$-modules such that $\hat{u}$ is a map of $\mathcal{O}_{\hat{X}}$-algebras, then, by 8.4.2, u is automatically a map of $\mathcal{O}_X$-algebras. The full faithfulness follows. If $\mathcal{A}$ is a finite $\mathcal{O}_{\hat{X}}$-algebra with proper support over $\hat{Y}$, then by 8.4.2, there exists a coherent $\mathcal{O}_X$-module A with proper support over Y such that $\hat{A} = \mathcal{A}$ as $\mathcal{O}_{\hat{X}}$-modules. But by 8.4.2 the maps $\mathcal{A} \otimes \mathcal{A} \to \mathcal{A}$ and $\mathcal{O}_{\hat{X}} \to \mathcal{A}$ giving the algebra structure on $\mathcal{A}$ uniquely algebraize to maps giving A a structure of $\mathcal{O}_X$-algebra such that $\hat{A} = \mathcal{A}$ as $\mathcal{O}_{\hat{X}}$-algebras.

COROLLARY 8.4.7. *Let X be a* proper *Y-scheme and let Z be a noetherian scheme, separated and of finite type over Y. Then the application*

$$\operatorname{Hom}_Y(X, Z) \to \operatorname{Hom}_{\hat{Y}}(\hat{X}, \hat{Z}) \ , \ f \mapsto \hat{f}$$

is bijective. In particular, the functor $X \mapsto \hat{X}$ from the category of proper *Y-schemes to the category of $\hat{Y}$-formal schemes is fully faithful.*

If $\hat{f} = \hat{g}$, the remark about the kernel of (8.1.6.4) shows (cf. [**EGAI**][10.9.4]) that $f = g$ in a neighborhood of X_0, hence everywhere since $X \to Y$ is proper (and, in particular, closed) and I is contained in the radical of A. To show the surjectivity, one applies 8.4.5 to the *graph* of a given morphism $\hat{X} \to \hat{Z}$, viewed as a closed formal subscheme of $(X \times_Y Z)^\wedge$.

REMARK 8.4.8. If, in 8.4.7, one drops the hypothesis of properness on X, the conclusion no longer holds in general. For example, if $X = Z = \operatorname{Spec} A[t]$, then $\hat{X} = \hat{Z} = \operatorname{Spf} A\{t\}$, and $\operatorname{Hom}_Y(X, Z) = A[t]$, while

$$\operatorname{Hom}_{\hat{Y}}(\hat{X}, \hat{Z}) = \operatorname{Hom}_{A,cont}(A\{t\}, A\{t\}) = A\{t\},$$

where $A\{t\} = A[t]^\wedge$ is the ring of restricted formal series $\sum a_n t^n$, i.e., such that a_n tends to 0 for the I-adic topology as n tends to infinity.

8.4.9. If X is a proper Y-scheme, $\hat{X}$ is a noetherian adic $\hat{Y}$-formal scheme, which is *proper* over $\hat{Y}$ (by which we mean that $X_0 = X \times_Y Y_0$ is proper over $Y_0 = \mathrm{Spec}(A/I)$). If $\mathcal{X}$ is a proper adic $\hat{Y}$-formal scheme, and if $\mathcal{X}$ is *algebraizable*, i.e., is of the form $\hat{X}$ for a proper Y-scheme, then by 8.4.7, X is unique (up to a unique isomorphism inducing the identity on $\hat{X}$). Deformation theory can produce proper adic $\hat{Y}$-formal schemes which are not algebraizable, cf. 8.5.24 (b). This, however, cannot happen in the *projective* formal case, as is shown by the next result, which is extremely useful.

THEOREM 8.4.10 ([**EGAIII1**, 5.4.5]). *Let $\mathcal{X} = \mathrm{colim}\, X_n$ be a proper, adic $\hat{Y}$-formal scheme, where $X_n = \mathcal{X} \times_{\hat{Y}} Y_n$. Let L be an invertible $\mathcal{O}_{\mathcal{X}}$-module such that $L_0 = L \otimes \mathcal{O}_{X_0} = L/IL$ is ample (so that X_0 is projective over Y_0). Then $\mathcal{X}$ is algebraizable, and if X is a proper Y-scheme such that $\hat{X} = \mathcal{X}$, then there exists an unique line bundle M on X such that $L = \hat{M}$, and M is ample (in particular, X is projective over Y).*

Using 8.4.3.2, choose n such that:

(i) $L_0^{\otimes n}$ is very ample, i.e., of the form $i_0^*\mathcal{O}_{P_0}(1)$ for a standard projective space $P_0 = \mathbb{P}^r_{Y_0}$ and a closed immersion $i_0 : X_0 \to P_0$.

(ii) $\Gamma(\mathcal{X}, L^{\otimes n}) \to \Gamma(X_0, L_0^{\otimes n})$ is surjective.

Using (ii), lift the canonical epimorphism $u_0 : \mathcal{O}_{X_0}^{r+1} \to L_0^{\otimes n}$ given by i_0 to an $\mathcal{O}_{\mathcal{X}}$-linear map $u : \mathcal{O}_{\mathcal{X}}^{r+1} \to L^{\otimes n}$. By Nakayama, each $u_k = u \otimes \mathcal{O}_{X_k} : \mathcal{O}_{X_k}^{r+1} \to L_k^{\otimes n}$ ($k \in \mathbb{N}$) is surjective, hence corresponds to a morphism $i_k : X_k \to P_k = \mathbb{P}^r_{Y_k}$ of Y_k-schemes. such that $L_k^{\otimes n} = i_k^*\mathcal{O}_{P_k}(1)$. By 8.4.4 (b) i_k is finite, hence a closed immersion by Nakayama. These closed immersions i_k form an inductive system $i_. : X_. \to P_.$, with cartesian squares of the type (8.1.5.2), hence define a closed formal subscheme (8.4.4 (a)) $i : \mathcal{X} \to \hat{P}$, where $\hat{P}$ is the completion of the standard projective space $P = \mathbb{P}^r_Y$ over $Y = \mathrm{Spec}\, A$, and $L^{\otimes n} = i^*\mathcal{O}_{\hat{P}}(1)$. By 8.4.5, there exists a (unique) closed subscheme $j : X \to P$ such that $\hat{X} = \mathcal{X}$. Moreover, by 8.4.2, there exists a (unique) line bundle M on X such that $L = \hat{M}$. Since $L^{\otimes n} = i^*\mathcal{O}_{\hat{P}}(1)$ and $(M^{\otimes n})^{\wedge} = \hat{M}^{\otimes n}$, we get $(M^{\otimes n})^{\wedge} = (j^*\mathcal{O}_P(1))^{\wedge}$, hence (by 8.4.2) $M^{\otimes n} = j^*\mathcal{O}_P(1))$, and therefore M is ample.

REMARKS 8.4.11.

(a) The main theorems in §8.2, 8.4 are analogous to the results of Serre on the comparison between algebraic and analytic geometry (GAGA). See [**Ser56**] and [**SGA1**, XII].

(b) The results of §8.2, 8.4 have been generalized by Knutson to algebraic spaces [**Knu71**, chap. V]. A generalization to *stacks* (Deligne-Mumford's stacks, or Artin's stacks [**Art74b**], [**LMB00**]) has been recently worked out. For a generalization of Zariski's main theorem (8.2.17), see [**LMB00**, 16.5]. A generalization to Artin stacks of the finiteness theorem for proper morphisms has been given by Faltings [**Fal03**], and, independently by Olsson [**Ols**], who deduces it from a Chow's lemma for Artin stacks. Olsson (loc. cit.) also generalizes Grothendieck's main existence theorem 8.4.2 to Artin stacks.

8.5. Applications to lifting problems

8.5.1. Let A be a local noetherian ring with maximal ideal $\mathbf{m}$ and residue field $k = A/\mathbf{m}$. Let $S = \mathrm{Spec}\, A$, with closed point $s = \mathrm{Spec}\, k$. Here is a prototype of lifting problems. Given a scheme X_0 of finite type over s, can one find a scheme X, of finite type and *flat* over S, lifting X_0, i.e., such that $X_s \simeq X_0$? For example, A could be a discrete valuation ring of mixed characteristic, with k of characteristic $p > 0$ and the fraction field K of characteristic zero, and a scheme X as above would provide a "lifting of X_0 to characteristic zero" (namely, the generic fiber X_η, $\eta = \operatorname{Spec} K$). Usually X_0 satisfies additional assumptions (e. g. properness, smoothness, etc.) and is sometimes endowed with additional structures (e. g. group structure), which are to be preserved in the lifting. We ignore this here for simplicity. Grothendieck's strategy to attack the problem consists of several steps.

(1) Try to lift X_0 to an inductive system of (flat and of finite type) schemes X_n such that $X_{n+1} \times_{S_{n+1}} S_n = X_n$. The closed immersion $S_n \to S_{n+1}$ is a thickening of order 1 (8.1.3): its ideal $\mathbf{m}^{n+1}\mathcal{O}_{S_{n+1}}$ is killed by $\mathbf{m}$, and *a fortiori* is of square zero. Suppose X_m has been constructed for $m \leq n$. To lift X_n to X_{n+1} over S_{n+1}, then one usually encounters an obstruction in a cohomology group of X_0, and when this obstruction vanishes, the set of isomorphism classes of such X_{n+1} is in bijection with another cohomology group of X_0. Automorphisms of a given X_{n+1} inducing the identity on X_n can also be described by a suitable cohomology group of X_0. Such a study is the object of *deformation theory.*

(2) Suppose that an inductive system $X.$ as in (1) has been found. It defines an adic (locally noetherian) formal scheme $\mathcal{X}$ over the completion $\hat{S} = \operatorname{Spf}\hat{A}$ of S at s, which is (by definition) flat and of finite type over $\hat{S}$ (8.1.5). The next problem is to *algebraize $\mathcal{X}$ over* $\operatorname{Spec}\hat{A}$, i.e., to find X of finite type over $\operatorname{Spec}\hat{A}$ such that $\hat{X} = \mathcal{X}$. Here one can try to apply the existence theorems of §4, assuming X_0 *proper*, in which case the algebraization is unique if it exists. The main tool – not to say the only one – is 8.4.10. For this we have first to assume that X_0 is *projective.* Let L_0 be an ample invertible sheaf on X_0. If such an L_0 can be chosen such that it lifts to $\mathcal{X}$, namely that there exists a projective system L_n of invertible sheaves on the X_n's such that $L_n = L_{n+1} \otimes \mathcal{O}_{X_n}$, then by 8.4.10, we are done: there exists a projective scheme X over $\operatorname{Spec}\hat{A}$ such that $\hat{X} = \mathcal{X}$. Supposing that L_m has been constructed for $m \leq n$, there is a cohomological obstruction to lifting L_n to L_{n+1}, similar to that alluded to in (1) and closely related to it.

(3) Having found X over $\operatorname{Spec}\hat{A}$, one cannot in general go further, i.e., descend X to $S = \operatorname{Spec} A$. But sometimes, in moduli problems, one encounters a situation where $X/\operatorname{Spec}\hat{A}$ enjoys a *versal* property. If moreover k is separably closed, then Artin's *approximation theory* [**Art69a**], [**Art74b**] usually enables us to descend X at least to the *henselization* S^h of S, i.e., find Z over S^h such that $X = \operatorname{Spec}\hat{A}\times_{S^h} Z$. In a sense, Artin's theory answers Grothendieck's question in [**Gro95a**, p. 15]: "Pour passer de résultats connus pour le complété d'un anneau local **à** des résultats correspondants pour cet anneau local lui-même, il faudrait un quatrième "théorème fondamental", dont l'énoncé définitif reste **à** trouver".

In this section we recall basic facts on deformation theory and give applications to problems related to (1) and (2).

A. Deformation of vector bundles.

8.5.2. The simplest deformation problem is the problem of deformation of *vector bundles*, i.e., *locally free sheaves of finite rank.* Let $i : X_0 \to X$ be a thickening of order one (8.1.3), defined by an ideal I of square zero. Let E_0 be a vector bundle on X_0. We want to "deform" (or "extend") E_0 over X, i.e., find a vector bundle E

on X such that $\mathcal{O}_{X_0} \otimes E = E/IE = E_0$. More precisely, by a *deformation* of E_0 over X we mean a pair of a vector bundle E on X and an $\mathcal{O}_X$-linear map $E \to i_* E_0$ inducing an isomorphism $i^*E \xrightarrow{\sim} E_0$. By a morphism $u : E' \to E$ of deformations we mean a morphism u such that $i^*u = Id_{E_0}$. Such a morphism is automatically an isomorphism.

THEOREM 8.5.3. *Let $i : X_0 \to X$ be as in 8.5.2.*

(a) Let E, F be vector bundles on X, $E_0 = i^*E$, $F_0 = i^*F$, and $u_0 : E_0 \to F_0$ be an $\mathcal{O}_{X_0}$-linear map. There is an obstruction

$$o(u_0, i) \in H^1(X_0, I \otimes \mathcal{H}om(E_0, F_0))$$

to the existence of an $\mathcal{O}_X$-linear map $u : E \to F$ extending u_0. When $o(u_0, i) = 0$, the set of u extending u_0 is an affine space under $H^0(X_0, I \otimes \mathcal{H}om(E_0, F_0))$.

(b) Let E_0 be a vector bundle on X_0. There is an obstruction

$$o(E_0, i) \in H^2(X_0, I \otimes \mathcal{E}nd(E_0))$$

whose vanishing is necessary and sufficient for the existence of a deformation E of E_0 over X. When $o(E_0, i) = 0$, the set of deformations of E_0 over X is an affine space under $H^1(X_0, \mathcal{E}nd(E_0) \otimes I)$, and the group of automorphisms of a given deformation E is identified by $a \mapsto a - Id$ with $H^0(X_0, \mathcal{E}nd(E_0) \otimes I)$.

The proof is elementary and would work in a more general context (ringed spaces or topoi). One first proves (a). The second assertion is clear. Moreover, extensions u of u_0 exist locally. Therefore we get a *torsor* P under $I \otimes \mathcal{H}om(E_0, F_0)$ on X_0, whose sections over an open subset U of X_0 are the $\mathcal{O}_X$-linear extensions of $u_0|U$. The class of P in $H^1(X_0, I \otimes \mathcal{H}om(E_0, F_0))$ is the obstruction $o(u_0, i)$. To prove (b), assume first, for simplicity, that X_0 (or X, this is equivalent) is *separated.* Choose $(\mathcal{U} = (U_i)_{i \in K}, (E_i)_{i \in K})$, where $\mathcal{U}$ is an affine open cover of X_0 and E_i a deformation to $X \cap U_i$ of $E_0|U_i$. Since X_0 is separated, $U_{ij} = U_i \cap U_j$ is affine, so by (a) one can find an isomorphism $g_{ij} : E_i|U_{ij} \xrightarrow{\sim} E_j|U_{ij}$ (inducing the identity on X_0). Such an isomorphism is unique up to the addition of $h_{ij} \in H^0(U_{ij}, I \otimes \mathcal{E}nd(E_0))$. Then

$$(i, j, k) \mapsto c_{ijk} = g_{ij} g_{ik}^{-1} g_{jk} - Id \in H^0(U_{ijk}, I \otimes \mathcal{E}nd(E_0))$$

is a 2-cocycle of $\mathcal{U}$ with values in $I \otimes \mathcal{E}nd(E_0)$, which is a coboundary dh, $h = (h_{ij})$ if and only if the g_{ij} can be modified into a gluing data for the (E_i), in other words if and only if E_0 can be deformed over X. Thus the class of c,

$$[c] = o(E_0, i) \in H^2(X_0, I \otimes \mathcal{E}nd(E_0))$$

is the desired obstruction, which can be checked to be independent of the choices. If E_1 and E_2 are two deformations of E_0 over X, then by (a) the local isomorphisms from E_1 to E_2 form a torsor under $I \otimes \mathcal{E}nd(E_0)$, whose class

$$[E_2] - [E_1] \in H^1(X_0, I \otimes \mathcal{E}nd(E_0))$$

depends only on the isomorphism classes $[E_i]$ of E_1 and E_2, and is zero if and only if $[E_1] = [E_2]$. One checks that this defines the desired affine structure on the set of isomorphism classes of deformations. That finishes the proof in the case X_0 is separated. In the general case, the data $(U_i), (E_i), (g_{ij})$ have to be replaced by data $(U_i), (E_i), (g_{ij}^{\alpha})$ where g_{ij}^{α} is an isomorphism from $E_i|U_{ij}^{\alpha}$ to $E_i j U_{ij}^{\alpha}$, for an open cover $(U_{ij}^{\alpha})_{\alpha \in A_{ij}}$ of U_{ij}. Then the g_{ij}^{α} provide a 2-cocycle of the *hypercovering* defined by

$(U_i), (U_{ij}^{\alpha})$ (cf. [**SGA4**, V 7]) whose cohomology class in $H^2(X_0, I \otimes \mathcal{E}nd(E_0))$ is the desired obstruction, and the rest of the proof goes on with minor modifications.

A more intrinsic way of presenting the proof is to use Giraud's language of *gerbes* [**Gir71**]. The deformations $U \mapsto \mathcal{E}(U)$ of E_0 over variable open subsets U of X_0 form a stack in groupoids, which is in fact a gerbe, i.e., has the following properties: two objects of $\mathcal{E}(U)$ are locally isomorphic, and for any U, there is an open cover (U_i) of U such that $\mathcal{E}(U_i)$ is nonempty. The sheaves of automorphisms of objects of $\mathcal{E}(U)$ form a global sheaf on X_0, namely $I \otimes \mathcal{E}nd(E_0)$, called the *band* ("lien") of the gerbe $\mathcal{E}$. The obstruction $o(E_0, i)$ is the *cohomology class* of $\mathcal{E}$. When this class is zero, the gerbe is *neutral*, which means that the choice of a global object E (a deformation of E_0 over X) identifies $\mathcal{E}$ with the gerbe of *torsors* under $I \otimes \mathcal{E}nd(E_0)$ (over variable open subsets of X_0). See [**Gir71**, VII 1.3.1] for a generalization of the preceding discussion to the case $GL(n)$) is replaced by a smooth group scheme G (and locally free sheaves of rank n by torsors under G).

REMARKS 8.5.4.

(a) The construction of the cocycle c in 8.5.3 (b) shows that if L_0, M_0 are *line bundles* on X_0, then

$$o(L_0 \otimes M_0, i) = o(L_0, i) + o(M_0, i)$$

in $H^2(X_0, I)$. Thus, on line bundles, the obstruction behaves like a first Chern class. In fact, the class $o(E_0, i)$ in 8.5.3 (b) can be viewed as a kind of *Atiyah class*, similar to that defined by Atiyah in [**Ati57**] to construct Chern classes in Hodge cohomology, see [**Ill71**, chap. IV, V] and [**KS04**, 1.4.1].

(b) In practice, the ideal I is killed by a bigger ideal J. More precisely, changing notations, let $X_0 \longrightarrow X_1 \xrightarrow{i_1} X_2$ be closed immersions, I (resp. J) the ideal of X_1 (resp. X_0) in X_2, and suppose that $I \cdot J = 0$. In particular, $I^2 = 0$ and I can be viewed not just as an $\mathcal{O}_{X_1}$-module, but as an $\mathcal{O}_{X_0}(= \mathcal{O}_{X_2}/J)$-module. For vector bundles E, F on X_2, the groups $H^q(X_1, I \otimes \mathcal{H}om(E_1, F_1))$ appearing in 8.5.3 (a), with i replaced by i_1 and $E_1 = i_1^*E$, $F_1 = i_1^*F$, can then be rewritten

$$H^q(X_1, I \otimes_{\mathcal{O}_{X_1}} \mathcal{H}om(E_1, F_1)) = H^q(X_0, I \otimes_{\mathcal{O}_{X_0}} \mathcal{H}om(E_0, F_0)),$$

with $E_0 = \mathcal{O}_{X_0} \otimes E$, $F_0 = \mathcal{O}_{X_0} \otimes F$. Similarly, for a vector bundle E_1 on X_1, the groups $H^q(X_1, I \otimes_{\mathcal{O}_{X_1}} \mathcal{E}nd(E_1))$ appearing in 8.5.3 (b) (with i replaced by i_1), can be rewritten

$$H^q(X_1, I \otimes_{\mathcal{O}_{X_1}} \mathcal{E}nd(E_1)) = H^q(X_0, I \otimes_{\mathcal{O}_{X_0}} \mathcal{E}nd(E_0)),$$

with $E_0 = \mathcal{O}_{X_0} \otimes E_1$.

COROLLARY 8.5.5. *Let A be a complete local noetherian ring, with maximal ideal $\mathbf{m}$ and residue field k. Let $S = \operatorname{Spec} A$, $\hat{S} = \operatorname{colim} S_n$, where $S_n = \operatorname{Spec} A/\mathbf{m}^{n+1}$. Let $\mathcal{X} = \operatorname{colim} X_n$, $X_n = S_n \times_{\hat{S}} \mathcal{X}$, be a flat adic locally noetherian formal scheme over $\hat{S}$ (8.1.5), and assume that $H^2(X_0, \mathcal{O}_{X_0}) = 0$. Then any line bundle L_0 on X_0 can be lifted to a line bundle L on $\mathcal{X}$. If moreover $H^1(X_0, \mathcal{O}_{X_0}) = 0$, then such a lifting L is unique up to a (nonunique) isomorphism (inducing the identity on L_0).*

Suppose that L_0 has been lifted to L_n on X_n. Let I_n be the ideal of X_n in X_{n+1}. By the flatness of $\mathcal{X}$ over $\hat{S}$, we have

$$I_n = \mathcal{O}_{X_0} \otimes_k \mathbf{m}^{n+1}/\mathbf{m}^{n+2}.$$

Taking 8.5.4 (b) into account, we see that the obstruction to lifting L_n to L_{n+1} on X_{n+1} lies in

$$H^2(X_0, I_n) = H^2(X_0, \mathcal{O}_{X_0}) \otimes_k \mathbf{m}^{n+1}/\mathbf{m}^{n+2} = 0,$$

whence the first assertion. For the second one, suppose L and L' are two liftings of L_0 on $\mathcal{X}$. Assume that an isomorphism $u_m : L_m \xrightarrow{\sim} L'_m$ has been constructed for $m \leq n$, with $u_0 = Id$. Then, since $H^1(X_0, \mathcal{O}_{X_0}) = 0$, by 8.5.3 (a) there is an isomorphism $u_{n+1} : L_{n+1} \xrightarrow{\sim} L'_{n+1}$ extending u_n, and $u = \lim u_n$ is an isomorphism from L to L' inducing the identity on L_0.

COROLLARY 8.5.6. *Let $\mathcal{X}$ be a proper, flat adic locally noetherian formal scheme over $\hat{S}$. Then:*

(a) If X/S is a proper scheme such that $\mathcal{X} = \hat{X}$, X is flat over S. Moreover, if $H^2(X_0, \mathcal{O}_{X_0}) = 0$, any line bundle L_0 on X_0 can be lifted to a line bundle L on X, which is unique (up to an isomorphism) if $H^1(X_0, \mathcal{O}_{X_0}) = 0$.

(b) If X_0 is projective *and an ample line bundle L_0 on X_0 can be lifted to a line bundle $\mathcal{L}$ on $\mathcal{X}$, there exists a projective and flat scheme X/S such that $\mathcal{X} = \hat{X}$ and an ample line bundle L on X such that $\hat{L} = \mathcal{L}$.*

Let us prove (a). Let x be a point of $X_0 = X_s$ ($s = S_0 = \operatorname{Spec} k$). For all $n \geq 0$, $\mathcal{O}_{X,x}/\mathbf{m}^{n+1}\mathcal{O}_{X,x} = \mathcal{O}_{X_n,x}$ is flat over $A_n = A/\mathbf{m}^{n+1}$, hence $\mathcal{O}_{X,x}$ is flat over A by the usual flatness criterion. As the set of points at which a morphism is flat is open, X is flat over S in a neighborhood of the special fibre X_s, hence everywhere since X is proper over S. The second assertion follows from 8.4.2 and 8.5.5. Assertion (b) follows from (a) and 8.4.10.

B. Deformation of smooth schemes.

8.5.7. We now turn to the problem of deforming schemes. Let $i : S_0 \to S$ be a thickening of order one (8.1.3), defined by an ideal I of square zero, and let X_0 be a *flat* scheme over S_0. By a *deformation* (or *lifting*) of X_0 over S we mean a cartesian square

(8.5.7.1)
$$\begin{array}{ccc} X_0 & \xrightarrow{j} & X \\ \downarrow & & \downarrow \\ S_0 & \longrightarrow & S \end{array}$$

with X flat over S. The flatness condition is expressed by the fact that the natural map

(8.5.7.2)
$$f_0^* I \to J,$$

where $f_0 : X_0 \to S_0$ is the structural morphism and J the ideal of X_0 in X, is an isomorphism. By a morphism of deformations we mean an S-morphism $u : X \to X'$ such that $uj = j'$ (where $j' : X_0 \to X'$). Such a morphism u is necessarily an isomorphism.

8.5.8. We will first discuss the smooth case, which is elementary. Let $f : X \to Y$ be a morphism of schemes. Recall that f is called *smooth* if f is locally of finite presentation (i.e., locally of finite type if Y is locally noetherian) and satisfies the equivalent conditions:

(i) f is flat and the geometric fibers $X_{\overline{y}}$ of X/Y are regular (here $\overline{y} \to y \in Y$ runs through the geometric points of Y, with $k(\overline{y})$ algebraically closed);

(ii) (*jacobian criterion*) for every point $x \in X$ there exist open affine neighborhoods U of x and V of $y = f(x)$ such that $f(U) \subset V$ and U is the closed subscheme of a standard affine space $\mathbb{A}^n_V = \operatorname{Spec} A[t_1, \cdots, t_n]$ (where $V = \operatorname{Spec A}$) defined by equations $g_1 = \cdots = g_r = 0$ ($g_i \in A[t_1, \cdots, t_n]$) such that $\operatorname{rk}(\partial g_i/\partial t_j)(x) = r$;

(iii) (*formal smoothness*) for every commutative square

$$\begin{array}{ccc} S_0 & \xrightarrow{g_0} & X \\ {\scriptstyle i}\downarrow & & \downarrow{\scriptstyle f} \\ S & \xrightarrow{h} & Y \end{array}$$

where i is a thickening of order 1, there exists, Zariski locally on S, a Y-morphism $g : S \to X$ extending g_0, i.e., such that $gi = g_0$.

(For the equivalence of these conditions and basic facts on smooth and étale morphisms, see [**BLR90**], [**Ill72a**], and [**EGAIV2**, §17], [**SGA1**, I, II, III] for a more comprehensive treatment.)

Suppose f is smooth. Then the sheaf of relative differentials $\Omega^1_{X/Y}$ is locally free of finite type, as well as the tangent sheaf

$$T_{X/Y} = \mathcal{H}om(\Omega^1_{X/Y}, \mathcal{O}_X).$$

Their common rank $r(x)$ at a point x of X is the dimension at x of the fiber $X_{f(x)}$, the *relative dimension* of X at x. It is a locally constant function of x. A morphism $f : X \to Y$ is called *étale* if f is smooth and of relative dimension zero at all points, in other words, f is smooth and $\Omega^1_{X/Y} = 0$, or, equivalently, f is flat, locally of finite presentation, and $\Omega^1_{X/Y} = 0$. Smoothness (resp. étaleness) is stable under composition and base change.

We will also need the definition of smoothness in the context of *formal schemes*. Let $\mathcal{Y} = \operatorname{colim} Y_n$ be a locally noetherian formal scheme, with the notations of 8.1.5. An adic morphism $f : \mathcal{X} \to \mathcal{Y}$ is called *smooth* if f is flat (8.1.5) and each X_n is smooth over Y_n (or, equivalently, by 8.5.8 (i), if X_0 is smooth over Y_0). We will refer to $\mathcal{X}$ as a *smooth formal scheme over $\mathcal{Y}$ lifting X_0*.

The main results about deformations of smooth schemes are summed up in the following theorem.

THEOREM 8.5.9.

(a) *Let X and Y be schemes over a scheme S, with Y smooth over S, and let $j : X_0 \to X$ be a closed subscheme defined by an ideal J of square zero. Let $g : X_0 \to Y$ be an S-morphism. There is an obstruction*

$$o(g, j) \in H^1(X_0, J \otimes_{\mathcal{O}_{X_0}} g^*T_{Y/S})$$

whose vanishing is necessary and sufficient for the existence of an S-morphism $h : X \to Y$ extending g, i.e., such that $hj = g$. When $o(g, j) = 0$, the set of extensions h of g is an affine space under $H^0(X_0, J \otimes_{\mathcal{O}_{X_0}} g^*T_{Y/S})$.

(b) *Let $i : S_0 \to S$ be a thickening of order one defined by an ideal I of square zero, and let X_0 be a smooth S_0-scheme. There is an obstruction*

$$o(X_0, i) \in H^2(X_0, f_0^*I \otimes T_{X_0/S_0})$$

(where $f_0 : X_0 \to S_0$ is the structural morphism) whose vanishing is necessary and sufficient for the existence of a deformation X of X_0 over S (8.5.7). When $o(X_0, i) = 0$, the set of isomorphism classes of such deformations is an affine space under $H^1(X_0, f_0^*I \otimes T_{X_0/S_0})$, and the group of automorphism of a fixed deformation is isomorphic to $H^0(X_0, f_0^*I \otimes T_{X_0/S_0})$. In particular, if X_0 is étale *over S_0, there exists a deformation X of X_0 over S, which is unique up to a unique isomorphism.*

Note that if X_0 is smooth (resp. étale), any deformation of X_0 over S is smooth (resp. étale). This follows from 8.5.8 (i).

The proof of 8.5.9 is similar to that of 8.5.3. One first proves (a). Since Y is smooth over S, an extension h of g exists locally on X_0. Moreover, two such extensions differ by an S-derivation of $\mathcal{O}_Y$ into g_*J, i.e., a section of $J \otimes_{\mathcal{O}_{X_0}} g^*T_{Y/S}$. Therefore, the extensions h over variable open subsets of X form a torsor on X under $J \otimes_{\mathcal{O}_{X_0}} g^*T_{Y/S}$, and $o(g, j)$ is the class of this torsor. To prove (b), one first observes that deformations of X_0 exist locally on X_0. This follows from 8.5.8 (ii) (lift the polynomials g_i's). Moreover, (a) implies that two deformations are locally isomorphic, and that, for any open subset U_0 of X_0, the sheaf of automorphisms of a deformation U of U_0 is identified by $a \mapsto a - Id$ with $f_0^*I \otimes T_{X_0/S_0}$. Therefore (cf. [**Gir71**, VII 1.2]) by associating to each open subset U_0 of X_0 the groupoid of deformations of U_0, we define a *gerbe* $\mathcal{G} = \mathcal{G}_{X_0}$ whose band is $f_0^*I \otimes T_{X_0/S_0}$. The class $o(X_0, i)$ of $\mathcal{G}$ in $H^2(X_0, f_0^*I \otimes T_{X_0/S_0})$ is the obstruction to the existence of an object of $\mathcal{G}(X_0)$, i.e., a deformation of X_0. When $o(X_0, i) = 0$, $\mathcal{G}$ is *neutral*, i.e., a (global) deformation X of X_0 exists. Once such an X has been chosen, one can identify $\mathcal{G}$ with the gerbe of torsors on X_0 under $f_0^*I \otimes T_{X_0/S_0}$ by associating to a deformation U of an open subset U_0 of X_0 the torsor of local isomorphisms between U and $X|U_0$. In particular, the set of isomorphism classes of deformations of X_0 is then identified to $H^1(X_0, f_0^*I \otimes T_{X_0/S_0})$.

As in the proof of 8.5.3 one can exhibit a 2-cocycle defining $o(X_0, i)$. Suppose, for simplicity, that X_0 is separated. Choose $(\mathcal{U} = ((U_0)_i)_{i \in K}, (U_i)_{i \in K})$ where $\mathcal{U}$ is an affine open cover of X_0 and U_i a deformation of $(U_0)_i$. Since X_0 is separated, $(U_0)_{ij} = (U_0)_i \cap (U_0)_j$ is affine, so by (a) there is an isomorphism of deformations $g_{ij} : U_i|(U_0)_{ij} \xrightarrow{\sim} U_j|(U_0)_{ij}$. Then

$$(i, j, k) \mapsto c_{ijk} = g_{ij} g_{ik}^{-1} g_{jk} - Id \in H^0((U_0)_{ijk}, f_0^*I \otimes T_{X_0/S_0})$$

is a 2-cocycle of $\mathcal{U}$ with values in $f_0^*I \otimes T_{X_0/S_0}$, whose class in $H^2(X_0, f_0^*I \otimes T_{X_0/S_0})$ represents $o(X_0, i)$.

REMARKS 8.5.10.

(a) The obstruction $o(X_0, i)$ satisfies the following functoriality property. Let $g_0 : Y_0 \to S_0$ be a smooth morphism and let $h_0 : X_0 \to Y_0$ be an S_0-morphism (so that $f_0 = g_0 h_0 : X_0 \to S_0$). Then $o(X_0, i)$ and $o(Y_0, i)$ have the same image in $H^2(X_0, f_0^*I \otimes h_0^*T_{Y_0/S_0})$ under the canonical maps

$$H^2(X_0, f_0^*I \otimes T_{X_0/S_0}) \to H^2(X_0, f_0^*I \otimes h_0^*T_{Y_0/S_0}) \leftarrow H^2(Y_0, g_0^*I \otimes T_{Y_0/S_0}).$$

Moreover, if X_0, Y_0 are smooth S_0-schemes, the obstruction $o(X_0 \times_{S_0} Y_0, i)$ to the lifting of $X_0 \times_{S_0} Y_0$ to S satisfies the formula

$$o(X_0 \times_{S_0} Y_0, i) = pr_1^* o(X_0, i) + pr_2^* o(Y_0, i),$$

where pr_1 (resp. pr_2) is the projection from $X_0 \times_{S_0} Y_0$ to X_0 (resp. Y_0) and pr_1^* is the composite

$$\begin{aligned} H^2(X_0, I \otimes T_{X_0/S_0}) &\to H^2(X_0 \times_{S_0} Y_0, I \otimes pr_1^* T_{X_0/S_0}) \\ &\to H^2(X_0 \times_{S_0} Y_0, I \otimes T_{X_0 \times_{S_0} Y_0}) \end{aligned}$$

of the functoriality map and the inclusion of the first direct summand, and similarly for pr_2^*.

The obstructions $o(g, j)$ satisfy a compatibility with respect to the composition of morphisms: in the situation of 8.5.9 (b), if X, Y, Z are smooth schemes over S, and $f_0 : X_0 \to Y_0$, $g_0 : Y_0 \to Z_0$ S_0-morphisms between their pull-backs to S_0, then the obstruction to lifting $h_0 = g_0 f_0$ to $h : X \to Z$ is the pull-back by g_0^* of the obstruction to lifting g_0 to $g : Y \to Z$.

(b) As in 8.5.4 (b), suppose $S_0 \longrightarrow S_1 \xrightarrow{i_1} S_2$ are closed immersions, where the ideal I of i_1 is killed by the ideal J of S_0 in S_2. Then, if $f_1 : X_1 \to S_1$ is a smooth morphism, the groups appearing in 8.5.9 (b) relative to the deformation of X_1 over S_2 can be rewritten $H^q(X_0, f_0^* I \otimes T_{X_0/S_0})$ where $f_0 : X_0 \to S_0$ is deduced from f_1 by base change.

C. Specialization of the fundamental group.

8.5.11. The combination of the existence theorem 8.4.2 with 8.5.3 and 8.5.9 has powerful applications. We will first discuss those pertaining to the *fundamental group*.

Let X be a locally noetherian scheme. By an *étale cover* ("revêtement étale" [**SGA1**, I 4.9]) of X we mean a *finite* and *étale* morphism $Y \to X$. A morphism $Y' \to Y$ of étale covers is defined as an X-morphism from Y' to Y. It is automatically an étale cover of Y. We denote by

(8.5.11.1) $$Et(X)$$

the category of étale covers of X. Suppose X is *connected* and fix a *geometric point* $\overline{x}$ of X, localized at some point x, i.e., a morphism $\operatorname{Spec} k(\overline{x}) \to \operatorname{Spec} k(x)$, with $k(\overline{x})$ a separably closed field. Then there is defined a profinite group

(8.5.11.2) $$\pi_1(X, \overline{x}),$$

called the *fundamental group* of X at $\overline{x}$, and an equivalence of categories

(8.5.11.3) $$Et(X) \xrightarrow{\sim} \{\pi_1(X, \overline{x}) - fsets\},$$

where $\{\pi_1(X, \overline{x}) - fsets\}$ denotes the category of finite sets on which $\pi_1(X, \overline{x})$ acts continuously [**SGA1**, V 7]. More precisely, the functor

$$Et(X) \to \{fsets\}, \quad Y \mapsto Y(\overline{x}) = Y_{\overline{x}},$$

associating to an étale cover Y of X the finite set of its points over $\overline{x}$, called *fiber functor at* $\overline{x}$, is pro-representable: there is a pro-object $P = (P_i)_{i \in I}$ of $Et(X)$, called a *universal (pro-) étale cover* of X, and an isomorphism

(8.5.11.4) $$\operatorname{Hom}(P, Y) = \operatorname{colim}_i \operatorname{Hom}(P_i, Y) \xrightarrow{\sim} Y(\overline{x})$$

functorial in $Y \in Et(X)$. The identity of P corresponds by (8.5.11.4) to a point $\xi \in P(\overline{x}) = \lim P_i(\overline{x})$, which in turn defines (8.5.11.4) by $(u : P \to Y) \mapsto u(\xi) \in Y(\overline{x})$. The P_i's which are *Galois*, i.e., are connected, nonempty and such that the natural map $\mathrm{Aut}(P_i) = \mathrm{Hom}(P_i, P_i) \to \mathrm{Hom}(P, P_i)(\simeq P_i(\overline{x}))$ is bijective form a cofinal system, and therefore we have

$$\mathrm{Hom}(P, P) = \mathrm{Aut}(P) = \lim_{i \in J} \mathrm{Aut}(P_i),$$

where J is the subset of I consisting of indices i for which P_i is Galois. The group opposite to the group $\mathrm{Aut}(P)$ of automorphisms of P is by definition $\pi_1(X, \overline{x})$. In other words, it is the group of automorphism of the fiber functor at $\overline{x}$. It acts continuously and functorially (on the left) on $Y(\overline{x})$, and this defines the equivalence (8.5.11.3). An étale cover Y is *connected* if and only if $\pi_1(X, \overline{x})$ acts transitively on $Y(\overline{x})$.

If $\overline{a} \to X$, $\overline{b} \to X$ are two geometric points, then, as X is connected, the fiber functors $F_{\overline{a}}$ at $\overline{a}$ and $F_{\overline{b}}$ at $\overline{b}$ are isomorphic [**SGA1**, V 5.6]. The choice of an isomorphism from $F_{\overline{a}}$ to $F_{\overline{b}}$ is called a *path* from $\overline{a}$ to $\overline{b}$. Such a path induces an isomorphism

$$\pi_1(X, \overline{a}) \xrightarrow{\sim} \pi_1(X, \overline{b}). \tag{8.5.11.5}$$

If X is not assumed to be connected, one defines the fundamental group of X at $\overline{x}$ as the fundamental group of the connected component containing x. The fundamental group is in a natural way a functor on geometrically pointed locally noetherian schemes. If $f : X \to Z$ is a morphism between *connected* locally noetherian schemes, then the inverse image functor

$$f^* : Et(Z) \to Et(X), \ Z' \mapsto X \times_Z Z'$$

is an equivalence if and only if the homomorphism

$$f_* : \pi_1(X, \overline{x}) \to \pi_1(Z, \overline{z})$$

is an isomorphism, where $\overline{z}$ is the geometric point $\overline{x} \to x \to z$ image of $\overline{x}$ by f. The homomorphism f_* is surjective if and only if the functor f^* is fully faithful, or equivalently, if for any connected étale cover Z' of Z, f^*Z' is connected [**SGA1**, V 6.9]. It is injective if and only if, for any étale cover X' of X, there exists an étale cover Z' of Z and a map from a connected component of f^*Z' to X' [**SGA1**, V 6.8].

The following result complements 8.2.18.

THEOREM 8.5.12. [**SGA1**, IX 1.10]

Let A be a complete local noetherian ring, with maximal ideal $\mathbf{m}$ and residue field k. Let $S = \mathrm{Spec}\, A$, $\hat{S} = \mathrm{Spf}\, A = \mathrm{colim}\, S_n$, where $S_n = \mathrm{Spec}\, A/\mathbf{m}^{n+1}$. Let X be a proper scheme over S. Then the inverse image functor

$$Et(X) \to Et(X_s),$$

where $s = S_0 = \mathrm{Spec}\, k$, is an equivalence. In other words, for any geometric point $\overline{x}$ of X_s, the natural homomorphism

$$\pi_1(X_s, \overline{x}) \to \pi_1(X, \overline{x})$$

is an isomorphism.

Let $\hat{X}$ be the formal completion of X along X_s, so that $\hat{X} = \operatorname{colim}_n X_n$, where $X_n = S_n \times_S X$, $S_n = \operatorname{Spec} A/\mathbf{m}^{n+1}$. Consider the natural morphisms

$$X_s \xrightarrow{i} \hat{X} \xrightarrow{j} X .$$

We have inverse image functors

$$Et(X) \xrightarrow{j^*} Et(\hat{X}) \xrightarrow{i^*} Et(X_s) ,$$

where $Et(\hat{X})$ denotes the category of étale covers of $\hat{X}$, i.e., of finite formal schemes $\mathcal{Y} = \operatorname{colim} Y_n$ over $\hat{X}$ (8.4.4 (b)) which are *étale*, i.e., such that Y_n is étale over X_n for all $n \geq 0$. By 8.5.9 (b), i^* is an equivalence. On the other hand, by 8.4.6, $(ji)^*$ is fully faithful. It remains to show that j^* is essentially surjective. Let $\mathcal{Y}$ be an étale cover of $\hat{X}$. By 8.4.6 there exists a unique scheme Y finite over X such that $\hat{Y} = \mathcal{Y}$. If y is a point of Y_s and $n \geq 0$, $\mathcal{O}_{Y,y}/\mathbf{m}^{n+1}\mathcal{O}_{Y,y} = \mathcal{O}_{Y_n,y}$ is flat over $\mathcal{O}_{X,x}/\mathbf{m}^{n+1}\mathcal{O}_{X,x} = \mathcal{O}_{X_n,x}$, where x is the image of y in X_s. Therefore Y is flat over X in a neighborhood of Y_s, and consequently flat over X since Y is proper over S. Moreover,

$$j^*\Omega^1_{Y/X} = \lim_n \Omega^1_{Y_n/X_n} = 0$$

since Y_n is étale over X_n. Hence, by 8.4.2, $\Omega^1_{Y/X} = 0$, and therefore Y is étale over X (8.5.8).

REMARK 8.5.13. It follows from Artin's approximation theorem that the conclusion of 8.5.12 still holds if A is only assumed to be henselian instead of complete, see [**SGA4-1/2**, IV 2.2]. Statements 8.2.18 and 8.5.12 are crucial in the proof of the *proper base change theorem* in étale cohomology ((loc. cit.) and [**SGA4**, XII]).

8.5.14. Theorem 8.5.12 is the starting point of Grothendieck's theory of *specialization for the fundamental group*[**SGA1**, X]. Let $f : X \to Y$ be a *proper* morphism of locally noetherian schemes, with connected geometric fibers. Let s and η be points of Y, such that $s \in \overline{\{\eta\}}$, $\overline{s}$ (resp. $\overline{\eta}$) a geometric point over s (resp. η), a (resp. b) a geometric point of $X_{\overline{s}}$ (resp. $X_{\overline{\eta}}$). Then there is defined (loc. cit. 2.1, 2.4) a homomorphism

$$\pi_1(X_{\overline{\eta}}, b) \to \pi_1(X_{\overline{s}}, a), \tag{8.5.14.1}$$

called the *specialization homomorphism*. This homomorphism is well defined up to an inner automorphism of the target. If Y is the spectrum of a henselian local noetherian ring A, with closed point s such that $s = \overline{s}$, (8.5.14.1) is the composition

$$\pi_1(X_{\overline{\eta}}, b) \to \pi_1(X, b) \xrightarrow{\sim} \pi_1(X, a) \xrightarrow{\sim} \pi_1(X_s, a),$$

where the first map is the functoriality map, the second one an isomorphism associated to a *path* from a to b (8.5.11) (such a path exists because the hypotheses imply, by 8.2.18, that X is connected), and the last one is the inverse of the isomorphism of 8.5.12, 8.5.13. The definition in the general case is more delicate, see (loc. cit.). It uses the fact that for a proper and connected scheme X over an algebraically closed field k, the fundamental group of X is invariant under algebraically closed extension of k (this fact is a (nontrivial) consequence of 8.5.12). Grothendieck's main result about (8.5.14.1) is the following theorem:

THEOREM 8.5.15. [**SGA1**, X 2.4, 3.8] *(Grothendieck's specialization theorem) Let $f : X \to Y$ be as in 8.5.14.*

(a) If f is flat and has geometrically reduced fibers (i.e., for any morphism $\overline{y} \to y \in Y$ with $\overline{y}$ the spectrum of an algebraically *closed field, $X_{\overline{y}}$ is reduced), then (8.5.14.1) is surjective;*

(b) If f is smooth and p is the characteristic exponent of s, then (8.5.14.1) induces an isomorphism on the largest prime to p quotients of the fundamental groups

$$\pi_1^{(p')}(X_{\overline{\eta}}, b) \xrightarrow{\sim} \pi_1^{(p')}(X_{\overline{s}}, a).$$

(We use the notation $\pi_1^{(p')}$ to denote the largest prime to p quotients; this notation has become more common than the notation $\pi_1^{(p)}$ used in (loc. cit.).)

Let us prove (a) in the case Y is the spectrum of a henselian local noetherian ring A, with algebraically closed residue field k and $s = \operatorname{Spec} k$ (the general case can be reduced to this one). We have to show that if Z is a connected étale cover of X, then $Z_{\overline{\eta}}$ is connected. Note that Z is again proper and flat over Y with geometrically reduced fibers. As Z is connected, so is the special fibre Z_s by 8.2.18. Therefore $H^0(Z_s, \mathcal{O}_{Z_s})$ is an artinian local k-algebra with residue field k. Since Z_s is reduced, $H^0(Z_s, \mathcal{O}_{Z_s}) = k$. The composition of the canonical maps

$$\mathcal{O}_Y \otimes k \to g_*\mathcal{O}_Z \otimes k \to H^0(Z_s, \mathcal{O}_{Z_s}) = k,$$

where $g : Z \to Y$ is the structural morphism, is the identity, in particular

$$g_*\mathcal{O}_Z \otimes k \to H^0(Z_s, \mathcal{O}_{Z_s})$$

is surjective. Since Z is flat over Y, it follows from 8.3.11 that this map is in fact an isomorphism and that $g_*\mathcal{O}_Z$ is free of rank 1 and its formation commutes with arbitrary base change, in other words, $g_*\mathcal{O}_Z = \mathcal{O}_Y$ holds *universally* (i.e., after any base change). In particular, $Z_{\overline{\eta}}$ is connected.

Let us sketch the proof of (b) with Y as in the proof of (a) above, supposing furthermore that A is a discrete valuation ring with uniformizing parameter π (see [**SGA1**, X 3] for details and [**OV00**] for a survey). By (a) the restriction functor $Et(X) \to Et(X_{\overline{\eta}})$ is fully faithful. Therefore it remains to show that any Galois étale cover V of $X_{\overline{\eta}}$ of group G, with G finite of order prime to p, is induced by a G-Galois étale cover of X. Thanks to 8.5.12, up to replacing Y by a finite extension, we may assume that V comes by base change from a G-étale cover V_0 of X_η. Let n be the order of G. Let $Y' = \operatorname{Spec} A'$, where $A' = A[t]/(t^n - \pi)$, and $X' = Y' \times_Y X$. $V' = \eta' \times_\eta V$ (η' the generic point of Y') the schemes deduced from X and V_0 by base change. Let R be the local ring of X at the generic point ζ of X_s. This is a discrete valuation ring, for which π is a uniformizing parameter. The local ring R' of X' at ζ is $R' = R[t]/(t^n - \pi)$. It then follows from Abhyankar's lemma that the restriction of V' to the generic point of $\operatorname{Spec} R'$ extends to an étale G-cover of $\operatorname{Spec} R'$, an therefore to an étale G-cover W' of an open subset U' of X' whose complement is a closed subset of X_s of codimension at least 1 in X_s, hence of codimension at least 2 in X'. Since X' is smooth over Y', hence regular, Zariski-Nagata's purity theorem implies that W' extends to an étale G-cover Z' of X'. By 8.5.12, Z' comes from an étale G-cover Z of X, which finishes the proof.

The argument sketched for the proof of (a) gives in fact the following result ([**SGA1**, 1.2], [**EGAIII2**, 7.8.6]):

PROPOSITION 8.5.16. *Let $f : X \to Y$ be a proper and flat morphism of locally noetherian schemes, having geometrically reduced fibers, and let*

$$X \to Y' \to Y$$

be its Stein factorization (8.2.11). Then Y' is an étale cover of Y, and its formation commutes with any base change. In particular f is cohomologically flat in degree zero (8.3.10), and the following conditions are equivalent:

(i) $f_*\mathcal{O}_X = \mathcal{O}_Y$;

(ii) the geometric fibers of f are connected.

REMARKS 8.5.17.

(a) Under the assumptions of 8.5.15 (a), i.e., for $f : X \to Y$ proper and flat, with geometrically reduced and connected fibers, the specialization homomorphism (8.5.14.1) has been extensively studied in the past few years, especially in the case of relative *curves*. See [**BLR00**] for a discussion of some aspects of this.

(b) A variant of the theory of the fundamental group in "logarithmic geometry" has been constructed by Fujiwara-Kato [**FK95**]. See [**Ill02**] for an introduction and [**Vid01**, I 2.2] for a generalization of Grothendieck's specialization theorem 8.5.15 in this context and an application [**Kis00**] to the action by outer automorphisms of the wild inertia on the prime to p fundamental group of varieties over local fields.

D. Curves.

8.5.18. We now turn to applications to liftings of curves. Let Y be a locally noetherian scheme. By a *curve* over Y we mean a morphism $f : X \to Y$ which is flat, separated and of finite type, with relative dimension 1. Assume f is *proper*. Then, for any coherent sheaf F on X, $R^qf_*F = 0$ for $q > 1$ by 8.2.10, and if moreover F is flat over Y, e. g. $F = \mathcal{O}_X$, the complex Rf_*F is perfect, of perfect amplitude in [0,1] (8.3.11). In general, f is cohomologically flat neither in degree 0 nor 1, as simple examples show [**Har77**, III 12.9.2]. However, if f has geometrically reduced fibers, f is cohomologically flat in degree 0 by 8.5.16, hence also in degree 1 by 8.3.11, i.e., $R^qf_*\mathcal{O}_X$ is locally free of finite type for all q. When, moreover, f has connected geometric fibers, so that $f_*\mathcal{O}_X = \mathcal{O}_Y$, the rank of the locally free sheaf $R^1f_*\mathcal{O}_X$ is called the *(arithmetic) genus* of the curve X over Y. If f is proper and smooth, then, by Grothendieck's duality theorem, the sheaves $R^qf_*\Omega^1_{X/Y}$ are also locally free of finite type, and there is defined a *trace map*

$$\mathrm{Tr} : R^1f_*\Omega^1_{X/Y} \to \mathcal{O}_Y,$$

and the pairing

$$R^qf_*\mathcal{O}_X \otimes R^{1-q}f_*\Omega^1_{X/Y} \to \mathcal{O}_Y$$

obtained by composing the natural pairing to $R^1f_*\Omega^1_{X/Y}$ with Tr is a perfect pairing between locally free sheaves of finite type. In particular, if f has connected geometric fibers and is of genus g, $f_*\Omega^1_{X/Y}$ is locally free or rank g. Finally, recall that any curve over a field is *quasi-projective*. See [**Har77**, III Ex. 5.8] for the case the curve is proper over an algebraically closed field (the general case can be reduced to this one).

The main result on liftings of curves is the following theorem:

THEOREM 8.5.19. [**SGA1**, III 7.3]

Let A be a complete local noetherian ring, with residue field k. Let $S = \operatorname{Spec} A$, $s = \operatorname{Spec} k$, and let X_0 be a projective and smooth scheme over s satisfying

$$\text{(i)} \qquad H^2(X_0, T_{X_0/s}) = 0.$$

Then there exists a proper and smooth formal scheme (8.5.8) $\mathcal{X}$ over $\hat{S}$ lifting X_0. If, in addition to (i), X_0 satisfies

$$\text{(ii)} \qquad H^2(X_0, \mathcal{O}_{X_0}) = 0,$$

then there exists a projective and smooth scheme X over S such that $X_s = X_0$.

Conditions (i) and (ii) are satisfied, for example, if X_0 is a proper and smooth curve over s. Note that, if X_0 is a proper, geometrically connected, smooth curve of genus g, then the same is true for the fibers of X over S.

Let $\hat{S} = \operatorname{Spf} A = \operatorname{colim} S_n$, where $S_n = \operatorname{Spec} A/\mathbf{m}^{n+1}$, $\mathbf{m}$ denoting the maximal ideal of A. Let us show that, under the assumption (i), there exists a (proper) and smooth formal scheme $\mathcal{X} = \operatorname{colim} X_n$ over $\hat{S}$ lifting X_0. Assume X_m, smooth over S_m, has been constructed for $m \leq n$ such that $X_m = S_m \times_{S_n} X_n$, and let $i_n : S_n \to S_{n+1}$ be the inclusion. Then, by 8.5.9, 8.5.10 (b), there is an obstruction

$$o(X_n, i_n) \in H^2(X_0, T_{X_0/s} \otimes \mathbf{m}^{n+1}/\mathbf{m}^{n+2})$$

to the existence of a smooth lifting X_{n+1} of X_n over S_{n+1}. But

$$H^2(X_0, T_{X_0/s} \otimes \mathbf{m}^{n+1}/\mathbf{m}^{n+2}) = H^2(X_0, T_{X_0/s}) \otimes \mathbf{m}^{n+1}/\mathbf{m}^{n+2},$$

which is zero by (i). This shows the existence of $\mathcal{X}$. As for he second assertion of 8.5.19, we deduce from 8.5.6 the existence of a projective and flat scheme X over S such that $X_s = X_0$. Then X is smooth over S at each point of X_s, hence in an open neighborhood of X_s, which has to be equal to X since X is proper over S.

By 8.5.19, *proper smooth curves in positive characteristic can be lifted to characteristic zero "without ramification"*: if k is of characteristic $p > 0$, one can take for A a *Cohen ring* for k, i.e., a complete discrete valuation ring with residue field k, fraction field of characteristic zero, and maximal ideal generated by p (the ring $W(k)$ of Witt vectors on k when k is perfect). Using this and the known structure of the (topological) fundamental group of compact Riemann surfaces, Grothendieck was able to deduce from the specialization theorem 8.5.15 the following results about the (algebraic) fundamental group of proper smooth curves in positive characteristic:

THEOREM 8.5.20 ([**SGA1**, X, 3.9, 3.10]). *Let k be an algebraically closed field of characteristic exponent p and let C be a proper, smooth and connected curve over k, of genus g. Let x be a rational point of C. Denote by Π_g the group defined by generators a_i, b_i ($1 \leq i \leq g$), subject to the relation $\prod_{1 \leq i \leq g}(a_i, b_i) = 1$, where $(a, b) := aba^{-1}b^{-1}$, and let $\widehat{\Pi_g}$ be its profinite completion. Then there exists a surjective homomorphism*

$$\widehat{\Pi_g} \to \pi_1(C, x),$$

inducing an isomorphism

$$\widehat{\Pi_g}^{(p')} \xrightarrow{\sim} \pi_1(C, x)^{(p')}$$

on the largest prime to p quotients.

Here is a sketch of the argument. One first treats the case where $k = \mathbb{C}$. Let C^{an} be the (compact, connected and of genus g) Riemann surface associated to C. By Riemann's existence theorem, the functor $C' \mapsto C'^{an}$ from the category of finite étale covers of C to that of finite étale covers of C^{an} is an equivalence. It follows that $\pi_1(C, x)$ is the profinite completion of $\pi_1(C^{an}, x)$. Topological arguments, using the representation of C^{an} as the quotient of a polygon with $4g$ edges $(e_i, e_i^{-1}, f_i, f_i^{-1})$ $(1 \leq i \leq g$ by the identification specified by the word $\prod_{1\leq i\leq g}(e_i, f_i)$ shows that $\pi_1(C^{an}, x) = \Pi_g$. So the result is proven in this case. The case where $p = 1$ is reduced to this one by standard limit arguments using the invariance of the (algebraic) fundamental group (of proper schemes) under arbitrary extension of algebraically closed fields. Finally, suppose $p \geq 2$. In 8.5.19, take $X_0 = C$ and $A = W(k)$ the ring of Witt vectors on k. Let X be a projective and smooth scheme over S such that $X_s = C$. Then, by 8.5.18, X/S is a projective and smooth curve with connected geometric fibers of genus g. The conclusion thus follows from the case $p = 1$ and 8.5.15.

REMARKS 8.5.21.

(a) If $g = 0$, then C is isomorphic to $\mathbb{P}^1_k$, hence simply connected (by Riemann-Hurwitz). More generally, all $\mathbb{P}^r_k$ are simply connected [**SGA1**, XI 1.1].
(b) If $g = 1$, then C is an *elliptic curve* and $\pi_1(C)$ is the *Tate module* of C, $T(C) = \lim {}_nC(k)$, where n runs through all integers ≥ 1, ${}_nC(k)$ denotes the kernel of the multiplication by n on $C(k)$, and for $m = nd$, ${}_mC(k)$ is sent to ${}_nC(k)$ by multiplication by d. More generally, if A is an abelian variety over k, then

$$\pi_1(A) = T(A),$$

where $T(-)$ is the Tate module, defined similarly [**Mum70**, IV 18].
(c) By 8.5.20, $\pi_1(C, x)$ is topologically of finite type. As Grothendieck observed in [**SGA1**, X 2.8], it seems unlikely that $\pi_1(C, x)$ could be topologically of finite presentation, but the question is still open. Using some Lefschetz type arguments for hyperplane sections, Grothendieck shows that more generally, for any proper connected scheme X over k, $\pi_1(X)$ is topologically of finite generation [**SGA1**, X, 2.9].
(d) There is a variant of the last assertion of 8.5.20 for *affine* curves. More precisely, let C be a proper, smooth and connected curve of genus $g \geq 0$ over k. Let n be an integer ≥ 1, $x_1, \cdots, x_n$ be distinct rational points of C, let $X = C - \{x_1, \cdots, x_n\}$, and pick a rational point x of X. Then there is an isomorphism

$$\widehat{\Pi}_{g,n}^{(p')} \xrightarrow{\sim} \pi_1(X, x)^{(p')},$$

where $\widehat{\Pi}_{g,n}^{(p')}$ is the prime to p quotient of the profinite completion of the (free) group $\Pi_{g,n}$ defined by generators a_i, b_i $(1 \leq i \leq g)$, s_i $(1 \leq i \leq n)$, subject to the relation $\prod_{1\leq i\leq g}(a_i, b_i)\prod_{1\leq i\leq n} s_i = 1$ [**SGA1**, XIII 2.12]. However, $\pi_1(X, x)$ is not topologically of finite type, even for $X = \mathbb{A}^1_k$. A finite group G is the Galois group of a connected étale cover of $\mathbb{A}^1_k$ if and only if its largest prime to p quotient is trivial (*Abhyankar's conjecture*, proven by Raynaud [**Ray94**]).
(e) For C proper, connected and smooth of genus $g \geq 2$, the (full) fundamental group of C encodes amazingly deep information about C. For example, let me mention the following striking result of Tamagawa:

THEOREM [**Tam04**]. *Let k be an algebraically closed field of characteristic $p > 0$, $A = \operatorname{Spec} k[[t]]$, and X a proper and smooth curve over S with connected geometric fibers of genus $g \geq 2$. Let $s = \operatorname{Spec} k$. Assume that the special fiber X_s can be defined over $\operatorname{Spec} k_0$, where k_0 is a finite subfield of k. Let $\overline{\eta}$ be a geometric point over the generic point η of S. Then, if the specialization homomorphism*

$$\pi_1(X_{\overline{\eta}}, b) \to \pi_1(X_s, a),$$

of (8.5.14.1) is an isomorphism, X is constant *over S, i.e., is isomorphic to $X_s \times_s S$.*

E. Abelian varieties.

8.5.22. Let me now come to liftings of abelian varieties. Let S be a scheme. Recall that an *abelian scheme* over S (*abelian variety* when S is the spectrum of a field) is an S-group scheme, which is proper and flat, and whose geometric fibers are reduced and irreducible. Let X be an abelian scheme over S. Then X is automatically *smooth* and *commutative*, see [**Mum70**, II 4] for the case S is the spectrum of an algebraically closed field, and [**Mum65**, 6.5] for the general case. It is also known that if S is normal, or even geometrically unibranch, X is *projective* over S [**Mur95**]. Counterexamples outside of these hypotheses have been given by Raynaud [**Ray70**].

Grothendieck has shown (unpublished) that abelian varieties admit formal liftings:

THEOREM 8.5.23. *Let $\hat{S} = \operatorname{Spf} A$ be as in 8.5.12, and let X_0 be an abelian variety over $s = \operatorname{Spec} k$. Then:*

(a) *There exists a proper and smooth formal scheme $\mathcal{X}$ over $\hat{S}$ such that $s \times_{\hat{S}} \mathcal{X} = X_0$ and a section e of $\mathcal{X}$ over $\hat{S}$ extending the unit section e_0 of X_0.*
(b) *Let $(\mathcal{X}, e)$ be a lifting of (X_0, e_0) over $\hat{S}$ as in (a), and let $X_n = S_n \times_{\hat{S}} \mathcal{X}$, with S_n as in 8.2.12. One can, in a unique way, inductively define a structure of abelian scheme on X_n over S_n having e_n as unit section and such that $X_n = S_n \times_{S_{n+1}} X_{n+1}$ as abelian schemes.*

Assuming that, for a fixed integer n, an abelian scheme X_n lifting X_0 has been constructed (with unit section e_n), we have to show that:

(i) there exists a smooth scheme X_{n+1} over S_{n+1} lifting X_n and a lifting e_{n+1} of e_n;

(ii) given a smooth lifting X_{n+1} of X_n as a scheme and a lifting e_{n+1} of e_n, there exists a unique group scheme structure on X_{n+1} over S_{n+1} lifting that of X_n over S_n and having e_{n+1} as unit section.

The proofs of (i) and (ii) are similar. In both cases one encounters an obstruction, which lives in a *nonzero* cohomology group. Using the *functoriality* (8.5.10 (a)) of the obstruction with respect to a suitable morphism, one shows that it is zero.

Let us sketch the proof of (i) (cf. [**Oor71**, p. 238]). Consider the obstruction

$$o(X_n) \in H^2(X_0, T_{X_0}) \otimes I$$

to the lifting of X_n to S_{n+1} (8.5.9 (b)), where we write T_{X_0} for $T_{X_0/s}$ and I for $\mathbf{m}^n/\mathbf{m}^{n+1}$ for brevity. Consider, too, the obstruction

$$o(X_n \times X_n) \in H^2(X_0 \times X_0, T_{X_0 \times X_0}) \otimes I$$

to the lifting of $X_n \times X_n$ to S_{n+1}. By the compatibility of obstructions with products (8.5.10 (a)) we have

$$o(X_n \times X_n) = pr_1^* o(X_n) + pr_2^* o(X_n). \tag{1}$$

Let

$$s : X_n \times X_n \to X_n \quad , \quad (x, y) \mapsto x + y$$

be the sum morphism. By functoriality of the obstructions (8.5.10 (a)), $o(X_n)$ and $o(X_n \times X_n)$ have the same image by the two maps

$$H^2(X_0, T_{X_0}) \otimes I \to H^2(X_0 \times X_0, s^* T_{X_0}) \otimes I \leftarrow H^2(X_0 \times X_0, T_{X_0 \times X_0}) \otimes I.$$

These two maps can be rewritten

$$H^2(X_0) \otimes t_{X_0} \otimes I \xrightarrow{s^* \otimes Id} H^2(X_0 \times X_0) \otimes t_{X_0} \otimes I \xleftarrow{Id \otimes s} H^2(X_0 \times X_0) \otimes t_{X_0 \times X_0} \otimes I$$

where we have written $H^*(-)$ instead of $H^*(-, \mathcal{O})$, and t means the tangent space at the origin, pull-back of the tangent bundle T by the unit section. In other words, we have

$$(s^* \otimes Id)(o(X_n)) = (Id \otimes s)(o(X_n \times X_n). \tag{2}$$

On the other hand, we know [**Ser59**, VII 21] that

$$H^q(X_0) = \Lambda^q H^1(X_0) \quad , \quad H^q(X_0 \times X_0) = \Lambda^q H^1(X_0 \times X_0),$$

and that

$$s^* : H^1(X_0) \to H^1(X_0 \times X_0) = H^1(X_0) \oplus H^1(X_0)$$

is the diagonal map. Choose a basis (e_i) $(1 \leq i \leq g)$ $(g = \dim X_0)$ for $H^1(X_0)$ and a basis ε_k $(1 \leq k \leq g)$ for t_{X_0} (actually, $H^1(X_0)$ and t_{X_0} are naturally dual to each other, and we could take dual bases, but we don't need this). Write

$$o(X_n) = \sum_{1 \leq i < j \leq g, 1 \leq k \leq g} a_{ij}^k e_i \wedge e_j \otimes \varepsilon_k,$$

with $a_{ij}^k \in I$. Let $e_i' = pr_1^* e_i$, $e_i'' = pr_2^* e_i$, $\varepsilon_k' = (\varepsilon_k, 0)$, $\varepsilon_k'' = (0, \varepsilon_k)$. By (1) we have

$$o(X_n \times X_n) = \sum a_{ij}^k e_i' \wedge e_j' \otimes \varepsilon_k' + \sum a_{ij}^k e_i'' \wedge e_j'' \otimes \varepsilon_k''. \tag{3}$$

Using that s^* is the diagonal map, hence sends e_i to $e_i' + e_i''$, we get

$$(s^* \otimes Id)(o(X_n)) = \sum a_{ij}^k (e_i' \wedge e_j' + e_i'' \wedge e_j' + e_i' \wedge e_j'' + e_i'' \wedge e_j'') \otimes \varepsilon_k.$$

Finally, since $s : t_{X_0 \times X_0} \to t_{X_0}$ sends ε_k' and ε_k'' to ε_k, we deduce from (2):

$$\sum a_{ij}^k (e_i' \wedge e_j' + e_i'' \wedge e_j' + e_i' \wedge e_j'' + e_i'' \wedge e_j'') \otimes \varepsilon_k = \sum a_{ij}^k (e_i' \wedge e_j' + e_i'' \wedge e_j'') \otimes \varepsilon_k.$$

Therefore

$$a_{ij}^k = 0$$

for all i, j, k, i.e., $o(X_n) = 0$. The existence of a lifting e_{n+1} of the unit section e_n is immediate.

For the proof of (ii), see [**Mum65**, 6.15]: one first shows that the obstruction to lifting the difference map $\mu_n : X_n \times_{S_n} X_n \to X_n$, $(x, y) \mapsto x - y$ is zero, using its compatibility (8.5.10 (a)) with composition with the diagonal map $x \mapsto (x, x)$ and the map $x \mapsto (x, 0)$; one normalizes the lifting of μ_n using e_{n+1} and one concludes by a rigidity lemma.

REMARKS 8.5.24.

(a) Using the arguments above, Grothendieck actually proved a more general and precise result than 8.5.23, namely: if $S_0 \to S$ is a closed immersion of *affine* schemes, defined by an ideal I of square zero, and if if X_0 is an abelian scheme over S_0, then there exists an abelian scheme X over S lifting X_0; moreover, the set of isomorphism classes of abelian schemes X over S lifting X_0 is an affine space under $\Gamma(S_0, t_{\hat{X}_0} \otimes t_{X_0} \otimes I)$, where $\hat{X}_0$ is the dual abelian scheme, and the group of automorphisms of any lifting X (inducing the identity on X_0) is zero. A different proof is given in [**Ill85**, A 1.1], using the theory of the cotangent complex, which provides an obstruction to the lifting of X_0 *as a flat commutative group scheme,* living in a cohomology group which is zero.

(b) Consider a formal abelian scheme $\mathcal{X} = \operatorname{colim} X_n$ as in 8.5.23 (b). It is not true in general that $\mathcal{X}$ is algebraizable: using the theory of formal moduli of abelian varieties, one can construct examples of nonalgebraizable $\mathcal{X}$ already for $k = \mathbb{C}$, $A = \mathbb{C}[[ts$ and $g = \dim X_0 = 2$. In contrast with the case of curves, it is indeed not always possible to lift an ample invertible sheaf L_0 on X_0 to $\mathcal{X}$ (or even to X_1). The step by step obstructions to such liftings lie in a group of the form $H^2(X_0, \mathcal{O}) \otimes I$, which is not zero for $g \geq 2$, and they can be nonzero.

On the other hand, Mumford has proven that *any abelian variety in positive characteristic can be lifted to characteristic zero* [**Mum69**]. More precisely, if k is an algebraically closed field of characteristic $p > 0$ and X_0 is an abelian variety over k, there exists a complete discrete valuation ring A having k as residue field and with fraction field of characteristic zero and a (projective) abelian scheme X over $\operatorname{Spec} A$ such that $X \otimes k = X_0$ (the ring A is a finite extension of the ring $W(k)$ of Witt vectors on k, which is in general ramified).

F. Surfaces.

8.5.25. Let Y be a locally noetherian scheme. By a *surface* over Y, we mean a scheme X over Y, which is flat, separated and of finite type and of relative dimension 2. We will be concerned only with *proper and smooth* surfaces. By a theorem of Zariski ([**Zar58**], [**Har70**]), a proper, smooth surface over a field is *projective.* In contrast with the case of curves and abelian varieties, there are proper, smooth surfaces over a field which do not lift formally. More precisely, let k be an algebraically closed field of characteristic $p > 0$. There are two kinds of nonliftability phenomena.

(a) *Nonliftability to* W_2. Let $W = W(k)$ be the ring of Witt vectors on k, $W_n = W/p^nW$ the ring of Witt vectors of length n. Let X_0 be a proper and smooth surface over k having nonclosed global differential forms of degree 1. Examples of such surfaces have been constructed by Mumford [**Mum61a**] and, later on, by Lang [**Lan79**], Raynaud and Szpiro (see [**Fos81**]). By a theorem of Deligne-Illusie [**DI87**, 2.4] this pathology prevents X_0 from being liftable to W_2.

(b) *Nonliftability to characteristic zero.* Improving a result of Serre [**Ser61**], Mumford [**Mum**] has constructed examples of proper and smooth surfaces X_0 over $s = \operatorname{Spec} k$ having the following property. Let A be *any* integral, complete local noetherian ring with residue field k and fraction field of characteristic zero. Then there exists no proper and smooth scheme X over $\operatorname{Spec} A$ such that $X_s = X_0$.

Using Hodge-Witt numbers, which are deep invariants of X_0 defined in terms of the de Rham-Witt complex, Ekedhal [**Eke86**, p. 114] observed that similar examples are provided by suitable Raynaud's surfaces as mentioned in (a).

The relation between phenomena of types (a) and (b) is not well understood.

8.5.26. Here are some results in the positive direction. As in 8.5.12, let A be a complete local noetherian ring, with maximal ideal $\mathbf{m}$ and residue field k. Let $S = \operatorname{Spec} A$, $\hat{S} = \operatorname{Spf} A = \operatorname{colim} S_n$, where $S_n = \operatorname{Spec} A/\mathbf{m}^{n+1}$. Let X_0 be a proper and smooth surface over $s = \operatorname{Spec} k$. Using 8.5.19 and the general results of [**Har77**, IV 2, 5] it is easy to see that if X_0 is *rational* or *ruled*, then X_0 lifts to a projective surface over S. On the other hand, we have seen that if X_0 is an *abelian* surface, then X_0 admits a formal smooth lifting $\mathcal{X}$ over $\hat{S}$. The same is true if X_0 is a *K3 surface*, i . e. a proper, smooth, connected surface such that $\Omega^2_{X_0/s}$ is trivial and $H^1(X_0, \mathcal{O}_{X_0}) = 0$. More precisely, we have the following result, due to Rudakov-Shafarevitch and Deligne:

THEOREM 8.5.27. [**Del81**, 1.8]
With the notations of 8.5.26, let X_0 be a K3 surface over an algebraically closed field k.

(a) There exists a proper and smooth formal scheme $\mathcal{X}$ over $\hat{S}$ lifting X_0.

(b) Let L_0 be an ample line bundle on X_0. Then there exists a complete discrete valuation ring R finite over the ring of Witt vectors $W(k)$, a proper and smooth scheme X over $T = \operatorname{Spec} R$ lifting X_0, and a lifting of L_0 to an ample line bundle L on X.

Let us prove (a). By a basic result of Rudakov-Shafarevitch [**RŠ76**] (see also [**Nyg79**] and [**LN80**]), we have

$$H^0(X_0, T_{X_O/k}) = 0.$$

Since $\Omega^2_{X_0/k}$ is trivial, we have $T_{X_O/k} = \Omega^1_{X_O/k}$, hence by Serre duality it follows that $H^2(X_0, T_{X_O/k}) = 0$. Therefore the conclusion follows from 8.5.19. The proof of (b) is much more difficult, since $H^2(X_0, \mathcal{O}) = k$ and one cannot apply 8.5.19. See [**Del81**] for details.

REMARKS 8.5.28.

(a) As in the case of abelian varieties, in the situation of 8.5.27 (a) it may happen that a given polarization of X_0 can't lift to $\mathcal{X}$, see [**Del81**, 1.6] for a more precise statement.

(b) For $p = 3$, M. Hirokado [**Hir99**] has constructed a Calabi-Yau threefold X_0/k (i.e., a smooth projective scheme of dimension 3 such that $\Omega^3_{X_0/k} \simeq \mathcal{O}_{X_0}$ and $H^1(X_0, \mathcal{O}_{X_0}) = H^2(X_0, \mathcal{O}_{X_0}) = 0$) having $b_3 = 0$, where

$$b_3 = \dim H^3(X_0, \mathbb{Q}_\ell), \ell \neq p.$$

By Hodge theory, such a scheme admits no smooth *projective* lifting to characteristic zero. This Calabi-Yau threefold is constructed from a blow-up of $\mathbb{P}^3_k$ by taking the quotient by a certain vector field. Thus, as Calabi-Yau threefolds can be considered as analogues of K3 surfaces, Deligne's result 8.5.27 does not extend to dimension 3.

G. Cotangent complex.

8.5.29. So far we have considered deformations of smooth morphisms only. To deal with more general morphisms, one must use the theory of the cotangent complex [**Ill71**]. For an extensive survey, see [**Ill72b**]. We will just give very brief indications.

Let $f : X \to Y$ be a morphism of schemes. The *cotangent complex* of f (or X/Y), denoted

$$L_{X/Y},$$

is a complex of $\mathcal{O}_X$-modules, concentrated in ≤ 0 degrees, defined as follows. The pair of functors: free $f^{-1}(\mathcal{O}_Y)$-algebra generated by a sheaf of sets, sheaf of sets underlying an $f^{-1}(\mathcal{O}_Y)$-algebra, defines a Godement style, standard simplicial $f^{-1}(\mathcal{O}_Y)$-algebra P, augmented to $\mathcal{O}_X$, whose components are free $f^{-1}(\mathcal{O}_Y)$-algebras over sheaves of sets, and such that the chain complex of the underlying augmented simplicial $f^{-1}(\mathcal{O}_Y)$-module is acyclic. Applying the functor Ω^1 (Kähler differentials) componentwise, one obtains a simplicial $f^{-1}(\mathcal{O}_Y)$-module $\Omega^1_{P/f^{-1}(\mathcal{O}_Y)} \otimes_P \mathcal{O}_X$, whose corresponding chain complex is $L_{X/Y}$. This complex has a natural augmentation to $\Omega^1_{X/Y}$, which defines an isomorphism $\mathcal{H}^0(L_{X/Y}) \xrightarrow{\sim} \Omega^1_{X/Y}$. Its components are flat $\mathcal{O}_X$-modules. It depends functorially on X/Y. Moreover, a sequence of morphisms $X \xrightarrow{f} Y \longrightarrow Z$ gives rise to a distinguished triangle in $D(X)$, called the *transitivity triangle*

$$f^* L_{Y/Z} \to L_{X/Z} \to L_{X/Y} \to .$$

Suppose f is a morphism locally of finite type between locally noetherian schemes. Then $L_{X/Y}$ is *pseudo-coherent* (8.3.6.1). If f is *smooth*, the augmentation $L_{X/Y} \to \Omega^1_{X/Y}$ is a quasi-isomorphism. If f is a *closed immersion*, defined by an ideal I, then there is a natural augmentation $L_{X/Y} \to I/I^2[1]$, which is a quasi-isomorphism when f is a *regular immersion*, i.e., is locally defined by a regular sequence; in this case I/I^2 is locally free. If f is *locally of complete intersection*, i.e., is locally (on X) the composition of a regular immersion and a smooth morphism, then $L_{X/Y}$ is perfect, of perfect amplitude in $[-1, 0]$ (8.3.6.3).

8.5.30. The relation between cotangent complex and deformation theory comes from the following fact. Let $f : X \to Y$ be a morphism of schemes. If $i : X \to X'$ is a closed immersion into a Y-scheme defined by an ideal I of square zero, I is a quasi-coherent module on X. We call i (or X') a *Y-extension of X by I*. For fixed I, these Y-extensions form an abelian group, which is shown to be canonically isomorphic to $Ext^1(L_{X/Y}, I)$. This isomorphism is functorial in I. Using the transitivity triangle (8.5.29), one easily deduces the following generalization of 8.5.9 (see [**Ill71**, III 2]):

THEOREM 8.5.31. *(a) Let X and Y be schemes over a scheme S, and let $j : X_0 \to X$ be a closed subscheme defined by an ideal J of square zero. Let $g : X_0 \to Y$ be an S-morphism. There is an obstruction*

$$o(g, j) \in \mathrm{Ext}^1(g^* L_{Y/S}, J)$$

whose vanishing is necessary and sufficient for the existence of an S-morphism $h : X \to Y$ extending g, i.e., such that $hj = g$. When $o(g, j) = 0$, the set of *extensions h of g is an affine space under* $\mathrm{Ext}^0(g^*L_{Y/S}, J) = \mathrm{Hom}(g^*\Omega^1_{Y/S}, J)$.

(b) Let $i : S_0 \to S$ be a thickening of order 1 defined by an ideal I of square zero, and let X_0 be a flat S_0-scheme. There is an obstruction

$$o(X_0, i) \in \mathrm{Ext}^2(L_{X_0/S_0}, f_0^*I)$$

(where $f_0 : X_0 \to S_0$ is the structural morphism) whose vanishing is necessary and sufficient for the existence of a deformation X of X_0 over S (8.5.7). When $o(X_0, i) = 0$, the set of isomorphism classes of such deformations is an affine space under $\mathrm{Ext}^1(L_{X_0/S_0}, f_0^*I)$, and the group of automorphisms of a fixed deformation is isomorphic to $\mathrm{Ext}^0(L_{X_0/S_0}, f_0^*I) = \mathrm{Hom}(\Omega^1_{X_0/S_0}, f_0^*I)$.

Here is an application to liftings of certain singular curves (generalizing the smooth case, dealt with in 8.5.19):

COROLLARY 8.5.32. *Let $S = \mathrm{Spec}\, A$ be as in 8.5.19. Let X_0 be a proper curve over s (8.5.18). We assume that X_0 is locally of complete intersection over s and is smooth over s outside a finite set of closed points. Then there exists a projective and flat curve X over S such that $X_s = X_0$.*

Note that such a lifting X is automatically locally of complete intersection over S ([**EGAIV3**, 11.3.8], [**EGAIV4**, 19.2.4]), and is smooth over S outside a finite subscheme (the nonsmoothness locus of X/S is closed and its special fiber is finite, hence is finite by 8.2.16).

As in the proof of 8.5.19, we first show that there exists a (proper) and flat formal scheme $\mathcal{X} = \mathrm{colim}\, X_n$ over $\hat{S}$ lifting X_0. Assume X_m, flat over S_m, has been constructed for $m \le n$ such that $X_m = S_m \times_{S_n} X_n$, and let $i_n : S_n \to S_{n+1}$ be the inclusion. Then, by 8.5.31, there is an obstruction

$$o(X_n, i_n) \in \mathrm{Ext}^2(L_{X_0/s}, \mathcal{O}_{X_0} \otimes \mathbf{m}^{n+1}/\mathbf{m}^{n+2}) = \mathrm{Ext}^2(L_{X_0/s}, \mathcal{O}_{X_0}) \otimes_k \mathbf{m}^{n+1}/\mathbf{m}^{n+2}$$

to the existence of a flat lifting X_{n+1} of X_n over S_{n+1}. Therefore it suffices to show

$$(*) \qquad \mathrm{Ext}^2(L_{X_0/s}, \mathcal{O}_{X_0}) = 0.$$

We have

$$\mathrm{Ext}^2(L_{X_0/s}, \mathcal{O}_{X_0}) = H^2(X_0, R\mathcal{H}om(L_{X_0/s}, \mathcal{O}_{X_0})).$$

Since $L_{X_0/s}$ is of perfect amplitude in $[-1, 0]$, $R\mathcal{H}om(L_{X_0/s}, \mathcal{O}_{X_0})$ is of perfect amplitude in $[0, 1]$, in particular,

$$\mathcal{E}xt^i(L_{X_0/s}, \mathcal{O}_{X_0}) = 0$$

for $i \neq 0, 1$. Hence it suffices to show

$$(1) \qquad H^2(X_0, \mathcal{H}om(L_{X_0/s}, \mathcal{O}_{X_0})) = 0,$$

$$(2) \qquad H^1(X_0, \mathcal{E}xt^1(L_{X_0/s}, \mathcal{O}_{X_0})) = 0.$$

(1) trivially holds because X_0 is of dimension 1. Since X_0 is smooth over s outside a finite closed subset Σ, $\mathcal{E}xt^1(L_{X_0/s}, \mathcal{O}_{X_0})$ is concentrated on Σ, which implies (2), hence $(*)$.

It remains to show that $\mathcal{X}$ is algebraizable to a projective scheme over S. If D is any effective divisor supported on the smooth locus of X_0 and meeting each irreducible component of X_0, then $\mathcal{O}_{X_0}(D)$ is ample, and since $H^2(X_0, \mathcal{O}_{X_0}) = 0$, the conclusion follows from 8.5.6.

REMARK 8.5.33. One can show that, under the assumptions of 8.5.32, the natural map $L_{X_0/s} \to \Omega^1_{X_0/s}$ is an isomorphism in $D(X_0)$.

8.6. Serre's examples

8.6.1. Let k be an algebraically closed field of characteristic $p > 0$, $n \geq 0$, $r \geq 1$ integers, G a finite group, and

$$\rho_0 : G \to PGL_{n+1}(k) \ (= GL_{n+1}(k)/k^*)$$

a representation. Let $P_0 = \mathbb{P}^n_k$. Since the group of k-automorphisms of P_0 is $PGL_{n+1}(k)$ [**Har77**, II 7.1.1], ρ_0 defines a (right) action of G on P_0. For $g \in G$, denote by Fix(g) the (closed) subscheme of fixed points of g (intersection of the graph of g and the diagonal in $P_0 \times_k P_0$). Let $Q_0 \subset P_0$ be the union of the Fix(g)'s for $g \neq e$. Consider the condition

$$r + \dim(Q_0) < n. \tag{8.6.1.1}$$

The starting point of Serre's construction is the following result [**Ser58**, Prop. 15]:

PROPOSITION 8.6.2. *Assume that (8.6.1.1) holds. Then there exists an integer $d_0 \geq 1$ such that, for any integer d divisible by d_0, one can find a smooth complete intersection $Y_0 = \mathrm{V}(h_1, \cdots, h_{n-r})$ of dimension r in P_0, with $\deg(h_i) = d$ for $1 \leq i \leq r$, which is stable under G, and on which G acts freely.*

By [**SGA2**, V 1.8] the action of G on P_0 is admissible, in particular, the quotient $Z_0 = P_0/G$ exists. The projection $f : P_0 \to Z_0$ is finite, and $(f_*\mathcal{O}_{P_0})^G = \mathcal{O}_{Z_0}$. By [**EGAII**, 6.6.4], Z_0 is projective (indeed, condition (II bis) of [**EGAII**, 6.5.1] is satisfied: as P_0 is normal, Z_0 is normal, too, as follows from the above formula for $\mathcal{O}_{Z_0}$). Choose an embedding $i : Z_0 \to \mathbb{P}^s_k$. Then $(if)^*\mathcal{O}_{\mathbb{P}^s_k}(1) = \mathcal{O}_{P_0}(d_0)$ for some integer $d_0 > 0$. For any integer $m \geq 1$, denote by $i_m : Z_0 \to \mathbb{P}^{N(m)}_k$ ($N(m) = \binom{s+m}{s} - 1$) the m-th multiple of i. Then $(i_m f)^*\mathcal{O}_{\mathbb{P}^{N(m)}_k}(1) = \mathcal{O}_{P_0}(d)$ where $d = md_0$. Since f is finite, $f(Q_0)$ is closed in Z_0 and $\dim(f(Q_0)) = \dim(Q_0)$. Since (8.6.1.1) holds, by a theorem of Bertini [**Jou79**, 6.11], there exists a linear subspace $L_0 = V(\ell_1, \cdots, \ell_{n-r})$ of $\mathbb{P}^{N(m)}_k$ of codimension $n - r$ (with $\deg(\ell_i) = 1$), such that $L_0 \cap Z_0$ is contained in $U_0 = Z_0 - f(Q_0)$ and L_0 is transversal to U_0. Since $f|U_0 : f^{-1}(U_0) \to U_0$ is étale, the homogeneous polynomials $h_i = (i_m f)^*\ell_i \in \Gamma(P_0, \mathcal{O}(d))$ $(1 \leq i \leq n - r)$ define a smooth complete intersection Y_0 in P_0, which is stable under G, and does not meet Q_0, hence on which G acts freely.

8.6.3. Let d and Y_0 be as in (8.6.2), with $d \geq 2$, and let

$$X_0 = Y_0/G$$

be the quotient of Y_0 by G. As G acts freely on Y_0, X_0/k is a smooth, projective scheme of dimension r, and the projection

$$f : Y_0 \to X_0$$

is an étale cover of group G [**SGA1**, V 2.3]. Moreover, since Y_0 is a complete intersection of dimension $r \geq 1$, Y_0 is *connected* [**Ser55**, no 78, Prop. 5].

The main point in Serre's construction is the following result.

PROPOSITION 8.6.4. *Assume $r \geq 3$, or $r = 2$, $(p, n+1) = 1$, and $p|d$. Let A be a complete local noetherian ring, with residue field k. Let $\mathcal{X}$ be a flat, formal scheme over A lifting X_0. Then $\mathcal{X}$ is algebraizable, i.e., (8.4.9) there exists a (unique) proper scheme X/A such that $\hat{X} = \mathcal{X}$. Moreover, X is projective and smooth over A and the representation ρ_0 (8.6.1) lifts to a representation*

$$\rho : G \to PGL_{n+1}(A) \ (= GL_{n+1}(A)/A^*).$$

The case $r \geq 3$ is dealt with in [**Ser61**]. The case $r = 2$ is due to Mumford [**Mum**].

By 8.5.9 (b), Y_0 lifts uniquely (up to a unique isomorphism) to a formal étale cover $\mathcal{Y} = \operatorname{colim} Y_m$ of $\mathcal{X} = \operatorname{colim} X_m$, i.e., such that Y_m is finite étale over X_m for all $m \geq 0$ (where $X_m = \mathcal{X} \otimes A/\mathbf{m}^{n+1}$). By 8.5.9 (b) again, the action of G on Y_0 extends (uniquely) to an action of G on $\mathcal{Y}$, making $\mathcal{Y}$ an étale Galois cover of $\mathcal{X}$ of group G (i.e., an inductive system of G-Galois étale covers $Y_m \to X_m$). Since $r \geq 2$, we have $H^1(Y_0, \mathcal{O}_{Y_0}) = 0$ and $H^0(Y_0, \mathcal{O}_{Y_0}) = k$ ([**Ser55**, no 78] or [**SGA7**, IX]), so 8.3.11.2 implies that $H^0(Y_m, \mathcal{O}_{X_m}) = A_m$ for all m.

Let $i : Y_0 \to P_0$ be the inclusion and $L_0 = \mathcal{O}_{Y_0}(1) = i^*\mathcal{O}_{P_0}(1)$. We shall show:

(*) L_0 lifts to an invertible sheaf $\mathcal{L}$ on $\mathcal{Y}$, unique up to a (nonunique) isomorphism (inducing the identity on L_0).

Assume first that $r \geq 3$. Then, by (loc. cit.), $H^2(Y_0, \mathcal{O}_{Y_0}) = 0$. Since $H^1(Y_0, \mathcal{O}_{Y_0}) = 0$, too, (*) is true by 8.5.5. Assume now that $r = 2$. Then it is no longer true that $H^2(Y_0, \mathcal{O}_{Y_0}) = 0$. To show that L_0 lifts (in which case it will lift uniquely as $H^1(Y_0, \mathcal{O}_{Y_0}) = 0$), Mumford argues as follows. We have

$$(**)_0 \qquad \Omega^2_{Y_0/k} = \mathcal{O}_{Y_0}(N),$$

with $N = (n-r)d - n - 1$. The hypotheses imply $(p, N) = 1$. Assume that, for $m \geq 0$, L_0 has been lifted to an invertible sheaf L_m on X_m, and the isomorphism $(**)_0$ lifted to an isomorphism

$$(**)_m \qquad L_m^{\otimes N} \simeq \Omega^2_{Y_m/A_m}.$$

Let $i_m : Y_m \to Y_{m+1}$ be the inclusion. Consider the obstruction $o(L_m, i_m)$ to lifting L_m to Y_{m+1} (8.5.3 (b)). By 8.5.4 (a), we have

$$o(L_m^{\otimes N}, i_m) = No(L_m, i_m).$$

Since $\Omega^2_{Y_{m+1}/A_{m+1}}$ lifts $\Omega^2_{Y_m/A_m}$, the isomorphism $(**)_m$ implies that $o(L_m^{\otimes N}, i_m) = 0$, hence $o(L_m, i_m) = 0$ as well, since p does not divide N. Hence L_m lifts to an invertible sheaf L_{m+1} on Y_{m+1}. Since $H^1(Y_0, \mathcal{O}_{Y_0}) = 0$, by 8.5.3 (a) the isomorphism $(**)_m$ lifts to an isomorphism $(**)_{m+1}$. Therefore L_0 lifts to an invertible sheaf $\mathcal{L}$ on $\mathcal{Y}$.

Since L_0 is ample, by 8.5.6 there exists a projective and flat scheme Y over A such that $\hat{Y} = \mathcal{Y}$ and an ample line bundle L on Y such that $\hat{L} = \mathcal{L}$. By [**EGAIII2**, 6.6.1], the norm $E_0 = N_{Y_0/X_0}L_0$ of L_0 is an ample line bundle on X_0. For $m \in \mathbb{N}$, let $E_m = N_{Y_m/X_m}L_m$ and $\mathcal{E} = \lim E_m$. Then $\mathcal{E}$ lifts E_0, so by 8.4.10 there exists a projective scheme X/A such that $\hat{X} = \mathcal{X}$ and an ample line bundle E on X such that $\hat{E} = \mathcal{E}$. By 8.5.6 (and the argument at the end of the proof of 8.5.19), X is smooth over A. Moreover, by 8.4.7, the étale Galois cover $\hat{Y} \to \hat{X}$ is deduced by completion of a (unique) étale Galois cover $Y \to X$ of group G, and by 8.4.2,

$E = N_{Y/X}L$.

It remains to show that ρ_0 lifts to A. Since $d \geq 2$, by ([FAC no 78] we have $H^0(Y_0, L_0) = k^{n+1}$, $H^1(Y_0, L_0) = 0$. Therefore, by 8.3.11.2, $H^0(Y_m, L_m) = A_m^{n+1}$ for all m, hence $H^0(Y, L) = A^{n+1}$ by 8.2.4. Let $g \in G$. Since $H^1(Y_0, \mathcal{O}_{Y_0}) = 0$, by 8.5.3 (a) and 8.4.2 there is an isomorphism $a(g) : L \xrightarrow{\sim} L$ above $g : Y \xrightarrow{\sim} Y$, i.e., an isomorphism $g^*L \xrightarrow{\sim} L$, unique up to an automorphism of L, such that $a(g)_0$ is the isomorphism $L_0 \to L_0$ above $g : Y_0 \xrightarrow{\sim} Y_0$ given by the action of g on Y_0 (which is well defined up to an automorphism of L_0). For g and h in G, we have $a(g)a(h) = a(gh)$ and $a(e) = Id$ up to an automorphism of L. Therefore we get a representation

$$\rho : G \to PGL(H^0(Y, L)) = PGL_{n+1}(A),$$

associating to g the automorphism $\rho(g)$ of $H^0(Y, L) = A^{n+1}$ induced by the pair $(g, a(g))$, which automorphism is well defined up to multiplication by an element of A^*. This representation lifts ρ_0.

8.6.5. Let now r and n be integers with $1 \leq r < n$, and let G be a group of type $(p, \cdots, p)$ of order p^s; i.e., $G \simeq \mathbb{F}_p^s$, with $s \geq n+1$. Assume moreover that $p \geq n+1$. Choose an *injective* homomorphism $h : G \to k$ (where k is considered as an additive group). Let $N = (u_{ij})$ be the nilpotent matrix of order $n+1$ defined by $u_{ij} = 1$ if $j = i+1$ and $u_{ij} = 0$ otherwise. For $g \in G$, let

$$\tilde{\rho}_0(g) = \exp(h(g)N) \in GL_{n+1}(k)$$

(which makes sense since $p \geq n+1$), and let $\rho_0(g)$ be the image of $\tilde{\rho}_0(g)$ in $PGL_{n+1}(k)$. We thus get a representation

$$\rho_0 : G \to PGL_{n+1}(k), \tag{8.6.5.1}$$

which is *faithful*, as h is injective. For any $g \in G$, $g \neq e$, $\mathrm{Fix}(g)$ consists of the single rational point $(1, 0, \cdots, 0)$ of P_0. In particular, $\dim Q_0 = 0$, with the notations of 8.6.1, so the condition (8.6.1.1) is satisfied.

PROPOSITION 8.6.6. *Assume that $p > n+1$. Let A be an integral local ring with residue field k and field of fractions K of characteristic zero. Then there exists no homomorphism $\rho : G \to PGL_{n+1}(A)$ lifting ρ_0 (8.6.5.1).*

The following argument is due to Serre (private communication). Suppose that such a homomorphism ρ exists. Since ρ_0 is injective, ρ is injective, too, and so is the composition, still denoted ρ, with the inclusion $PGL_{n+1}(A) \to PGL_{n+1}(K)$. Since K is of characteristic zero and p does not divide $n+1$, this representation lifts to a (faithful) representation $\rho' : G \to SL(V)$, where $V = K^{n+1}$. As G is commutative and K is of characteristic zero, up to extending the scalars to a finite extension of K, V decomposes into a sum

$$V = \bigoplus_{1 \leq i \leq n+1} V_i$$

of 1-dimensional subspaces stable under G, corresponding to characters $\chi_i : G \to \mathrm{Aut}(V_i) = K^*$, whose product is 1. The kernel Z of ρ is the intersection of the kernels H_i of χ_i, for $1 \leq i \leq n$. Each χ_i is a homomorphism from G to $\mu_p(K)$, so can be viewed as a linear form φ_i on G considered as a vector space over $\mathbb{F}_p$. Since $\sum \varphi_i = 0$ and G is of dimension $s \geq n+1$ over $\mathbb{F}_p$, Z cannot be zero, which contradicts the faithfulness of ρ.

COROLLARY 8.6.7. *Let r, n be integers such that $2 \leq r < n$ and $p > n+1$. Let $G = \mathbb{F}_p^s$, with $s \geq n+1$. There exists a smooth, projective complete intersection Y_0 of dimension r in P_0, stable under the action of G on P_0 defined by the representation ρ_0 constructed in (8.6.5.1), and on which G acts freely, and such that the smooth, projective scheme $X_0 = Y_0/G$ has the following property. Let A be an integral, complete, local noetherian ring with residue field k and field of fractions K of characteristic zero. Then there exists no formal scheme $\mathcal{X}$, flat over A, lifting X_0.*

Let d_0 be an integer ≥ 1 having the properties stated in 8.6.2. If $r \geq 3$, take any nonzero multiple $d \geq 2$ of d_0, and if $r = 2$, take any nonzero multiple d of d_0 which is divisible by p. By 8.6.2, choose a smooth, complete intersection Y_0 in P_0, of degree $(d, \cdots, d)$, stable under the action of G on P_0 defined by the representation ρ_0 constructed in (8.6.5.1), and on which G acts freely. Let $X_0 = Y_0/G$. Assume that there exists a formal scheme $\mathcal{X}$, flat over A, lifting X_0. Since $p > n+1$ and, if $r = 2$, p divides d, the assumptions of 8.6.4 are satisfied, and its conclusion, together with 8.6.6, yields a contradiction.

The minimal examples are obtained for $r = 2$, $n = 3$, $s = 4$, $p = 5$. (In [**Ser61**], the minimal ones were for $r = 3$, $n = 4$, $s = 5$, $p = 7$.)

REMARK 8.6.8. Let X_0 be the scheme considered in 8.6.7. Let A be a complete, local noetherian ring with residue field k, which is the base of a *formal versal deformation* $\mathcal{X}$ of X_0 [**Sch68**]. Such a ring A is a W-algebra which is formally of finite type, where $W = W(k)$ is the ring of Witt vectors on k. Let $K_0 = W[1/p]$ be the fraction field of W. It follows from 8.6.7 that $A \otimes_W K_0 = 0$, in other words *there exists an integer $n_0 \geq 1$ such that $p^{n_0} A = 0$.* Otherwise, one could find an integral closed subscheme $T = \operatorname{Spec} B$ of $\operatorname{Spec} A$ with generic point of characteristic zero. By pulling back $\mathcal{X}$ to $\operatorname{Spf} B$, we would obtain a contradiction.

8.7. A letter of Serre

In the letter below, Serre proves that in fact $n_0 = 1$. His argument also shows that in 8.6.7 it suffices to assume $s \geq 2$ instead of $s \geq n+1$.

Paris, le 11/10/03

Cher Illusie,

Voici la démonstration du fait que la variété formelle de modules que tu sais est “tuée par p”.

Notations – Je considère un anneau local A de corps résiduel k de caractéristique p. Soit $n > 1$, avec $n \leq p$. On s'intéresse à un sous-groupe G de $GL(n,k)$, de type (p,p), et ayant la propriété suivante : pour tout $s \in G$, $s \neq 1$, le noyau de $s-1$ est de dimension 1 (autrement dit, on peut représenter s par un bloc de Jordan de longueur n).

THÉORÈME 1. *Si G est relevable dans $GL(n,A)$, on a $p = 0$ dans A.*

THÉORÈME 2. *Si $p > n$, et si G est relevable dans $PGL(n,A)$, il est relevable dans $SL(n,A)$* (de sorte que l'on a $p = 0$ dans A d'après le th. 1).

Bien sûr, c'est le th. 2 qui est utile pour les variétés formelles de modules, le premier cas intéressant étant $n = 4$; le th. 2 s'applique alors si $p \geq 5$.

Le th. 2 est presque immédiat : si z est un élément d'ordre p de $PGL(n,A)$, on peut le relever de façon unique en un élément z' d'ordre p de $SL(n,A)$. En effet, on choisit un relèvement z'' de s dans $GL(n, A)$; on a $z''^p = c$, avec $c \in A^*$. Si $d = \det(z'')$, on a $d^p = c^n$ et comme n n'est pas divisible par p, ceci montre que c est de la forme u^p avec $u \in A^*$. L'élément $z' = z''/u$ appartient à $SL(n, A)$ et est un relèvement de z. On applique ceci aux différents éléments de G (vu comme sous-groupe de $PGL(n, A)$). Les relèvements obtenus commutent entre eux (cela résulte de l'unicité du relèvement) et donnent un plongement de G dans $SL(n, A)$. Il reste à voir que l'image de ce "G" dans $SL(n, k)$ est bien le groupe G de départ, mais c'est clair, car ledit G est visiblement contenu dans $SL(n, k)$.

Passons aux choses sérieuses, i.e., à la *démonstration du th. 1.*

On suppose G relevé dans $GL(n, A)$ et on désire montrer que $p = 0$ dans A. Soit $\mathbf{m} = \operatorname{Ker}(A \to k)$ l'idéal maximal de A. Si l'on démontre que p appartient à l'idéal $p\mathbf{m}$, il en résultera (par un Nakayama évident) que $p = 0$. En d'autres termes, il suffit de prouver que l'image de p dans $A/p\mathbf{m}$ est 0. *Nous pouvons donc supposer que* $p\mathbf{m} = 0$.

Soit (s, s') un ensemble générateur de G, vu comme sous-groupe de $GL(n, A)$. J'écrirai s sous la forme $s = 1 + e$; en notant par $\underline{e}$ la réduction modulo $\mathbf{m}$ de e, on peut supposer que $\underline{e}$ est la matrice nilpotente type de rang $n-1$ (un seul bloc de Jordan). Le polynôme caractéristique de $\underline{e}$ est T^n, et c'est aussi son polynôme minimal. Le polynôme caractéristique de e est de la forme $T^n + a_1 T^{n-1} + \cdots + a_n$, avec $a_i \in \mathbf{m}$. Comme les $\underline{e}^i$ sont linéairement indépendants sur k pour $i = 0, \cdots, n-1$, il en est de même des e^i sur A, de sorte que l'anneau $A[e]$ est libre de rang n sur A, avec pour base $1, e, \cdots, e^{n-1}$. De plus, *le A-module A^n est libre de rang* 1 *sur* $A[e]$. Ceci entraîne que *tout endomorphisme de A^n qui commute à e appartient à* $A[e]$. Ceci s'applique en particulier au second générateur s' de G : on peut écrire s' comme un polynôme en e à coefficients dans A :

$$s' = a_0 + a_1 e + \cdots + a_{n-1} e^{n-1},$$

avec $a_i \in A$. Je vais maintenant exploiter la relation $s'^p = 1$ pour obtenir une relation entre les a_i. De façon générale, si x est un élément de $A[e]$, je noterai $t_0(x), t_1(x), \cdots$, les coefficients de x dans la base $1, e, \cdots, e^{n-1}$ de $A[e]$.

PROPOSITION. *On a*

$$t_1(x^p) = p t_1(x)(t_0(x)^{p-1} - t_1(x)^{p-1}).$$

Admettons pour un instant cet énoncé. Si on l'applique à $x = s'$, compte tenu de $s'^p = 1$, on a $t_1(x)^p = 0$, d'où $pu = 0$, où u est donné par

$$u = a_1(a_0^{p-1} - a_1^{p-1}).$$

Mais on connaît les images dans k de a_0 et a_1. Celle de a_0 est évidemment 1, et celle de a_1 est un élément t de k *qui n'est pas dans* $\mathbb{Z}/p\mathbb{Z}$ (si cette image était égale à j, avec $0 \leq j < p$, l'élément $s^{-j}s'$ de G serait tel que $\dim \operatorname{Ker}(1 - s^{-j}s') > 1$). On a donc $t - t^p \neq 0$ dans k, ce qui signifie que u *est inversible*, et que l'équation $pu = 0$ entraîne $p = 0$, comme on le désirait.

Reste à démontrer la proposition ci-dessus. Cela peut se faire par un développement multinomial brutal. Cela conduit à un fatras d'indices. Je vais suivre une méthode plus douce. Posons $f(x) = t_1(x^p)$.

LEMME. *Si $y \in A[e]$ est un multiple de e^2, on a $f(x+y) = f(x)$.*

On écrit $(x+y)^p = x^p + v + y^p$, où v est une somme de termes de la forme py', avec y' divisible par e^2. Or, si y' est multiple de e^2, on a $t_1(y') = 0$ (mod. $\mathbf{m}$) : et, d'autre part, le fait que $s^p = 1$ signifie que $(1+e)^p = 1$, i.e., $e^p = -\sum \binom{p}{i} e^i$, où la sommation porte sur les indices i tels que $0 < i < p$. Comme tous les $\binom{p}{i}$ sont divisibles par p, on voit que e^p est divisible par p, et $e^{2p} = 0$ (car $p^2 = 0$ dans A, vu que $p\mathbf{m} = 0$). On a donc $y^p = 0$. D'où le résultat cherché.

Ceci fait, pour prouver la proposition, on peut éliminer les termes de x en $e^2, e^3, \cdots$, i.e., supposer que x est de la forme $a + be$. On calcule alors $f(x) = t_1(x^p)$ par la même méthode que ci-dessus : on écrit

$$x^p = a^p + pa^{p-1}be + \cdots + b^p e^p,$$

d'où

$$f(x) = t_1(a^p) + pt_1(a^{p-1}be) + pt_1(\cdots) + t_1(b^p e^p).$$

Les valeurs respectives de ces termes sont :

$$0, pa^{p-1}b, 0, \cdots, 0, -pb^p$$

(noter que $t_1(e^p) = -p$ à cause de la formule $e^p = -\sum_{1 \leq i \leq p-1} \binom{p}{i} e^i$ donnée ci-dessus).

On obtient donc bien

$$f(x) = pa^{p-1}b - pb^p,$$

comme on le désirait.

Ouf, cqfd, petit carré, etc.

Bien à toi,
J-P. Serre

Part 5

The Picard scheme

Steven L. Kleiman

CHAPTER 9

The Picard scheme

9.1. Introduction

On any ringed space X, the isomorphism classes of invertible sheaves form a group; it is denoted by $\mathrm{Pic}(X)$ and called the (absolute) *Picard group.* Suppose X is a "projective variety"; in other words, X is an integral scheme that is projective over an algebraically closed field k. Then, as is proved in these notes, the group $\mathrm{Pic}(X)$ underlies a natural k-scheme, which is a disjoint union of quasi-projective schemes, and the operations of multiplying and of inverting are given by k-maps. This scheme is denoted by $\mathbf{Pic}_{X/k}$ and called the *Picard scheme.* It is reduced in characteristic zero, but not always in positive characteristic. When X varies in an algebraic family, correspondingly, $\mathbf{Pic}_{X/k}$ does too.

The Picard scheme was introduced in 1962 by Grothendieck. He sketched his theory in two Bourbaki talks, nos. 232 and 236, which were reprinted along with his commentaries in [**FGA**]. But Grothendieck advanced an old subject, which was actively being developed by many others at the time. Nevertheless, Grothendieck's theory was revolutionary, both in concept and in technique.

In order to appreciate Grothendieck's contribution fully, we have to review the history of the Picard scheme. Reviewing this history serves as well to introduce and to motivate Grothendieck's theory. Furthermore, the history is rich and fascinating, and it is a significant part of the history of algebraic geometry.

So let us now review the history of the Picard scheme up to 1962. We need only summarize and elaborate on scattered parts of Brigaglia, Ciliberto, and Pedrini's article [**BCP04**] and of the author's article [**Kle04**]. Both articles give many precise references to the original sources and to the secondary literature; so few references are given here.

The Picard scheme has roots in the 1600s. Over the course of that century, the Calculus was developed, through the efforts of many individuals, in order to design lenses, to aim cannons, to make clocks, to hang cables, and so on. Thus interest arose in the properties of functions appearing as indefinite integrals.

Notably, in 1694, James Bernoulli analyzed the way rods bend, and was led to introduce the "lemniscate," a figure eight with equation $(x^2 + y^2)^2 = a^2(x^2 - y^2)$ where a is nonzero. In polar coordinates, he found the arc length s to be given by

$$s = \int_0^r \frac{a^2\, dr}{\sqrt{a^4 - r^4}}.$$

He surmised that s cannot be expressed in terms of the elementary functions. Similar integrals had already arisen in attempts to rectify elliptical orbits; so these integrals became known as "elliptic integrals."

In 1698, James's brother, John, recalled there are algebraic relations among the arguments of sums and differences of logarithms and of the inverse trigonometric functions. Then he showed that, similarly, given two arcs from the origin on the cubical parabola $y = x^3$, their lengths differ by the length of a certain third such arc. And he posed the problem of finding more cases of this phenomenon.

Sure enough, between 1714 and 1720, Fagnano found, in an ad hoc manner, similar relations for the cords and arcs of ellipses, hyperbolas, and lemniscates. In turn, Fagnano's work led Euler in 1757 to discover the "addition formula"

$$\int_0^{x_1} \frac{dx}{\sqrt{1-x^4}} \pm \int_0^{x_2} \frac{dx}{\sqrt{1-x^4}} = \int_0^{x_3} \frac{dx}{\sqrt{1-x^4}}$$

where the variables x_1, x_2, x_3 must satisfy the symmetric relation

$$x_1^4x_2^4x_3^4 + 2x_1^4x_2^2x_3^2 + 2x_1^2x_2^4x_3^2 + 2x_1^2x_2^2x_3^4 \\ + x_1^4 + x_2^4 + x_3^4 - 2x_1^2x_2^2 - 2x_1^2x_3^2 - 2x_2^2x_3^2 = 0.$$

In 1759, Euler generalized this formula to some other elliptic integrals. Specifically, Euler found the sum or difference of two to be equal to a certain third plus an elementary function. Moreover, he expressed regret that he could handle only square roots and fourth powers, but not higher roots or powers.

In 1826, Abel made a great advance: he discovered an addition theorem of sweeping generality. It concerns certain algebraic integrals, which soon came to be called "Abelian integrals." They are of the following form:

$$\psi x := \int_{x_0}^{x} R(x, y)\, dx$$

where x is an independent complex variable, R is a rational function, and $y = y(x)$ is an integral algebraic function; that is, y is the implicit multivalued function defined by an irreducible equation of the form

$$f(x, y) := y^n + f_1(x)y^{n-1} + \cdots + f_n(x) = 0$$

where the $f_i(x)$ are polynomials in x.

Let p be the genus of the curve $f = 0$, and let $h_1, \ldots, h_\alpha$ be rational numbers. Then Abel's addition theorem asserts that

$$h_1\psi x_1 + \cdots + h_\alpha \psi x_\alpha = v + \psi x_1' + \cdots + \psi x_p'$$

where v is an elementary function of the independent variables $x_1, \ldots, x_\alpha$ and where $x_1', \ldots, x_p'$ are algebraic functions of them. More precisely, v is a complex-linear combination of one algebraic function of $x_1, \ldots, x_\alpha$ and of logarithms of others; moreover, $x_1', \ldots, x_p'$ work for every choice of ψx. Lastly, p is minimal: given algebraic functions $x_1', \ldots, x_{p-1}'$ of $x_1, \ldots, x_p$, there exists an integral ψx such that, for any elementary function v,

$$\psi x_1 + \cdots + \psi x_p \neq v + \psi x_1' + \cdots + \psi x_{p-1}'.$$

Abel finished his 61-page manuscript in Paris, and submitted it in person on 30 October 1826 to the Royal Academy of Sciences, which appointed Cauchy and Legendre as referees. However, the Academy did not publish it until 1841, long after Abel's death from tuberculosis on 6 April 1829.

Meanwhile, Abel feared his manuscript was lost forever. So in Crelle's Journal, **3** (1828), he summarized his general addition theorem informally. Then he treated in detail a major special case, that in which $f(x, y) := y^2 - \varphi(x)$ where $\varphi(x)$ is a square-free polynomial of degree $d \geq 1$. In particular, Abel found

$$p = \begin{cases} (d-1)/2, & \text{if } d \text{ is odd;} \\ (d-2)/2, & \text{if } d \text{ is even.} \end{cases}$$

Thus, if $d \geq 5$, then $p \geq 2$, and so Euler's formula does not extend.

With Jacobi's help, Legendre came to appreciate the importance of this case. To it, Legendre devoted the third supplement to his long treatise on elliptic integrals, which are recovered when $d = 3, 4$. For $d \geq 5$, the integrals share many of the same formal properties. So Legendre termed them "ultra-elliptiques."

Legendre sent a copy of the supplement to Crelle for review on 24 March 1832, and Crelle asked Jacobi to review it. Jacobi translated "ultra-elliptiques" by "hyperelliptischen," and the prefix "hyper" has stuck. In his cover letter, Legendre praised Abel's addition theorem, calling it, in the immortal words of Horace's Ode 3, XXX.1, "a monument more lasting than bronze" (monumentum aere perennius). In his review, Jacobi said that the theorem would be a most noble monument were it to acquire the name **Abel's Theorem.** And it did!

Jacobi was inspired to give, a few months later, the first of several proofs of Abel's Theorem in the hyperelliptic case. Furthermore, he posed the famous problem, which became known as the "Jacobi Inversion Problem." He asked, "what, in the general case, are those functions whose inverses are Abelian integrals, and what does Abel's theorem show about them?"

Jacobi solved the inversion problem when $d = 5, 6$. Namely, he formed

$$\psi x := \int_{x_0}^{x} \frac{dx}{\sqrt{\varphi(x)}} \text{ and } \psi_1 x := \int_{x_0}^{x} \frac{x\,dx}{\sqrt{\varphi(x)}},$$

and he set

$$\psi x + \psi y = u \text{ and } \psi_1 x + \psi_1 y = v.$$

He showed $x + y$ and xy are single-valued functions of u and v with four periods.

Some historians have felt Abel had this inversion in mind, but ran out of time. At any rate, in 1827, Abel had originated the idea of inverting elliptic integrals, obtaining what became known as "elliptic functions." He and Jacobi studied them extensively. Moreover, Jacobi introduced "theta functions" as an aid in the study; they were generalized by Riemann in 1857, and used to solve the inversion problem in arbitrary genus.

Abel's paper on hyperelliptic integrals fills twelve pages. Eight are devoted to a computational proof of a key intermediate result. A half year later, in Crelle's Journal, **4** (1829), Abel published a 2-page paper with a conceptual proof of this result for any Abelian integral ψx. The result says that $\psi x_1 + \cdots + \psi x_\mu$ is equal to an elementary function v if $x_1, \ldots, x_\mu$ are not independent, but are the abscissas of the variable points of intersection of the curve $f = 0$ with a second plane curve that varies in a linear system — although this geometric formulation is Clebsch's.

In each of the first two papers, Abel addressed two more, intermediate questions: First, when is the sum $\psi x_1 + \cdots + \psi x_\mu$ constant? Second, what is the number α of x_i that can vary independently? Remarkably, the answers involve the genus p.

For hyperelliptic integrals, Abel found $\psi x_1 + \cdots + \psi x_\mu$ is constant if ψx is a linear combination of the following p integrals:

$$\int_{x_0}^{x} \frac{dx}{\sqrt{\varphi(x)}}, \int_{x_0}^{x} \frac{x\,dx}{\sqrt{\varphi(x)}}, \ldots, \int_{x_0}^{x} \frac{x^{p-1}\,dx}{\sqrt{\varphi(x)}}.$$

Here $x_1, \ldots, x_\mu$ are the abscissas of the variable points of intersection of the curve $y^2 = \varphi(x)$ and the curve $\theta_1(x)y = \theta_2(x)$, whose coefficients vary, but $\theta_2(x)$ and $\varphi(x)$ retain a fixed common factor $\varphi_1(x)$. Furthermore, Abel found

$$\mu - \alpha \geq p; \tag{9.1.1}$$

equality does not always hold, but can be achieved, given d and α, by choosing the degrees of $\theta_1(x)$ and $\varphi_1(x)$ appropriately.

Suppose $\mu - \alpha = p$. Then

$$\psi x_1 + \cdots + \psi x_\alpha = v - (\psi x_{\alpha+1} + \cdots + \psi x_{\alpha+p})$$

where $x_{\alpha+1}, \ldots, x_{\alpha+p}$ are algebraic functions of $x_1, \ldots, x_\alpha$. Similarly, given any $x'_1, \ldots, x'_{\alpha'}$, we get

$$\psi x'_1 + \cdots + \psi x'_{\alpha'} + \psi x_{\alpha+1} + \cdots + \psi x_{\alpha+p} = v' - (\psi x''_1 + \cdots + \psi x''_p).$$

Subtract this formula from the one above, and set $V := v - v'$. The result is

$$\psi x_1 + \cdots + \psi x_\alpha - \psi x'_1 - \cdots - \psi x'_{\alpha'} = V + \psi x''_1 + \cdots + \psi x''_p,$$

namely, the addition theorem with $h_i = \pm 1$. This result is essentially Abel's main theorem on hyperelliptic integrals.

In his Paris manuscript, Abel addressed the two intermediate questions for an arbitrary f. However, his computations are more involved, and his results, less definitive. He found constancy holds when ψx is of the form

$$\psi x := \int_{x_0}^{x} \frac{h(x,y)}{\partial f/\partial y}\, dx \tag{9.1.2}$$

where $\deg h \leq \deg f - 3$. Also, h must satisfy certain linear conditions; namely, h must vanish suitably everywhere $\partial f/\partial y$ does on the curve $f = 0$, at finite distance and at infinity. Abel took the maximum number of independent h as the genus p.

Furthermore, Abel found that there exists an $i \geq 0$ such that

$$\mu - \alpha = p - i. \tag{9.1.3}$$

This equation does not contradict (9.1.1) as the two α's differ; in (9.1.1), the linear system of intersections is incomplete, whereas in (9.1.3), the system is complete.

Abel's ideas have been clarified and completed over the course of time through the efforts of many. Doubtless, Riemann made the greatest contribution in his revolutionary 1857 paper on Abelian functions. In his thesis of 1851, he had developed a way of extending complex analysis to a multivalued function y of a single variable x by viewing y as a single-valued function on an abstract multisheeted covering of the x-plane, the "Riemann surface" of y. In 1857, he treated the case where the surface is compact, and showed this case is precisely the case where y is algebraic.

Riemann defined the genus p topologically, essentially as half the first Betti number of the surface. However, the term "genus" is not Riemann's, but Clebsch's. Clebsch introduced it in 1865 to signal his aim of using p in order to classify algebraic curves. And he showed that every curve of genus 0 is birationally equivalent to a line, and every curve of genus 1, to a nonsingular plane cubic.

Also in 1865, Clebsch gave an algebro-geometric formula for the genus p of a plane curve: if the curve has degree d and, at worst, δ nodes and κ cusps, then

$$p = (d-1)(d-2)/2 - \delta - \kappa.$$

The next year, Clebsch and Gordan employed this formula to prove the birational invariance of p; they determined how d, δ, and κ change.

Plainly, birationally equivalent curves have homeomorphic Riemann surfaces, and so the same genus p. But Clebsch was no longer satisfied in just showing the consequences of Riemann's work. He now wanted to establish the theory of Abelian integrals on the basis of the algebraic theory of curves as developed by Cayley, Salmon, and Sylvester. At the time, Riemann's theory was strange and suspect; there was, as yet, no theory of manifolds, and no proof of the Dirichlet principle. Clebsch's efforts led to a sea change in algebraic geometry, which turned toward the study of birational invariants.

Riemann defined an integral to be of the "first kind" if it is finite everywhere. He proved these integrals form a vector space of dimension p. Furthermore, each can be expressed in the form (9.1.2) provided the curve $f = 0$ has, at worst, double points; if so, the linear conditions on h just require h to vanish at these double points. In 1874, Brill and M. Noether generalized this result to ordinary m-fold points: h must vanish to order $m - 1$. They termed such h "adjoints." Meanwhile, starting with Kronecker in 1858 and Noether in 1871 and continuing through Muhly and Zariski in 1938, many algebraic geometers developed corresponding ways of reducing the singularities of a given plane curve by means of birational transformations.

Euler noted the integral $\int_0^x dx/\sqrt{1-x^4}$ has a "modulus of multivaluedness" like that of the inverse trigonometric functions. Abel noted an arbitrary Abelian integral has a similar ambiguity, but viewed it as a sort of constant of integration, and avoided it by keeping the domain small. Riemann clarified the issue completely. He proved every integral of the first kind has $2p$ "periods," which are numbers that generate all possible changes in the value of the integral arising from changes in the path of integration.

Riemann, in effect, did as follows. He fixed a basis $\psi_1 x, \ldots, \psi_p x$ of the integrals of the first kind, and he fixed a homology basis of $2p$ paths. Then, inside the vector space $\mathbb{C}^p$, he formed the lattice $\mathbb{L}$ generated by the $2p$ corresponding p-vectors of periods. And he proved the quotient is a p-dimensional complex torus

$$J := \mathbb{C}^p/\mathbb{L}.$$

Later J was termed the "Jacobian" to honor Jacobi's work on inversion.

Let C be the curve $f = 0$, or better, the associated Riemann surface. Let $C^{(\mu)}$ be its μ-fold symmetric product. Riemann, in effect, formed the following map:

$$\Psi_\mu \colon C^{(\mu)} \to J \text{ given by } \Psi_\mu(x_1, \ldots, x_\mu) = \left(\sum_{i=1}^{\mu} \psi_1 x_i, \ldots, \sum_{i=1}^{\mu} \psi_p x_i\right).$$

This map Ψ_μ is rather important. It has been called the "Abel–Jacobi map" and the "Abel map." The latter name is historically more correct and shorter, so better.

Riemann, in effect, studied the fibers of the Abel map Ψ_μ. He proved that, if two divisors $x_1 + \cdots + x_\mu$ and $x'_1 + \cdots + x'_\mu$ are linearly equivalent, then

$$\Psi_\mu(x_1, \ldots, x_\mu) = \Psi_\mu(x'_1, \ldots, x'_\mu).$$

Riemann called this result "Abel's Addition Theorem," and cited Jacobi's 1832 proof of it in the hyperelliptic case.

The converse of this result holds too. But Abel did not recognize it, and it lies, at best, between the lines of Riemann's paper. The converse was first explicitly stated by Clebsch in 1864, and first proved in full generality some time later by Weierstrass. In 1913, in Weyl's celebrated book on Riemann surfaces, Weyl combined the result and its converse under the heading of Abel's Theorem. Ever since then, most mathematicians have done the same, even though Weyl explained it is not historically correct to do so.

Together, the above result and its converse imply the fiber $\Psi_\mu^{-1}\Psi_\mu(x_1, \ldots, x_\mu)$ is the complete linear system determined by $x_1 + \cdots + x_\mu$. Its dimension is just Abel's α, the number of x_i that can vary independently in the system. Furthermore, in effect, Riemann rediscovered Abel's formula (9.1.3), and in 1864, Roch identified i as the number of independent adjoints vanishing on $x_1, \ldots, x_\mu$. In 1874, Brill and Noether, inspired by Clebsch, gave the first algebro-geometric treatment of the formula, whose statement they named the "Riemann–Roch Theorem."

Finally, Riemann treated the inversion problem. In effect, he proved that the Abel map Ψ_p is biholomorphic on a certain saturated Zariski open subset $U \subset C^{(p)}$; namely, U is the complement of the image of $C^{(p-1)}$ in $C^{(p)}$ under the map

$$(x_2, \ldots, x_p) \mapsto (x_0, x_2, \ldots, x_p).$$

The inverse map $\Psi_p U \to U$ can be expressed using the coordinate functions on $C^{(p)}$, so in terms of functions on $\Psi_p U$. Since $J := \mathbb{C}^p/\mathbb{L}$, these functions can be lifted to an open subset of $\mathbb{C}^p$, and then continued to meromorphic functions on $\mathbb{C}^p$ with $2p$ periods. Riemann termed these special functions "Abelian functions."

Two years later, in 1859, Riemann proved that every meromorphic function F in p variables has at most $2p$ independent period vectors. Those F with exactly $2p$ soon became known as "Abelian functions." They were studied by many, including Weierstrass, Frobenius, and Poincaré. In particular, in 1869, Weierstrass observed that not every F comes from a curve.

Form the set K of all "Abelian functions" whose group of periods contains a given lattice $\mathbb{L}$ in $\mathbb{C}^p$ of rank $2p$. It turns out K is a field of transcendence degree p over $\mathbb{C}$. Hence K is the field of rational functions on a p-dimensional projective algebraic variety A, which is parameterized on a Zariski open set by p of them. There are many such A, and all were called "Abelian varieties" at first. In 1919, Lefschetz proved there is a distinguished A, whose underlying set can be identified with $\mathbb{C}^p/\mathbb{L}$ in a natural way, and he restricted the term "Abelian variety" to it.

Not only do the points of an Abelian variety A form a group, but the operations of adding and of inverting are given by polynomials. Thus A is a complete algebraic group, or an "Abelian variety" in Weil's sense of 1948. Weil proved each such abstract Abelian variety is commutative. Earlier, in 1889, Picard had, in effect, proved this commutativity in the case of a surface.

Every connected projective algebraic group is parameterized globally by Abelian functions with a common lattice of periods. This fact was proved by Picard for surfaces in 1889 and, assuming the group is commutative, in any dimension in 1895. His proof was completed at certain points of analysis in 1903 by Painlevé. Thus the two definitions of Abelian variety agree, Lefschetz's and Weil's; however, Weil worked in arbitrary characteristic.

In the case of C above, its Jacobian J is thus an Abelian variety. Moreover, J is the quotient of $C^{(\mu)}$ for any $\mu \geq p$ by linear equivalence. So J and Ψ_μ are defined by integrals, but given by polynomials! And addition on J corresponds to addition of divisors. Therefore, J and Ψ_μ can be constructed algebro-geometrically just by forming the quotient. Severi attributed this construction to Castelnuovo.

In 1905, Castelnuovo generalized the construction to surfaces. To set the stage, he reviewed the case of curves, calling it very well known (notissimo). His work is a milestone in the history of irregular surfaces, which began in 1868.

In 1868, Clebsch generalized Abel's formula (9.1.2) to a surface $f(x, y, z) = 0$ with ordinary singularities and no point at infinity on the z-axis; in other words, $f = 0$ is a general projection of a smooth surface. Clebsch showed that every double integral of the first kind is of the form

$$\iint \frac{h(x, y, z)}{\partial f/\partial z}\, dx\, dy$$

where $\deg h = \deg f - 4$ and h vanishes when $\partial f/\partial z$ does. The number of independent integrals became known as the "geometric genus" and denoted by p_g.

In 1870, M. Noether found an algebro-geometric proof that p_g is a birational invariant, as conjectured by Clebsch. In 1871, Cayley found a formula for the expected number of independent h, later called the "arithmetic genus" and denoted by p_a. He observed that, if $f = 0$ is a ruled surface over a base curve of genus p, then $p_a = -p$, but $p_g = 0$. Later in 1871, Zeuthen used Cayley's formula to give an algebro-geometric proof that p_a too is a birational invariant.

In 1875, Noether explained the unexpected discrepancy between p_g and p_a: the vanishing conditions on h need not be independent; in any case, $p_g \geq p_a$. He noted that, if the surface $f = 0$ in 3-space is smooth or rational, then $p_g = p_a$. It was expected that equality usually holds; so when it did, $f = 0$ was termed "regular." The difference $p_g - p_a$ gives a quantitative measure of the failure of $f = 0$ to be regular; so $p_g - p_a$ was termed its "irregularity."

In 1884, Picard studied, on the surface $f = 0$, simple integrals

$$\int P(x, y, z)\, dx + \int Q(x, y, z)\, dy$$

that are closed, or $\partial P/\partial y = \partial Q/\partial x$; they became known as "Picard integrals." And q was used to denote the number of independent Picard integrals of the first kind. Picard noted that, if $f = 0$ is smooth, then $q = 0$. In 1894, Humbert proved that, if $q = 0$, then every algebraic system of curves is contained in a linear system.

Inspired by Humbert's result, in 1896, Castelnuovo proved that, if $p_g - p_a = 0$, then every algebraic system of curves on $f = 0$ is contained in a linear system under a certain restriction. In 1899, Enriques removed the restriction. For a modern version of this result and of its converse, which together characterize regular surfaces, see Exercises 9.5.16 and 9.5.17.

In 1897, Castelnuovo fixed a linear system of curves on the surface $f = 0$, and studied the "characteristic" linear system cut out on a general member by the other members. Let δ be the amount, termed the "deficiency," by which the dimension of the characteristic system falls short of the dimension of its complete linear system. Castelnuovo proved that

$$\delta \leq p_g - p_a,$$

and equality holds for the system cut out by the surfaces of suitably high degree.

In February 1904, Severi extended Castelnuovo's work. Severi took a complete algebraic system of curves, without repetition, on the surface $f = 0$, say with parameter space Σ of dimension R. Let $\sigma \in \Sigma$ be a general point, and C_σ the corresponding curve. Say C_σ moves in a complete linear system of dimension r. Now, to each tangent direction at $\sigma \in \Sigma$, Severi associated, in an injective fashion, a member of the complete characteristic system on C_σ. Thus he got an R-dimensional "characteristic" linear subsystem. The complete system has dimension $r+\delta$. Hence,

$$R \le r + p_g - p_a.$$

A few months later, Enriques and, shortly afterward, Severi gave proofs that, if C_σ is sufficiently positive, then equality holds above; in other words, then the characteristic system of Σ is complete. Both proofs turned out to have serious gaps, as Severi himself observed in 1921. Meanwhile, in 1910, Poincaré gave an analytic construction of a family with $R = p_g - p_a$ and $r = 0$. It follows formally, by means of the Riemann–Roch Theorem for surfaces, that whenever C_σ is sufficiently positive, the characteristic system is complete. After Severi's criticism, it became a major open problem to find a purely algebro-geometric treatment of this issue. But see Corollary 9.5.5 and Remark 9.5.18: the solution finally came forty years later with Grothendieck's systematic use of nilpotents!

In mid January 1905, Severi proved that $p_g - p_a \ge q$ and that $p_g - p_a = b - q$ where b is the number of independent Picard integrals of the second kind, which is equal to the first Betti number. Simultaneously and independently, Picard too proved that $p_g - p_a = b - q$.

A week later and more fully that May and June, Castelnuovo took the last step in this direction. He fixed C_σ sufficiently positive, and formed the quotient, P say, of Σ modulo linear equivalence. So P is projective, and $\dim P = p_g - p_a$. Furthermore, since two sufficiently positive curves sum to a third, it follows that P is independent of the choice of C_σ, and is a commutative group variety. Hence P is an Abelian variety by the general theorem of Picard, completed by Painlevé, mentioned above. Hence P is parameterized by $\dim P$ Abelian functions. Castelnuovo proved they induce independent Picard integrals on $f = 0$. Therefore, $p_g - p_a \le q$. Thus Castelnuovo obtained the Fundamental Theorem of Irregular Surfaces:

$$\dim P = p_g - p_a = q = b/2. \tag{9.1.4}$$

For a modern discussion of the result, see Remark 9.5.15 and Exercise 9.5.16. In 1905, the term "Abelian variety" was not yet in use, and so, naturally enough, Castelnuovo termed P the "Picard variety" of the surface $f = 0$.

Picard also studied Picard integrals of the third kind on the surface $f = 0$. In 1901, he proved that there is a smallest integer ϱ such that any $\varrho + 1$ curves are the logarithmic curves of some such integral. On the basis of this result, in 1905, Severi proved, in effect, that ϱ is the rank of the group of all curves modulo algebraic equivalence. In 1952, Néron proved the result in arbitrary characteristic. So the group is now called the "Néron–Severi group," and ϱ is called the "Picard number."

In 1908 and 1910, Severi studied, in effect, the torsion subgroup of the Néron–Severi group, notably proving it is finite. In 1957, Matsusaka proved this finiteness in arbitrary characteristic. However, there is no special name for this subgroup or for its order. For more about them and ϱ, see Corollary (9.6.17) and Remark (9.6.19).

The impetus to work in arbitrary characteristic came from developments in number theory. In 1921, E. Artin developed, in his thesis, an analogue of the Riemann Hypothesis, in effect, for a hyperelliptic curve over a prime field of odd characteristic. In 1929, F. K. Schmidt generalized Artin's work to all curves over all finite fields, and recast it in the geometric style of Dedekind and Weber. In 1882, they had viewed a curve as the set of discrete valuation rings in a finitely generated field of transcendence degree 1 over $\mathbb{C}$, and they had given an abstract algebraic treatment of the Riemann–Roch Theorem. Schmidt observed that their treatment works with little change in arbitrary characteristic, and he used the Riemann–Roch Theorem to prove that Artin's zeta function satisfies a natural functional equation.

In 1936, Hasse proved Artin's Riemann hypothesis in genus 1 using an analogue of the theory of elliptic functions. Then he and Deuring noted that to extend the proof to higher genus would require developing a theory of correspondences between curves analogous to that developed by Hurwitz and others. This work inspired Weil to study the fixed points of the Frobenius correspondence, and led to his announcement in 1940 and to his two great proofs in 1948 of Artin's Riemann hypothesis for the zeta function of an arbitrary curve and also to his proof of the integrality of his analogue of Artin's L-functions of 1923 and 1930.

First, in 1946, Weil carefully rebuilt the foundation of algebraic geometry from scratch. Following in the footsteps of E. Noether, van der Waerden, and Schmidt, Weil took a variable coefficient field of arbitrary characteristic inside a fixed algebraically closed coordinate field of infinite transcendence degree. Then he formed "abstract" varieties by patching pieces of projective varieties, and said when these varieties are "complete." Finally, he developed a calculus of cycles.

In 1948, Weil published two exciting monographs. In the first, he reproved the Riemann–Roch theorem for (smooth complete) curves, a theorem he regarded as fundamental (see [**Wei79**, I, p. 562, top; II, p. 541, top]). Then he developed an elementary theory of correspondences between curves, which included Castelnuovo's theorem of 1906 on the positive definiteness of the equivalence defect of a correspondence. Of course, Castelnuovo's proof was set over $\mathbb{C}$, but "its translation into abstract terms was essentially a routine matter once the necessary techniques had been created," as Weil put it in his 1954 ICM talk. Finally, Weil derived the Riemann hypothesis.

In the second monograph, Weil established the abstract theory of Abelian varieties. He constructed the Jacobian J of a curve C of genus p by patching together copies of an open subset of the symmetric product $C^{(p)}$. Then taking a prime l different from the characteristic, he constructed, out of the points on J of order l^n for all $n \geq 1$, an l-adic representation of the ring of correspondences, equivalent to the representation on the first cohomology group of C. Finally, he proved the positive definiteness of the trace of this representation, reproved the Riemann hypothesis for the zeta function, and completed the proof of his analogue of Artin's conjectured integrality for L-functions of number fields.

Weil left open two questions: Do a curve and its Jacobian have the same coefficient field? Is every Abelian variety projective? Both questions were soon answered in the affirmative by Chow and Matsusaka. However, there has remained some general interest in constructing nonprojective varieties and in finding criteria for projectivity. Furthermore, Weil was led in 1956 to study the general question of descent of the coefficient field, and this work in turn inspired Grothendieck's general descent theory, which he sketched in [**FGA**, no. 236].

In 1949, Weil published his celebrated conjectures about the zeta function of a variety of arbitrary dimension. Weil did not explicitly explain these conjectures in terms of a hypothetical cohomology theory, but such an explanation lies between the lines of his paper. Furthermore, it was credited to him explicitly in Serre's 1956

"Mexico paper" [**Ser58**, p. 24] and in Grothendieck's 1958 ICM talk.

In his talk, Grothendieck announced that he had found a new approach to developing the desired "Weil cohomology." He wrote: "it was suggested to me by the connections between sheaf-theoretic cohomology and cohomology of Galois groups on the one hand, and the classification of unramified coverings of a variety on the other ..., and by Serre's idea that a 'reasonable' algebraic principal fiber space ..., if it is not locally trivial, should become locally trivial on some covering unramified over a given point." This is the announcement of Grothendieck topology.

In 1960, Grothendieck and Dieudonné [**EGAI**, p. 6] listed the titles of the chapters they planned to write. The last one, Chapter XIII, is entitled "Cohomologie de Weil." The next-to-last is entitled "Schémas abéliens et schémas de Picard." Earlier, at the end of his 1958 ICM talk, Grothendieck had listed five open problems; the fifth is to construct the Picard scheme.

In 1950, Weil published a remarkable note on Abelian varieties. For each complete normal variety X of any dimension in any characteristic, he said there ought to be two associated Abelian varieties, the "Picard" variety P and the "Albanese" variety A, with certain properties, discussed just below. He explained he had complete proofs for smooth complex X, and "sketches" in general. Soon all was proved.

Weil's sketches rest on two criteria for linear equivalence, developed in 1906 by Severi and reformulated in the 1950 note by Weil. He announced proofs of them in 1952, and published the details in 1954. For some more information, see [**Zar71**, p. 120] and Remark 9.5.8. In Weil's commentaries on his 1954 paper, he wrote: "Ever since 1949, I considered the construction of an algebraic theory of the Picard variety as the task of greatest urgency in abstract algebraic geometry."

The properties are these. First, P parameterizes the linear equivalence classes of divisors on X. And there exists a map $X \to A$ that is "universal" in the sense that every map from X to an Abelian variety factors through it. In his commentaries on the note, Weil explained that P had been introduced and named by Castelnuovo; so historically speaking, it would be justified to name P after him, but it was better not to tamper with common usage. By contrast, Weil chose to name A after Albanese in order to honor his work in 1934 viewing A as a quotient of a symmetric power of X, although A had been introduced and studied in 1913 by Severi.

Second, if X is an Abelian variety, then X is equal to the Picard variety of P; so each of X and P is the "dual" of the other. If X is arbitrary, then A and P are dual Abelian varieties; in fact, the universal map $X \to A$ induces the canonical isomorphism from the Picard variety of A onto P. If X is a curve, then both A and P coincide with the Jacobian of X, and the universal map $X \to A$ is just the Abel map; in other words, the Jacobian is "autodual." This autoduality can be viewed as an algebro-geometric statement of Abel's theorem and its converse for integrals of the first kind. For some more information, see Remarks 9.5.24–9.5.26.

In 1951, Matsusaka gave the first algebraic construction of P. However, he had to extend the ground field because he applied Weil's results: one of the equivalence criteria, and the construction of the Jacobian. Both applications involve the "generic curve," which is the section of X by a generic linear space of complementary dimension. In 1952, Matsusaka gave a different construction; it does not require extending the ground field, but requires X to be smooth.

Both constructions are like Castelnuovo's in that they begin by constructing a complete algebraic system of sufficiently positive divisors, and then form the quotient modulo linear equivalence. To parameterize the divisors, Matsusaka used the theory of "Chow coordinates," which was developed by Chow and van der Waerden in 1938 and refined by Chow contemporaneously. In 1952, Matsusaka also used this theory to form the quotient. In the same paper, he gave the first construction of A, again using the Jacobian of the generic curve, but he did not relate A and P.

In 1954, Chow published a construction of the Jacobian similar to Matsusaka's second construction of P. Chow had announced it in 1949, and both Weil and Matsusaka had referred to it in the meantime. In 1955, Chow constructed A and P by a new procedure; he took the "image" and the "trace" of the Jacobian of a generic curve. Moreover, he showed that the universal map $X \to A$ induces an isomorphism from the Picard variety of A onto P.

In a course at the University of Chicago, 1954–55, Weil gave a more complete and elegant treatment, based on the "see-saw principle," which he adapted from Severi, and on his own Theorem of the Square and Theorem of the cube. This treatment became the core of Lang's 1959 book, "Abelian.Varieties." The idea is to construct A first using the generic curve, and then to construct P as a quotient of A modulo a finite subgroup; thus there is no need for Chow coordinates.

In 1959 and 1960, Nishi and Cartier independently established the duality between A and P in full generality.

Between 1952 and 1957, Rosenlicht published a remarkable series of papers, which grew out of his 1950 Harvard thesis. It was supervised by Zariski, who had studied Abelian functions and algebraic geometry with Castelnuovo, Enriques, and Severi in Rome from 1921 to 1927. Notably Rosenlicht generalized to a curve with arbitrary singularities the notions of linear equivalence and differentials of the first kind. Then he constructed a "generalized Jacobian" over $\mathbb{C}$ by integrating and in arbitrary characteristic by patching. It is not an Abelian variety, but an extension of the Jacobian of the normalized curve by an affine algebraic group.

Rosenlicht cited Severi's 1947 monograph, "Funzioni Quasi Abeliane," where the generalized Jacobian was discussed for the first time, but only for curves with double points. In turn, Severi traced the history of the corresponding theory of quasi-Abelian functions back to Klein, Picard, Poincaré, and Lefschetz.

In 1956, Igusa established the compatibility of specializing a curve with specializing its generalized Jacobian in arbitrary characteristic when the general curve is smooth and the special curve has at most one node. Igusa explained that, in 1952, Néron had studied the total space of such a family of Jacobians, but had not explicitly analyzed the special fiber. Igusa's approach is, in spirit, like Castelnuovo's, Chow's, and Matsusaka's before him and Grothendieck's after him. But Grothendieck went considerably further: he proved compatibility with specialization for a family of varieties of arbitrary dimension with arbitrary singularities, both in equicharacteristic and in mixed.

In 1960, Chevalley constructed a Picard variety for any normal variety X using locally principal divisorial cycles. Cartier had already focused on these cycles in his 1958 Paris thesis. But Chevalley said he would call them simply "divisors," and we follow suit, although they are now commonly called "Cartier divisors."

First, Chevalley constructed a "strict" Albanese variety; it is universal for regular maps (morphisms) into Abelian varieties. Then he took its Picard variety to be that of X. He noted his Picard and Albanese varieties need not be equal to those of a desingularization of X. By contrast, Weil's P and A are birational invariants, and his universal map $X \to A$ is a rational map, which is defined wherever X is smooth. In 1962, Seshadri generalized Chevalley's construction to a variety with arbitrary singularities, thus recovering Rosenlicht's generalized Jacobian.

Back in 1924, van der Waerden initiated the project of rebuilding the whole foundation of algebraic geometry on the basis of commutative algebra. His goal was to develop a rigorous theory of Schubert's enumerative geometry, as called for by Hilbert's fifteenth problem. Van der Waerden drew on Elimination Theory, Ideal Theory, and Field Theory as developed in the schools of Kronecker, of Dedekind, and of Hilbert. Van der Waerden originated, notably, the algebraic notion of specialization as a replacement for the topological notion of continuity.

In 1934, as Zariski wrote his book [**Zar71**], he lost confidence in the clarity, precision, and completeness of the algebraic geometry of his Italian teachers. He spent a couple of years studying the algebra of E. Noether and Krull, and aimed to reduce singularities rigorously. He introduced three algebraic tools: normalization, valuation theory, and completion. He worked extensively with the rings obtained by localizing affine coordinate domains at arbitrary primes over arbitrary fields. And, in 1944, he put a topology on the set of all valuation rings in a field of algebraic functions, and used the property that any open covering has a finite subcovering.

In 1949, Weil saw that the "Zariski topology" can be put on his abstract varieties, simplifying the old exposition and suggesting the construction of new objects, such as locally trivial fiber spaces. In his paper of 1950 on Abelian varieties, he noted that line bundles correspond to linear equivalence classes of divisors, and predicted that line bundles would play a role in the theory of quasi-Abelian functions.

In 1955, Serre provided abstract algebraic geometry with a very powerful new tool: sheaf cohomology. Given a variety equipped with the Zariski topology, he assembled the local rings into the stalks of a "structure" sheaf. Then he developed a cohomology theory of coherent sheaves, analogous to the one that he and Cartan and Kodaira and Spencer had just developed in complex analytic geometry, and had so successfully applied to establish and to generalize previous work on complex algebraic varieties.

About the same time, a general theory of abstract algebraic geometry was developed by Chevalley. He did not use sheaves and cohomology, but did work with what he called "schemes," obtained by patching "affine schemes"; an affine scheme is the set of rings obtained by localizing a finitely generated domain over an arbitrary field. Nevertheless, he soon returned to a more traditional theory of "varieties" when he worked on the theory of algebraic groups.

In January 1954, Chevalley lectured on schemes at Kyoto University. His lectures inspired Nagata in [**Nag56**] to generalize the theory by replacing the coefficient field by a Dedekind domain. But Nagata used Zariski's term "model," not Chevalley's term "scheme." Earlier, at the 1950 ICM, Weil had recalled Kronecker's dream of an algebraic geometry over the integers; however, Nagata did not cite Weil's talk, and likely was not motivated by it.

In the fall of 1955, Chevalley lectured on schemes over fields at the Séminaire Cartan–Chevalley, and Grothendieck was there. By February 1956 (see [**CS01**, p. 32]), he was patching the spectra of arbitrary Noetherian rings, and studying the cohomology of Cartier's "quasi-coherent" sheaves. There is good reason for the added generality: nilpotents allow better handling of higher-order infinitesimal deformations, of inseparability in positive characteristic, and of passage to the completion; quasi-coherent sheaves have the technical convenience of coherent sheaves without their cumbersome finiteness. By October 1958 (see [**CS01**, p. 63]), Grothendieck and Dieudonné had begun the gigantic program of writing EGA — rebuilding once again the foundation of algebraic geometry in order to provide a more flexible framework, more powerful methods, and a more refined theory.

Also in 1958, there appeared two other papers, which discuss objects similar to Chevalley's schemes: Kähler published a 400 page foundational monograph, which introduced general base changes, and Chow and Igusa published a four-page note, which proved the Künneth Formula for coherent sheaves. The two works are mentioned briefly in [**CS01**, p. 101], in [**EGAI**, p. 8] and in [**EGAG**, p. 6]; however, they seem to have had little or no influence on Grothendieck.

Finally, in 1961–1962, Grothendieck constructed the Hilbert scheme and the Picard scheme. The construction is a technical masterpiece, showing the tremendous power of Grothendieck's new tools. In particular, a central role is played by the theory of flatness. It was introduced by Serre in 1956 as a formal device for use in comparing algebraic functions and analytic functions. Then Grothendieck developed the theory extensively, for he recognized that flatness is the technical condition that best expresses continuity across a family.

Grothendieck [**FGA**, p. 221-1] saw Hilbert schemes as "destined to replace" Chow coordinates. However, he [**FGA**, p. 221-7] had to appeal to the theory of Chow coordinates for a key finiteness result: in projective space, the subvarieties of given degree form a bounded family. A few years later, Mumford [**Mum66**, Lects. 14, 15] gave a simple direct proof of this finiteness; his proof introduced an important new tool, now known as "Castelnuovo–Mumford" regularity. In a slightly modified form, this tool plays a central role in the proofs of the finiteness theorems for the Picard scheme, which are addressed in Chapter 6 below.

In spirit, Grothendieck's construction of the Picard scheme is like Castelnuovo's and Matsusaka's. He began with the component Σ of the Hilbert scheme determined by a sufficiently positive divisor. Then he formed the quotient; in fact, he did so twice for diversity. First, he used "quasi-sections"; second, and more elegantly, he used the Hilbert scheme of Σ.

Grothendieck's definition of the two schemes is yet a greater contribution than his construction of them. He defined them by their functors of points. These schemes are universal parameter spaces; so they receive a map from a scheme T just when T parameterizes a family of subschemes or of invertible sheaves, respectively, and this map is unique.

What is a universal family? The answer seems obvious when we use schemes. But Chow coordinates parameterize positive cycles, not subschemes. And even in the analytic theory of the Picard variety P, there was some question about the sense in which P parameterizes divisor classes. Indeed, the American Journal of Math., **74** (1952), contains three papers on P. First, Igusa constructed P, but left universality unsettled. Then Weil and Chow settled it with different arguments.

A functor of points, or "representable functor," is not an arbitrary contravariant functor from schemes to sets. It is determined locally, so is a sheaf. But it suffices to represent a sheaf locally, as the patching is determined. Thus to construct the Hilbert scheme, the first step is to check that the Hilbert functor is, in fact, a sheaf for the Zariski topology, that is, a "Zariski sheaf."

The naive Picard functor is not a Zariski sheaf. So the first step is to localize it, or form the associated sheaf. This time, the Zariski topology is not fine enough. However, a representable functor is an fpqc sheaf by a main theorem of Grothendieck's descent theory. In practice, it is enough to localize for the étale Grothendieck topology or for the fppf, and these localizations are more convenient to work with. The localizations of the Picard functor are discussed in Chapter 2.

The next step is to cover the localized Picard functor by representable Zariski open subfunctors. This step is elementary. But it is technically involved, more so than any other argument in these notes. It is carried out in the proof of the main result, Theorem 9.4.8. Each subfunctor is represented by a quotient of an open subscheme of the Hilbert scheme. Thus the Picard scheme is constructed.

In sum, Grothendieck's method of representable functors is like Descartes's method of coordinate axes: simple, yet powerful. Here is one hallmark of genius!

In the notes that follow, our primary aim is to develop in detail most of Grothendieck's original theory of the Picard scheme basically by filling in his sketch in [**FGA**]. Our secondary aim is to review in brief much of the rest of the theory developed by Grothendieck and by others. We review the secondary material in a series of scattered remarks. The remarks refer to each other and to the primary discussion, but the primary discussion never refers to the remarks. So the remarks may be safely ignored in a first reading.

Notably, the primary discussion does not develop Grothendieck's method of "relative representability." Indeed, the details would take us too far afield. On the other hand, were we to use the method, we could obtain certain existence theorems and finiteness theorems in greater generality by reducing to the cases that we do handle. Consequently, in Sections 4–6, a number of results just assume the Picard scheme exists, rather than assume hypotheses guaranteeing it does. However, we do discuss the method and its applications in Remark 9.4.18 and in other remarks.

These notes also contain many exercises, which call for working examples and constructing proofs. Unlike the remarks, these exercises are an integral part of the primary discussion, which not only is enhanced by them, but also is based in part on them. Furthermore, the exercises are designed to foster comprehension. The answers involve no new concepts or techniques. The exercises are meant to be easy; if a part seems to be hard, then some review and reflection may be in order. However, all the answers are worked out in detail in Appendix 9.6.

These notes assume familiarity with the basic algebraic geometry developed in Chapters II and III of Hartshorne's popular textbook [**Har77**], and assume familiarity, but to a lesser extent, with the foundational material developed in Grothendieck and Dieudonné's monumental reference books [**EGAI**] to [**EGAG**]. In addition, these notes assume familiarity with basic Grothendieck topology, descent theory, and Hilbert-scheme theory, such as that explained on pp. 129–147, 199–201, and 215–221 in Bosch, Lütkebohmert, and Raynaud's welcome survey book [**BLR90**]; this material and more was introduced by Grothendieck in three Bourbaki talks, nos. 190, 212, and 221, which were reprinted in [**FGA**] and are still worth reading.

Of course, when specialized results are used below, precise references are provided.

Throughout these notes, we work only with locally Noetherian schemes, just as Grothendieck did in [**FGA**]. Shortly afterward, Grothendieck promoted the elimination of this restriction, through a limiting process that reduces the general case to the Noetherian case. Ever since, it has been common to make this reduction. However, the process is elementary and straightforward. Using it here would only be distracting.

Given two locally Noetherian schemes lying over a third, the (fibered) product is not necessarily locally Noetherian. Consequently, there are minor technical difficulties in working with the fpqc topology on the category of locally Noetherian schemes. On the other hand, in practice, there is no need to use the fpqc topology. Therefore, its use has been eliminated from these notes.

Throughout, we work with a **fixed** separated map of finite type

$$f\colon X \to S.$$

For convenience, when given an S-scheme T, we set

$$X_T := X \times_S T$$

and denote the projection by $f_T\colon X_T \to T$. Also, when given a T-scheme T' and given quasi-coherent sheaves $\mathcal{N}$ on T and $\mathcal{M}$ on X_T, we denote the pullback sheaves by $\mathcal{N}|\,T'$ or $\mathcal{N}_{T'}$ and by $\mathcal{M}|\,X_{T'}$ or $\mathcal{M}_{T'}$.

Given an S-scheme P, we call an S-map $T \to P$ a *T-point* of P, and we denote the set of all T-points by $P(T)$. As T varies, the sets $P(T)$ form a contravariant functor on the category of S-schemes, called the *functor of points* of P.

Section 2 introduces and compares the four common relative Picard functors, the likely candidates for the functor of points of the Picard scheme. They are simply the functor $T \mapsto \mathrm{Pic}(X_T)/\mathrm{Pic}(T)$ and its associated sheaves in the Zariski topology, the étale topology, and the fppf topology. Section 3 treats relative effective (Cartier) divisors on X/S and the relation of linear equivalence. We prove these divisors are parameterized by an open subscheme of the Hilbert scheme of X/S. Furthermore, we study the subscheme parameterizing the divisors whose fibers are linearly equivalent, and prove it is of the form $\mathbf{P}(\mathcal{Q})$ where $\mathcal{Q}$ is a certain coherent sheaf on S.

Section 4 begins the study of the Picard scheme $\mathbf{Pic}_{X/S}$ itself. Notably, we prove Grothendieck's main theorem: $\mathbf{Pic}_{X/S}$ exists when X/S is projective and flat and its geometric fibers are integral. Then we work out Mumford's example showing the necessity of the integrality hypothesis. Section 5 studies $\mathbf{Pic}^0_{X/S}$, which is the union of the connected components of the identity element of the fibers of $\mathbf{Pic}^0_{X/S}$. In particular, we compute the tangent space at the identity of each fiber. It is remarkable how much we can prove *formally* about $\mathbf{Pic}^0_{X/S}$. Section 6 proves the two deeper finiteness theorems. They concern $\mathbf{Pic}^\tau_{X/S}$, which consists of the points with a multiple in $\mathbf{Pic}^0_{X/S}$, and $\mathbf{Pic}^\varphi_{X/S}$, whose points represent the invertible sheaves with a given Hilbert polynomial φ.

Finally, there are two appendices. Appendix A contains detailed answers to all the exercises. Appendix B develops basic divisorial intersection theory, which is used freely throughout Section 6; the treatment is short, simple, and elementary.

These notes owe much to comments made on earlier drafts by Allen Altman, Ethan Cotterill, Eduardo Esteves, Rebecca Lehman, Joseph Lipman, Jean-Pierre Serre, and Angelo Vistoli. The author is very grateful.

9.2. The several Picard functors

Our first job is to identify a likely candidate for the functor of points of the Picard scheme. In fact, there are several reasonable such Picard functors, and each one is more likely to be representable than the preceding. In this section, they all are formally introduced and compared.

DEFINITION 9.2.1. The *absolute Picard functor* Pic_X is the functor from the category of (locally Noetherian) S-schemes T to the category of abelian groups defined by the formula

$$\mathrm{Pic}_X(T) := \mathrm{Pic}(X_T)$$

where $Pic(X_T)$ just means the group of isomorphism classes of invertible sheaves on X_T, that is, on $X \times_S T$.

The absolute Picard functor is a "prepared presheaf" in this sense: given any family of S-schemes T_i, we have

$$\mathrm{Pic}_X(\coprod T_i) = \prod \mathrm{Pic}_X(T_i).$$

Hence, given a covering family $\{T_i \to T\}$ in the Zariski topology, the étale topology, or any other Grothendieck topology on the category of S-schemes, there is no harm, when we consider the sheaf associated to Pic_X, in working simply with the single map $T' \to T$ where $T' := \coprod T_i$. And doing so lightens the notation, making for easier reading. Therefore, we do so throughout, calling $T' \to T$ simply a *covering* in the given topology.

The absolute Picard functor Pic_X is never a separated presheaf in the Zariski topology. Indeed, take an S-scheme T that carries an invertible sheaf $\mathcal{N}$ such that $f_T^*\mathcal{N}$ is nontrivial. (For example, take $T := \mathbf{P}^1_X$ and $\mathcal{N} := \mathcal{O}_T(1)$. Then the diagonal map $X \to X \times X$ induces a section g of f_T; that is, $gf_T = 1_T$. Hence $f_T^*\mathcal{N}$ is nontrivial.) Now, there exists a Zariski covering $T' \to T$ such that the pullback $\mathcal{N}|T'$ is trivial; here, T' is simply the disjoint union of the subsets in a suitable ordinary open covering of T. So the pullback $f_{T'}^*\mathcal{N}|X_{T'}$ is trivial too. Thus $\mathrm{Pic}_X(T)$ has a nonzero element whose restriction is zero in $\mathrm{Pic}_X(T')$.

According to descent theory, every representable functor is a sheaf in the Zariski topology — in fact, in the étale and fppf topologies as well. Therefore, the absolute Picard functor Pic_X is never representable. So, in the hope of obtaining a representable functor that differs as little as possible from Pic_X, we now define a sequence of three successively more promising relative functors.

DEFINITION 9.2.2. The *relative Picard functor* $\mathrm{Pic}_{X/S}$ is defined by

$$\mathrm{Pic}_{X/S}(T) := \mathrm{Pic}(X_T)/\,\mathrm{Pic}(T).$$

Denote its associated sheaves in the Zariski, étale, and fppf topologies by

$$\mathrm{Pic}_{(X/S)\,(\mathrm{zar})},\ \mathrm{Pic}_{(X/S)\,(\mathrm{\acute{e}t})},\ \mathrm{Pic}_{(X/S)\,(\mathrm{fppf})}\,.$$

We now have a sequence of five Picard functors, and each one maps naturally into the next. So each one maps naturally into any of its successors. If the latter functor is one of the three just displayed, then it is a sheaf in the indicated topology; in fact, it is the sheaf associated to any one of its predecessors, and the map between them is the natural map from a presheaf to its associated sheaf, as is easy to check. In particular, the three displayed sheaves are the sheaves associated to the absolute Picard functor Pic_X, as well as to the relative Picard functor $\mathrm{Pic}_{X/S}$.

Since $\mathrm{Pic}_{X/S}$ is not a priori a sheaf, it is remarkable that it is representable so often in practice.

Note that, for every S-scheme T, each T-point of $\mathrm{Pic}_{(X/S)\,(\mathrm{fppf})}$, or element of $\mathrm{Pic}_{(X/S)\,(\mathrm{fpqc})}(T)$, is represented by an invertible sheaf $\mathcal{L}'$ on $X_{T'}$ for some fppf-covering $T' \to T$. Moreover, there must be an fppf-covering $T'' \to T' \times_T T'$ such that the two pullbacks of $\mathcal{L}'$ to $X_{T''}$ are isomorphic — or to put it informally, the restrictions of $\mathcal{L}'$ must agree on a covering of the overlaps.

Furthermore, a second such sheaf $\mathcal{L}_1$ on X_{T_1} represents the same T-point if and only if there is an fppf-covering $T_1' \to T_1 \times_T T'$ such that the pullbacks of $\mathcal{L}'$ and $\mathcal{L}_1$ to $X_{T_1'}$ are isomorphic. Technically, this condition includes the preceding one, which concerns the case where $T_1 = T'$ and $\mathcal{L}_1 = \mathcal{L}'$ since $\mathcal{L}'$ must represent the same T-point as itself. Of course, similar considerations apply to the Picard functors for the Zariski and étale topologies as well.

EXERCISE 9.2.3. Given an S-scheme T of the form $T = \mathrm{Spec}(A)$ where A is a local ring, show that the natural maps are isomorphisms

$$\mathrm{Pic}_X(A) \xrightarrow{\sim} \mathrm{Pic}_{X/S}(A) \xrightarrow{\sim} \mathrm{Pic}_{(X/S)\,(\mathrm{zar})}(A)$$

where $\mathrm{Pic}_X(A) := \mathrm{Pic}_X(T)$, where $\mathrm{Pic}_{X/S}(A) := \mathrm{Pic}_{X/S}(T)$, and so forth.

Assume A is Artin local with algebraically closed residue field. Show

$$\mathrm{Pic}_X(A) \xrightarrow{\sim} \mathrm{Pic}_{(X/S)\,(\text{ét})}(A).$$

Assume A is an algebraically closed field k. Show

$$\mathrm{Pic}_X(k) \xrightarrow{\sim} \mathrm{Pic}_{(X/S)\,(\mathrm{fppf})}(k).$$

EXERCISE 9.2.4. Show that the natural map

$$\mathrm{Pic}_{(X/S)\,(\mathrm{zar})} \to \mathrm{Pic}_{(X/S)\,(\text{ét})}$$

need not be an isomorphism. Specifically, take X to be the following curve in the real projective plane:

$$X : u^2 + v^2 + w^2 = 0 \text{ in } \mathbf{P}^2_{\mathbb{R}}.$$

Then X has no $\mathbb{R}$-points, but over the complex numbers $\mathbb{C}$, there is an isomorphism

$$\varphi\colon X_{\mathbb{C}} \xrightarrow{\sim} \mathbf{P}^1_{\mathbb{C}}.$$

Show that $\varphi^*\mathcal{O}(1)$ defines an element of $\mathrm{Pic}_{(X/\mathbb{R})\,(\text{ét})}(\mathbb{R})$, which is not in the image of $\mathrm{Pic}_{(X/\mathbb{R})\,(\mathrm{zar})}(\mathbb{R})$.

The main result of this section is the following comparison theorem. It identifies two useful conditions: the first guarantees that the first three relative functors can be viewed as subfunctors of the fourth; together, the two conditions guarantee that all four functors coincide. The second condition has three successively weaker forms. Before we can prove the theorem, we must develop some theory.

THEOREM 9.2.5 (Comparison). Assume $\mathcal{O}_S \xrightarrow{\sim} f_*\mathcal{O}_X$ holds universally; that is, for any S-scheme T, the comorphism of f_T is an isomorphism, $\mathcal{O}_T \xrightarrow{\sim} f_{T*}\mathcal{O}_{X_T}$.

1. Then the natural maps are injections:

$$\mathrm{Pic}_{X/S} \hookrightarrow \mathrm{Pic}_{(X/S)\,(\mathrm{zar})} \hookrightarrow \mathrm{Pic}_{(X/S)\,(\text{ét})} \hookrightarrow \mathrm{Pic}_{(X/S)\,(\mathrm{fppf})}\,.$$

2. All three maps are isomorphisms if also f has a section; the latter two maps are isomorphisms if also f has a section locally in the Zariski topology; and the last map is an isomorphism if also f has a section locally in the étale topology.

EXERCISE 9.2.6. Assume $\mathcal{O}_S \xrightarrow{\sim} f_*\mathcal{O}_X$ holds universally. Using Theorem 9.2.5, show that its four functors have the same geometric points; in other words, for every algebraically closed field k containing the residue class field of a point of S, the k-points of all four functors are, in a natural way, the same. Show, in fact, that these k-points are just the elements of $\operatorname{Pic}(X_k)$.

What if $\mathcal{O}_S \xrightarrow{\sim} f_*\mathcal{O}_X$ does not necessarily hold universally?

LEMMA 9.2.7. Assume $\mathcal{O}_S \xrightarrow{\sim} f_*\mathcal{O}_X$. Then the functor $\mathcal{N} \mapsto f^*\mathcal{N}$ is fully-faithful from the category $\mathcal{C}$ of locally free sheaves of finite rank on S to that on X. The essential image is formed by the sheaves $\mathcal{M}$ on X such that (i) the image $f_*\mathcal{M}$ is in $\mathcal{C}$ and (ii) the natural map $f^*f_*\mathcal{M} \to \mathcal{M}$ is an isomorphism.

PROOF. (Compare with [**EGAIV4**, 21.13.2].) For any $\mathcal{N}$ in $\mathcal{C}$, there is a string of three natural isomorphisms

$$\mathcal{N} \xrightarrow{\sim} \mathcal{N} \otimes f_*\mathcal{O}_X \xrightarrow{\sim} \mathcal{N} \otimes f_*f^*\mathcal{O}_S \xrightarrow{\sim} f_*f^*\mathcal{N}. \tag{9.2.7.1}$$

The first isomorphism arises by tensor product with the comorphism of f; this comorphism is an isomorphism by hypothesis. The second isomorphism arises from the identification $\mathcal{O}_X = f^*\mathcal{O}_S$. The third arises from the projection formula.

For any $\mathcal{N}'$ in $\mathcal{C}$, also $Hom(\mathcal{N}, \mathcal{N}')$ is in $\mathcal{C}$. Hence, (9.2.7.1) yields an isomorphism

$$Hom(\mathcal{N}, \mathcal{N}') \xrightarrow{\sim} f_*f^*\, Hom(\mathcal{N}, \mathcal{N}').$$

Now, since $\mathcal{N}$ and $\mathcal{N}'$ are locally free of finite rank, the natural map

$$f^*\, Hom(\mathcal{N}, \mathcal{N}') \to Hom(f^*\mathcal{N}, f^*\mathcal{N}')$$

is an isomorphism locally, so globally. Hence there is an isomorphism of groups

$$\operatorname{Hom}(\mathcal{N}, \mathcal{N}') \xrightarrow{\sim} \operatorname{Hom}(f^*\mathcal{N}, f^*\mathcal{N}').$$

In other words, $\mathcal{N} \mapsto f^*\mathcal{N}$ is fully-faithful.

Finally, the essential image consists of the sheaves $\mathcal{M}$ that are isomorphic to those of the form $f^*\mathcal{N}$ for some $\mathcal{N}$ in $\mathcal{C}$. Given such an $\mathcal{M}$ and $\mathcal{N}$, there is an isomorphism $f_*\mathcal{M} \simeq \mathcal{N}$ owing to (9.2.7.1). Hence $f_*\mathcal{M}$ is in $\mathcal{C}$, and $f^*f_*\mathcal{M} \to \mathcal{M}$ is an isomorphism locally, so globally; thus (i) and (ii) hold. Conversely, if (i) and (ii) hold, then $\mathcal{M}$ is, by definition, in the essential image. □

PROOF OF PART 1 OF THEOREM 9.2.5. Given $\lambda \in \operatorname{Pic}_{X/S}(T)$, represent λ by an invertible sheaf $\mathcal{L}$ on X_T. Suppose λ maps to 0 in $\operatorname{Pic}_{(X/S)\,(\mathrm{fppf})}(T)$. Then there exist an fppf covering $p\colon T' \to T$ and an isomorphism $p_X^*\mathcal{L} \simeq f_{T'}^*\mathcal{N}'$ for some invertible sheaf $\mathcal{N}'$ on T'. Hence Lemma 9.2.7 implies that $f_{T'*}p_X^*\mathcal{L} \simeq \mathcal{N}'$. Now, p is flat, so $p^*f_{T*}\mathcal{L} \xrightarrow{\sim} f_{T'*}p_X^*\mathcal{L}$. So $p^*f_{T*}\mathcal{L} \simeq \mathcal{N}'$. Hence $f_{T*}\mathcal{L}$ is invertible and the natural map $f_T^*f_{T*}\mathcal{L} \to \mathcal{L}$ is an isomorphism, as both statements hold after pullback via p, which is faithfully flat. Therefore, $\lambda = 0$. Thus $\operatorname{Pic}_{X/S} \hookrightarrow \operatorname{Pic}_{(X/S)\,(\mathrm{fppf})}$.

The rest is formal. Indeed, take the last injection, and form the associated sheaves in the Zariski topology. This operation is exact by general (Grothendieck) topology, and $\operatorname{Pic}_{(X/S)\,(\mathrm{fppf})}$ remains the same, as it is already a Zariski sheaf. Thus $\operatorname{Pic}_{(X/S)\,(\mathrm{zar})} \hookrightarrow \operatorname{Pic}_{(X/S)\,(\mathrm{fppf})}$. Similarly, $\operatorname{Pic}_{(X/S)\,(\text{ét})} \hookrightarrow \operatorname{Pic}_{(X/S)\,(\mathrm{fppf})}$.

Alternatively, we can avoid the use of Lemma 9.2.7 by starting from the fact that $\operatorname{Pic}_{(X/S)\,(\mathrm{fppf})}$ is the sheaf associated to Pic_X, rather than to $\operatorname{Pic}_{X/S}$. This way, we may assume $\mathcal{N}' = \mathcal{O}_{T'}$. Then $f_{T'*}p_X^*\mathcal{L} \simeq \mathcal{N}'$ because $\mathcal{O}_S \xrightarrow{\sim} f_*\mathcal{O}_X$ holds universally. We now proceed just as before. □

DEFINITION 9.2.8. Assume f has a section g, so $fg = 1$. Let T be an S-scheme, and $\mathcal{L}$ a sheaf on X_T. Then a *g-rigidification* of $\mathcal{L}$ is the choice of an isomorphism $u\colon \mathcal{O}_T \xrightarrow{\sim} g_T^*\mathcal{L}$, assuming one exists.

LEMMA 9.2.9. *Assume f has a section g, and let T be an S-scheme. Form the group of isomorphism classes of pairs $(\mathcal{L}, u)$ where $\mathcal{L}$ is an invertible sheaf on X_T and u is a g-rigidification of $\mathcal{L}$. Then this group is carried isomorphically onto $\mathrm{Pic}_{X/S}(T)$ by the homomorphism ρ defined by $\rho(\mathcal{L}, u) := \mathcal{L}$.*

PROOF. Given λ in $\mathrm{Pic}_{X/S}(T)$, represent λ by an invertible sheaf $\mathcal{M}$ on X_T. Set $\mathcal{L} := \mathcal{M} \otimes (f_T^* g_T^* \mathcal{M})^{-1}$. Then $\mathcal{L}$ too represents λ. Also $g_T^*\mathcal{L} = g_T^*\mathcal{M} \otimes g_T^* f_T^* g_T^* \mathcal{M}^{-1}$. Now, $g_T^* f_T^* = 1$ as $fg = 1$. Hence the natural isomorphism $g_T^*\mathcal{M} \otimes (g_T^*\mathcal{M})^{-1} \xrightarrow{\sim} \mathcal{O}_T$ induces a g-rigidification of $\mathcal{L}$. Thus ρ is surjective.

To prove ρ is injective, let $(\mathcal{L}, u)$ represent an element of its kernel. Then there exist an invertible sheaf $\mathcal{N}$ on T and an isomorphism $v\colon \mathcal{L} \xrightarrow{\sim} f_T^*\mathcal{N}$. Set $w := g_T^* v \circ u$, so $w\colon \mathcal{O}_T \xrightarrow{\sim} g_T^*\mathcal{L} \xrightarrow{\sim} \mathcal{N}$. Now, a map of pairs is just a map w' of the first components such that $g_T^* w'$ is compatible with the two g-rigidifications. So $v\colon (\mathcal{L}, u) \xrightarrow{\sim} (f_T^*\mathcal{N}, w)$ and $f_T^* w\colon (\mathcal{O}_{X_T}, 1) \xrightarrow{\sim} (f_T^*\mathcal{N}, w)$. Thus ρ is injective. □

LEMMA 9.2.10. *Assume f has a section g, and assume $\mathcal{O}_S \xrightarrow{\sim} f_*\mathcal{O}_X$ holds universally. Let T be an S-scheme, $\mathcal{L}$ an invertible sheaf on X_T, and u a g-rigidification of $\mathcal{L}$. Then every automorphism of the pair $(\mathcal{L}, u)$ is trivial.*

PROOF. An automorphism of $(\mathcal{L}, u)$ is just an automorphism $v\colon \mathcal{L} \xrightarrow{\sim} \mathcal{L}$ such that $g_T^* v \circ u\colon \mathcal{O}_T \xrightarrow{\sim} g_T^*\mathcal{L} \xrightarrow{\sim} g_T^*\mathcal{L}$ is equal to u. But then $g_T^* v = 1$. Now,

$$v \in \mathrm{Hom}(\mathcal{L}, \mathcal{L}) = \mathrm{H}^0(\mathcal{H}om(\mathcal{L}, \mathcal{L})) = \mathrm{H}^0(\mathcal{O}_{X_T}) = \mathrm{H}^0(\mathcal{O}_T);$$

the middle equation holds since the natural map $\mathcal{O}_{X_T} \to \mathcal{H}om(\mathcal{L}, \mathcal{L})$ is locally an isomorphism, so globally one, and the last equation holds since $\mathcal{O}_T \xrightarrow{\sim} f_*\mathcal{O}_{X_T}$. But $g_T^* v = 1$. Therefore, $v = 1$. □

PROOF OF PART 2 OF THEOREM 9.2.5. Suppose f has a section g. Owing to Part 1, it suffices to prove that every $\lambda \in \mathrm{Pic}_{(X/S)\,(\mathrm{fppf})}(T)$ lies in $\mathrm{Pic}_{X/S}(T)$. Represent λ by a $\lambda' \in \mathrm{Pic}_{X/S}(T')$ where $T' \to T$ is an fppf covering. Then there is an fppf covering $T'' \to T' \times_T T'$ such that the two pullbacks of λ' to $X_{T''}$ are equal. We may assume $T'' \xrightarrow{\sim} T' \times_T T'$ because $\mathrm{Pic}_{X/S}$ is separated for the fppf topology, again owing to Part 1.

Owing to Lemma 9.2.9, we may represent λ' by a pair $(\mathcal{L}', u')$ where $\mathcal{L}'$ is an invertible sheaf on $X_{T'}$ and u' is a g-rigidification of $\mathcal{L}'$. Furthermore, on $X_{T' \times T'}$, there is an isomorphism v' from the pullback of $(\mathcal{L}', u')$ via the first projection onto the pullback via the second.

Consider the three projections $X_{T' \times T' \times T'} \to X_{T' \times T'}$. Let v'_{ij} denote the pullback of v' via the projection to the ith and jth factors. Then $v_{13}'^{-1} v'_{23} v'_{12}$ is an automorphism of the pullback of $(\mathcal{L}', u')$ via the first projection $X_{T' \times T' \times T'} \to X_{T'}$. So, owing to Lemma 9.2.10, this automorphism is trivial. Hence $(\mathcal{L}', u')$ descends to a pair $(\mathcal{L}, u)$ on X_T. Therefore, λ lies in $\mathrm{Pic}_{X/S}(T)$.

The rest is formal. Indeed, suppose that there is a Zariski covering $T' \to T$ such that $f_{T'}$ has a section. Then, by the above, $\mathrm{Pic}_{(X/S)\,(\mathrm{zar})}\,|T' \xrightarrow{\sim} \mathrm{Pic}_{(X/S)\,(\mathrm{fppf})}\,|T'$. Hence, by general (Grothendieck) topology, $\mathrm{Pic}_{(X/S)\,(\mathrm{zar})} \xrightarrow{\sim} \mathrm{Pic}_{(X/S)\,(\mathrm{fppf})}$ since both source and target are sheaves in the Zariski topology and a map of sheaves is an isomorphism if it is so locally. Similarly, $\mathrm{Pic}_{(X/S)\,(\text{ét})} \xrightarrow{\sim} \mathrm{Pic}_{(X/S)\,(\text{fppf})}$ if f has a section locally in the étale topology. □

REMARK 9.2.11. There is another way to prove Theorem 9.2.5. This way is more sophisticated, and yields more information, which we won't need. Here is the idea.

Recall [**Har77**, Ex. III, 4.5, p. 224] that, for any ringed space R, there is a natural isomorphism

$$\mathrm{Pic}(R) = \mathrm{H}^1(R, \mathcal{O}_R^*). \tag{9.2.11.1}$$

Now, given any S-scheme T, form the presheaf $T' \mapsto \mathrm{H}^1(X_{T'}, \mathcal{O}_{X_{T'}}^*)$ on T. Its associated sheaf is [**Har77**, Prp. III, 8.1, p. 250] simply $\mathrm{R}^1 f_{T*}\mathcal{O}_{X_T}^*$. Therefore,

$$\mathrm{Pic}_{(X/S)\,(\mathrm{zar})}(T) = \mathrm{H}^0(T, \mathrm{R}^1 f_{T*}\mathcal{O}_{X_T}^*). \tag{9.2.11.2}$$

Consider the Leray spectral sequence [**God58**, Thm. II, 4.17.1, p. 201]

$$\mathrm{E}_2^{pq} := \mathrm{H}^p(T, \mathrm{R}^q f_{T*}\mathcal{O}_{X_T}^*) \Longrightarrow \mathrm{H}^{p+q}(X_T, \mathcal{O}_{X_T}^*),$$

and form its exact sequence of terms of low degree [**God58**, Thm. I, 4.5.1, p. 82]

$$\begin{aligned} 0 \to \mathrm{H}^1(T, f_{T*}\mathcal{O}_{X_T}^*) \to \mathrm{H}^1(X_T, \mathcal{O}_{X_T}^*) \to \mathrm{H}^0(T, \mathrm{R}^1 f_{T*}\mathcal{O}_{X_T}^*) \\ \to \mathrm{H}^2(T, f_{T*}\mathcal{O}_{X_T}^*) \to \mathrm{H}^2(X_T, \mathcal{O}_{X_T}^*). \end{aligned} \tag{9.2.11.3}$$

If $\mathcal{O}_S \xrightarrow{\sim} f_*\mathcal{O}_X$ holds universally, then $\mathrm{H}^1(T, \mathcal{O}_T^*) \xrightarrow{\sim} \mathrm{H}^1(T, f_{T*}\mathcal{O}_{X_T}^*)$. Hence the beginning of (9.2.11.3) becomes

$$0 \to \mathrm{Pic}(T) \to \mathrm{Pic}(X_T) \to \mathrm{Pic}_{(X/S)\,(\mathrm{zar})}(T).$$

Thus $\mathrm{Pic}_{X/S} \hookrightarrow \mathrm{Pic}_{(X/S)\,(\mathrm{zar})}$.

If also f has a section g, then g induces, for each p, a left inverse of the map $\mathrm{H}^p(T, f_{T*}\mathcal{O}_{X_T}^*) \to \mathrm{H}^p(X_T, \mathcal{O}_{X_T}^*)$ induced by f. So the latter is injective. Hence, $\mathrm{H}^1(X_T, \mathcal{O}_{X_T}^*) \to \mathrm{H}^0(T, \mathrm{R}^1 f_{T*}\mathcal{O}_{X_T}^*)$ is surjective. Thus $\mathrm{Pic}_{X/S} \xrightarrow{\sim} \mathrm{Pic}_{(X/S)\,(\mathrm{zar})}$.

The preceding argument works for the étale and fppf topologies too with little change. First of all, consider the functor

$$\mathbb{G}_{\mathrm{m}}(T) := \mathrm{H}^0(T, \mathcal{O}_T^*).$$

Let u be an indeterminate. Then $\mathbb{G}_{\mathrm{m}}(T)$ is representable by the S-scheme

$$\mathbb{G}_{\mathrm{m}} := \mathrm{Spec}(\mathcal{O}_S[u, u^{-1}]).$$

Indeed, giving an S-map $T \to \mathbb{G}_{\mathrm{m}}$ is the same as giving an $\mathrm{H}^0(S, \mathcal{O}_S)$-homomorphism from $\mathrm{H}^0(S, \mathcal{O}_S)[u, u^{-1}]$ to $\mathrm{H}^0(T, \mathcal{O}_T)$, and giving such a homomorphism is the same as assigning to u a unit in $\mathrm{H}^0(T, \mathcal{O}_T)$. Now, since $\mathbb{G}_{\mathrm{m}}(T)$ is representable, it is a sheaf for the étale and fppf topologies.

Grothendieck's generalization [**FGA**, p. 190-16] of Hilbert's Theorem 90 asserts the formula

$$\mathrm{Pic}(T) = \mathrm{H}^1(T, \mathbb{G}_{\mathrm{m}})$$

where the H^1 can be computed in either the étale or fppf topology. The proof is simple, and similar to the proof of (9.2.11.1). The H^1 can be computed as a Čech group. And, essentially by definition, for a covering $T' \to T$, a Čech cocycle with values in $\mathbb{G}_{\mathrm{m}}$ amounts to descent data on $\mathcal{O}_{T'}$. The data is effective by descent theory, and the resulting sheaf on T is invertible since the covering is faithfully flat.

In the present context, the exact sequence (9.2.11.3) becomes

$$(9.2.11.4)\quad 0 \to \mathrm{H}^1(T, f_{T*}\mathbb{G}_m) \to \mathrm{H}^1(X_T, \mathbb{G}_m) \to \mathrm{H}^0(T, \mathrm{R}^1 f_{T*}\mathbb{G}_m) \to \mathrm{H}^2(T, f_{T*}\mathbb{G}_m) \to \mathrm{H}^2(X_T, \mathbb{G}_m).$$

Furthermore, the proof of (9.2.11.2) yields, for example in the fppf topology,

$$\mathrm{Pic}_{(X/S)\,(\mathrm{fppf})}(T) = \mathrm{H}^0(T, \mathrm{R}^1 f_{T*}\mathbb{G}_m).$$

If $\mathcal{O}_S \xrightarrow{\sim} f_*\mathcal{O}_X$ holds universally, then it follows from the definitions that $f_{T*}\mathbb{G}_m = \mathbb{G}_m$. Hence the beginning of (9.2.11.4) becomes

$$0 \to \mathrm{Pic}(T) \to \mathrm{Pic}(X_T) \to \mathrm{Pic}_{(X/S)\,(\mathrm{fppf})}(T).$$

Thus $\mathrm{Pic}_{X/S} \hookrightarrow \mathrm{Pic}_{(X/S)\,(\mathrm{fppf})}$. And $\mathrm{Pic}_{X/S} \xrightarrow{\sim} \mathrm{Pic}_{(X/S)\,(\mathrm{fppf})}$ if also f has a section. As before, the rest of Theorem 9.2.5 follows formally.

The étale group $\mathrm{H}^2(T, \mathbb{G}_m)$ was studied extensively by Grothendieck [**DIX68**, pp. 46–188]. He showed that it gives one of two significant generalizations of the Brauer group of central simple algebras over a field. The other is the group of Azumaya algebras on T. He denoted the latter by $\mathrm{Br}(T)$ and the former by $\mathrm{Br}'(T)$. Hence [**DIX68**, pp. 127–128], if $\mathcal{O}_S \xrightarrow{\sim} f_*\mathcal{O}_X$ holds universally, then (9.2.11.4) becomes

$$(9.2.11.5)\qquad 0 \to \mathrm{Pic}(T) \to \mathrm{Pic}(X_T) \to \mathrm{Pic}_{(X/S)\,(\text{ét})}(T) \to \mathrm{Br}'(T) \to \mathrm{Br}'(X_T);$$

in particular, the obstruction to representing an element of $\mathrm{Pic}_{(X/S)\,(\text{ét})}(T)$ by an invertible sheaf on X_T is given by an element of $\mathrm{Br}'(T)$, which maps to 0 in $\mathrm{Br}'(X_T)$.

Using the smoothness of $\mathbb{G}_m$ as an S-scheme, Grothendieck [**DIX68**, p. 180] proved that the natural homomorphisms are isomorphisms from the étale groups $\mathrm{H}^p(T, \mathbb{G}_m)$ to the corresponding fppf groups. Hence, if $\mathcal{O}_S \xrightarrow{\sim} f_*\mathcal{O}_X$ holds universally, then it follows from (9.2.11.4) via the Five Lemma that, whether f has a section locally in the étale topology or not,

$$\mathrm{Pic}_{(X/S)\,(\text{ét})} \xrightarrow{\sim} \mathrm{Pic}_{(X/S)\,(\mathrm{fppf})}\,.$$

Nevertheless, when a discussion is set in the greatest possible generality, it is common to work with $\mathrm{Pic}_{(X/S)\,(\mathrm{fppf})}$ and call it *the* Picard functor.

9.3. Relative effective divisors

Grothendieck constructed the Picard scheme by taking a suitable family of effective divisors and forming the quotient modulo linear equivalence. This section develops the basic theory of these notions.

9.3.1 (Effective divisors). A closed subscheme $D \subset X$ is called an *effective (Cartier) divisor* if its ideal $\mathcal{I}$ is invertible. Given an $\mathcal{O}_X$-module $\mathcal{F}$ and $n \in \mathbb{Z}$, set

$$\mathcal{F}(nD) := \mathcal{F} \otimes \mathcal{I}^{\otimes -n}.$$

In particular, $\mathcal{O}_X(-D) = \mathcal{I}$. So the inclusion $\mathcal{I} \hookrightarrow \mathcal{O}_X$ yields, via tensor product with $\mathcal{O}_X(D)$, an injection $\mathcal{O}_X \hookrightarrow \mathcal{O}_X(D)$, which, in turn, corresponds to a global section of $\mathcal{O}_X(D)$. This section is not arbitrary since it corresponds to an injection. Sections corresponding to injections are termed *regular.*

Conversely, given an arbitrary invertible sheaf $\mathcal{L}$ on X, let $\mathrm{H}^0(X, \mathcal{L})_{\mathrm{reg}}$ denote the subset of $\mathrm{H}^0(X, \mathcal{L})$ consisting of the regular sections, those corresponding to injections $\mathcal{L}^{-1} \hookrightarrow \mathcal{O}_X$. And let $|\mathcal{L}|$ denote the set of effective divisors D such that $\mathcal{O}_X(D)$ is, in some way, isomorphic to $\mathcal{L}$. For historical reasons, $|\mathcal{L}|$ is called the *complete linear system* associated to $\mathcal{L}$ (but $|\mathcal{L}|$ needn't be a $\mathbf{P}^n$ if X isn't integral).

EXERCISE 9.3.2. Under the conditions of (9.3.1), establish a canonical isomorphism

$$\mathrm{H}^0(X,\mathcal{L})_{\mathrm{reg}}/\mathrm{H}^0(X,\mathcal{O}_X^*) \xrightarrow{\sim} |\mathcal{L}|.$$

DEFINITION 9.3.3. A *relative effective divisor* on X/S is an effective divisor $D \subset X$ that is S-flat.

LEMMA 9.3.4. Let $D \subset X$ be a closed subscheme, $x \in D$ a point, and $s \in S$ its image. Then the following statements are equivalent:

(i) The subscheme $D \subset X$ is a relative effective divisor at x (that is, in a neighborhood of x).
(ii) The schemes X and D are S-flat at x, and the fiber D_s is an effective divisor on X_s at x.
(iii) The scheme X is S-flat at x, and the subscheme $D \subset X$ is cut out at x by one element that is regular (a nonzerodivisor) on the fiber X_s.

PROOF. For convenience, set $A := \mathcal{O}_{S,s}$ and denote its residue field by k. In addition, set $B := \mathcal{O}_{X,x}$ and $C := \mathcal{O}_{D,x}$. Then $B \otimes_A k = \mathcal{O}_{X_s,x}$.

Assume (i), and let's prove (ii). By hypothesis, D is an effective divisor at x. So there is a regular element $b \in B$ that generates the ideal of D. Multiplication by b defines a short exact sequence

$$0 \to B \to B \to C \to 0.$$

In turn, this sequence induces the following exact sequence:

$$\mathrm{Tor}_1^A(B,k) \to \mathrm{Tor}_1^A(B,k) \to \mathrm{Tor}_1^A(C,k) \to B \otimes k \to B \otimes k.$$

By hypothesis, D is S-flat at x; hence, $\mathrm{Tor}_1^A(C,k) = 0$. So $B \otimes k \to B \otimes k$ is injective. Its image is the ideal of D_s. Thus D_s is an effective divisor.

Since $\mathrm{Tor}_1^A(C,k) = 0$, the map $\mathrm{Tor}_1^A(B,k) \to \mathrm{Tor}_1^A(B,k)$ is surjective. This map is given by multiplication by b, and b lies in the maximal ideal of B. Also, $\mathrm{Tor}_1^A(B,k)$ is a finitely generated B-module. Hence, $\mathrm{Tor}_1^A(B,k) = 0$ by Nakayama's lemma. Therefore, by the local criterion of flatness [**SGA1**, Thm. 5.6, p. 98] or [**OB72**, Thm. 6.1, p. 73], B is A-flat; in other words, X is S-flat at x. Thus (ii) holds.

Assume (ii). To prove (iii), denote the ideal of D in B by I, and that of D_s in $B \otimes k$ by I'. Take an element $b \in I$ whose image b' in $B \otimes k$ generates I'; such a b exists because D_s is an effective divisor at x by hypothesis. For the same reason, b' is regular. It remains to prove b generates I.

Consider the short exact sequence

$$0 \to I \to B \to C \to 0.$$

By hypothesis, C is A-flat. Hence the map $I \otimes k \to B \otimes k$ is injective. Its image is I', which is generated by b'. So the image of b in $I \otimes k$ generates it. Hence, by Nakayama's lemma, b generates I. Thus (iii) holds.

Assume (iii). To prove (i), again denote the ideal of D in B by I. By hypothesis, I is generated by an element b whose image b' in $B \otimes k$ is regular. We have to prove b is regular and C is A-flat.

The exact sequence $0 \to I \to B \to C \to 0$ yields this one:

$$\operatorname{Tor}_1^A(B,k) \to \operatorname{Tor}_1^A(C,k) \to I \otimes k \to B \otimes k. \tag{9.3.4.1}$$

The last map is injective for the following reason. Since $I = Bb$, multiplication by b induces a surjection $B \to I$, so a surjection $B \otimes k \to I \otimes k$. Consider the composition

$$B \otimes k \to I \otimes k \to B \otimes k.$$

It is given by multiplication by b', so is injective because b' is regular. Hence $B \otimes k \xrightarrow{\sim} I \otimes k$. Therefore, $I \otimes k \to B \otimes k$ is injective.

By hypothesis, B is A-flat. So $\operatorname{Tor}_1^A(B,k) = 0$. Hence the exactness of (9.3.4.1) implies $\operatorname{Tor}_1^A(C,k) = 0$. Therefore, by the local criterion, C is A-flat.

Since B and C are A-flat and $0 \to I \to B \to C \to 0$ is exact, also I is A-flat.

Define K by the exact sequence $0 \to K \to B \to I \to 0$. Then the sequence

$$0 \to K \otimes k \to B \otimes k \to I \otimes k \to 0$$

is exact since I is A-flat. But $B \otimes k \xrightarrow{\sim} I \otimes k$. Hence $K \otimes k = 0$. Therefore, $K = 0$ by Nakayama's lemma. But K is the kernel of multiplication by b on B. So b is regular. Thus (i) holds. □

EXERCISE 9.3.5. Let D and E be relative effective divisors on X/S, and $D+E$ their sum. Show $D+E$ is a relative effective divisor too.

DEFINITION 9.3.6. Define a functor $\operatorname{Div}_{X/S}$ by the formula

$$\operatorname{Div}_{X/S}(T) := \{\text{relative effective divisors } D \text{ on } X_T/T\}.$$

Note $\operatorname{Div}_{X/S}$ is indeed a functor. Namely, given a relative effective divisor D on X_T/T and an arbitrary S-map $p\colon T' \to T$, we have to see the T'-flat closed subscheme $D_{T'} \subset X_{T'}$ is an effective divisor. So let $\mathcal{I}$ denote the ideal of D. Since D is T-flat, $p_{X_T}^*\mathcal{I}$ is equal to the ideal of $D_{T'}$. But, since $\mathcal{I}$ is invertible, so is $p_{X_T}^*\mathcal{I}$. Thus $D_{T'}$ is a (relative) effective divisor.

THEOREM 9.3.7. *Assume X/S is projective and flat. Then $\operatorname{Div}_{X/S}$ is representable by an open subscheme $\mathbf{Div}_{X/S}$ of the Hilbert scheme $\mathbf{Hilb}_{X/S}$.*

PROOF. Set $H := \mathbf{Hilb}_{X/S}$, and let $W \subset X \times H$ be the universal (closed) subscheme, and $q\colon W \to H$ the projection. Let V denote the set of points $w \in W$ at which W is an effective divisor. Plainly V is open in W. Set $Z := q(W - V)$. Then Z is closed because q is proper. Set $U := H - Z$. Then U is open, and $q^{-1}U$ is an effective divisor in $X \times U$. In fact, since q is flat, $q^{-1}U$ is a relative effective divisor in $X \times U/U$.

It remains to show that U represents $\operatorname{Div}_{X/S}$. So let T be an S-scheme, and $D \subset X_T/T$ a relative effective divisor. By the universal property of the pair (H,W), there exists a unique map $g\colon T \to H$ such that $g_X^{-1}W = D$. We have to show that g factors through U.

For each $t \in T$, the fiber D_t is an effective divisor as it is obtained by base change (or owing to Lemma 9.3.4). But $D_t = W_{g(t)} \otimes k_t$ where k_t is the residue field of t. So $W_{g(t)}$ too is a divisor, as a field extension is faithfully flat. But $X \times H$ and W are H-flat. Hence W is, by Lemma 9.3.4, a relative effective divisor along the fiber over $g(t)$. So $g(t) \in U$. But U is open. Hence g factors through U. □

EXERCISE 9.3.8. Assume $f\colon X \to S$ is flat and is projective Zariski locally over S. Assume its fibers are curves, that is, of *pure* dimension 1. Given $m \geq 1$, let $\mathrm{Div}^m_{X/S}$ be the functor whose T-points are the relative effective divisors D on X_T/T with fibers D_t of degree m.

Show the $\mathrm{Div}^m_{X/S}$ are representable by open and closed subschemes of finite type $\mathbf{Div}^m_{X/S} \subset \mathbf{Div}_{X/S}$, which are disjoint and cover.

Let $X_0 \subset X$ be the subscheme where X/S is smooth. Show $X_0 = \mathbf{Div}^1_{X/S}$.

Let X_0^m be the m-fold S-fibered product. Show there is a natural S-map

$$\alpha\colon X_0^m \to \mathbf{Div}^m_{X/S},$$

which is given on T-points by $\alpha(\Gamma_1, \ldots, \Gamma_m) = \sum \Gamma_i$.

REMARK 9.3.9. Consider the map α of Exercise 9.3.8. Plainly α is compatible with permuting the factors of X_0^m. Hence α factors through the symmetric product $X_0^{(m)}$. In fact, α induces an isomorphism

$$X_0^{(m)} \xrightarrow{\sim} \mathbf{Div}^m_{X_0/S}.$$

This isomorphism is treated in detail in [**Del73**, Prp. 6.3.9, p. 437] and in outline in [**BLR90**, Prp. 3, p. 254].

9.3.10 (The module $\mathcal{Q}$). Assume $f\colon X \to S$ is proper, and let $\mathcal{F}$ be a coherent $\mathcal{O}_X$-module flat over S. Recall from [**EGAIII2**, 7.7.6] that there exist a coherent $\mathcal{O}_S$-module $\mathcal{Q}$ and an isomorphism of functors in the quasi-coherent $\mathcal{O}_S$-module $\mathcal{N}$:

$$q\colon \mathcal{H}om(\mathcal{Q}, \mathcal{N}) \xrightarrow{\sim} f_*(\mathcal{F} \otimes f^*\mathcal{N}). \tag{9.3.10.1}$$

The pair $(\mathcal{Q}, q)$ is unique, up to unique isomorphism, and by [**EGAIII2**, 7.7.9], forming it commutes with changing the base, in particular, with localizing.

Fix $s \in S$ and assume $S = \mathrm{Spec}(\mathcal{O}_{S,s})$. Note that the following conditions are equivalent:

(i) the $\mathcal{O}_S$-module $\mathcal{Q}$ is free (or equivalently, projective);
(ii) the functor $\mathcal{N} \mapsto f_*(\mathcal{F} \otimes f^*\mathcal{N})$ is right exact;
(iii) for all $\mathcal{N}$, the natural map is an isomorphism, $f_*(\mathcal{F}) \otimes \mathcal{N} \xrightarrow{\sim} f_*(\mathcal{F} \otimes f^*\mathcal{N})$;
(iv) the natural map is a surjection, $\mathrm{H}^0(X, \mathcal{F}) \otimes k_s \twoheadrightarrow \mathrm{H}^0(X_s, \mathcal{F}_s)$.

Indeed, the equivalence of (i)–(iii) is elementary and straightforward. Moreover, (iv) is a special case of (iii). Conversely, (iv) implies (iii) by [**EGAIII2**, 7.7.10] or [**OB72**, Cor. 5.1 p. 72]; this useful implication is known as the "property of exchange."

In addition, (i)–(iv) are implied by the following condition:

(v) the first cohomology group of the fiber vanishes, $\mathrm{H}^1(X_s, \mathcal{F}_s) = 0$.

Indeed, (v) implies that $\mathrm{R}^1 f_*(\mathcal{F} \otimes f^*\mathcal{N}) = 0$ for all $\mathcal{N}$ by [**EGAIII2**, 7.5.3] or [**OB72**, Cor. 2.1 p. 68]; in turn, this vanishing implies (ii) owing to the long exact sequence of higher direct images.

EXERCISE 9.3.11. Assume $f\colon X \to S$ is proper and flat, and its geometric fibers are reduced and connected. Show $\mathcal{O}_S \xrightarrow{\sim} f_*\mathcal{O}_X$ holds universally.

DEFINITION 9.3.12. Let $\mathcal{L}$ be an invertible sheaf on X. Define a subfunctor $\mathrm{LinSys}_{\mathcal{L}/X/S}$ of $\mathrm{Div}_{X/S}$ by the formula

$$\mathrm{LinSys}_{\mathcal{L}/X/S}(T) := \{\text{relative effective divisors } D \text{ on } X_T/T \text{ such that } \mathcal{O}_{X_T}(D) \simeq \mathcal{L}_T \otimes f_T^*\mathcal{N} \text{ for some invertible sheaf } \mathcal{N} \text{ on } T\}.$$

Notice the similarity of this definition with that, Definition 9.2.2, of $\mathrm{Pic}_{X/S}$: both definitions work with isomorphism classes of invertible sheaves on X_T modulo $\mathrm{Pic}(T)$, in the hope of producing a representable functor. Here, this hope is fulfilled under suitable hypotheses on f, according to the next theorem.

THEOREM 9.3.13. *Assume X/S is proper and flat, and its geometric fibers are integral. Let $\mathcal{L}$ be an invertible sheaf on X, and let $\mathcal{Q}$ be the $\mathcal{O}_S$-module associated in Subsection 9.3.10 to $\mathcal{F} := \mathcal{L}$. Set $L := \mathbf{P}(\mathcal{Q})$. Then L represents $\mathrm{LinSys}_{\mathcal{L}/X/S}$.*

PROOF. Let $D \in \mathrm{LinSys}_{\mathcal{L}/X/S}(T)$. Say $\mathcal{O}_{X_T}(D) \simeq \mathcal{L}_T \otimes f_T^*\mathcal{N}$. Then $\mathcal{N}$ is determined up to isomorphism. Indeed, let $\mathcal{N}'$ be a second choice. Then

$$\mathcal{L}_T \otimes f_T^*\mathcal{N} \simeq \mathcal{L}_T \otimes f_T^*\mathcal{N}'.$$

So $f_T^*\mathcal{N} \simeq f_T^*\mathcal{N}'$ since $\mathcal{L}$ is invertible. Hence $\mathcal{N} \simeq \mathcal{N}'$ by Lemma 9.2.7, which applies to $f_T \colon X_T \to T$ since $\mathcal{O}_S \xrightarrow{\sim} f_*\mathcal{O}_X$ holds universally by Exercise 9.3.11.

Say D is defined by $\sigma \in \mathrm{H}^0(X_T, \mathcal{L}_T \otimes f_T^*\mathcal{N})$. Now, forming $\mathcal{Q}$ commutes with changing the base to T, and so (9.3.10.1) becomes

$$\mathcal{H}om(\mathcal{Q}_T, \mathcal{N}) \xrightarrow{\sim} f_{T*}(\mathcal{L}_T \otimes f_T^*\mathcal{N}). \tag{9.3.13.1}$$

Hence σ corresponds to a map $u \colon \mathcal{Q}_T \to \mathcal{N}$.

Let $t \in T$. Since D is a relative effective divisor on X_T/T, its fiber D_t is a divisor on X_t by Lemma 9.3.4. Since D_t is defined by $\sigma_t \in \mathrm{H}^0(X_t, \mathcal{L}|X_t)$, necessarily $\sigma_t \neq 0$. But σ_t corresponds to $u \otimes k_t$, so $u \otimes k_t \neq 0$. Now, $\mathcal{N}$ is invertible, so $\mathcal{N} \otimes k_t$ is a k_t-vector space of dimension 1. So $u \otimes k_t$ is surjective. Hence, by Nakayama's lemma, u is surjective at t. But t is arbitrary. So u is surjective everywhere.

Therefore, $u \colon \mathcal{Q}_T \to \mathcal{N}$ defines an S-map $p \colon T \to L$ by [**EGAII**, 4.2.3]. Since $(\mathcal{N}, u)$ is determined up to isomorphism, a second choice yields the same p.

Plainly, this construction is functorial in T. Thus we obtain a map of functors,

$$\Lambda \colon \mathrm{LinSys}_{\mathcal{L}/X/S}(T) \to L(T).$$

Let us prove Λ is an isomorphism.

Let $p \in L(T)$, so $p \colon T \to L$ is an S-map. Then p arises from a surjection $u \colon \mathcal{Q}_T \twoheadrightarrow \mathcal{N}$; namely, $u = p^*\alpha$ where $\alpha \colon \mathcal{Q}_L \twoheadrightarrow \mathcal{O}(1)$ is the tautological map. Moreover, there is only one such pair $(\mathcal{N}, u)$ up to isomorphism.

Via the isomorphism in (9.3.13.1), the surjection u corresponds to a global section $\sigma \in \mathrm{H}^0(X_T, \mathcal{L}_T \otimes f_T^*\mathcal{N})$. Let $t \in T$. Then $u \otimes k_t$ is surjective, so $u \otimes k_t \neq 0$. But $u \otimes k_t$ corresponds to $\sigma_t \in \mathrm{H}^0(X_t, \mathcal{L}|X_t)$, so $\sigma_t \neq 0$. But X_t is integral since the geometric fibers of X/S are integral by hypothesis. Hence σ_t is regular.

The section σ defines a map $(\mathcal{L}_T \otimes f_T^*\mathcal{N})^{-1} \to \mathcal{O}_{X_T}$. Its image is the ideal of a closed subscheme $D \subset X$, which is cut out locally by one element; moreover, on the fiber X_t, this element corresponds to σ_t, so is regular. Hence D is a relative effective divisor on X_T/T by Lemma 9.3.4. In fact, $D \in \mathrm{LinSys}_{\mathcal{L}/X/S}(T)$. Plainly, D is the only such divisor corresponding to $(\mathcal{N}, u)$, so mapping to p under Λ.

Thus Λ is an isomorphism. In other words, L represents $\mathrm{LinSys}_{\mathcal{L}/X/S}$. □

EXERCISE 9.3.14. Under the conditions of Theorem 9.3.13, show that there exists a natural relative effective divisor W on X_L/L such that

$$\mathcal{O}_{X_L}(W) = \mathcal{L}_L \otimes f_L^*\mathcal{O}_L(1).$$

Furthermore, W possesses the following universal property: given any S-scheme T and any relative effective divisor D on X_T/T such that $\mathcal{O}_{X_T}(D) \simeq \mathcal{L}_T \otimes f_T^*\mathcal{N}$ for some invertible sheaf $\mathcal{N}$ on T, there exist a unique S-map $w\colon T \to L$ such that $(1 \times w)^{-1}W = D$.

9.4. The Picard scheme

This section proves Grothendieck's main theorem about the Picard scheme, which asserts its existence if X/S is projective and flat and its geometric fibers are integral; in fact, the functor $\mathrm{Pic}_{(X/S)\,(\text{ét})}$ is representable. The proof involves Grothendieck's method of using functors to prescribe patching. The basic theory is developed in [**EGAG**, Ch. 0, Sct. 4.5, pp. 102–107], and is applied in [**EGAG**, Ch. 1, Sct. 9, pp. 354–401] to the construction of Grassmannians and related parameter schemes. The present construction involves the basic theory, but is more sophisticated and more complicated because it works not simply with the Zariski topology, but also with the étale topology.

DEFINITION 9.4.1. If any of the four relative Picard functors of Definition 9.2.2 is representable, then the representing scheme is called the *Picard scheme* and denoted by $\mathbf{Pic}_{X/S}$. Moreover, we say simply that the Picard scheme $\mathbf{Pic}_{X/S}$ exists.

Notice that, although there are four relative Picard functors, there is at most one Picard scheme. Of course, if any functor is representable, then the representing scheme is uniquely determined, up to a unique isomorphism that preserves the identification of the given functor with the functor of points of the representing scheme. But here, there is more to the story.

Indeed, for example, say $\mathrm{Pic}_{(X/S)\,(\text{ét})}$ is representable by $\mathbf{Pic}_{X/S}$. Then, by descent theory, $\mathrm{Pic}_{(X/S)\,(\text{ét})}$ is already a sheaf in the fppf topology; so it is equal to its associated sheaf $\mathrm{Pic}_{(X/S)\,(\text{fppf})}$. Hence, $\mathrm{Pic}_{(X/S)\,(\text{fppf})}$ too is representable by $\mathbf{Pic}_{X/S}$. On the other hand, $\mathrm{Pic}_{X/S}$ may or may not be representable; however, if it is, then it, as well, must be representable by $\mathbf{Pic}_{X/S}$.

EXERCISE 9.4.2. Assume $\mathbf{Pic}_{X/S}$ exists, and $\mathcal{O}_S \xrightarrow{\sim} f_*\mathcal{O}_X$ holds universally. Let T be an S-scheme, and $\mathcal{L}$ an invertible sheaf on X_T. Show that there exist a subscheme $N \subset T$ and an invertible sheaf $\mathcal{N}$ on N with these three properties: first, $\mathcal{L}_N \simeq f_N^*\mathcal{N}$; second, given any S-map $t\colon T' \to T$ such that $\mathcal{L}_{T'} \simeq f_{T'}^*\mathcal{N}'$ for some invertible sheaf $\mathcal{N}'$ on T', necessarily t factors through N and $\mathcal{N}' \simeq t^*\mathcal{N}$; and third, N is a closed subscheme if $\mathbf{Pic}_{X/S}$ is separated. Show also that the first two properties determine N uniquely and $\mathcal{N}$ up to isomorphism.

EXERCISE 9.4.3. Assume $\mathbf{Pic}_{X/S}$ exists. An invertible sheaf $\mathcal{P}$ on $X \times \mathbf{Pic}_{X/S}$ is called a *universal sheaf*, or *Poincaré sheaf*, if $\mathcal{P}$ possesses the following property: given any S-scheme T and any invertible sheaf $\mathcal{L}$ on X_T, there exists a unique S-map $h\colon T \to \mathbf{Pic}_{X/S}$ such that, for some invertible sheaf $\mathcal{N}$ on T,

$$\mathcal{L} \simeq (1 \times h)^*\mathcal{P} \otimes f_T^*\mathcal{N}.$$

Show that a universal sheaf $\mathcal{P}$ exists if and only if $\mathbf{Pic}_{X/S}$ represents $\mathrm{Pic}_{X/S}$.

Assume $\mathcal{O}_S \xrightarrow{\sim} f_*\mathcal{O}_X$ holds universally. Show that, if $\mathcal{P}$ exists, then it is unique up to tensor product with the pullback of a unique invertible sheaf on $\mathbf{Pic}_{X/S}$.

Show that, if also f has a section, then a universal sheaf $\mathcal{P}$ exists.

Find an example where no universal sheaf $\mathcal{P}$ exists.

EXERCISE 9.4.4. Assume $\mathbf{Pic}_{X/S}$ exists, and let S' be an S-scheme. Show that $\mathbf{Pic}_{X_{S'}/S'}$ exists too, and in fact, that

$$\mathbf{Pic}_{X_{S'}/S'} = \mathbf{Pic}_{X/S} \times_S S'.$$

Thus forming the Picard scheme commutes with changing the base.

Find an example where $\mathbf{Pic}_{X_{S'}/S'}$ represents $\mathrm{Pic}_{X_{S'}/S'}$, but $\mathbf{Pic}_{X/S}$ does not represent $\mathrm{Pic}_{X/S}$.

EXERCISE 9.4.5. Assume $\mathbf{Pic}_{X/S}$ exists, and either it represents $\mathrm{Pic}_{(X/S)\,(\mathrm{fppf})}$ or $\mathcal{O}_S \xrightarrow{\sim} f_*\mathcal{O}_X$ holds universally. Show the scheme points of $\mathbf{Pic}_{X/S}$ correspond, in a natural bijective fashion, to the classes of invertible sheaves $\mathcal{L}$ on the fibers of X/S. A class is, by definition, represented by an $\mathcal{L}$ on an X_k where k is a field containing the residue field k_s of a (scheme) point $s \in S$; an $\mathcal{L}'$ on an $X_{k'}$ represents the same class if and only if there is a third field k'' containing the other two such that $\mathcal{L}|X_{k''} \simeq \mathcal{L}'|X_{k''}$.

DEFINITION 9.4.6. The *Abel map* is the natural map of functors

$$A_{X/S}(T)\colon \mathrm{Div}_{X/S}(T) \to \mathrm{Pic}_{X/S}(T)$$

defined by sending a relative effective divisor D on X_T/T to the sheaf $\mathcal{O}_{X_T}(D)$. The target $\mathrm{Pic}_{X/S}$ may be replaced by any of its associated sheaves. If $\mathbf{Pic}_{X/S}$ exists, then the term *Abel map* may refer to the corresponding map of schemes

$$\mathbf{A}_{X/S}\colon \mathbf{Div}_{X/S} \to \mathbf{Pic}_{X/S}\,.$$

EXERCISE 9.4.7. Assume X/S is proper and flat with integral geometric fibers. Assume $\mathbf{Pic}_{X/S}$ exists, and denote it by P. View $\mathbf{Div}_{X/S}$ as a P-scheme via the Abel map. Assume a universal sheaf $\mathcal{P}$ exists, and let $\mathcal{Q}$ be the sheaf on P associated to $\mathcal{P}$ as in Subsection 9.3.10. Show $\mathbf{P}(\mathcal{Q}) = \mathbf{Div}_{X/S}$ as P-schemes.

THEOREM 9.4.8 (Main). Assume $f\colon X \to S$ is projective Zariski locally over S, and is flat with integral geometric fibers.

(1) Then $\mathbf{Pic}_{X/S}$ exists, is separated and locally of finite type over S, and represents $\mathrm{Pic}_{(X/S)\,(\text{ét})}$.

(2) If also S is Noetherian and X/S is projective, then $\mathbf{Pic}_{X/S}$ is a disjoint union of open subschemes, each an increasing union of open quasi-projective S-schemes.

PROOF. By [**EGAG**, (0, 4.5.5), p. 106], it is a local matter on S to represent a Zariski sheaf on the category of S-schemes. Moreover, it is also a local matter on S to prove that an S-scheme is separated and locally of finite type. Hence, in order to prove (1), we may assume S is Noetherian and X/S is projective.

Plainly an S-scheme is separated if it is a disjoint union of separated open subschemes, or if it is an increasing union of separated open subschemes. Hence (1) follows from (2).

To prove (2), owing to Yoneda's lemma, we may view the category of schemes as a full subcategory of the category of functors by identifying a scheme T with its functor of points. Denote this functor too by T in order to lighten the notation.

And say that the functor is a scheme, as well as that it is representable. Also, set

$$P := \mathrm{Pic}_{(X/S)\,(\text{ét})}\,.$$

Note $P(T) = \mathrm{Hom}(T, P)$.

Given a polynomial $\varphi \in \mathbb{Q}[n]$, let $P^\varphi \subset P$ be the étale subsheaf associated to the presheaf whose T-points are represented by the invertible sheaves $\mathcal{L}$ on X_T such that we have

$$\chi(X_t, \mathcal{L}_t^{-1}(n)) = \varphi(n) \text{ for all } t \in T. \tag{9.4.8.1}$$

Notice that this presheaf is well defined, because (9.4.8.1) remains valid after any base change $p\colon T' \to T$; indeed, for any $t' \in T'$, for any i, and for any n, we have

$$\mathrm{H}^i\big(X_{t'}, \mathcal{L}_{t'}^{-1}(n)\big) = \mathrm{H}^i\big(X_{p(t')}, \mathcal{L}_{p(t')}^{-1}(n)\big) \otimes_{k_t} k_{t'}$$

because cohomology commutes with flat base change by [**Har77**, Prp. III, 9.3, p. 255]. Hence P^φ is well defined too.

Fix a map $T \to P$, and represent it by means of an étale covering $p\colon T' \to T$ and an invertible sheaf $\mathcal{L}'$ on $X_{T'}$. Consider the subset $T'^\varphi \subset T'$ defined as follows:

$$T'^\varphi := \{\, t' \in T' \mid \chi\big(X_{t'}, \mathcal{L}_{t'}^{-1}(n)\big) = \varphi(n)\,\}.$$

Then T'^φ is open by [**EGAIII2**, 7.9.11].

Set $T^\varphi := p(T'^\varphi)$. Then $T^\varphi \subset T$ is open as $T'^\varphi \subset T'$ is open and p is étale.

Moreover, $T'^\varphi = p^{-1}(T^\varphi)$. Indeed, let $t' \in p^{-1}(T^\varphi)$. Say $p(t') = p(t'_1)$ where $t'_1 \in T'^\varphi$. Now, there is an étale covering $T'' \to T' \times_T T'$ such that the two pullbacks of $\mathcal{L}'$ to $X_{T''}$ are isomorphic. Let $t'' \in T''$ have image $t' \in T'$ under the first map $T'' \to T'$ and have the image $t'_1 \in T'$ under the second map. Then

$$\chi\big(X_{t'}, \mathcal{L}_{t'}^{-1}(n)\big) = \chi\big(X_{t''}, \mathcal{L}_{t''}^{-1}(n)\big) = \chi\big(X_{t'_1}, \mathcal{L}_{t'_1}^{-1}(n)\big) = \varphi(n).$$

Hence $t' \in T'^\varphi$. Thus $T'^\varphi \supset p^{-1}(T^\varphi)$. Therefore, $T'^\varphi = p^{-1}(T^\varphi)$.

Furthermore, T^φ is (represents) the fiber product of functors $P^\varphi \times_P T$. Indeed, to see they have the same R-points, let $r\colon R \to T$ be a map; form $R' := R \times_T T'$ and $r'\colon R' \to T'$. Suppose r factors through T^φ. Then r' factors through T'^φ. So $R' \to T' \to P$ factors through P^φ essentially by definition. Now, $R' \to R$ is an étale covering. Hence $R \to T \to P$ factors through P^φ since P^φ is an étale sheaf.

Conversely, suppose $R \to T \to P$ factors through P^φ. Then $R \to P$ is defined by means of an étale covering $R'' \to R$ and an invertible sheaf $\mathcal{L}''$ on $X_{R''}$ such that $\chi(X_u, {\mathcal{L}''}_u^{-1}(n)) = \varphi(n)$ for all $u \in R''$. Since both $\mathcal{L}''$ and $\mathcal{L}'$ define $R \to P$, there is an étale covering $R''' \to R'' \times_R R'$ such that the pullbacks of $\mathcal{L}''$ and $\mathcal{L}'$ to $X_{R'''}$ are isomorphic. Hence the image of $r'\colon R' \to T'$ lies in T'^φ. But the latter is open. Hence r' factors through it. Therefore, $r\colon R \to T$ factors through T^φ. Thus T^φ and $P^\varphi \times_P T$ have the same R-points.

Let φ vary. Plainly the T'^φ are disjoint and cover T'. So the T^φ are disjoint and cover T. Hence, by a general result [**EGAG**, (0, 4.5.4), p. 103], if the P^φ are (representable by) schemes, then P is their disjoint union. Thus it remains to represent each P^φ by an increasing union of open quasi-projective S-schemes.

Fix φ. Given $m \in \mathbb{Z}$, let $P_m^\varphi \subset P^\varphi$ be the étale subsheaf associated to the presheaf whose T-points are represented by the $\mathcal{L}$ on X_T such that, in addition to (9.4.8.1), we have

$$\mathrm{R}^i f_{T*}\mathcal{L}(n) = 0 \text{ for all } i \geq 1 \text{ and } n \geq m. \tag{9.4.8.2}$$

Notice that this presheaf is well defined, because (9.4.8.2) remains valid after any base change $p\colon T' \to T$, as is shown next.

First, let's see that (9.4.8.2) is equivalent to the following condition:

$$\mathrm{H}^i(X_t, \mathcal{L}_t(n)) = 0 \text{ for all } i \geq 1, \text{ all } n \geq m, \text{ and all } t \in T. \tag{9.4.8.3}$$

Indeed, (9.4.8.3) implies (9.4.8.2); in fact, given any i, t and n, if $\mathrm{H}^i(X_t, \mathcal{L}_t(n)) = 0$, then $\mathrm{R}^i f_{T*}\big(\mathcal{L}(n) \otimes f_T^*\mathcal{N}\big)_t = 0$ for all quasi-coherent $\mathcal{N}$ on T by [**EGAIII2**, 7.5.3] or [**OB72**, Cor. 2.1 p. 68].

Conversely, assume (9.4.8.2). Fix t and n. Let's proceed by descending induction on i to prove $\mathrm{H}^i(X_t, \mathcal{L}_t(n))$ vanishes. It vanishes for $i \gg 1$ by Serre's Theorem [**EGAIII1**, 2.2.2]. Suppose it vanishes for some $i \geq 2$. Then $\mathrm{R}^i f_{T*}\big(\mathcal{L}(n) \otimes f_T^*\mathcal{N}\big)_t$ vanishes for all quasi-coherent $\mathcal{N}$ on T, as just noted. So $\mathrm{R}^{i-1} f_{T*}\big(\mathcal{L}(n) \otimes f_T^*\mathcal{N}\big)_t$ is right exact in $\mathcal{N}$ owing to the long exact sequence of higher direct images. Therefore, by general principles, there is a natural isomorphism of functors

$$\mathrm{R}^{i-1} f_{T*}(\mathcal{L}(n))_t \otimes \mathcal{N}_t \xrightarrow{\sim} \mathrm{R}^{i-1} f_{T*}\big(\mathcal{L}(n) \otimes f_T^*\mathcal{N}\big)_t.$$

Since (9.4.8.2) holds, both source and target vanish. Taking $\mathcal{N} := k_t$ yields the vanishing of $\mathrm{H}^{i-1}(X_t, \mathcal{L}_t(n))$. Thus (9.4.8.2) implies (9.4.8.3).

Finally, for any $t' \in T'$, any i, and any n, we have

$$\mathrm{H}^i\big(X_{t'}, \mathcal{L}_{t'}(n)\big) = \mathrm{H}^i\big(X_{p(t')}, \mathcal{L}_{p(t')}(n)\big) \otimes_{k_t} k_{t'}$$

because cohomology commutes with flat base change. So (9.4.8.3) remains valid after the base change $p\colon T' \to T$; whence, (9.4.8.2) does too. Thus the presheaf is well defined, and so P_m^φ is too.

Arguing much as we did for $P^\varphi \times_P T$, we find, given a map $T \to P^\varphi$, that, as m varies, the products $P_m^\varphi \times_{P^\varphi} T$ form a nested sequence of open subschemes of T, whose union is T. In the argument, the key change is in proving openness. In place of [**EGAIII2**, 7.9.11], we use the following part of Serre's Theorem [**EGAIII1**, 2.2.2]: given a coherent sheaf $\mathcal{F}$ on a projective scheme over a Noetherian ring A, there are only finitely many $i \geq 1$ and $n \geq m$ such that $H^i(\mathcal{F}(n))$ is nonzero, and all these nonzero A-modules are finitely generated. Hence, if there is a prime $\mathbf{p}$ of A such that $H^i(\mathcal{F}(n))_{\mathbf{p}} = 0$ for all $i \geq 1$ and $n \geq m$, then there is an $a \notin \mathbf{p}$ such that $H^i(\mathcal{F}(n))_a = 0$ for all $i \geq 1$ and $n \geq m$.

Therefore, again by [**EGAG**, (0, 4.5.4), p. 103], it suffices to represent each P_m^φ by a quasi-projective S-scheme.

Fix φ and m. Set $\varphi_0(n) := \varphi(m+n)$. Then there is an isomorphism of functors $P_m^\varphi \xrightarrow{\sim} P_0^{\varphi_0}$, which is defined as follows. First, define an endomorphism ε of $\mathrm{Pic}_{X/S}$ by sending an invertible sheaf $\mathcal{L}$ on an X_T to $\mathcal{L}(m)$. Plainly ε is an automorphism. So ε induces an automorphism ε^+ of the associated sheaf P. Plainly ε^+ carries P_m^φ onto $P_0^{\varphi_0}$. Thus it suffices to represent $P_0^{\varphi_0}$ by a quasi-projective S-scheme.

The function $s \mapsto \chi(X_s, \mathcal{O}_{X_s}(n))$ is locally constant on S by [**EGAIII2**, 7.9.11]. Hence we may assume it is constant by replacing S by an open and closed subset. Set $\psi(n) := \chi(X_s, \mathcal{O}_{X_s}(n))$.

Consider the Abel map $A_{X/S}\colon \mathrm{Div}_{X/S} \to P$. Note that $\mathrm{Div}_{X/S}$ is a scheme, in fact, an open subscheme of the Hilbert scheme $\mathbf{Hilb}_{X/S}$, by Theorem 9.3.7. Form the product $P_0^{\varphi_0} \times_P \mathrm{Div}_{X/S}$. It is a scheme Z, in fact, an open subscheme of $\mathrm{Div}_{X/S}$, by what was proved above. Set $\theta(n) := \psi(n) - \varphi_0(n)$. Plainly Z lies in $\mathbf{Hilb}_{X/S}^{\theta}(n)$, which is projective over S. Hence Z is quasi-projective.

Let's now prove the projection $\alpha\colon Z \to P_0^{\varphi_0}$ is a surjection of étale sheaves. In other words, given a T and a $\lambda \in P_0^{\varphi_0}(T)$, we have to find an étale covering $T_1 \to T$ and a $\lambda_1 \in Z(T_1)$ such that $\alpha(\lambda_1) \in P_0^{\varphi_0}(T_1)$ is equal to the image of λ.

Represent λ by means of an étale covering $p\colon T' \to T$ and an invertible sheaf $\mathcal{L}'$ on $X_{T'}$. Virtually by definition, the product $T' \times_{P_0^{\varphi_0}} Z$ is equal to $\mathrm{LinSys}_{\mathcal{L}'/X_{T'}/T'}$. So, by Theorem 9.3.13, this product is equal to $\mathbf{P}(\mathcal{Q})$ where $\mathcal{Q}$ is the $\mathcal{O}_{T'}$-module associated to $\mathcal{L}'$ as in Subsection 9.3.10. Now, $m = 0$, so $\mathrm{H}^1(X_t, \mathcal{L}_t) = 0$ for all $t \in T'$ owing to (9.4.8.3). Since (v) implies (i) in Subsection 9.3.10, therefore $\mathcal{Q}$ is locally free.

Hence $\mathbf{P}(\mathcal{Q})$ is smooth over T'. So there exist an étale covering $T_1 \to T'$ and a T'-map $T_1 \to \mathbf{P}(\mathcal{Q})$ by [**EGAIV4**, 17.16.3 (ii)]. Then the composition $T_1 \to \mathbf{P}(\mathcal{Q}) \to Z \to P_0^{\varphi_0}$ is equal to the composition $T_1 \to T' \to T \to P_0^{\varphi_0}$. In other words, the map $T_1 \to Z$ is a $\lambda_1 \in Z(T_1)$ such that $\alpha(\lambda_1) \in P_0^{\varphi_0}(T_1)$ is equal to the image of λ. Since the composition $T_1 \to T' \to T$ is an étale covering, α is thus a surjection of étale sheaves.

Plainly, the map $\alpha\colon Z \to P_0^{\varphi_0}$ is defined by the invertible sheaf associated to the universal relative effective divisor on X_Z/Z. So taking $T := Z$ and $T' := T$ above, we conclude that the product $Z \times_{P_0^{\varphi_0}} Z$ is a scheme and that the first projection is smooth and projective. Therefore, the theorem now results from the following general lemma. □

LEMMA 9.4.9. Let $\alpha\colon Z \to P$ be a map of étale sheaves, and set $R := Z \times_P Z$. Assume α is a surjection, Z is representable by a quasi-projective S-scheme, R is representable by an S-scheme, and the first projection $R \to Z$ is representable by a smooth and proper map. Then P is representable by a quasi-projective S-scheme, and α is representable by a smooth map.

PROOF. To lighten the notation, let Z and R denote the corresponding schemes as well. Since the structure map $Z \to S$ is quasi-projective, it is separated; whence, the first projection $Z \times_S Z \to Z$ is too. But the first projection $R \to Z$ is proper, and factors naturally through a map $h\colon R \to Z \times_S Z$. Hence h is proper. But h is a monomorphism; that is, h is injective on T-points. Therefore, h is a closed embedding by [**EGAIV3**, 8.11.5].

Plainly, for each S-scheme T, the subset $R(T) \subset Z(T) \times_{S(T)} Z(T)$ is the graph of an equivalence relation. Also, the map of schemes $R \to Z$ is flat and proper, and Z is a quasi-projective S-scheme. It follows that there exist a quasi-projective S-scheme Q and a faithfully flat and projective map $Z \to Q$ such that $R = Z \times_Q Z$. (In other words, a flat and proper equivalence relation on a quasi-projective scheme is effective.) In fact, R defines a map from Z to the Hilbert scheme $\mathbf{Hilb}_{Z/S}$, and its graph lies in the universal scheme as a closed subscheme, which descends to $Q \subset \mathbf{Hilb}_{Z/S}$; for more details, see [**AK80**, Thm. (2.9), p. 70].

Since $Z \to Q$ is flat, it is smooth if (and only if) its fibers are smooth. But these fibers are, up to extension of the ground field, the same as those of $R \to Z$. And $R \to Z$ is smooth by hypothesis. Thus $Z \to Q$ is smooth.

It remains to see that Q represents P. First, observe that $Z \to Q$ is (represents) a surjection of étale sheaves. Indeed, given an element of $Q(T)$, set $A := Z \times_Q T$. Then $A \to T$ is smooth. So there exist an étale covering $T' \to T$ and a T-map $T' \to A$ by [**EGAIV4**, 17.16.3 (ii)]. Then $T' \to A \to Z$ is an element of $Z(T')$, which maps to the element of $Q(T')$ induced by the given element of $Q(T)$.

Since $Z \to Q$ is a surjection of étale sheaves, Q is, in the category of étale sheaves, the coequalizer of the pair of maps $R \rightrightarrows Z$ by Exercise 9.4.10 below, which is an elementary exercise in general Grothendieck topology. By the same exercise, P too is the coequalizer of this pair of maps. But, in any category, the coequalizer is unique up to unique isomorphism. Thus Q represents P. □

EXERCISE 9.4.10. Given a map of étale sheaves $F \to G$, show it is a surjection if and only if G is the coequalizer of the pair of maps $F \times_G F \rightrightarrows F$.

EXERCISE 9.4.11. Assume X/S is projective and flat, its geometric fibers are integral, and S is Noetherian. Let $Z \subset \mathbf{Pic}_{X/S}$ be a subscheme of finite type. Show Z is quasi-projective.

EXERCISE 9.4.12. Assume $f\colon X \to S$ is projective Zariski locally over S, and is flat with integral geometric fibers. First, show that, if a universal sheaf $\mathcal{P}$ exists, then the Abel map $\mathbf{A}_{X/S}\colon \mathbf{Div}_{X/S} \to \mathbf{Pic}_{X/S}$ is projective Zariski locally over S.

Second, show that, in general, $\mathbf{A}_{X/S}\colon \mathbf{Div}_{X/S} \to \mathbf{Pic}_{X/S}$ is proper. Proceed by reducing this case to the first: just use $f\colon X \to S$ itself to change the base.

EXERCISE 9.4.13. Assume $f\colon X \to S$ is flat and projective Zariski locally over S. Assume its geometric fibers X_k are integral curves of arithmetic genus $\dim \mathrm{H}^1(\mathcal{O}_{X_k})$ at least 1. Let $X_0 \subset X$ be the open subscheme where X/S is smooth. Show there is a natural closed embedding $A\colon X_0 \hookrightarrow \mathbf{Pic}_{X/S}$.

EXAMPLE 9.4.14. In Theorem 9.4.8, the geometric fibers of f are assumed to be integral. This condition is needed not only for the proof to work, but also for the statement to hold. The matter is clarified by the following example, which is attributed to Mumford and is described in [**FGA**, p. 236-1] and in [**BLR90**, p. 210].

Let $\mathbb{R}[\![t]\!]$ be the ring of formal power series, S its spectrum. Let $X \subset \mathbf{P}^2_S$ be the subscheme with inhomogeneous equation $x^2 + y^2 = t$, and $f\colon X \to S$ the structure map. The generic fiber X_σ is a nondegenerate conic. The special fiber X_0 is a pair of conjugate lines; X_0 is irreducible and geometrically connected. So f is flat by the implication (iii)⇒(i) of Lemma 9.3.4 with $D := X$. And $\mathcal{O}_S \xrightarrow{\sim} f_*\mathcal{O}_X$ holds universally by Exercise 9.3.11. However, as we'll see, $\mathbf{Pic}_{X/S}$ does *not* exist!

On the other hand, set $S' := \mathbb{C}[\![t]\!]$ and $X' := X \times S'$. Then $f_{S'}$ has sections: for example, one section g' is defined by setting

$$x := \sqrt{-1} \text{ and } y := \sqrt{1+t} = 1 + (1/2)t - (1/8)t^2 + \cdots .$$

Hence all four relative Picard functors of X'/S' are equal by the Comparison Theorem, Theorem 9.2.5. Furthermore, as we'll see, $\mathbf{Pic}_{X'/S'}$ exists, but is *not* separated!

In fact, $\mathbf{Pic}_{X'/S'}$ is a disjoint union of isomorphic open nonseparated subschemes S'_n for $n \in \mathbb{Z}$. Each S'_n is obtained from S' by repeating the origin infinitely often; more precisely, to form S'_n, take a copy $S'_{a,b}$ of S' for each pair $a, b \in \mathbb{Z}$ with $a+b = n$, and glue the $S'_{a,b}$ together off their closed points. Each $S'_{a,b}$ parameterizes a different degeneration of the invertible sheaf of degree n on the generic fiber X'_σ; the degenerate sheaf has degree a on the first line, and degree b on the second. Also, complex conjugation interchanges the two lines, so interchanges $S'_{a,b}$ and $S'_{b,a}$.

Suppose $\mathbf{Pic}_{X/S}$ exists. Then $\mathbf{Pic}_{X/S} \times_S S' = \mathbf{Pic}_{X'/S'}$ by Exercise 9.4.4. Since the closed points of $S'_{a,b}$ and $S'_{b,a}$ are conjugate, they map to the same point of $\mathbf{Pic}_{X/S}$. This point lies in an affine open subscheme U. So the two closed points lie in the preimage U' of U in $\mathbf{Pic}_{X'/S'}$. But U' is affine since U, S and S' are affine. However, if $a \neq b$, then the two closed points are distinct, and so lie in no common affine U'. We have a contradiction. Thus $\mathbf{Pic}_{X/S}$ cannot exist.

Finally, let's prove $\mathrm{Pic}_{X'/S'}$ is representable by $\amalg S'_n$. First, note X' is regular; in fact, in $\mathbf{P}^2_{\mathbb{C}} \times \mathbf{A}^1_{\mathbb{C}}$, the equation $x^2 + y^2 = t$ defines a smooth surface. So on X' every reduced curve is an effective divisor. In particular, consider these three: the line $L : x = \sqrt{-1}y$, $t = 0$, the line $M : x = -\sqrt{-1}y$, $t = 0$, and the image A of the section g of $f_{S'}$ defined above.

Set $\mathcal{P}'_{a,b} := \mathcal{O}_{X'}(bL + nA)$ where $n = a + b$. The restriction to the generic fiber $\mathcal{P}'_{a,b}|\, X'_\sigma$ has degree n since $L \cap X'_\sigma$ is empty and A is the image of a section. And $\mathcal{P}'_{a,b}|\, M$ has degree b since $A \cap M$ is empty and $L \cap M$ is a reduced $\mathbb{C}$-rational point. And $\mathcal{P}'_{a,b}|L$ has degree a since, in addition, $\mathcal{O}_{X'}(L) \otimes \mathcal{O}_{X'}(M) \simeq \mathcal{O}_{X'}$ as the ideal of $L + M$ is generated by t. Lastly, every invertible sheaf on S' is trivial; fix a g-rigidification $\mathcal{O}_{S'} \xrightarrow{\sim} g^*\mathcal{P}'$, and use it to identify the two sheaves.

On $X' \times_{S'} \amalg S'_n$, form an invertible sheaf $\mathcal{P}'$ by placing $\mathcal{P}'_{a,b}$ on $S'_{a,b}$; plainly, the $\mathcal{P}'_{a,b}$ patch together. It now suffices to show this: given any S'-scheme T and any invertible sheaf $\mathcal{L}$ on on X'_T, there exist a unique S'-map $q\colon T \to \amalg S'_n$ and some invertible sheaf $\mathcal{N}$ on T such that $(1 \times q)^*\mathcal{P}' \simeq \mathcal{L} \otimes f_T^*\mathcal{N}$.

Replace $\mathcal{L}$ by $\mathcal{L} \otimes f_T^* g_T^* \mathcal{L}^{-1}$. Then $g_T^*\mathcal{L} = \mathcal{O}_T$ since $g_T^* f_T^* = 1$. Hence, if q and $\mathcal{N}$ exist, then necessarily $\mathcal{N} \simeq \mathcal{O}_T$ since $g_T^*(1 \times q)^*\mathcal{P}' = q^*g^*\mathcal{P}'$ and $g^*\mathcal{P}' = \mathcal{O}_{S'}$.

Plainly, we may assume T is connected. Then the function $s \mapsto \chi(X'_t, \mathcal{L}_t)$ is constant on T by [**EGAIII2**, 7.9.5]. Set $n := \chi(X'_t, \mathcal{L}_t) - 1$.

Fix a, b with $a + b = n$. Set

$$\mathcal{M} := \mathcal{L}^{-1} \otimes (1 \times \tau)^*\mathcal{P}'_{a,b} \text{ and } \mathcal{N} := f_{T*}\mathcal{M}.$$

Plainly $g_T^*\mathcal{M} = \mathcal{O}_T$. Form the natural map $u\colon f_T^*\mathcal{N} \to \mathcal{M}$.

Let T_σ be the generic fiber of the structure map $\tau\colon T \to S'$. Then $T_\sigma \subset T$ is open. Let $t \in T_\sigma$. Then X'_t is a nondegenerate plane conic with a rational point A_t. So $X'_t \simeq \mathbf{P}^1_{k_t}$. Hence $\mathcal{L}_t \simeq \mathcal{O}_{X'_t}(nA_t)$. So $\mathcal{M}_t \simeq \mathcal{O}_{X'_t}$. Hence $\mathrm{H}^1(X'_t, \mathcal{M}_t) = 0$ and $\mathrm{H}^0(X'_t, \mathcal{M}_t) = k_t$ by Serre's explicit computation [**EGAIII1**, 2.1.12].

Therefore, $\mathcal{N}$ is invertible at t, and forming $\mathcal{N}$ commutes with passing to X'_t, owing to the theory in Subsection 9.3.10. So forming $u\colon f_T^*\mathcal{N} \to \mathcal{M}$ commutes with passing to X'_t too. But u is an isomorphism on X'_t. Hence u is surjective along X'_t by Nakayama's lemma. But both source and target of u are invertible along X'_t. Hence u is an isomorphism along X'_t, so over T_σ as $t \in T_\sigma$ is arbitrary. Now, $g_T^*\mathcal{M} = \mathcal{O}_T$ and $g_T^* f_T^* = 1$. Hence $\mathcal{N}|T_\sigma = \mathcal{O}_{T_\sigma}$. Therefore, $\mathcal{L}|X'_{T_\sigma} = (1 \times \tau)^*\mathcal{P}'_{a\,b}|X'_{T_\sigma}$.

Let $T'_{a,b}$ be the set of $t \in T$ such that $\tau(t) \in S'$ is the closed point and $\mathcal{L}|L_t$ has degree a and $\mathcal{L}|\, M_t$ has degree b. Fix $t \in T'_{a,b}$. Then $\mathcal{M}|L_t$ has degree 0, so $\mathcal{M}|L_t \simeq \mathcal{O}_{L_t}$. Similarly, $\mathcal{M}|M_t \simeq \mathcal{O}_{M_t}$. Consider the natural short exact sequence

$$0 \to \mathcal{O}_{L_t}(-1) \to \mathcal{O}_{X'_t} \to \mathcal{O}_{M_t} \to 0.$$

Twist by $\mathcal{M}$ and take cohomology. Thus $\mathrm{H}^1(X'_t, \mathcal{M}_t) = 0$ and $\mathrm{H}^0(X'_t, \mathcal{M}_t) = k_t$. Hence, on X'_t, the map u becomes a map $\mathcal{O}_{X'_t} \to \mathcal{M}_t$. This map is surjective as it is surjective after restriction to L_t and to M_t. So this map is an isomorphism.

As above, we conclude u is an isomorphism on an open neighborhood V' of X'_t. Set $W' := f_{T'}(X'_T - V')$. Since f is proper, W' is open. But $f^{-1}W' \subset V'$. So u is an isomorphism over W'. Hence $\mathcal{O}_{X'_{t'}} \xrightarrow{\sim} \mathcal{M}_{t'}$ for all $t' \in W'$. So $W' \subset T'_{a,b} \cup T_\sigma$.

Set $T_{a,b} := T'_{a,b} \cup T_\sigma$. Then $T_{a,b} \subset T$ is open as it contains a neighborhood of each of its points. Furthermore, u is an isomorphism over $T_{a,b}$. Hence $\mathcal{N}|T_\sigma = \mathcal{O}_{T_\sigma}$, again since $g_T^*\mathcal{M} = \mathcal{O}_T$ and $g_T^* f_T^* = 1$. Therefore, $\mathcal{L}|X'_{T_{a,b}} = (1 \times \tau)^* \mathcal{P}'_{a,b}|X'_{T_{a,b}}$.

Let $q_{a,b} \colon T_{a,b} \to \amalg S'_n$ be the composition of the structure map $\tau \colon T \to S'$ and the inclusion of S' in S'_n as $S'_{a,b}$. Plainly $(1 \times q_{a,b})^*\mathcal{P}' = \mathcal{L}|X'_{T_{a,b}}$. Plainly, as a and b and n vary, the $q_{a,b}$ patch to a map $q \colon T \to \amalg S'_n$ such that $(1 \times q)^*\mathcal{P}' = \mathcal{L}$. Plainly this map q is the only S'-map q such that $(1 \times q)^*\mathcal{P}' \simeq \mathcal{L} \otimes f_T^*\mathcal{N}$ for some invertible sheaf $\mathcal{N}$ on T. Thus $\amalg S'_n$ represents $\mathrm{Pic}_{X'/S'}$, and $\mathcal{P}'$ is a universal sheaf.

EXERCISE 9.4.15. Assume $X = \mathbf{P}(\mathcal{E})$ where $\mathcal{E}$ is a locally free sheaf on S and is everywhere of rank at least 2. Show $\mathbf{Pic}_{X/S}$ exists, and represents $\mathrm{Pic}_{X/S}$; in fact, $\mathbf{Pic}_{X/S} = \mathbb{Z}_S$ where $\mathbb{Z}_S$ stands for the disjoint union of copies of S indexed by $\mathbb{Z}$.

EXERCISE 9.4.16. Consider the curve X of Exercise 9.2.4. Show $\mathbf{Pic}_{X/\mathbb{R}} = \mathbb{Z}_\mathbb{R}$.

PROPOSITION 9.4.17. *Assume that* $\mathbf{Pic}_{X/S}$ *exists and represents* $\mathrm{Pic}_{(X/S)\,(\mathrm{fppf})}$. *Then* $\mathbf{Pic}_{X/S}$ *is locally of finite type.*

PROOF. Set $P := \mathrm{Pic}_{(X/S)\,(\mathrm{fppf})}$. Owing to [**EGAIV3**, 8.14.2], we just need to check the following condition. For every filtered inverse system of affine S-schemes T_i, the natural map is a bijection:

$$\varinjlim P(T_i) \xrightarrow{\sim} P\bigl(\varprojlim T_i\bigr).$$

To check it is injective, set $T := \varprojlim T_i$. Fix i and let $\lambda_i \in P(T_i)$. Represent λ_i by a sheaf $\mathcal{L}'_i$ on $X_{T'_i}$ where $T'_i \to T_i$ is an fppf covering. Set $T' := T'_i \times_{T_i} T$. Set $\mathcal{L}' := \mathcal{L}_i|X_{T'}$. Let λ be the image of λ_i in $P(T)$. Then $\mathcal{L}'$ represents λ.

Suppose $\lambda = 0 \in P(T)$. Then there exists an fppf covering $T'' \to T'$ such that $\mathcal{L}'|X_{T''} \simeq \mathcal{O}_{X_{T''}}$. It follows from [**EGAIV3**, 8.8.2, 8.10.5(vi), 11.2.6, 8.5.2(i), and 8.5.2.4] that there exist a $j \geq i$ and an fppf covering $T''_j \to T'_j$ with $T'_j := T'_i \times_{T_j} T$ such that $\mathcal{L}'|X_{T''_j} \simeq \mathcal{O}_{X_{T''_j}}$. So λ_i maps to $0 \in P(T_j)$. Thus the map is injective.

Surjectivity can be proved similarly. □

REMARK 9.4.18. There are three important existence theorems for $\mathbf{Pic}_{X/S}$, which refine Theorem 9.4.8. They were proved soon after it, and each involves new ideas.

First, Mumford proved the following generalization of Theorem 9.4.8, and it fits in nicely with Example 9.4.14.

Theorem 9.4.18.1. *Assume X/S is projective and flat, and its geometric fibers are reduced and connected; assume the irreducible components of its ordinary fibers are geometrically irreducible. Then* $\mathbf{Pic}_{X/S}$ *exists.*

Mumford stated this theorem at the bottom of Page viii in [**Mum66**]. He said the proof is a generalization of that [**Mum66**, Lects. 19–21] in the case where S is the spectrum of an algebraically closed field and X is a smooth surface. That proof is based on his theory of independent 0-cycles. This theory is further developed in [**AK79**, pp. 23–28] and used to prove [**AK79**, Thm. (3.1)], which asserts the existence of a natural compactification of $\mathbf{Pic}_{X/S}$ when X/S is flat and locally projective with integral geometric fibers.

On the other hand, Grothendieck [**FGA**, p. 236-1] attributed to Mumford a slightly different theorem, in which neither the geometric fibers nor the ordinary fibers are assumed connected (see [**BLR90**, p. 210] also). Grothendieck said the proof is based on a refinement of Mumford's construction of quotients, and referred to the forthcoming notes of a Harvard seminar of Mumford and Tate's, held in the spring of 1962.

Mumford was kind enough, in November 2003, to mail the present author his personal folder of handwritten notes from the seminar; the folder is labeled, "Groth–Mumford–Tate," and contains notes from talks by each of the three, and notes written by each of them. Virtually all the content has appeared elsewhere; Mumford's contributions appeared in Mumford's books [**Mum65**], [**Mum66**], and [**Mum70**].

The notes contain, in Mumford's hand, a precise statement of the theorem and a rough sketch of the proof. This statement too is slightly different from that of Theorem 9.4.18.1: he crossed out the hypothesis that the geometric fibers are connected; and he made the weaker assumption that the ordinary fibers are connected. The proof is broadly like his proof in [**Mum66**, Lects. 19–21].

Second, Grothendieck proved this theorem of "generic representability."

Theorem 9.4.18.2. *Assume X/S is proper, and S is integral. Then there exists a nonempty open subset $V \subset S$ such that $\mathbf{Pic}_{X_V/V}$ exists, represents $\mathrm{Pic}_{(X_V/V)\,(\mathrm{fppf})}$, and is a disjoint union of open quasi-projective subschemes.*

A particularly important special case is covered by the following corollary.

Corollary 9.4.18.3. *Assume S is the spectrum of a field k, and X is complete. Then $\mathbf{Pic}_{X/k}$ exists, represents $\mathrm{Pic}_{(X/k)\,(\mathrm{fppf})}$, and is a disjoint union of open quasi-projective subschemes.*

Before Grothendieck discovered Theorem 9.4.18.2, he [**FGA**, Cor. 6.6, p. 232-17] proved Corollary 9.4.18.3 assuming X/k is projective. To do so, he developed a method of "relative representability," by which Theorem 9.4.8 implies the existence of the Picard scheme in other cases. The method incorporates a "dévissage" of Oort's [**Oor62**, §6] ; the latter yields the Picard scheme as an extension of a group subscheme of $\mathbf{Pic}_{X_{\mathrm{red}}/k}$ by an affine group scheme.

In [**FGA**, Rem. 6.6, p. 232-17], Grothendieck said it is "extremely plausible" that Corollary 9.4.18.3 holds in general, and can be proved by extending the method of relative representability, so that it covers the case of a surjective map $X' \to X$ with X' projective, such as the map provided by Chow's lemma [**EGAII**, Thm. 5.6.1]. Furthermore, he conjectured, in [**FGA**, Rem. 6.6, p. 232-17], that, for any surjective map $X' \to X$ between proper schemes over a field, the induced map on Picard schemes is affine. He said he was led to the conjecture by considerations of "nonflat descent," a version of descent theory where the maps are not required to be flat, but the objects are.

Thanking Grothendieck for help, Murre [**Mur64**, (II.15)] gave the first proof of the heart of Corollary 9.4.18.3: if X/k is proper, then $\mathbf{Pic}_{X/k}$ exists and is locally of finite type. Murre did not use the method of relative representability. Rather, he identified seven conditions [**Mur64**, (I.2.1)] that are necessary and sufficient for the representability of a functor from schemes over a field to Abelian groups. Then he showed the seven are satisfied by the Picard functor localized in the fpqc topology. To handle the last two conditions, he used Chow's lemma and Theorem 9.4.8.

In the meantime, Grothendieck had proved Theorem 9.4.18.2, according to Murre [**Mur64**, p. 5]. Later, the proof appeared in two parts. The first part established a key intermediate result, the following theorem.

Theorem 9.4.18.4. *Assume X/S is proper. Let $\mathcal{F}$ be a coherent sheaf on X, and $S_{\mathcal{F}} \subset S$ the subfunctor of all S-schemes T such that $\mathcal{F}_T$ is T-flat. Then $S_{\mathcal{F}}$ is representable by an unramified S-scheme of finite type.*

Murre sketched Grothendieck's proof of this theorem in [**Mur95**, Cor. 1, p. 294-11]. The proof involves identifying and checking eight conditions that are necessary and sufficient for the representability of a functor by a separated and unramified S-scheme locally of finite type. As in Murre's proof of Corollary 9.4.18.3, a key step is to show the functor is "pro-representable": there exist certain natural topological rings, and if the functor is representable by Y, then these rings are the completions of the local rings at the points of Y that are closed in their fibers over S.

In the second part of Grothendieck's proof of Theorem 9.4.18.2, the main result is the following theorem of relative representability.

Theorem 9.4.18.5. *Assume S is integral. Let $X' \to X$ be a surjective map of proper S-schemes. Then there is a nonempty open subset $V \subset S$ such that the map*

$$\mathrm{Pic}_{(X_V/V)\,(\mathrm{fppf})} \to \mathrm{Pic}_{(X'_V/V)\,(\mathrm{fppf})}$$

is representable by quasi-affine maps of finite type.

In other words, for every S-scheme T and map $T \to \mathrm{Pic}_{(X'_V/V)\,(\mathrm{fppf})}$, the fibered product $\mathrm{Pic}_{(X_V/V)\,(\mathrm{fppf})} \times_{\mathrm{Pic}_{(X'_V/V)\,(\mathrm{fppf})}} T$ is representable and its projection to T is a quasi-affine map of finite type between schemes.

Raynaud [**Ray71**, Thm. 1.1] gave Grothendieck's proof of Theorem 9.4.18.5; the main ingredients are, indeed, Oort dévissage and nonflat descent. As a first consequence, Raynaud [**Ray71**, Cor. 1.2] derived Theorem 9.4.18.2. In order to show $\mathbf{Pic}_{X/S}$ is a disjoint union of open quasi-projective subschemes, he used a finiteness theorem for $\mathbf{Pic}_{X/S}$, which Grothendieck had stated under (v) in [**FGA**, p. C-08], and which is proved below as Theorem 9.6.16. As a second consequence of Theorem 9.4.18.5, Raynaud [**Ray71**, Cors. 1.5] established Grothendieck's conjecture that, over a field, a surjective map of proper schemes induces an affine map on Picard schemes.

The third important existence theorem for $\mathbf{Pic}_{X/S}$ is due to M. Artin, whose work greatly clarifies the situation. Artin proved $\mathbf{Pic}_{X/S}$ exists when it should, but not as a scheme. Rather, it exists as a more general object, called an "algebraic space," which is closer in nature to a (complex) analytic space.

Grothendieck [**FGA**, Rem. 5.2, p. 232-13] had said: "it is not ruled out that $\mathbf{Pic}_{X/S}$ exists whenever X/S is proper and flat and such that $\mathcal{O}_S \xrightarrow{\sim} f_*\mathcal{O}_X$ holds universally. At least, this statement is proved in the context of analytic spaces when X/S is, in addition, projective." Mumford's example, Example 9.4.14, shows the statement is false for schemes; Artin's theorem shows the statement holds for algebraic spaces.

Algebraic spaces were introduced by Artin [**Art69b**, §1], and the theory developed by himself and his student Knutson [**Knu71**]. The spaces are constructed by gluing together schemes along open subsets that are isomorphic locally in the étale topology. Over $\mathbb{C}$, these open sets are locally analytically isomorphic; so an algebraic space is a kind of complex analytic space. In general and more formally, an algebraic space is the quotient in the category of étale sheaves of a scheme divided by an étale equivalence relation.

Artin [**Art69a**, Thm. 3.4, p. 35], proceeding in the spirit of Grothendieck and Murre, identified five conditions on a functor that are necessary and sufficient for it to be a locally separated algebraic space that is locally of finite type over a field or over an excellent Dedekind domain. In the proof of necessity, a key new ingredient is Artin's approximation theorem; it implies that the topological rings given by pro-representability can be algebraized. The resulting schemes are then glued together via an étale equivalence relation.

By checking that the five conditions hold, Artin [**Art69a**, Thm. 7.3, p. 67] proved the following theorem.

Theorem 9.4.18.6. *Let $f\colon X \to S$ be a flat, proper, and finitely presented map of algebraic spaces. Assume forming $f_*\mathcal{O}_X$ commutes with changing S (or f is "cohomologically flat in dimension zero"). Then $\mathbf{Pic}_{X/S}$ exists as an algebraic space, which is locally of finite presentation over S.*

Since f is finitely presented, the statement can be reduced to the case where S is locally of finite type over a field or over an excellent Dedekind domain. Artin's proof of Theorem 9.4.18.6 is direct: it involves no reduction whatsoever to Theorem 9.4.8.

On the other hand, it is not known, in general, whether Theorem 9.4.18.6 implies Theorem 9.4.8, which asserts $\mathbf{Pic}_{X/S}$ is a scheme. However, more is known over a *field.* Indeed, given an algebraic space that is locally of finite type over a field, Artin [**Art69a**, Lem. 4.2, p. 43] proved this: if the space is a sheaf of groups, then it is a group scheme. Thus Theorem 9.4.18.6 implies the heart of Corollary 9.4.18.3.

Thus, in Theorem 9.4.18.6, the fibers over S of the algebraic space $\mathbf{Pic}_{X/S}$ are schemes; they are the Picard schemes of the fibers X_s, though the X_s need not be schemes. In particular, if S and X are schemes — so the X_s are too — then the $\mathbf{Pic}_{X_s/k_s}$ are schemes. Furthermore, then they form a family; its total space $\mathbf{Pic}_{X/S}$ is an algebraic space, but not necessarily a scheme.

Remark 9.4.19. In a body of work of some importance, the theory of the Picard scheme was developed in two nonstandard cases: one without properness, the other without flatness. The first is the case of the punctured spectrum of a Noetherian local algebra over a field; the resulting scheme is known as the *local Picard scheme.* The second is the case of a scheme that is proper, but not flat, over a complete local ring. Remarkably, the two cases are closely related, and results in both cases are essential ingredients in the only available proof of the following purely algebraic statement, known as *Samuel's conjecture*: given a complete local UFD A with algebraically closed residue field, the power series ring in one variable $A[[t]]$ is also a UFD.

Mumford began the work in his celebrated paper [**Mum61b**, § 2(a), pp. 13–16]. He worked analytically at a singular point P of a normal complex algebraic surface F. He formed the local ring of holomorphic functions $\mathcal{O}^h_{F,P}$, and proved that its ideal classes form an analytic group. He did so by choosing a desingularization $\pi\colon F' \to F$ and by analyzing $(\mathrm{R}^1\pi_*\mathcal{O}^{h*}_{F'})_P$. He showed the latter is an analytic group variety, whose points represent the classes of analytic divisors defined along the exceptional locus $\pi^{-1}P$ modulo the divisors of meromorphic functions. Then he formed the quotient of this group modulo the subgroup generated by the classes of the components of $\pi^{-1}P$, and showed this quotient parameterizes the ideal classes of $\mathcal{O}^h_{F,P}$. Mumford entitled § 2(a) "Local analytic Picard varieties and unique factorization," yet he said nothing about factorization. But, he did suggest his work might be algebraized by using the common completion of $\mathcal{O}^h_{F,P}$ and $\mathcal{O}_{F,P}$.

Grothendieck said in [**SGA2**, pp. 189–193] that, in that paper, Mumford had raised the following "fundamental problem": given the punctured spectrum U of a complete local ring A with residue field k, to construct a projective system of locally algebraic groups G_n/k, and a canonical isomorphism $\mathrm{Pic}(U) = G(k)$ where $G(k) := \varprojlim G_n(k)$. Again giving Mumford credit for the idea, Grothendieck said that one way to attack the problem is as follows.

Suppose U is regular and nonempty, and $\mathrm{Spec}\, A$ has a desingularization X isomorphic to U over U. Then $\mathrm{Pic}(U) = \mathrm{Pic}(X)/D$ where D is the subgroup generated by the components of the exceptional locus. Also, the existence theorem of formal geometry [**EGAIII1**, 5.1.6, p. 150] yields $\mathrm{Pic}(X) = \varprojlim \mathrm{Pic}(X_n)$ where X_n is the fiber defined by the $(n+1)$-st power of the maximal ideal of A. If A contains a field of representatives, also denoted by k, then $\mathrm{Pic}(X_n) = \mathbf{Pic}_{X_n/k}(k)$ at least if k is separably closed. Hence, in this case, $\mathbf{Pic}_{X_n/k}/D$ works as G_n, and the system is essentially constant. However, it depends on the choice of field of representatives, which is unique, according to [**EGAIV1**, 19.8.7, p. 115], if k is perfect of positive characteristic or is a number field, but not otherwise.

Grothendieck expanded on his ideas in an exchange of letters between May 1969 and March 1970 with Lipman, who kindly made copies available to the author. In particular, Grothendieck observed that, in the preceding case, the transition maps $G_n \to G_{n-1}$ are affine by Oort dévissage [**Ray71**, Prp. 3.1, p. 604], also see Remark 9.4.18. So the limit is a group scheme G by the theory in [**EGAIV3**, §8.2]. Furthermore, the "Lie algebra" of G, the tangent space $\mathrm{T}_0\, G$, is equal to $\mathrm{H}^1(\mathcal{O}_X)$. And, to eliminate the assumption that U is regular and $\mathrm{Spec}\, A$ is desingularizable, Grothendieck suggested looking for a proper surjective map $f\colon X \to \mathrm{Spec}(A)$ that restricts to an isomorphism over U and that induces a surjection $\mathrm{Pic}(X) \twoheadrightarrow \mathrm{Pic}(U)$.

Grothendieck also addressed the case where A is of mixed characteristic, but k is perfect. He suggested replacing A by an arbitrary Artin ring and X by a proper A-scheme, and then representing the fppf sheaf P associated to the functor $B \mapsto \mathrm{Pic}(X \otimes_A A(B))$ where B is a k-algebra and $A(B)$ is the Greenberg algebra. If the Picard scheme $\mathbf{Pic}_{X/A}$ exists, then its "Greenberg realization" represents P. But $\mathbf{Pic}_{X/A}$ may fail to exist; Lipman gave an example in [**Lip76**, p. 16]. To represent P in general, Grothendieck suggested checking Artin's "handy" necessary and sufficient conditions, which are mentioned above in Remark 9.4.18.

Lipman took up Grothendieck's program, and represented P in [**Lip76**]. But Lipman found it preferable, though equivalent, to work with the fpqc sheaf associated to the functor $B \mapsto \mathrm{Pic}(X_B)$ where $X_B := X \otimes_{\mathbf{W}(k)} \mathbf{W}(B)$ and where $\mathbf{W}(k)$ and $\mathbf{W}(B)$ are the rings of Witt vectors. Furthermore, rather than check Artin's conditions, Lipman argued as follows. For each $m \geq 0$, he formed the mth power of the nilradical of $\mathcal{O}_X$, the corresponding subscheme X_m of X, and the functor P_m obtained on replacing X by X_m. Then X_1 is just X_{red}, which is a k-scheme. It follows that $P_1 = \mathrm{Pic}_{(X_{\mathrm{red}}/k)\,(\mathrm{fppf})}$; so P_1 is representable by Corollary 9.4.18.3. To pass from P_{m-1} to P_m, Lipman developed an appropriate version of Oort dévissage. But $X_m = X$ for $m \gg 0$. Thus P is representable. Finally, Lipman studied the natural map $\mathrm{Pic}(X_B) \to P(B)$ when B is perfect. And he computed the dimension λ of the Lie algebra of P; interestingly, $\lambda \geq \mathrm{length}\, H^1(O_X)$, but '$>$' can hold.

Lipman went on to treat formal schemes $\mathfrak{X}$ in [**Lip74**, §2, §3]. He fixed a complete Noetherian local ring A whose residue class field k is perfect of arbitrary characteristic p. If $p = 0$, then he set $\mathbf{W}(k) := k$, and fixed a map $\mathbf{W}(k) \to A$; if $p > 0$, then, as before, $\mathbf{W}(k)$ denotes the Witt vectors, and there is a canonical map $\mathbf{W}(k) \to A$. Lipman considered an $\mathfrak{X}$ that is the direct limit of a nested sequence of Noetherian closed subschemes X_m proper over A. Given a k-algebra B, set $\mathfrak{W}_B := \mathrm{Spf}(\mathbf{W}(B))$ and $\mathfrak{X}_B := \mathfrak{X} \times_{\mathfrak{W}_k} \mathfrak{W}_B$. Lipman formed the fpqc sheaf P associated to the functor $B \mapsto \mathrm{Pic}(\mathfrak{X}_B)$, and proved P is representable by $\varprojlim G_n$ where, if $p = 0$, then $G_n := \mathbf{Pic}_{X_n/k}$ and, if $p > 0$, then G_n is the representing scheme of the preceding paragraph with $X := X_n$. In this passage to the limit, it is helpful to use the fpqc topology; possibly, it is even necessary when $\mathfrak{X}$ is the completion of a proper A-scheme X along the closed fiber, unless X is generically finite over $\mathrm{Spec}(A)$. In addition, Lipman proved that $\mathrm{Pic}(\mathfrak{X}_K) = P(K)$ when K is an algebraically closed field, and observed that $\mathrm{Pic}(\mathfrak{X}_K) = \mathrm{Pic}(X_K)$ when $\mathfrak{X}$ is the completion of an X.

Boutot devoted his thesis [**Bou78**] to the local Picard scheme. He fixed an arbitrary field k and a Noetherian local k-algebra A with residue field k. He defined the local Picard functor Q directly as the étale sheaf associated to the functor $B \mapsto \mathrm{Pic}(\widetilde{U}_B)$ where B is a k-algebra, U is the punctured spectrum of A, and $\widetilde{U}_B$ is the preimage of U in the spectrum of the Henselized product $A\widetilde{\otimes}_k B$. He proved Q is representable and has Lie algebra $\mathrm{H}^1(\mathcal{O}_U)$ if $\mathrm{H}^1(\mathcal{O}_U)$ is finite dimensional and $\mathrm{depth}(A) \geq 2$, for example, if the completion of A is normal and $\dim A \geq 3$. He proceeded by checking Artin's conditions using a variant of the Ramanujam–Samuel Theorem [**EGAIV4**, 21.14.1] and by using Elkik–Artin approximation theory to algebraize the formal deformations. Furthermore, given any proper A-scheme X, he formed an analogue of $\mathrm{Pic}(\mathfrak{X}_B)$, namely, the subgroup of $\mathrm{Pic}(A\widetilde{\otimes}_k B)$ of sheaves whose restrictions are trivial locally over $\widetilde{U}_B$; then, he studied the associated étale sheaf, and related it to the local Picard functor.

The above theory supplied the tools needed to establish Samuel's conjecture. In [**Lip75**, pp. 534, 537], Lipman explained the philosophy, which is largely due to Grothendieck and contained in his letters. Here is the idea. Set $R := A[[t]]$. If $\dim A = 1$, then A is regular; plainly, then R is regular, whence a UFD by the Auslander–Buchsbaum Theorem [**EGAIV4**, Thm. 21.11.1, p. 302]. Assume $\dim A \geq 2$. Since A is a UFD, its ideal class group $\mathrm{Cl}(A)$ is trivial. Suppose first U is regular; then $\mathrm{Cl}(A) = \mathrm{Pic}(U)$ since A is normal and $\dim A \geq 2$. Since k is algebraically closed, $\mathrm{Pic}(U) = Q(k)$ where Q is the local Picard scheme, assuming it exists. Thus Q has only one scheme point. Hence $Q(k[[t]]) = 0$. Therefore, $\mathrm{Pic}(U') = 0$ where U' is the punctured spectrum of R. So $\mathrm{Cl}(R) = 0$ since R is normal and $\dim R \geq 2$. Thus R is a UFD.

Suppose U is not necessarily regular, and let $I \subset R$ be a divisorial ideal. Then there is an open set $V \subset \mathrm{Spec}(A)$ whose complement is of codimension at least 2 and over which the restriction of the sheaf $\widetilde{I}$ is invertible. Furthermore, Boutot proved in [**Bou78**, Prp. 2.4, p. 143] that there is a projective birational map $\varphi\colon X \to \mathrm{Spec}(A)$ such that $\varphi^{-1}V \to V$ is an isomorphism and $\widetilde{I} \mid \varphi^{-1}V$ extends to an invertible sheaf on X. Lipman used Boutot's result and his own theory of the Picard scheme of the completion $\mathfrak{X}$ of X, rather than the theory of the local Picard scheme, to establish Samuel's conjecture, fully as stated above, in [**Lip74**, §1]. In the introduction, Lipman discussed related work of Scheja, Storch, and Danilov, and he devoted [**Lip75**] to an extensive survey of unique factorization in complete local rings. In particular, he noted that Danilov's form [**Dan70**, p. 228] of Samuel's conjecture remains open: $A[\![t]\!]$ is a UFD if the strict Henselization of A is a UFD.

Samuel himself assumed that A is complete, but not that k is algebraically closed. In this form, the conjecture is false! Salmon gave the first counterexample, see [**Lip75**, p. 537]: A is the completion of the local ring at the vertex of the cone over a smooth plane cubic E with a single rational point e over a field k. The local Picard scheme is just E. Therefore, A is a UFD, but $A \otimes_k \bar{k}$ is not, where $\bar{k}$ is the algebraic closure. Now, $E(k[\![t]\!]) \neq 0$ since $\widehat{\mathcal{O}}_{E,e} = k[\![t]\!]$. Furthermore, it follows from the sequence 9.2.11.5 that the Brauer group $\mathrm{Br}'(k[\![t]\!])$ houses the obstruction to representing an element of $E(k[\![t]\!])$ by an invertible sheaf on the blowup of $\mathrm{Spec}(R)$, so by an element of $\mathrm{Cl}(R)$. Now, $\mathrm{Br}'(k[\![t]\!]) \xrightarrow{\sim} \mathrm{Br}'(k)$ owing to [**Mil80**, Rem. 3.11(a), p. 116]. And $\mathrm{Br}'(k) = 0$ by Tsen's Theorem [**FK88**, Prp. 1.15, p. 15] when k is a function field in one variable over an algebraically closed field. Thus, in this case, $A[\![t]\!]$ is not a UFD. A more direct analysis of essentially the same example, but valid for more k, was given by Danilov in [**Dan70**, § 1].

9.5. The connected component of the identity

Having treated the existence of the Picard scheme $\mathbf{Pic}_{X/S}$, we now turn to its structure. In this section, we study the union $\mathbf{Pic}^0_{X/S}$ of the connected components of the identity element, $\mathbf{Pic}^0_{X_s/k_s}$, for $s \in S$. We establish a number of basic properties, especially when S is the spectrum of a field.

It is remarkable how much we can prove about $\mathbf{Pic}^0_{X/S}$ formally, or nearly so, from general principles. Notably, we can do without the finiteness theorems proved in the next section. In order to emphasize the formal nature and corresponding generality of the arguments, most of the results are stated with the general hypothesis that $\mathbf{Pic}_{X/S}$ exists instead of with specific hypotheses that imply it exists.

LEMMA 9.5.1. *Let k be a field, and G a group scheme locally of finite type. Let G^0 denote the connected component of the identity element e.*

(1) Then G is separated.

(2) Then G is smooth if it has a geometrically reduced open subscheme.

(3) Then G^0 is an open and closed group subscheme of finite type; it is geometrically irreducible; and forming it commutes with extending k.

PROOF. Since e is a k-point, it is closed. Define a map $\alpha\colon G \times G \to G$ on T-points by $\alpha(g,h) := gh^{-1}$. Then $\alpha^{-1}e \subset G \times G$ is a closed subscheme. Its T-points are just the pairs (g,g) for $g \in G(T)$; so $\alpha^{-1}e$ is the diagonal. Thus (1) holds.

Suppose G has a geometrically reduced open subscheme V. To prove G is smooth, we may replace k by its algebraic closure. Then V contains a nonempty smooth open subscheme W. Furthermore, given any two closed points g, $h \in G$, there is an automorphism of G that carries g to h, namely, multiplication by $g^{-1}h$. Taking $g \in W$, we conclude that G is smooth at h. Thus (2) holds.

Consider (3). By definition, G^0 is the largest connected subspace containing e. But the closure of a connected subspace is plainly connected; so G^0 is closed. By [**EGAI**, 6.1.9], in any locally Noetherian topological space, the connected components are open. Thus G^0 is open too.

Since G^0 is connected and has a k-point, G^0 is geometrically connected by [**EGAIV2**, 4.5.14]. Thus forming G^0 commutes with extending k.

Furthermore, $G^0 \times G^0$ is connected by [**EGAIV2**, 4.5.8]. So $\alpha(G^0 \times G^0) \subset G$ is connected, and contains e, so lies in G^0. Thus G^0 is a subgroup.

To prove G^0 is geometrically irreducible and quasi-compact, we may replace k by its algebraic closure. Then $G^0_{\text{red}} \times G^0_{\text{red}}$ is reduced by [**EGAIV2**, 4.6.1]. Hence α induces a map from $G^0_{\text{red}} \times G^0_{\text{red}}$ into G^0_{red}. So G^0_{red} is a subgroup. Thus we may replace G^0 by G^0_{red}.

Since G^0 is reduced and k is algebraically closed, G^0 contains a nonempty smooth affine open subscheme U. Take arbitrary k-points $g \in U$ and $h \in G^0$. Then $hg^{-1}U$ is smooth and open, and contains h. Hence G^0 is irreducible locally at h. But h is arbitrary, and G^0 is connected. So G^0 is irreducible by [**EGAI**, 6.1.10].

Since G^0 is irreducible, its open subschemes U and hU meet. So their intersection contains a k-point g_1 since k is algebraically closed. Then $g_1 = hh_1$ for some a k-point $h_1 \in U$. Then $h = g_1h_1^{-1}$. But $h \in G^0$ is arbitrary. Hence $\alpha(U \times U) = G^0$. Now, $U \times U$ is affine, so quasi-compact. Hence G^0 is quasi-compact. By hypothesis, G is locally of finite type. Hence G^0 is of finite type. Thus (3) holds. □

REMARK 9.5.2. Let G be a group scheme of finite type over a field, G_{red} its reduction. Then G_{red} need not be a group subscheme, because $G_{\text{red}} \times G_{\text{red}}$ need not be reduced. Waterhouse [**Wat79**, p. 53] gives two conditions equivalent to reducedness when G is finite in Exercise 9, and he gives a counterexample in Exercise 10.

PROPOSITION 9.5.3. *Assume S is the spectrum of a field k. Assume $\mathbf{Pic}_{X/k}$ exists and represents $\mathrm{Pic}_{(X/k)\,(\mathrm{fppf})}$. Then $\mathbf{Pic}_{X/k}$ is separated, and it is smooth if it has a geometrically reduced open subscheme. Furthermore, the connected component of the identity $\mathbf{Pic}^0_{X/k}$ is an open and closed group subscheme of finite type; it is geometrically irreducible; and forming it commutes with extending k.*

PROOF. Proposition 9.4.17 and Lemma 9.5.1 imply this result formally. □

THEOREM 9.5.4. Assume S is the spectrum of a field k. Assume X/k is projective and geometrically integral. Then $\mathbf{Pic}^0_{X/k}$ exists and is quasi-projective. If also X/k is geometrically normal, then $\mathbf{Pic}^0_{X/k}$ is projective.

PROOF. Theorem 9.4.8 implies $\mathbf{Pic}_{X/k}$ exists and represents $\mathrm{Pic}_{(X/k)\,(\text{ét})}$, so $\mathrm{Pic}_{(X/k)\,(\mathrm{fppf})}$. Hence $\mathbf{Pic}^0_{X/k}$ exists and is of finite type by Proposition 9.5.3 (in fact, here Proposition 9.4.17 is logically unnecessary since $\mathbf{Pic}_{X/k}$ is locally of finite type by Theorem 9.4.8). So $\mathbf{Pic}^0_{X/k}$ is quasi-projective by Exercise 9.4.11.

Suppose X is also geometrically normal. Since $\mathbf{Pic}^0_{X/k}$ is quasi-projective, to prove it is projective, it suffices to prove it is proper. By Lemma 9.5.1, forming $\mathbf{Pic}^0_{X/k}$ commutes with extending k. And by [**EGAIV2**, 2.7.1(vii)], a k-scheme is complete if (and only if) it is after extending k. So we may, and do, assume k is algebraically closed.

Recall the structure theorem of Chevalley and Rosenlicht for algebraic groups, or reduced connected group schemes of finite type over k; see [**Con02**, Thm. 1.1, p. 3]. The theorem says that every algebraic group is an extension of an Abelian variety (or complete algebraic group) by a linear (or affine) algebraic group. Recall also that every solvable linear algebraic k-group is triangularizable (the Lie–Kolchin theorem); so, if it's nontrivial, then it contains a copy of the multiplicative group or of the additive group; see [**Bor69**, (10.5) and (10.2)]. Now, $\mathbf{Pic}_{X/k}$ is commutative, so solvable. Hence it suffices to show that, if T denotes the affine line minus the origin, then every k-map $t\colon T \to (\mathbf{Pic}^0_{X/k})_{\rm red}$ is constant.

Since k is algebraically closed, X/k has a section. So t arises from an invertible sheaf $\mathcal{L}$ on $X \times T$ by the Comparison Theorem, Theorem 9.2.5. Since $X \times T$ is integral, there is a divisor D such that $\mathcal{O}(D) = \mathcal{L}$ by [**Har77**, Ex. II, 6.15, p. 145].

Form the projection $p\colon X \times T \to X$. Restrict $\mathcal{L}$ to its generic fiber. This restriction is trivial as T is an open subset of the line. Therefore, there exists a rational function φ on $X \times T$ such that $(\varphi) + D$ restricts to the trivial divisor. Let $s\colon X \to X \times T$ be a section. Set $E := s^*((\varphi) + D)$; then E is a well-defined divisor on X. Plainly, p^*E and $(\varphi)+D$ coincide as cycles; whence, they coincide as divisors by [**AK70**, Prp. (3.10), p. 139], since $X \times T$ is normal. Therefore, $\mathcal{L} = p^*\mathcal{O}(E)$. Thus $t\colon T \to \mathbf{Pic}_{X/k}$ is constant. □

COROLLARY 9.5.5. Assume S is the spectrum of an algebraically closed field k. Assume X is projective and integral. Set $P := \mathbf{Pic}^0_{X/k}$, and let $\mathcal{P}$ be the restriction to X_P of a Poincaré sheaf. Then a Poincaré family W exists; by definition, W is a relative effective divisor on X_P/P such that

$$\mathcal{O}_{X_P}\big(W - (W_0 \times P)\big) \simeq \mathcal{P} \otimes f_P^*\mathcal{N}$$

where W_0 is the fiber over $0 \in P$ and where $\mathcal{N}$ is an invertible sheaf on P.

PROOF. Note that P exists and is quasi-projective by Theorem 9.5.4 and that $\mathcal{P}$ exists by Exercise 9.3.11 and Exercise 9.4.3. Since P is Noetherian, Serre's Theorem [**EGAIII1**, 2.2.1] implies there is an N such that $R^i f_{P*}\mathcal{P}(n) = 0$ for all $i > 0$ and $n \geq N$. Recall that (9.4.8.2) implies (9.4.8.3); similarly, $\mathrm{H}^i(\mathcal{P}_t(n)) = 0$ for all $t \in P$. Fix an $n \geq N$ such that $\dim \mathrm{H}^0(\mathcal{O}_X(n)) > \dim P$.

Say $\lambda \in \mathbf{Pic}_{X/k}$ represents $\mathcal{O}_X(n)$. Form the automorphism of $\mathbf{Pic}_{X/k}$ of multiplication by λ. Plainly, P is carried onto the connected component, P' say, of λ. Let $q\colon P \xrightarrow{\sim} P'$ be the induced isomorphism. Let $\mathcal{P}'$ be the restriction to $X_{P'}$ of a Poincaré sheaf. Plainly, $(1 \times q)^*\mathcal{P}' \simeq \mathcal{P}(n) \otimes f_P^*\mathcal{N}$ for some invertible sheaf $\mathcal{N}$.

By Exercise 9.4.7, there is a coherent sheaf $\mathcal{Q}$ on P such that $\mathbf{P}(\mathcal{Q}) = \mathbf{Div}_{X/S}$. Moreover, $\mathcal{Q}|P'$ is locally free of rank $\dim \mathrm{H}^0(\mathcal{O}_X(n))$ owing to Subsection 9.3.10. So $\mathcal{Q}|P'$ is of rank at least $1 + \dim P$. Now, there is an m such that the sheaf $\mathcal{H}om(\mathcal{Q}|P', \mathcal{O}_P)(m)$ is generated by finitely many global sections; so a general linear combination of them vanishes nowhere by a well-known lemma [**Mum66**, p. 148] attributed to Serre. Hence there is a surjection $\mathcal{Q}|P' \twoheadrightarrow \mathcal{O}_P(m)$. Correspondingly, there is a P'-map $h'\colon P' \to \mathbf{P}(\mathcal{Q}|P')$; in other words, h' is a section of the restriction over P' of the Abel map $\mathbf{A}_{X/S}\colon \mathbf{Div}_{X/S} \to \mathbf{Pic}_{X/S}$.

Let $W' \subset X_{P'}$ be the pullback under $1 \times h'$ of the universal relative effective divisor. Then $\mathcal{O}_{X_{P'}}(W') = \mathcal{P}'$ since h' is a section of $\mathbf{A}_{X/S}|P'$. So, in particular, $\mathcal{O}_X(W'_\lambda) = \mathcal{O}_X(n)$. Set $W := (1 \times q)^{-1}W'$. Plainly, W is a Poincaré family. □

REMARK 9.5.6. More generally, Theorem 9.5.4 holds whenever X/k is proper, whether X is integral or not. The proof is essential the same, but uses Corollary 9.4.18.3 in place of Theorem 9.4.8. In fact, the proof of the quasi-projectivity assertion is easier, and does not require Proposition 9.5.3, since $\mathbf{Pic}^0_{X/k}$ is given as contained in a quasi-projective scheme. On the other hand, the proof of the projectivity assertion requires an additional step: the reduction, when k is algebraically closed, to the case where X is irreducible. Here is the idea: since X

is normal, X is the disjoint union of its irreducible components X_i, and plainly $\mathbf{Pic}_{X/k} = \prod \mathbf{Pic}_{X_i/k}$; hence, if the $\mathbf{Pic}^0_{X_i/k}$ are complete, so is $\mathbf{Pic}^0_{X/k}$.

Chow [**Cho57**, Thm. p. 128] proved every algebraic group — indeed, every homogeneous variety — is quasi-projective. Hence, in Lemma 9.5.1 , if G is reduced, then G^0 is quasi-projective. In characteristic 0, remarkably G is smooth, so reduced; this result is generally attributed to Cartier, and is proved in [**Mum66**, p. 167].

It follows that Theorem 9.5.4 holds in characteristic 0 whenever X is a proper algebraic k-space. Indeed, the quasi-projectivity assertion holds in view of the preceding discussion and of the discussion at the end of Remark 9.4.18. The proof of the projectivity assertion has one more complication: it is necessary to work with an étale covering $U \to X$ where U is a scheme and with an invertible sheaf $\mathcal{L}$ on $U \times T$. However, the proof shows that, in arbitrary characteristic, if X is a proper and normal algebraic k-space, then $\mathbf{Pic}^0_{X/k}$ is proper.

EXERCISE 9.5.7. Assume X/S is projective and smooth, its geometric fibers are irreducible, and S is Noetherian. Using the Valuative Criterion [**Har77**, Thm. 4.7, p. 101] rather than the Chevalley–Rosenlicht structure theorem, prove that a closed subscheme $Z \subset \mathbf{Pic}_{X/S}$ is projective over S if it is of finite type.

REMARK 9.5.8. There are three interesting alternative proofs of the second assertion of Theorem 9.5.4. The first alternative uses Exercise 9.5.7. It was sketched by Grothendieck [**FGA**, p. 236-12], and runs basically as follows. Proceed by induction on the dimension r of X/k. If $r = 0$, then $\mathbf{Pic}^0_{X/k}$ is S, so trivially projective. If $r = 1$, then $\mathbf{Pic}^0_{X/k}$ is projective by Exercise 9.5.7.

Suppose $r \geq 2$. As in the proof of the theorem, reduce to the case where k is algebraically closed. Let Y be a general hyperplane section of X. Then Y too is normal [**Sei50**, Thm. 7′, p. 376]. Plainly the inclusion $\varphi\colon Y \hookrightarrow X$ induces a map

$$\varphi^* : \mathbf{Pic}^0_{X/k} \to \mathbf{Pic}^0_{Y/k} .$$

By induction, $\mathbf{Pic}^0_{Y/k}$ is projective. So $\mathbf{Pic}^0_{X/k}$ is projective too if φ^* is finite.

In order to handle φ^*, Grothendieck suggested using a version of the "known equivalence criteria." In this connection, he [**FGA**, p. 236-02] announced that [**SGA2**] contains some key preliminary results, which must be combined with the existence theorems for the Picard scheme. In fact, [**SGA2**, Cor. 3.6, p. 153] does directly imply φ^* is injective for $r \geq 3$. Hence φ^* is generically finite. So φ^* is finite since it is homogeneous.

For any $r \geq 2$, [**Kle73**, Lem. 3.11, p. 639, and Rem. 3.12, p. 640] assert $\ker \varphi^*$ is finite and unipotent, so trivial in characteristic 0; however, the proofs in [**Kle73**] require the compactness of $\mathbf{Pic}^0_{X/k}$. It would be good to have a direct proof of this assertion also when $r = 2$, a proof in the spirit of [**SGA2**].

In characteristic 0, Mumford [**Mum67**, p. 99] proved as follows $\ker \varphi^*$ vanishes. Suppose not. First of all, $\ker \varphi^*$ is reduced by Cartier's theorem. So $\ker \varphi^*$ contains a point λ of order $n > 1$. And λ defines an unramified Galois cover X'/X with group $\mathbb{Z}/n$. Set $Y' := Y \times_X X'$. Then Y' is a disjoint union of n copies of Y because $\lambda \in \ker \varphi^*$. On the other hand, Y' is ample since Y is; hence, Y' is connected (by Corollary B.29 for example). We have a contradiction. Thus $\ker \varphi^*$ vanishes.

In characteristic 0, the injectivity of φ^* also follows from the Kodaira Vanishing Theorem. Indeed, as just noted, $\ker \varphi^*$ is reduced. So φ^* is injective if and only if its differential is zero. Now, this differential is, owing to Theorem 9.5.11 below, equal to the natural map

$$\mathrm{H}^1(\mathcal{O}_X) \to \mathrm{H}^1(\mathcal{O}_Y).$$

Its kernel is equal to $\mathrm{H}^1(\mathcal{O}_X(-1))$ owing to the long exact sequence of cohomology.

In characteristic 0, if X is smooth, then $\mathrm{H}^1(\mathcal{O}_X(-1))$ vanishes by the Kodaira Vanishing Theorem. If also $r = 2$, then the dual group $\mathrm{H}^1(\Omega^2_X(1))$ vanishes by the theorem on the regularity of the adjoint system, which was proved by Picard in 1906 using Abelian integrals and by Severi in 1908 using algebro-geometric methods. For more information, see [**Zar71**, pp. 181, 204–206] and [**Mum67**, pp. 94–97].

For any $r \geq 2$, there are, as Grothendieck [**FGA**, p. 236-12] suggested, finitely many smooth irreducible curves $Y_i \subset X$ such that the induced map is injective:

$$\mathbf{Pic}^0_{X/k} \to \prod \mathbf{Pic}^0_{Y_i/k}\,.$$

So, again, since the $\mathbf{Pic}^0_{Y_i/k}$ are projective, $\mathbf{Pic}^0_{X/k}$ is projective too.

To find the Y_i, use the final version of the "equivalence criteria" proved by Weil [**Wei54**, Cor. 2, p. 159]; it says in other words that, if W/T is the family of smooth 1-dimensional linear-space sections of X (or even a nonempty open subfamily), then the induced map is injective:

$$\mathbf{Pic}^0_{X/k}(k) \to \mathbf{Pic}^0_{W/T}(T).$$

For each finite set F of k-points of T, let K_F be the kernel of the map

$$\mathbf{Pic}^0_{X/k} \to \textstyle\prod_{t\in F} \mathbf{Pic}_{W_t/k}\,.$$

Since K_F is closed, we may assume, by Noetherian induction, that K_F contains no strictly smaller K_G. Suppose K_F has a nonzero k-point. It yields a nonzero T-point of $\mathrm{Pic}_{W/T}$, so a nonzero k-point of $\mathbf{Pic}_{W_t/k}$ for some k-point t of T. Let G be the union of F and $\{t\}$. Then K_G is strictly smaller than K_F. So $K_F = 0$. Take the Y_i to be the W_t for $t \in F$.

In characteristic 0 or if $r = 2$, another way to finish is to take a desingularization $\vartheta\colon X' \to X$. Since X is normal, a divisor D on X is principal if ϑ^*D is principal; hence, the induced map

$$\vartheta^* : \mathbf{Pic}^0_{X/k} \to \mathbf{Pic}^0_{X'/k}$$

is injective. But $\mathbf{Pic}^0_{X'/k}$ is projective by Exercise 9.5.7. So $\mathbf{Pic}^0_{X/k}$ is projective too.

The second alternative proof of Theorem 9.5.4 is similar to the proof in Answer 9.5.7. (It too may be due to Grothendieck — see [**EGAIV4**, 21.14.4, iv] — but was indicated to the author by Mumford in a private communication in 1974). Again we reduce to the case where k is algebraically closed. Then, using a more refined form of the Valuative Criterion (obtained modifying [**Har77**, Ex. 4.11, p. 107] slightly), we need only check this statement: given a k-scheme T of the form $T = \mathrm{Spec}(C)$ where C is a complete discrete valuation ring with algebraically closed residue field k_0, and with fraction field K say, and given a divisor D on X_K, its closure D' is a divisor on X_T.

This statement follows from the Ramanujam–Samuel theorem [**EGAIV4**, 21.14.1], a result in commutative algebra. Apply it taking B to be the local ring of a closed point of X_T, and A to be the local ring of the image point in X_{k_0}. The hypotheses hold because A and B share the residue field k_0. The completion $\widehat{A}$ is a normal domain by a theorem of Zariski's [**ZS60**, Thm. 32, p. 320]. And $A \to B$ is formally smooth because C is a formal power series ring over k_0 by a theorem of

Cohen's [**ZS60**, Thm. 32, p. 320].

The third alternative proof is somewhat like the second, but involves some geometry instead of the Ramanujam–Samuel theorem; see [**AK74**, Thm. 19, p. 138]. Moreover, C need not be complete, just discrete, and k_0 need not be algebraically closed. Here is the idea. Let $E \subset X_{k_0}$ be the closed fiber of D'/T. In $\mathbf{Hilb}_{X_L/L}$, form the sets U and V parameterizing the divisorial cycles linearly equivalent to those of the form D_L+H and E_L+H as H ranges over the divisors whose associated sheaves are algebraically equivalent to $\mathcal{O}_{X_L}(n)$ for a suitably large n.

It can be shown that U and V are dense open subsets of the same irreducible component of $\mathbf{Hilb}_{X_L/L}$. Hence they have a common point. Let I and J be the ideals of D and E. Then there are invertible sheaves $\mathcal{L}$ and $\mathcal{M}$ on X_L such that $I_L \otimes \mathcal{L}$ and $J_L \otimes \mathcal{M}$ are isomorphic. Since I_L is invertible, so is J_L. Hence so is J. Thus E is a divisor, as desired.

More generally, if X is not necessarily projective, but is simply complete and normal, then $\mathbf{Pic}^0_{X/k}$ is still complete, whether X is a scheme or algebraic space. Indeed, the original proof and its second alternative work without essential change. The first and third alternatives require X to be projective. However, it is easy to see as follows that this case implies the general case.

Namely, we may assume k is algebraically closed. By Chow's lemma, there is a projective variety Y and a birational map $\gamma\colon Y \to X$. Since X and Y are normal, a divisor D on X is the divisor of a function h if and only if γ^*D is the divisor of γ^*h. Hence the induced map $\mathbf{Pic}^0_{X/k} \to \mathbf{Pic}^0_{Y/k}$ is injective. It follows, as above, that $\mathbf{Pic}^0_{X/k}$ is complete since $\mathbf{Pic}^0_{Y/k}$ is.

DEFINITION 9.5.9. Assume S is the spectrum of a field k. Let $\mathcal{L}$ and $\mathcal{N}$ be invertible sheaves on X. Then $\mathcal{L}$ is said to be *algebraically equivalent* to $\mathcal{N}$ if, for some n and all i with $1 \le i \le n$, there exist a connected k-scheme of finite type T_i, geometric points s_i, t_i of T_i with the same field, and an invertible sheaf $\mathcal{M}_i$ on X_{T_i} such that

$$\mathcal{L}_{s_1} \simeq \mathcal{M}_{1,s_1},\ \mathcal{M}_{1,t_1} \simeq \mathcal{M}_{2,s_2},\ \ldots,\ \mathcal{M}_{n-1,t_{n-1}} \simeq \mathcal{M}_{n,s_n},\ \mathcal{M}_{n,t_n} \simeq \mathcal{N}_{t_n}.$$

PROPOSITION 9.5.10. *Assume that S is the spectrum of a field k. Assume that $\mathbf{Pic}_{X/k}$ exists and represents $\mathrm{Pic}_{(X/k)\,(\mathrm{fppf})}$. Let $\mathcal{L}$ be an invertible sheaf on X, and $\lambda \in \mathbf{Pic}_{X/k}$ the corresponding point. Then $\mathcal{L}$ is algebraically equivalent to $\mathcal{O}_X$ if and only if $\lambda \in \mathbf{Pic}^0_{X/k}$.*

PROOF. Suppose $\mathcal{L}$ is algebraically equivalent to $\mathcal{O}_X$, and use the notation of Definition 9.5.9. Then $\mathcal{M}_i$ defines a map $\tau_i\colon T_i \to \mathbf{Pic}_{X/k}$. Now, $\mathcal{M}_{n,t_n} \simeq \mathcal{O}_{X_{t_n}}$. So $\tau_n(t_n) \in \mathbf{Pic}^0_{X/k}$. Suppose $\tau_i(t_i) \in \mathbf{Pic}^0_{X/k}$. Then $\tau_i(T_i) \subset \mathbf{Pic}^0_{X/k}$ since T_i is connected. So $\tau_i(s_i) \in \mathbf{Pic}^0_{X/k}$. But $\mathcal{M}_{i,s_i} \simeq \mathcal{M}_{i-1,t_{i-1}}$. So $\tau_{i-1}(t_{i-1}) \in \mathbf{Pic}^0_{X/k}$. Descending induction yields $\tau_1(s_1) \in \mathbf{Pic}^0_{X/k}$. But $\mathcal{M}_{1,s_1} \simeq \mathcal{L}_{s_1}$. Thus $\lambda \in \mathbf{Pic}^0_{X/k}$.

Conversely, suppose $\lambda \in \mathbf{Pic}^0_{X/k}$. The inclusion $\mathbf{Pic}^0_{X/k} \hookrightarrow \mathbf{Pic}_{X/k}$ is defined by an invertible sheaf $\mathcal{M}$ on X_T for some fppf covering $T \to \mathbf{Pic}^0_{X/k}$. Let t_1, t_2 be geometric points of T lying over λ, $0 \in \mathbf{Pic}^0_{X/k}$. Let $T_1, T_2 \subset T$ be irreducible components containing t_1, t_2, and $T_1', T_2' \subset \mathbf{Pic}^0_{X/k}$ their images. The latter contain open subsets because an fppf map is open by [**EGAIV2**, 2.4.6]. Since $\mathbf{Pic}^0_{X/k}$ is irreducible by Lemma 9.5.1 , these open subsets contain a common point. Say it is the image of geometric points t_1, s_2 of T_1, T_2. Then $\mathcal{M}_{1,t_1} \simeq \mathcal{M}_{2,s_2}$ by Exercise 9.2.6. Set $\mathcal{M}_i := \mathcal{M}|T_i$. Thus $\mathcal{L}$ is algebraically equivalent to $\mathcal{O}_X$. □

THEOREM 9.5.11. Assume S is the spectrum of a field k. Assume $\mathbf{Pic}_{X/k}$ exists and represents $\mathrm{Pic}_{(X/k)\,(\text{ét})}$. Let $\mathrm{T}_0\,\mathbf{Pic}_{X/k}$ denote the tangent space at 0. Then

$$\mathrm{T}_0\,\mathbf{Pic}_{X/k} = \mathrm{H}^1(\mathcal{O}_X).$$

PROOF. Let P be any k-scheme locally of finite type, $e \in P$ a rational point, A its local ring, and $\mathbf{m}$ its maximal ideal. Usually, by the "tangent space" $\mathrm{T}_e P$ is meant the Zariski tangent space $\mathrm{Hom}(\mathbf{m}/\mathbf{m}^2, k)$. However, as in differential geometry, $\mathrm{T}_e P$ may be viewed as the vector space of k-derivations $\delta\colon A \to k$. Indeed, $\delta(\mathbf{m}^2) = 0$; so δ induces a linear map $\mathbf{m}/\mathbf{m}^2 \to k$. Conversely, every such linear map arises from a δ, and δ is unique because $\delta(1) = 0$.

Let k_ε be the ring of "dual numbers," the ring obtained from k by adjoining an element ε with $\varepsilon^2 = 0$. Then any (fixed) derivation δ induces a local homomorphism of k-algebras $u\colon A \to k_\varepsilon$ by $u(a) := \overline{a} + \delta(a)\varepsilon$ where $\overline{a} \in k$ is the residue of a. Conversely, every such u arises from a unique δ.

On the other hand, to give a u is the same as to give a k-map t_ε from the "free tangent vector" $\mathrm{Spec}(k_\varepsilon)$ to P such that the image of t_ε has support at e. Denote the set of t_ε by $P(k_\varepsilon)_e$. Thus, as sets,

$$T_e P = P(k_\varepsilon)_e. \tag{9.5.11.1}$$

The vector space structure on $T_e P$ transfers as follows. Given $a \in k$, define a k-algebra homomorphism $\mu_a\colon k_\varepsilon \to k_\varepsilon$ by $\mu_a \varepsilon := a\varepsilon$. Now, let δ be a derivation, and u the corresponding homomorphism. Then $a\delta$ corresponds to $\mu_a u$. Thus multiplication by a transfers as $P(\mu_a)\colon P(k_\varepsilon)_e \to P(k_\varepsilon)_e$.

Let $k_{\varepsilon,\varepsilon'}$ denote the ring obtained from k_ε by adjoining an element ε' with $\varepsilon\varepsilon' = 0$ and $(\varepsilon')^2 = 0$. Define a homomorphism $\sigma_1\colon k_{\varepsilon,\varepsilon'} \to k_\varepsilon$ by $\varepsilon \mapsto \varepsilon$ and $\varepsilon' \mapsto 0$. Define another $\sigma_2\colon k_{\varepsilon,\varepsilon'} \to k_\varepsilon$ by $\varepsilon \mapsto 0$ and $\varepsilon' \mapsto \varepsilon$. The σ_i induce a map of sets

$$\pi\colon P(k_{\varepsilon,\varepsilon'})_e \to P(k_\varepsilon)_e \times P(k_\varepsilon)_e$$

where $P(k_{\varepsilon,\varepsilon'})_e$ is the set of maps with image supported at e. Plainly π is bijective.

Define a third homomorphism $\sigma\colon k_{\varepsilon,\varepsilon'} \to k_\varepsilon$ by $\varepsilon \mapsto \varepsilon$ and $\varepsilon' \mapsto \varepsilon$. Given two derivations δ, δ', define a homomorphism $v\colon A \to k_{\varepsilon,\varepsilon'}$ by $v(a) := \overline{a} + \delta(a)\varepsilon + \delta'(a)\varepsilon'$. Then $\delta + \delta'$ corresponds to σv. Therefore, addition on $T_e P$ transfers as

$$\alpha\colon P(k_\varepsilon)_e \times P(k_\varepsilon)_e \to P(k_\varepsilon)_e \text{ where } \alpha := P(\sigma)\pi^{-1}.$$

Suppose P is a group scheme, $e \in P$ the identity. The natural ring homomorphism $\rho\colon k_\varepsilon \to k$ induces a group homomorphism $P(\rho)\colon P(k_\varepsilon) \to P(k)$. Plainly

$$P(k_\varepsilon)_e = \ker P(\rho). \tag{9.5.11.2}$$

The left side $\mathrm{T}_0 P$ is a vector space; the right side is a group. Does addition on the left match multiplication on the right? Yes, indeed! We know α is the addition map. We must show α is also the multiplication map. Let us do so.

Since π and $P(\sigma)$ arise from ring homomorphisms, both are group homomorphisms. Now $\alpha := P(\sigma)\pi^{-1}$. Hence α is a group homomorphism too. So

$$\alpha(m, n) = \alpha(m, e) \cdot \alpha(e, n)$$

where $e \in P(k_\varepsilon)_e$ is the identity. So we have to show $\alpha(m, e) = m$ and $\alpha(e, n) = n$.

Consider the inclusion $\iota\colon k_\varepsilon \to k_{\varepsilon,\varepsilon'}$. Plainly $\sigma_2\iota\colon k_\varepsilon \to k_{\varepsilon'}$ factors through $\rho\colon k_\varepsilon \to k$. Hence $P(\sigma_2)P(\iota)(m) = e$ for any $m \in P(k_\varepsilon)_e$ due to Formula (9.5.11.2). On the other hand, $\sigma_1\iota$ is the identity of k_ε. Thus $\pi P(\iota)(m) = (m, e)$

Plainly $\sigma\iota\colon k_\varepsilon \to k_\varepsilon$ is also the identity of k_ε. So $P(\sigma)P(\iota)(m) = m$. Hence

$$\alpha(m, e) = P(\sigma)\pi^{-1}\pi P(\iota)(m) = m.$$

Similarly $\alpha(e, n) = n$. Thus α is the multiplication map.

Take $P := \mathbf{Pic}_{X/k}$, so $e = 0$. Then Formulas (9.5.11.1) and (9.5.11.2) yield

$$\mathrm{T}_0\,\mathbf{Pic}_{X/k} = \ker\bigl(\mathbf{Pic}_{X/k}(k_\varepsilon) \to \mathbf{Pic}_{X/k}(k)\bigr). \tag{9.5.11.3}$$

To compute this kernel, set $X_\varepsilon := X \otimes_k k_\varepsilon$, and form the truncated exponential sequence of sheaves of Abelian groups:

$$0 \to \mathcal{O}_X \to \mathcal{O}_{X_\varepsilon}^* \to \mathcal{O}_X^* \to 1,$$

where the first map takes a local section b to $1 + b\varepsilon$. This sequence is split by the map $\mathcal{O}_X^* \to \mathcal{O}_{X_\varepsilon}^*$ defined by $a \mapsto a + 0\cdot\varepsilon$. Hence taking cohomology yields this split exact sequence of Abelian groups:

$$0 \to \mathrm{H}^1(\mathcal{O}_X) \to \mathrm{H}^1(\mathcal{O}_{X_\varepsilon}^*) \to \mathrm{H}^1(\mathcal{O}_X^*) \to 1.$$

However, $\mathbf{Pic}_{X/k}$ represents $\mathrm{Pic}_{(X/k)\,(\text{ét})}$, which is the sheaf associated to the presheaf $T \mapsto \mathrm{H}^1(\mathcal{O}_{X_T}^*)$. So there is a natural commutative square of groups

$$\begin{array}{ccc} \mathrm{H}^1(\mathcal{O}_{X_\varepsilon}^*) & \longrightarrow & \mathrm{H}^1(\mathcal{O}_X^*) \\ \downarrow & & \downarrow \\ \mathbf{Pic}_{X/k}(k_\varepsilon) & \to & \mathbf{Pic}_{X/k}(k). \end{array} \tag{9.5.11.4}$$

Hence, there is an induced homomorphism between the horizontal kernels. Owing to Formula (9.5.11.3), this homomorphism is an additive map

$$v\colon \mathrm{H}^1(\mathcal{O}_X) \to \mathrm{T}_0\,\mathbf{Pic}_{X/k}\,.$$

Let $a \in k$. On $T_0\,\mathbf{Pic}_{X/k}$, scalar multiplication by a is, owing to the discussion after Formula (9.5.11.1), the map induced by $\mu_a\colon k_\varepsilon \to k_\varepsilon$. Now, μ_a induces an endomorphism of the above square. At the top, it arises from the map of sheaves of groups $\mathcal{O}_X^* \to \mathcal{O}_X^*$ defined by $\varepsilon \mapsto a\varepsilon$. So the induced endomorphism of $\mathrm{H}^1(\mathcal{O}_X)$ is scalar multiplication by a. Thus v is a map of k-vector spaces.

Square (9.5.11.4) maps to the corresponding square obtained by making a field extension K/k. Since the kernels are vector spaces, there is an induced square

$$\begin{array}{ccc} \mathrm{H}^1(\mathcal{O}_X) \otimes_k K & \longrightarrow & \mathrm{H}^1(\mathcal{O}_{X_K}) \\ {\scriptstyle v\otimes_k K}\downarrow & & \downarrow \\ \mathrm{T}_0\,\mathbf{Pic}_{X/k} \otimes_k K & \to & \mathrm{T}_0\,\mathbf{Pic}_{X_K/K}\,. \end{array}$$

The two horizontal maps are isomorphisms. Hence, if the right-hand map is an isomorphism, so is v. Thus we may assume k is algebraically closed.

In Square (9.5.11.4), the two vertical maps are isomorphisms by Exercise 9.2.3 since $\mathbf{Pic}_{X/k}$ represents $\mathrm{Pic}_{(X/k)\,(\text{ét})}$ and k is algebraically closed. Therefore, v too is an isomorphism, as desired. □

REMARK 9.5.12. There is a relative version of Theorem 9.5.11. Namely, assume $\mathbf{Pic}_{X/S}$ exists, represents $\mathrm{Pic}_{(X/S)\,(\text{ét})}$, and is locally of finite type, but let S be arbitrary. Then $\mathrm{R}^1 f_*\mathcal{O}_X$ is equal to the normal sheaf of $\mathbf{Pic}_{X/S}$ along the identity section, its "Lie algebra"; the latter is simply the dual of the restriction to this section of the sheaf of relative differentials. For more information, see the recent treatment [**LLR04**, § 1] and the references it cites.

COROLLARY 9.5.13. Assume S is the spectrum of a field k. Assume $\mathbf{Pic}_{X/k}$ exists and represents $\mathrm{Pic}_{(X/k)\,(\text{ét})}$. Then

$$\dim \mathbf{Pic}_{X/k} \leq \dim \mathrm{H}^1(\mathcal{O}_X).$$

Equality holds if and only if $\mathbf{Pic}_{X/k}$ is smooth at 0; if so, then $\mathbf{Pic}_{X/k}$ is smooth of dimension $\dim \mathrm{H}^1(\mathcal{O}_X)$ everywhere.

PROOF. Plainly we may assume k is algebraically closed. Then, given any closed point $\lambda \in \mathbf{Pic}_{X/k}$, there is an automorphism of $\mathbf{Pic}_{X/k}$ that carries 0 to λ, namely, "multiplication" by λ. So $\mathbf{Pic}_{X/k}$ has the same dimension at λ as at 0, and $\mathbf{Pic}_{X/k}$ is smooth at λ if and only if it is smooth at 0.

By general principles, $\dim_0 \mathbf{Pic}_{X/k} \leq \dim \mathrm{T}_0\, \mathbf{Pic}_{X/k}$, and equality holds if and only if $\mathbf{Pic}_{X/k}$ is regular at 0. Moreover, $\mathbf{Pic}_{X/k}$ is regular at 0 if and only if it is smooth at 0 since k is algebraically closed. Therefore, the corollary results from Theorem 9.5.11. □

COROLLARY 9.5.14. Assume S is the spectrum of a field k. Assume $\mathbf{Pic}_{X/k}$ exists and represents $\mathrm{Pic}_{(X/k)\,(\text{ét})}$. If k is of characteristic 0, then $\mathbf{Pic}_{X/k}$ is smooth of dimension $\dim \mathrm{H}^1(\mathcal{O}_X)$ everywhere.

PROOF. Since k is of characteristic 0, any group scheme locally of finite type over k is smooth by Cartier's theorem [**Mum66**, Thm. 1, p. 167]. So the assertion follows from Corollary 9.5.13. □

REMARK 9.5.15. Over a field k of positive characteristic, $\mathbf{Pic}_{X/k}$ need not be smooth, even when X is a connected smooth projective surface. Examples were constructed by Igusa [**Igu55**] and Serre [**Ser58**, n° 20].

On the other hand, Mumford [**Mum66**, Lect. 27, pp. 193–198] proved $\mathbf{Pic}_{X/k}$ is smooth if and only if all of Serre's Bockstein operations β_i vanish; here

$$\beta_1 : \mathrm{H}^1(\mathcal{O}_X) \to \mathrm{H}^2(\mathcal{O}_X) \text{ and } \beta_i\colon \ \ker \beta_{i-1} \to \operatorname{cok} \beta_{i-1} \text{ for } i \geq 2.$$

In fact, the tangent space to $\mathbf{Pic}_{X_{\mathrm{red}}/k}$ is the subspace of $\mathrm{H}^1(\mathcal{O}_X)$ given by

$$\mathrm{T}_0\, \mathbf{Pic}_{X_{\mathrm{red}}/k} = \bigcap \ker \beta_i.$$

Moreover, here X need not be smooth or 2-dimensional.

The examples illustrate further pathologies. Set

$$g := \dim \mathbf{Pic}_{X/k},\ h^{0,1} := \dim \mathrm{H}^1(\mathcal{O}_X), \text{ and } h^{1,0} := \dim \mathrm{H}^0(\Omega^1_X).$$

In Igusa's example, $g = 1$, $h^{0,1} = 2$, and $h^{1,0} = 2$; in Serre's, $g = 0$, $h^{0,1} = 1$, and $h^{1,0} = 0$. Moreover, Igusa had just proved that, in any event, $g \leq h^{1,0}$.

By contrast, in characteristic 0, Serre's Comparison Theorem [**Har77**, Thm. 2.1, p. 440] says that $h^{0,1}$ and $h^{1,0}$ can be computed by viewing X as a complex analytic manifold. Hence Hodge Theory yields

$$h^{0,1} = h^{1,0} \text{ and } h^{0,1} + h^{1,0} = b$$

where b is the first Betti number; see [**Zar71**, p. 200]. Therefore, the following exercise now yields the Fundamental Theorem of Irregular Surfaces (9.1.4).

EXERCISE 9.5.16. Assume S is the spectrum of a field k. Assume X is a projective, smooth, and geometrically irreducible surface. According to the original definitions as stated in modern terms, the "geometric genus" of X is the number $p_g := \dim \mathrm{H}^0(\Omega_X^2)$; its "arithmetic genus" is the number $p_a := \varphi(0) - 1$ where $\varphi(n)$ is the polynomial such that $\varphi(n) = \dim \mathrm{H}^0(\Omega_X^2(n))$ for $n \gg 0$; and its "irregularity" q is the difference between the two genera, $q := p_g - p_a$.

Show $\dim \mathbf{Pic}_{X/k} \leq q$, with equality in characteristic 0.

EXERCISE 9.5.17. Assume S is the spectrum of an algebraically closed field k. Assume X is projective and integral. Set $q := \dim \mathrm{H}^1(\mathcal{O}_X)$.

Show $q = 0$ if and only if every algebraic system of curves is "contained completely in a linear system." The latter condition means just that, given any relative effective divisor D on X_T/T where T is a connected k-scheme, there exist invertible sheaves $\mathcal{L}$ on X and $\mathcal{N}$ on T such that $\mathcal{O}_{X_T}(D) \simeq \mathcal{L}_T \otimes f_T^*\mathcal{N}$. The condition may be put more geometrically: in the notation of Exercise 9.3.14, it means there is a map, necessarily unique, $w\colon T \to L$ such that $(1 \times w)^{-1}W = D$.

In characteristic 0, show $q = 0$ if the condition holds for all smooth such T.

REMARK 9.5.18. Assume S is the spectrum of a field k. Assume X/k is projective and geometrically integral. Let $D \subset X$ be an effective divisor, and $\mathcal{N}_D$ its normal sheaf. Let $\delta \in \mathbf{Div}_{X/k}$ be the point representing D, and $\lambda \in \mathbf{Pic}_{X/k}$ the point representing $\mathcal{O}_X(D)$.

Then the tangent space at δ is given by the formula

$$\mathrm{T}_\delta \,\mathbf{Div}_{X/k} = \mathrm{H}^0(\mathcal{N}_D), \tag{9.5.18.1}$$

which respects the vector space structure of each side. This formula can be proved with a simple elementary computation; see [**Mum66**, Cor., p. 154].

Form the fundamental exact sequence of sheaves

$$0 \to \mathcal{O}_X \to \mathcal{O}_X(D) \to \mathcal{N}_D \to 0,$$

and consider its associated long exact sequence of cohomology groups

$$\begin{aligned} 0 \longrightarrow \mathrm{H}^0(\mathcal{O}_X) &\to \mathrm{H}^0(\mathcal{O}_X(D)) \to \mathrm{H}^0(\mathcal{N}_D) \\ \xrightarrow{\partial^0} \mathrm{H}^1(\mathcal{O}_X) &\to \mathrm{H}^1(\mathcal{O}_X(D)) \xrightarrow{u} \mathrm{H}^1(\mathcal{N}_D) \xrightarrow{\partial^1} \mathrm{H}^2(\mathcal{O}_X). \end{aligned}$$

Another elementary computation shows that the boundary map ∂^0 is equal to the tangent map of the Abel map, $\mathrm{T}_\delta \,\mathbf{Div}_{X/k} \to T_\lambda \,\mathbf{Pic}_{X/k}$; see [**Mum66**, Prp., p. 165].

By definition, D is said to be "semiregular" if the boundary map ∂^1 is injective. Plainly, it is equivalent that $u = 0$. So it is equivalent that $\dim \mathrm{H}^0(\mathcal{N}_D) = R$ where

$$R := \dim \mathrm{H}^1(\mathcal{O}_X) + \dim \mathrm{H}^0(\mathcal{O}_X(D)) - 1 - \dim \mathrm{H}^1(\mathcal{O}_X(D)).$$

Semiregularity was recognized in 1944 by Severi as precisely the right positivity condition for the old (1904) theorem of the completeness of the characteristic system, although Severi formulated the condition in an equivalent dual manner.

In its modern formulation, the theorem of the completeness of the characteristic system asserts that $\mathbf{Div}_{X/k}$ is smooth of dimension R at δ if and only if D is semiregular, provided the characteristic is 0, or more generally, $\mathbf{Pic}_{X/k}$ is smooth. Indeed, Formula (9.5.18.1) says that the characteristic system of $\mathbf{Div}_{X/k}$ on D is always equal to the complete linear system of the invertible sheaf $\mathcal{N}_D$. But $\mathbf{Div}_{X/k}$ can have a nilpotent or a singularity at δ. So, in effect, Enriques and Severi had simply sought conditions guaranteeing $\mathbf{Div}_{X/k}$ is smooth of dimension R at δ.

The first purely algebraic discussion of the theorem was made by Grothendieck [**FGA**, Sects. 221-5.4 to 5.6], and he proved it in the two most important cases. Specifically, he noted that, if $\mathrm{H}^1(\mathcal{O}_X(D)) = 0$, then the Abel map is smooth; see the end of the proof of Theorem 9.4.8. Hence $\mathbf{Div}_{X/k}$ is smooth at δ if and only if $\mathbf{Pic}_{X/k}$ is smooth; furthermore, $\mathbf{Pic}_{X/k}$ is smooth in characteristic zero by Cartier's theorem. Grothendieck pointed out that this case had been treated with transcendental means by Kodaira in 1956.

Grothendieck also observed $\mathrm{H}^1(\mathcal{N}_D)$ houses the obstruction to deforming D in X. Hence, if this group vanishes, then $\mathbf{Div}_{X/k}$ is smooth at δ in any characteristic. Mumford [**Mum66**, pp. 157–159] explicitly worked out the obstruction and its image under ∂^1. If $\partial^1 = 0$, so if D is semiregular, then $\mathbf{Div}_{X/k}$ is smooth if this image vanishes. Inspired by work of Kodaira and Spencer in 1959, Mumford used the exponential in characteristic 0 and proved the image vanishes. Mumford did not use Cartier's theorem; so the latter results from taking a D with $\mathrm{H}^1(\mathcal{O}_X(D)) = 0$.

A purely algebraic proof of the full completeness theorem is given in [**Kle73**, Thm., p. 307]. This proof was inspired by Kempf's (unpublished) thesis. The proof does not use obstruction theory, but only simple formal properties of a scheme of the form $\mathbf{P}(\mathcal{Q})$ where $\mathcal{Q}$ arises from an invertible sheaf $\mathcal{F}$ as in Subsection 9.3.10. This proof works, more generally, if S is arbitrary and if X/S is projective and flat and has integral geometric fibers. Here D is the divisor on the geometric fiber through δ. Then provided $\mathbf{Pic}_{X/S}$ is smooth, $\mathbf{Div}_{X/S}$ is smooth of relative dimension R at δ if and only if D is semiregular.

There is a celebrated example, valid over an algebraically closed field k of any characteristic, where $\mathbf{Div}_{X/k}$ is nonreduced at δ. The example was discovered by Severi and Zappa in the 1940s, and is explained in [**Mum66**, pp. 155–156]. Here is the idea. Let C be an elliptic curve, and $0 \to \mathcal{O}_C \to \mathcal{E} \to \mathcal{O}_C \to 0$ the nontrivial extension; set $X := \mathbf{P}(\mathcal{E})$.

Let D be the section of X/C defined by $\mathcal{E} \to \mathcal{O}_C$. Then $\mathcal{N}_D = \mathcal{O}_C$ by [**Har77**, Prp. 2.8, p. 372]; so $\dim T_\delta \mathbf{Div}_{X/k} = 1$. However, δ is an isolated point. Otherwise, the connected component of δ contains a second closed point. And it represents a curve D' algebraically equivalent to D. So $\deg \mathcal{O}_D(D') = \deg \mathcal{N}_D = 0$. Hence D and D' are disjoint. Let F be a fiber. Then $\deg \mathcal{O}_F(D) = \deg \mathcal{O}_F(D') = 1$. Hence D' is a second section. Therefore, $\mathcal{E}$ is decomposable, a contradiction.

PROPOSITION 9.5.19. *Assume* $\mathbf{Pic}_{X/S}$ *exists and represents* $\mathrm{Pic}_{(X/S)\,(\text{ét})}$. *Let* $s \in S$ *be a point such that* $\mathrm{H}^2(\mathcal{O}_{X_s}) = 0$. *Then there exists an open neighborhood of* s *over which* $\mathbf{Pic}_{X/S}$ *is smooth.*

PROOF. By the Semicontinuity Theorem [**EGAIII2**, 7.7.5-I)], there exists an open neighborhood U of s such that $\mathrm{H}^2(\mathcal{O}_{X_t}) = 0$ for all $t \in U$. Replace S by U.

By [**EGAIII1**, **0**-10.3.1, p. 20], there is a flat local homomorphism from $\mathcal{O}_s$ into a Noetherian local ring B whose residue field is algebraically closed. By [**EGAIV2**, 6.8.3], if f_B is smooth, then $f\colon X \to S$ is smooth along X_s. Replace S by Spec(B).

By the Infinitesimal Criterion for Smoothness [**SGA1**, p. 67], it suffices to show this: given any S-scheme T of the form $T = \mathrm{Spec}(A)$ where A is an Artin local ring that is a finite $\mathcal{O}_s$-algebra and given any closed subscheme $R \subset T$ whose ideal $\mathcal{I}$ has square 0, every R-point of $\mathbf{Pic}_{X/S}$ lifts to a T-point.

The residue field of A is a finite extension of k_s, which is algebraically closed; so the two fields are equal. Hence the R-point is defined by an invertible sheaf on

X_R by Exercise 9.2.3. So we want to show invertible sheaves on X_R lift to X_T.

Since $\mathcal{I}^2 = 0$, we can form the truncated exponential sequence

$$0 \to f_T^*\mathcal{I} \to \mathcal{O}_{X_T}^* \to \mathcal{O}_{X_R}^* \to 1.$$

It yields the following exact sequence:

$$\mathrm{H}^1(\mathcal{O}_{X_T}^*) \to \mathrm{H}^1(\mathcal{O}_{X_R}^*) \to \mathrm{H}^2(f_T^*\mathcal{I}).$$

Hence it suffices to show $\mathrm{H}^2(f_T^*\mathcal{I}) = 0$.

Since T is affine, $\mathrm{H}^2(f_T^*\mathcal{I}) = H^0(\mathrm{R}^2 f_{T*} f_T^*\mathcal{I})$ owing to [**Har77**, Prp. 8.5, p. 251]. But $\mathrm{H}^2(\mathcal{O}_{X_t}) = 0$ for all $t \in S$. Hence $\mathrm{R}^2(f_T^*\mathcal{I}) = 0$ by the Property of Exchange [**EGAIII2**, 7.7.5 II and 7.7.10]. Thus $\mathrm{H}^2(f_T^*\mathcal{I}) = 0$, as desired. □

PROPOSITION 9.5.20. *Assume* $\mathbf{Pic}_{X/S}$ *exists, and represents* $\mathrm{Pic}_{(X/S)\,(\mathrm{fppf})}$. *For* $s \in S$, *assume all the* $\mathbf{Pic}^0_{X_s/k_s}$ *are smooth of the same dimension. Then* $\mathbf{Pic}_{X/S}$ *has an open group subscheme* $\mathbf{Pic}^0_{X/S}$ *of finite type whose fibers are the* $\mathbf{Pic}^0_{X_s/k_s}$. *Furthermore, if* S *is reduced, then* $\mathbf{Pic}^0_{X/S}$ *is smooth over* S. *Moreover, if all the* $\mathbf{Pic}^0_{X_s/k_s}$ *are complete and if* $\mathbf{Pic}_{X/S}$ *is separated over* S, *then* $\mathbf{Pic}^0_{X/S}$ *is closed in* $\mathbf{Pic}_{X/S}$ *and proper over* S.

PROOF. First off, $\mathbf{Pic}_{X/S}$ is locally of finite type by Proposition 9.4.17. Now, for every $s \in S$, the schemes $\mathbf{Pic}_{X/S} \otimes k_s$ and $\mathbf{Pic}_{X_s/k_s}$ coincide by Exercise 9.4.4. And the $\mathbf{Pic}^0_{X_s/k_s}$ are smooth of the same dimension by hypothesis. So the $\mathbf{Pic}^0_{X_s/k_s}$ form an open subscheme $\mathbf{Pic}^0_{X/S}$ of $\mathbf{Pic}_{X/S}$ and the structure map $\sigma\colon \mathbf{Pic}^0_{X/S} \to S$ is universally open by [**EGAIV3**, 15.6.3 and 15.6.4]. Furthermore, if S is reduced, then σ is flat by [**EGAIV3**, 15.6.7], so smooth by [**EGAIV4**, 17.5.1].

Define a map $\alpha\colon \mathbf{Pic}^0_{X/S} \times_S \mathbf{Pic}^0_{X/S} \to \mathbf{Pic}_{X/S}$ by $\alpha(g,h) := gh^{-1}$. Then α factors through the open subscheme $\mathbf{Pic}^0_{X/S}$ because forming α commutes with passing to the fibers. Hence $\mathbf{Pic}^0_{X/S}$ is a subgroup

To prove $\mathbf{Pic}^0_{X/S}$ is of finite type, we may work locally on S, and so assume S is Noetherian. Since $\mathbf{Pic}_{X/S}$ is locally of finite type, we need only prove $\mathbf{Pic}^0_{X/S}$ is quasi-compact.

Let $V \subset S$ be a nonempty affine open subscheme, and $U \subset \sigma^{-1}V$ another. Then $\sigma U \subset S$ is open since σ is an open map. Set $U' := \sigma^{-1}\sigma U$. Then U' is open, and α restricts to a map $\alpha'\colon U \times_V U \to U'$. In fact, α' is surjective because its geometric fibers are surjective by an argument at the end of the proof of Lemma 9.5.1. Now, $U \times_V U$ so quasi-compact. Hence U' is quasi-compact.

Set $T := S - \sigma U$. By Noetherian induction, we may assume $\sigma^{-1}T$ is quasi-compact. Therefore, $U' \cup \sigma^{-1}T$ is quasi-compact. But it is equal to $\mathbf{Pic}^0_{X/S}$. Thus $\mathbf{Pic}^0_{X/S}$ is quasi-compact, as desired.

Moreover, if all the $\mathbf{Pic}^0_{X_s/k_s}$ are complete and if $\mathbf{Pic}_{X/S}$ is separated over S, then $\mathbf{Pic}^0_{X/S}$ is proper over S by [**EGAIV3**, 15.7.11]. Finally, consider the inclusion map $\mathbf{Pic}^0_{X/S} \hookrightarrow \mathbf{Pic}_{X/S}$. It is proper as $\mathbf{Pic}^0_{X/S}$ is proper and $\mathbf{Pic}_{X/S}$ is separated; hence, $\mathbf{Pic}^0_{X/S}$ is closed. □

REMARK 9.5.21. Assume the characteristic is 0 and $f\colon X \to S$ is smooth and proper. Then, in the last two propositions, more can be said, as Vistoli explained to the author in early May 2004. These additions result from Part (i) of Theorem (5.5) on p. 123 in Deligne's article [**Del68**] (which uses Hodge Theory on p. 121). Deligne's theorem asserts that, under the present conditions, all the sheaves $\mathrm{R}^q f_* \Omega^p_{X/S}$ are locally free of finite rank, and forming them commutes with changing the base.

In Proposition 9.5.19, *$\mathbf{Pic}_{X/S}$ is smooth over the connected component S_0 of s.* Indeed, Deligne's theorem implies $\mathrm{R}^2 f_* \mathcal{O}_{X/S} | S_0$ vanishes since it is locally free, so of constant rank, and its formation commutes with passage to every fiber, in particular, to that over $s \in S_0$. Hence, $\mathrm{H}^2(\mathcal{O}_{X_t}) = 0$ for every $t \in S_0$. Therefore, Proposition 9.5.19, as it stands, implies $\mathbf{Pic}_{X/S}$ is smooth over S_0.

In Proposition 9.5.20, *there is no need to assume all the $\mathbf{Pic}^0_{X_s/k_s}$ are smooth of the same dimension.* Indeed, all the $\mathbf{Pic}^0_{X_s/k_s}$ are smooth of dimension $\dim \mathrm{H}^1(\mathcal{O}_{X_s})$ by Corollary 9.5.13. But $\dim \mathrm{H}^1(\mathcal{O}_{X_s})$ is constant on each connected component of S owing to Deligne's theorem.

Furthermore, if $f\colon X \to S$ is smooth and projective Zariski locally over S, then $\mathbf{Pic}^0_{X/S}$ is smooth whether or not S is reduced. Indeed, we may assume S is of finite type over $\mathbb{C}$; the reduction is standard, and sketched by Deligne at the beginning of his proof of his Theorem (5.5). By the Infinitesimal Criterion for Smoothness [**SGA1**, p. 67], it suffices to show this: given any S-scheme T of the form $T = \mathrm{Spec}(A)$ where A is an Artin local ring that is a finite $\mathbb{C}$-algebra and given any closed subscheme $R \subset T$, every R-point of $\mathbf{Pic}^0_{X/S}$ lifts to a T-point. But, $\mathbf{Pic}^0_{X/S}$ is open in $\mathbf{Pic}_{X/S}$ by the above argument. Hence it suffices to show every R-point of $\mathbf{Pic}^0_{X/S}$ lifts to a T-point of $\mathbf{Pic}_{X/S}$.

The residue field of A is a finite extension of $\mathbb{C}$; so the two fields are equal. Hence the R-point is defined by an invertible sheaf on X_R by Exercise 9.2.3. So we want to show every invertible sheaf $\mathcal{L}$ on X_R lifts to X_T. By Serre's Comparison Theorem [**Har77**, Thm. 2.1, p. 440], it suffices to lift $\mathcal{L}$ to an analytic invertible sheaf since $f_T\colon X_T \to T$ is projective. So pass now to the analytic category.

Let X_0 denote the closed fiber of X_T, and form the exponential sequence:

$$0 \to \mathbb{Z}_{X_0} \to \mathcal{O}_{X_0} \to \mathcal{O}^*_{X_0} \to 0.$$

Consider the class of $\mathcal{L}_{X_0}$ in $\mathrm{H}^1(\mathcal{O}^*_{X_0})$. It maps to 0 in $\mathrm{H}^2(\mathbb{Z}_{X_0})$ because this group is discrete and $\mathcal{L}$ defines an R-point of $\mathbf{Pic}^0_{X/S}$.

Form the exponential map $\mathcal{O}_{X_R} \to \mathcal{O}^*_{X_R}$, form its kernel $\mathcal{Z}$, and form the natural map $\kappa\colon \mathcal{Z} \to Z_{X_0}$. Then κ is bijective. Indeed, let a be a local section of $\mathcal{Z}$. Then $1 + a + a^2/2 + \cdots = 1$. Set $u := 1 + a/2 + \cdots$. Then $au = 0$. Suppose a maps to the local section 0 of $\mathcal{O}_{X_0}$. Then a is nilpotent. So u is invertible. Hence $a = 0$. Thus κ is injective. But κ is obviously surjective. Thus κ is bijective.

Consider the class λ of $\mathcal{L}$ in $\mathrm{H}^1(\mathcal{O}^*_{X_R})$. It follows that λ maps to 0 in $\mathrm{H}^2(\mathcal{Z})$. Hence λ comes from a class γ in $\mathrm{H}^1(\mathcal{O}_{X_R})$. By Deligne's theorem, γ lifts to a class γ' in $\mathrm{H}^1(\mathcal{O}_{X_T})$. The image of γ' in $\mathrm{H}^1(\mathcal{O}^*_{X_T})$ gives the desired lifting of $\mathcal{L}$ to X_T.

EXAMPLE 9.5.22. The following example complements Proposition 9.5.20, and was provided, in early May 2004, by Vistoli. The example shows *$\mathbf{Pic}^0_{X/S}$ can be smooth and proper over S and open and closed in $\mathbf{Pic}_{X/S}$, although $\mathbf{Pic}_{X/S}$ isn't smooth.* In fact, $f\colon X \to S$ is smooth and projective, its geometric fibers are integral, and S is a smooth curve over an algebraically closed field k of arbitrary characteristic. Furthermore, $\mathbf{Pic}_{X/S}$ has a component that is a reduced k-point; so it is not smooth over S, nor even flat.

To construct $f\colon X \to S$, set $P := \mathbb{P}^3_k$ and fix $d \geq 4$. Set $\mathcal{Q} := \mathrm{H}^0(\mathcal{O}_P(d))^*$ where the '$*$' means dual. Set $H := \mathbb{P}(\mathcal{Q})$. Then H represents the functor (on

k-schemes T) whose T-points are the T-flat closed subschemes of P_T whose fibers are surfaces of degree d by Exercise 9.5.17 and Theorem 9.3.13. Let $W \subset P \times H$ be the universal subscheme. Its ideal $\mathcal{O}_{P\times H}(-W)$ is equal to the tensor product of the pullbacks of $\mathcal{O}_P(-d)$ and $\mathcal{O}_H(-1)$ by Exercise 9.3.14. Set $N := \dim H$.

Let G be the Grassmannian of lines in P, and $L \subset P \times G$ the universal line. Let $\pi\colon L \to G$ be the projection. Form the exact sequence of locally free sheaves

$$0 \to \mathcal{K} \to \pi_*\mathcal{O}_{P\times G}(d) \to \pi_*\mathcal{O}_L(d) \to 0,$$

which defines $\mathcal{K}$; the right-hand map is surjective, since forming it commutes with passing to the fibers, and on the fibers, it is plainly surjective. Set $I := \mathbf{P}(\mathcal{K}^*)$. Then I is smooth, irreducible, and of dimension $4 + N - (d+1)$, or $N - d + 3$.

Note that $\pi_*\mathcal{O}_{P\times G}(d) = \mathcal{Q}_G^*$. So there is a surjection $\mathcal{Q}_G \to \mathcal{K}^*$, and it induces a closed embedding $I \subset G \times H$. Furthermore, given a T-point of $G \times H$, it lies in I if and only if $L_T \subset W_T$. Indeed, the latter means the ideal of $W_T \subset P_T$ maps to 0 in $\mathcal{O}_{L_T}$; in other words, the composition

$$\mathcal{O}_{P_T}(-d) \otimes \mathcal{O}_H(-1)_{P_T} \to O_{P_T} \to \mathcal{O}_{L_T}$$

vanishes. Equivalently, the composition

$$\mathcal{O}_H(-1)_T \to g_{T*}\mathcal{O}_{P_T}(d) \to g_{T*}\mathcal{O}_{L_T}(d)$$

vanishes. But the first map is equal to the natural map $\mathcal{O}_H(-1)_T \to \mathcal{Q}_T^*$. So it is equivalent that this map factors through $\mathcal{K}_T$, or that $\mathcal{Q}_T \to \mathcal{O}_H(1)_T$ factors through $\mathcal{K}_T^*$. Equivalently, the T-point of $G \times H$ lies in I. Thus I is the graph of the incidence correspondence.

Let $J \subset H$ be the image of I. Note $J \neq H$ since $\dim I = \dim H - d + 3$ and $d \geq 4$. Consider the open set $U \subset H$ over which $W \to H$ is smooth. Then $U \cap J$ is nonempty. Indeed, choose coordinates w, x, y, z for P. Then $U \cap J$ contains the point representing the surface $\{w^d - x^d = y^d - z^d\}$ if $p \nmid d$ or the surface $\{wx^{d-1} = y^{d-1}\}$ if $p \mid d$, because, in either case, the surface is smooth and contains a line, either $\{w = x,\ y = z\}$ or $\{w = 0,\ y = 0\}$.

Let $s \in U \cap J$ be a simple k-point. Take a line $S \subset H$ through s and transverse to J. Replace S by $S \cap U$. Let $X \subset W$ be the preimage of S, and $f\colon X \to S$ the induced map. Then f is smooth and projective, and its geometric fibers are integral. Hence $\mathbf{Pic}_{X/S}$ exists, is separated, and represents $\mathrm{Pic}_{(X/S)\,(\text{ét})}$ by Theorem 9.4.8. Moreover, $\mathrm{H}^1(\mathcal{O}_{X_t}) = 0$ for each $t \in S$; hence, Corollary 9.5.13 implies $\mathbf{Pic}_{X_t/k_t}$ is reduced and discrete. In particular, $\mathbf{Pic}^0_{X_t/k_t}$ is smooth, of dimension 0, and complete. Therefore, Proposition 9.5.20 implies $\mathbf{Pic}^0_{X/S}$ is smooth and proper over S and open and closed in $\mathbf{Pic}_{X/S}$

Since $s \in J$, the surface $X_s \subset P$ contains a line M. And X_s is smooth since $s \in U$; hence M is a divisor. Say $\mathcal{O}_{X_s}(M)$ defines the k-point $\mu \in \mathbf{Pic}_{X/S}$. View μ as a reduced closed subscheme. Then μ is a connected component of its fiber $\mathbf{Pic}_{X_s/k}$ because, as just noted, this fiber is reduced and discrete. It remains to prove μ is a connected component of $\mathbf{Pic}_{X/S}$.

Suppose not, and let's find a contradiction. Let Q be the connected component of $\mu \in \mathbf{Pic}_{X/S}$. Let k_ε be the ring of "dual numbers," and set $T := \mathrm{Spec}(k_\varepsilon)$. Then there is a closed embedding $T \hookrightarrow \mathbf{Pic}_{X/S}$ supported at μ. However, T does not embed into the fiber $\mathbf{Pic}_{X_s/k}$ because the latter is reduced and discrete. Hence the structure map $\mathbf{Pic}_{X/S} \to S$ embeds T into S.

The embedding $T \hookrightarrow \mathbf{Pic}_{X/S}$ corresponds to an invertible sheaf $\mathcal{M}$ on X_T by Exercise 9.2.3 since k is algebraically closed. Moreover, $\mathcal{M}|X_s \simeq \mathcal{O}_{X_s}(M)$ since the embedding is supported at μ.

Note $\mathrm{H}^1(\mathcal{O}_{X_s}(M)) = 0$. Indeed, by Serre duality [**Har77**, Cor. 7.7, p. 244, and Cor. 7.12, p. 246], it suffices to show $\mathrm{H}^1(\omega_{X_s}(-M)) = 0$. Form the sequence

$$0 \to \omega_{X_s}(-M) \to \omega_{X_s} \to \omega_{X_s}|M \to 0.$$

Now, $\omega_{X_s} \simeq \mathcal{O}_{X_s}(d-4)$ since $\omega_{X_s} \simeq \omega_P \otimes \mathcal{O}_{X_s}(X_s)$ by [**Har77**, Prp. 8.20, p. 182] and $\omega_P \simeq \mathcal{O}_P(-4)$ by [**Har77**, Eg. 8.20.1, p. 182]. But $\mathrm{H}^1(\mathcal{O}_{X_s}(d-4)) = 0$ because $X_s \subset P$ is a hypersurface, and $\mathrm{H}^0(\mathcal{O}_{X_s}(d-4)) \to \mathrm{H}^0(\mathcal{O}_M(d-4))$ is surjective because $\mathrm{H}^0(\mathcal{O}_P(d-4)) \to \mathrm{H}^0(\mathcal{O}_M(d-4))$ is. Thus $\mathrm{H}^1(\mathcal{O}_{X_s}(M)) = 0$.

Therefore, $\mathrm{H}^0(\mathcal{M}) \otimes k \to \mathrm{H}^0(\mathcal{O}_{X_s}(M))$ is surjective by the implication (v)$\Rightarrow$(iv) of Subsection 9.3.10. So the section of $\mathcal{O}_{X_s}(M)$ defining M extends to a section of $\mathcal{M}$. The extension defines a relative effective divisor on X_T, which restricts to M, owing to the implication (iii)$\Rightarrow$(i) of Lemma 9.3.4. It follows that the embedding $T \hookrightarrow H$ factors through I, so through J. However, J and S meet transversally at s; whence, T cannot embed into $J \cap S$. Here is the desired contradiction. The discussion is now complete.

EXERCISE 9.5.23. Assume X/S is projective and flat, its geometric fibers are integral curves of arithmetic genus p_a, and S is Noetherian. Show the "generalized Jacobians" $\mathbf{Pic}^0_{X_s/k_s}$ form a smooth quasi-projective family of relative dimension p_a. And show this family is projective if and only if X/S is smooth.

REMARK 9.5.24. Assume X is an Abelian S-scheme of relative dimension g; that is, X is a smooth and proper S-group scheme with geometrically connected fibers of dimension g. Then X needn't be projective Zariski locally over S.

Indeed, according to Raynaud [**Ray66**, Rem. 8.c, p. 1315], Grothendieck found two such examples: one where S is reduced and 1-dimensional, and another where S is the spectrum of the ring of dual numbers of a field of characteristic 0. In [**Ray70**, Ch. XII], Raynaud gave detailed constructions of similar examples.

Assume X/S is projective, and S is Noetherian. Then $\mathbf{Pic}^0_{X/S}$ exists, and is also a projective Abelian S-scheme of relative dimension g. Set $X^* := \mathbf{Pic}^0_{X/S}$.

Indeed, $\mathbf{Pic}_{X/S}$ exists, represents $\mathrm{Pic}_{(X/S)\,(\text{ét})}$, and is locally of finite type by the main theorem, Theorem 9.4.8. For $s \in S$, the $\mathbf{Pic}^0_{X_s/k_s}$ are smooth and proper of dimension g by [**Mum70**, § 13]. Hence X^* exists by Proposition 9.5.20, and is projective by Exercise 9.5.7. Finally, X^* is smooth owing to a more sophisticated version of the proof of Proposition 9.5.19; see [**Mum65**, pp. 117–118].

There exists a universal sheaf $\mathcal{P}$ on $X \times \mathbf{Pic}_{X/S}$ by Exercise 9.4.3 since f has a section g, namely, the identity section. Normalize $\mathcal{P}$ by tensoring it with $f^*_{X^*} g^*_{X^*} \mathcal{P}$. Then its restriction to $X \times X^*$ defines a map, which is a "duality" isomorphism

$$\pi \colon X \xrightarrow{\sim} X^{**}.$$

Indeed, forming π commutes with changing S, and π's geometric fibers are isomorphisms by [**Mum70**, Cor., p. 132]. But X and X^{**} are proper over S, and X is flat. Therefore, π is an isomorphism by [**EGAIII1**, 4.6.7].

REMARK 9.5.25. Assume S is the spectrum of an algebraically closed field k, and X is normal, integral, and projective. Then $\mathbf{Pic}^0_{X/k}$ is irreducible and projective by Proposition 9.5.3 and Theorem 9.5.4. Set $P := (\mathbf{Pic}^0_{X/k})_{\mathrm{red}}$ and $A := \mathbf{Pic}^0_{P/k}$.

Then P is plainly an Abelian variety; whence, A is an Abelian variety too by Remark 9.5.24.

Fix a point $x \in X(k)$. Let B be an Abelian variety, and set $B^* := \mathbf{Pic}^0_{B/k}$. Then B^* is an Abelian variety too, and there is a canonical isomorphism $B \xrightarrow{\sim} B^{**}$ by Remark 9.5.24. Let $\xi\colon X \to X \times B^*$ be the map defined by $0 \in B^*(k)$, and let $\beta\colon B^* \to X \times B^*$ be the map defined by x.

Consider a map $a\colon B^* \to P$ such that $a(0) = 0$. By the Comparison Theorem, Theorem 9.2.5, the map a corresponds to an invertible sheaf $\mathcal{L}$ on $X \times B^*$ such that $\xi^*\mathcal{L} \simeq \mathcal{O}_X$. Normalize $\mathcal{L}$ by tensoring it with $(\beta^*\mathcal{L})_X$. Then $\mathcal{L}$ defines a map $b\colon X \to B$ such that $b(x) = 0$.

Reversing the preceding argument, we see that every such b arises from a unique map $a\colon B^* \to \mathbf{Pic}_{X/k}$ such that $a(0) = 0$. Since B^* is integral, a factors through P. Thus the maps $a\colon B^* \to P$ and $b\colon X \to B$ are in bijective correspondence. Plainly, this correspondence is compatible with maps $b'\colon B \to B'$ such that $b'(0) = 0$. In particular, 1_P corresponds to a natural map $u\colon X \to A$ such that $u(x) = 0$, and every map $b\colon X \to B$ factors uniquely through u.

REMARK 9.5.26. Assume X/S is projective and smooth, its geometric fibers are connected curves of genus $g > 0$, and S is Noetherian. Set $J := \mathbf{Pic}^0_{X/S}$; it exists and is a projective Abelian S-scheme by Exercise 9.5.23. Set $J^* := \mathbf{Pic}^0_{J/S}$; it exists, is a projective Abelian scheme, and is "dual" to J by Remark 9.5.24.

Suppose X has an invertible sheaf $\mathcal{L}$ whose fibers $\mathcal{L}_s$ are of degree 1. Define an associated "Abel" map

$$A_{\mathcal{L}}\colon X \to J$$

directly on T-points as follows. Given $t\colon T \to X$, its graph subscheme $\Gamma_t \subset X \times T$ is a relative effective divisor; see Answer 9.4.13. Use $\mathcal{L}_T \otimes \mathcal{O}_{X_T}(-\Gamma)$ to define $A_{\mathcal{L}}(t)$.

Then $A_{\mathcal{L}}$ induces, via pullback, an "autoduality" isomorphism

$$A^*_{\mathcal{L}}\colon J^* \xrightarrow{\sim} J.$$

This isomorphism is independent of the choice of $\mathcal{L}$; in fact, it exists even if no $\mathcal{L}$ does. These facts are proved in [**EGK02**, Thm. 2.1, p. 595]. In fact, a more general autoduality result is proved: it applies to the natural compactification of J, which parameterizes torsion-free sheaves, when the geometric fibers of X are not necessarily smooth, but integral with double points at worst. And the proof starts from scratch, recovering the original case of a single smooth curve over a field.

REMARK 9.5.27. Assume X/S is proper and flat. Assume its geometric fibers are curves, but not necessarily integral. Then there are two remarkable theorems asserting the existence of $\mathbf{Pic}_{X/S}$ as an algebraic space and of $\mathbf{Pic}^0_{X/S}$ as a separated S-scheme. These theorems are important in the theory of Néron models; so in [**BLR90**, Sect. 9.4], their proofs are sketched, and the original papers, cited.

One theorem is due to Raynaud. He assumes, in addition, that S is the spectrum of a discrete valuation ring, that X is normal, and that $\mathcal{O}_S \xrightarrow{\sim} f_*\mathcal{O}_X$ holds. Furthermore, given any geometric fiber of X/S, he measures the lengths of the local rings at the generic points of its irreducible components, and he assumes their greatest common divisor is 1. Then he proves the above existence assertions.

The other theorem is due to Deligne. Instead, he assumes, in addition, X/S is semi-stable; that is, its geometric fibers are reduced and connected, and have, at worst, ordinary double points. Then he proves, in addition, $\mathbf{Pic}^0_{X/S}$ is smooth and quasi-projective; in fact, it carries a canonical S-ample invertible sheaf.

9.6. The torsion component of the identity

This section establishes the two main finiteness theorem for $\mathbf{Pic}_{X/S}$, when X/S is projective and its geometric fibers are integral. The first theorem asserts the finiteness of the torsion component $\mathbf{Pic}^{\tau}_{X/S}$, an open and closed group subscheme. By definition, it consists of the points with a multiple in the connected component $\mathbf{Pic}^{0}_{X/S}$, which was studied in the previous section.

The second theorem asserts the finiteness of a larger sort of subset $\mathbf{Pic}^{\varphi}_{X/S}$. Its points represent the invertible sheaves with a given Hilbert polynomial φ. The section starts by developing numerical characterizations of $\mathbf{Pic}^{\tau}_{X/S}$, or rather of the corresponding invertible sheaves, when S is the spectrum of an algebraically closed field. This development assumes some familiarity with basic intersection theory, which is developed in Appendix B.

DEFINITION 9.6.1. Assume S is the spectrum of a field. Let $\mathcal{L}$ and $\mathcal{N}$ be invertible sheaves on X. Then $\mathcal{L}$ is said to be *τ-equivalent* to $\mathcal{N}$ if, for some nonzero m depending on $\mathcal{L}$ and $\mathcal{N}$, the mth power $\mathcal{L}^{\otimes m}$ is algebraically equivalent to $\mathcal{N}^{\otimes m}$.

In addition, $\mathcal{L}$ is said to be *numerically equivalent* to $\mathcal{N}$ if, for every complete curve $Y \subset X$, the corresponding intersection numbers are equal:

$$\textstyle\int c_1\mathcal{L}\cdot[Y] = \int c_1\mathcal{N}\cdot[Y].$$

It is sufficient, by additivity, to take Y to be complete and integral. It is then equivalent that $\deg \mathcal{L}_Y = \deg \mathcal{N}_Y$ or that $\deg \mathcal{L}_{Y'} = \deg \mathcal{N}_{Y'}$ where Y' is the normalization of Y, because, in any event,

$$\textstyle\int c_1\mathcal{L}\cdot[Y] = \deg \mathcal{L}_Y = \deg \mathcal{L}_{Y'}.$$

DEFINITION 9.6.2. Assume S is the spectrum of a field. Let Λ be a family of invertible sheaves on X. Then Λ is said to be *bounded* if there exist an S-scheme T of finite type and an invertible sheaf $\mathcal{M}$ on X_T such that, given $\mathcal{L} \in \Lambda$, there exists a geometric point t of T such that $\mathcal{L}_t \simeq \mathcal{M}_t$.

THEOREM 9.6.3. Assume S is the spectrum of an algebraically closed field, and X is projective. Let $\mathcal{L}$ be an invertible sheaf on X. Then the following conditions are equivalent:

(a) The sheaf $\mathcal{L}$ is τ-equivalent to $\mathcal{O}_X$.
(b) The sheaf $\mathcal{L}$ is numerically equivalent to $\mathcal{O}_X$.
(c) The family $\{\mathcal{L}^{\otimes p} \mid p \in \mathbb{Z}\}$ is bounded.
(d) For every coherent sheaf $\mathcal{F}$ on X, we have $\chi(\mathcal{F}\otimes\mathcal{L}) = \chi(\mathcal{F})$.
(e) For every closed integral curve $Y \subset X$, we have $\chi(\mathcal{L}_Y) = \chi(\mathcal{O}_Y)$.
(f) For every integer p, the sheaf $\mathcal{L}^{\otimes p}(1)$ is ample.

If X is irreducible, then all the conditions above are equivalent to the following one:

(g) For every pair of integers p, n, we have $\chi(\mathcal{L}^{\otimes p}(n)) = \chi(\mathcal{O}_X(n))$.

If X is irreducible of dimension $r \geq 2$, then all the conditions above are equivalent to the following one:

(h) Setting $\ell := c_1\mathcal{L}$ and $h := c_1\mathcal{O}_X(1)$, we have $\int \ell h^{r-1} = 0$ and $\int \ell^2 h^{r-2} = 0$.

PROOF. Let us proceed by establishing the following implications:

$$(c) \Longrightarrow (a) \Longrightarrow (d) \Longrightarrow (e) \Longleftrightarrow (b) \Longrightarrow (c) \Longrightarrow (f) \Longrightarrow (b);$$
$$(d) \Longrightarrow (g) \Longrightarrow (h) \Longrightarrow (b);\text{ and } (g) \Longrightarrow (b) \text{ if } \dim X = 1.$$

Assume (c). Then, by definition, there exist an S-scheme T of finite type and an invertible sheaf $\mathcal{M}$ on X_T such that, given $p \in \mathbb{Z}$, there exists a geometric point t of T such that $\mathcal{L}_t^{\otimes p} = \mathcal{M}_t$. Apply the Pigeonhole Principle: say $\mathcal{L}^{\otimes p_1}$ and $\mathcal{L}^{\otimes p_2}$ belong to the same connected component of T, but $p_1 \neq p_2$. Set $m := p_1 - p_2$. Then $\mathcal{L}^{\otimes m}$ is algebraically equivalent to $\mathcal{O}_X$. Thus (a) holds.

Furthermore, for each $t \in T$, there exists an n such that $\mathcal{M}_t(n)$ is ample by [**EGAII**, 4.5.8]. So t has a neighborhood U such that $\mathcal{M}_u(n)$ is ample for every $u \in U$ by [**EGAIII1**, 4.7.1]. Since T is quasi-compact, T is covered by finitely many of the U. Let N be the product of the corresponding n. Then $\mathcal{M}_t(N)$ is ample for every $t \in T$. In particular, $\mathcal{L}^{\otimes p}(N)$ is ample for every $p \in \mathbb{Z}$. So $\left(\mathcal{L}^{\otimes q}(1)\right)^{\otimes N}$ is ample for every $q \in \mathbb{Z}$. Thus (f) holds.

Assume (a). The function $n \mapsto \chi(\mathcal{F} \otimes \mathcal{L}^{\otimes n})$ is a polynomial. To prove it is constant, we may replace $\mathcal{L}$ by $\mathcal{L}^{\otimes m}$ for any nonzero m. Thus we may assume $\mathcal{L}$ is algebraically equivalent to $\mathcal{O}_X$. So let T be a connected S-scheme, and $\mathcal{M}$ an invertible sheaf on X_T. Then for a fixed n, as $t \in T$ varies, the function $t \mapsto \chi(\mathcal{F} \otimes \mathcal{L}_t^{\otimes n})$ is constant by [**EGAIII2**, 7.9.5]. It follows that (d) holds.

Assume (d). Taking $\mathcal{F} := \mathcal{O}_Y$, we get (e). Taking $\mathcal{F} := \mathcal{L}^{\otimes p}(n)$, we get $\chi(\mathcal{L}^{\otimes(p+1)}(n)) = \chi(\mathcal{L}^{\otimes p}(n))$. Thus whether or not X is irreducible, (g) holds,.

Assume (g). Then X is irreducible, say of dimension r. Set $\ell := c_1\mathcal{L}$ and $h := c_1\mathcal{O}_X(1)$. Write

$$\chi(\mathcal{L}^{\otimes p}(n)) = \sum_{0 \leq i,\, j \leq r} a_{ij} \binom{p+i}{i} \binom{n+j}{j}$$

where $a_{ij} = \int \ell^i h^j$ if $i + j = r$. Then (g) implies $a_{ij} = 0$ if $i \geq 1$. If $r \geq 2$, then (h) follows. Suppose $r = 1$. Then $\int \ell = 0$. Now, X_{red} is the only closed integral curve contained in X, and $\int \ell$ is a multiple of $\int \ell \cdot [X_{\text{red}}]$. Thus (b) holds.

Conditions (e) and (b) are equivalent since $\chi(\mathcal{L}_Y) = \deg(\mathcal{L}_Y) + \chi(\mathcal{O}_Y)$ by Riemann's Theorem.

Assume (f). Then for every closed integral curve $Y \subset X$, we have

$$0 \leq \deg(\mathcal{L}_Y^{\otimes p}(1)) = p \deg(\mathcal{L}_Y) + \deg(\mathcal{O}_Y(1))$$

for every integer p. So $\deg(\mathcal{L}_Y) = 0$. Thus (b) holds.

Assume (h). Then X is irreducible of dimension $r \geq 2$. To prove (b), plainly we may replace X by its reduction. We proceed by induction on r. If $r = 2$, then (b) holds by the Hodge Index Theorem.

Suppose $r \geq 3$. Given a complete integral curve $Y \subset X$, take n so that the twisted ideal $\mathcal{I}_{Y,X}(n-1)$ is generated by its global sections. View these sections as sections of $\mathcal{O}_X(n-1)$. Then they define a linear system that is free of base points on $X - Y$. So the global sections of $\mathcal{I}_Y(n)$ define a linear system that is very ample on $X - Y$. In particular, this system maps $X - Y$ onto a variety of dimension at least 2; in other words, the system is not "composite with a pencil."

Therefore, the generic member H_η is geometrically irreducible by [**Zar58**, Thm. I.6.3, p. 30]. In the first instance, we must apply the cited theorem to the induced system on the normalization of X, and we conclude that the preimage of

H_η is geometrically irreducible. But then H_η is too. Therefore, by [**EGAIV3**, 9.7.7], a general member H is irreducible.

Set $\ell_1 := c_1\mathcal{L}_H$ and $h_1 := c_1\mathcal{O}_H(1)$. Then, by the Projection Formula,

$$\textstyle\int \ell_1 h_1^{r-2} = n\int \ell h^{r-1} = 0 \text{ and } \int \ell_1^2 h_1^{r-3} = n\int \ell^2 h^{r-2} = 0.$$

So by induction, $\mathcal{L}_H$ is numerically equivalent to $\mathcal{O}_X$ on H. But $Y \subset H$ since H arises from a section of $\mathcal{I}_Y(n)$. Hence, by the Projection Formula,

$$\textstyle\int \ell \cdot [Y] = \int \ell_1 \cdot [Y] = 0.$$

Thus (b) holds.

Finally, assume (b). By Lemma 9.6.6 below, there is an m such that, if $\mathcal{N}$ is an invertible sheaf on X numerically equivalent to $\mathcal{O}_X$, then $\mathcal{N}$ is m-regular. So $\mathcal{N}(m)$ is generated by its global sections, and its higher cohomology groups vanish.

Set $\varphi(n) := \chi(\mathcal{O}_X(n))$ and $M := \varphi(m)$. Then $\dim \mathrm{H}^0(\mathcal{N}(m)) = M$, since $\chi(\mathcal{N}(n)) = \varphi(n)$ also by Lemma 9.6.6 below. Set $\mathcal{F} := \mathcal{O}_X(-m)^{\oplus M}$. Then $\mathcal{N}$ is a quotient of $\mathcal{F}$.

Set $T := \mathbf{Quot}^{\varphi}_{\mathcal{F}/X/k}$. Then T is of finite type. Let $\mathcal{M}$ be the universal quotient. Then there exists a k-point $t \in T$ such that $\mathcal{N} = \mathcal{M}_t$. Let $U \subset X \times T$ be the open set on which $\mathcal{M}$ is invertible. Let $R \subset T$ be the image of the complement of U. Then R is closed. Replace T by $T - R$, and $\mathcal{M}$ by its restriction. Then $t \in T$ still. Thus the invertible sheaves on X numerically equivalent to $\mathcal{O}_X$ form a bounded family. In particular, (c) holds. □

EXERCISE 9.6.4. Consider the preceding paragraph, the last one in the proof of Theorem 9.6.3. Using $\mathbf{Div}_{X/k}$ instead of $\mathbf{Quot}^{\varphi}_{\mathcal{F}/X/k}$, give another proof that the invertible sheaves $\mathcal{N}$ on X numerically equivalent to $\mathcal{O}_X$ form a bounded family.

LEMMA 9.6.5. *Assume S is the spectrum of an algebraically closed field, and X is projective of dimension r. Let $\mathcal{F}$ be a coherent sheaf on X. Then there is a number $B_{\mathcal{F}}$ such that, if $\mathcal{L}$ is any invertible sheaf on X numerically equivalent to $\mathcal{O}_X$, then*

$$\dim \mathrm{H}^0(\mathcal{L}\otimes\mathcal{F}(n)) \le B_{\mathcal{F}}\tbinom{n+r}{r} \text{ for all } n \ge 0.$$

PROOF. Suppose $r = 0$. Then $\mathcal{L}\otimes\mathcal{F}(n) = \mathcal{F}$. So we may take $B_{\mathcal{F}} = \dim \mathrm{H}^0(\mathcal{F})$.

Suppose $r \ge 1$. Given a short exact sequence $0 \to \mathcal{F}' \to \mathcal{F} \to \mathcal{F}'' \to 0$, we have

$$\dim \mathrm{H}^0(\mathcal{L}\otimes\mathcal{F}(n)) \le \dim \mathrm{H}^0(\mathcal{L}\otimes\mathcal{F}'(n)) + \dim \mathrm{H}^0(\mathcal{L}\otimes\mathcal{F}''(n)).$$

So given $B_{\mathcal{F}'}$ and $B_{\mathcal{F}''}$, we can take $B_{\mathcal{F}} := B_{\mathcal{F}'} + B_{\mathcal{F}''}$.

Say $X = \mathrm{Proj}(A)$ and $\mathcal{F} = \widetilde{M}$ where M is a finitely generated graded A-module. Then there is a filtration by graded submodules

$$M =: M_q \supset M_{q-1} \supset \cdots \supset M_1 \supset M_0 := 0$$

such that $M_{i+1}/M_i \simeq (A/P_i)[p_i]$ where P_i is a homogeneous prime for each i. It follows that we may assume X is integral and $\mathcal{F} = \mathcal{O}_X(p)$.

Let $\mathcal{L}$ be numerically equivalent to $\mathcal{O}_X$. Set $\ell := c_1\mathcal{L}$ and $h := c_1\mathcal{O}_X(1)$. Suppose $\mathcal{L}(p)$ has a nonzero section. It defines a divisor D, possibly 0. Hence

$$\textstyle 0 \le \int h^{r-1}[D] = \int h^{r-1}\,\ell + p\int h^r.$$

But $\int h^{r-1}\,\ell = 0$ because h^{r-1} is represented by a curve since $r \ge 1$. And $\int h^r > 0$. Hence $p \ge 0$. Thus $\mathrm{H}^0(\mathcal{L}(-1)) = 0$.

Let H be a hyperplane section of X. Then there is an exact sequence

$$0 \to \mathcal{L}(n-1) \to \mathcal{L}(n) \to \mathcal{L}_H(n) \to 0.$$

By induction on r, we may assume there is a number B such that

$$\dim \mathrm{H}^0(\mathcal{L}_H(n)) \le B\binom{n+r-1}{r-1} \text{ for all } n \ge 0;$$

moreover, B works for every $\mathcal{L}$. Hence

$$\dim \mathrm{H}^0(\mathcal{L}(n)) - \dim \mathrm{H}^0(\mathcal{L}(n-1)) \le B\binom{n+r-1}{r-1} \text{ for all } n \ge 0.$$

But $\mathrm{H}^0(\mathcal{L}(-1)) = 0$. Since $\binom{n+r-1}{r} + \binom{n+r-1}{r-1} = \binom{n+r}{r}$, induction on n yields

$$\dim \mathrm{H}^0(\mathcal{L}(n)) \le B\binom{n+r}{r} \text{ for all } n \ge 0.$$

Recall $\mathcal{F} = \mathcal{O}_X(p)$. If $p \le 0$, then $\mathcal{F} \subset \mathcal{O}_X$; so we may take $B_{\mathcal{F}} := B$. But if $p \ge 0$, then $\binom{p+n+r}{r} \le \binom{p+r}{r}\binom{n+r}{r}$ since every monomial of degree $p+n$ is the product of one of degree p and one of degree n; so we may take $B_{\mathcal{F}} := B\binom{p+r}{r}$. □

LEMMA 9.6.6. Assume S is the spectrum of an algebraically closed field, and X is projective. Then there is an integer m such that, if $\mathcal{L}$ is any invertible sheaf on X numerically equivalent to $\mathcal{O}_X$, then $\mathcal{L}$ is m-regular, and

$$\chi(\mathcal{L}(n)) = \chi(\mathcal{O}_X(n)) \text{ for all } n.$$

PROOF. Set $r := \dim(X)$, and proceed by induction on r. If $r = 0$, then both assertions are trivial. So assume $r \ge 1$.

First, let us establish the asserted equation. Given an $\mathcal{L}$, fix $q \ge 1$ such that $\mathcal{L}(q)$ is very ample. Take effective divisors F and G such that

$$\mathcal{O}_X(F) = \mathcal{O}_X(q) \text{ and } \mathcal{O}_X(G) = \mathcal{L}(q).$$

For every p, plainly $\mathcal{L}_F^{\otimes p}$ and $\mathcal{L}_G^{\otimes p}$ are numerically equivalent to $\mathcal{O}_F$ and $\mathcal{O}_G$.

Form $0 \to \mathcal{O}_X(-F) \to \mathcal{O}_X \to \mathcal{O}_F \to 0$ and $0 \to \mathcal{O}_X(-G) \to \mathcal{O}_X \to \mathcal{O}_G \to 0$. Tensor them with $\mathcal{L}^{\otimes p}(n+q)$. We get

$$\begin{aligned} &0 \to \mathcal{L}^{\otimes p}(n) \to \mathcal{L}^{\otimes p}(n+q) \to \mathcal{L}_F^{\otimes p}(n+q) \to 0, \\ &0 \to \mathcal{L}^{\otimes p-1}(n) \to \mathcal{L}^{\otimes p}(n+q) \to \mathcal{L}_G^{\otimes p}(n+q) \to 0. \end{aligned} \tag{9.6.6.1}$$

Apply $\chi(\bullet)$ and subtract. We get

$$\chi(\mathcal{L}^{\otimes p}(n)) - \chi(\mathcal{L}^{\otimes p-1})(n)) = \chi(\mathcal{L}_G^{\otimes p}(n+q)) - \chi(\mathcal{L}_F^{\otimes p}(n+q)).$$

By induction, the right hand side varies as a polynomial in n, which is independent of p. Hence there are polynomials $\varphi_1(n)$ and $\varphi_0(n)$ such that

$$\chi(\mathcal{L}^{\otimes p}(n)) = \varphi_1(n)p + \varphi_0(n). \tag{9.6.6.2}$$

Suppose $\varphi_1 \ne 0$. Say $\varphi_1(n) \ne 0$ for all $n \ge n_1$.

By induction, there is an integer n_2 such that $\mathcal{L}_F^{\otimes p}$ is n_2-regular for every p. So $\mathrm{H}^i(\mathcal{L}_F^{\otimes p}(n)) = 0$ for $i \ge 1$, for $n \ge n_2 - i$, and for every p. Hence, owing to Sequence (9.6.6.1), there is an isomorphism

$$\mathrm{H}^i(\mathcal{L}^{\otimes p}(n)) \xrightarrow{\sim} \mathrm{H}^i(\mathcal{L}^{\otimes p}(n+q)) \text{ for } i \ge 2, \text{ for } n \ge n_2, \text{ and for every } p.$$

Now, for each $i \ge 2$, each p, and each n, there is a $j \ge 0$ such that

$$\mathrm{H}^i(\mathcal{L}^{\otimes p}(n+jq)) = 0$$

by Serre's Theorem. Therefore,

$$\mathrm{H}^i(\mathcal{L}^{\otimes p}(n)) = 0 \text{ for } i \geq 2, \text{ for } n \geq n_2 \text{ and for every } p.$$

Hence $\mathrm{H}^0(\mathcal{L}^{\otimes p}(n)) \geq \chi(\mathcal{L}^{\otimes p}(n))$ for $n \geq n_2$ and every p. Take $n := \max(n_2, n_1)$. Owing to Equation (9.6.6.2), then $\mathrm{H}^0(\mathcal{L}^{\otimes p}(n)) \to \infty$ as $p \to \infty$ if $\varphi_1(n) > 0$ or as $p \to -\infty$ if $\varphi_1(n) < 0$. However, by Lemma (9.6.5), there is a number B such that $\mathrm{H}^0(\mathcal{L}^{\otimes p}(n)) \leq B$ for any p. This contradiction means $\varphi_1 = 0$. Hence Equation (9.6.6.2) yields $\chi(\mathcal{L}(n)) = \chi(\mathcal{O}_X(n))$ for all n, as desired.

Finally, in order to prove there is an m such that every $\mathcal{L}$ numerically equivalent to $\mathcal{O}_X$ is m-regular, we must modify Mumford's original work [**Mum66**, Lect. 14] because these $\mathcal{L}$ are not ideals. However, as Mumford himself points out [**Mum66**, pp. 102–103], the hypothesis that his sheaf $\mathcal{I}$ is an ideal enters only through the bound $\dim \mathrm{H}^0(\mathcal{I}(n)) \leq \binom{n+r}{r}$. Plainly, this bound can be replaced by the bound $B_{\mathcal{F}}$ with $\mathcal{F} := \mathcal{O}_X$ of Lemma 9.6.5. Of course, in addition, we must use the fact we just proved, that all the $\mathcal{L}$ have the same Hilbert polynomial. □

EXERCISE 9.6.7. Assume S is the spectrum of an algebraically closed field, and X is projective and integral of dimension $r \geq 1$. Set $h := c_1\mathcal{O}_X(1)$. Let $\mathcal{L}$ be an invertible sheaf on X, and set $\ell := c_1\mathcal{L}$. Say

$$\chi(\mathcal{L}(n)) = \textstyle\sum_{0\leq i\leq r} a_i\binom{n+i}{i} \text{ and } a := \int \ell h^{r-1}.$$

Suppose $a < a_r$. Show $\dim \mathrm{H}^0(\mathcal{L}(n)) \leq a_r\binom{n+r}{r}$ for all $n \geq 0$. Furthermore, modifying Mumford's work [**Mum66**, pp. 102–103] slightly, show there is a polynomial Φ_r depending only on r such that $\mathcal{L}$ is m-regular with $m := \Phi_r(a_0, \ldots, a_{r-1})$.

In general, show there is a polynomial Ψ_r depending only on r such that $\mathcal{L}$ is m-regular with $m := \Psi_r(a_0, \ldots, a_r; a)$.

DEFINITION 9.6.8. Let G/S be a group scheme. For $n > 0$, let $\varphi_n \colon G \to G$ denote the nth power map. Then G^τ is the set defined by the formula

$$G^\tau := \textstyle\bigcup_{n>0} \varphi_n^{-1} G^0$$

where G^0 is the union of the connected components of the identity G_s^0 for $s \in S$.

LEMMA 9.6.9. Let k be a field, and G a commutative group scheme locally of finite type. Then G^τ is an open group subscheme, and forming it commutes with extending k. Moreover, if G^τ is quasi-compact, then it is closed and of finite type.

PROOF. By Lemma 9.5.1, G^0 is an open and closed group subscheme of finite type, and forming it commutes with extending k. Now, the nth power map φ_n is continuous, and forming it commutes with extending k; also, φ_n is a homomorphism since G is commutative. So G^τ is the filtered union of the open and closed group subschemes $\varphi_n^{-1}G^0$, and forming them commutes with extending k. Hence G^τ is an open group subscheme, and forming it commutes with extending k. Moreover, if G^τ is quasi-compact, then G^τ is the union of finitely many of the $\varphi_n^{-1}G^0$, and so G^τ is closed; also, then G^τ is of finite type since G is locally of finite type. □

EXERCISE 9.6.10. Let k be a field, and G a commutative group scheme locally of finite type. Let $H \subset G$ be a group subscheme of finite type. Show $H \subset G^\tau$. (Thus, if G^τ is of finite type, then it is the largest group subscheme of finite type.)

EXERCISE 9.6.11. Assume S is the spectrum of a field k. Assume $\mathbf{Pic}_{X/k}$ exists and represents $\mathrm{Pic}_{(X/k)\,(\mathrm{fppf})}$. Let $\mathcal{L}$ be an invertible sheaf on X, and $\lambda \in \mathbf{Pic}_{X/k}$ the corresponding point. Show $\mathcal{L}$ is τ-equivalent to $\mathcal{O}_X$ if and only if $\lambda \in \mathbf{Pic}^{\tau}_{X/k}$.

PROPOSITION 9.6.12. *Assume S is the spectrum of a field k. Assume $\mathbf{Pic}_{X/k}$ exists and represents $\mathrm{Pic}_{(X/k)\,(\mathrm{fppf})}$. Then $\mathbf{Pic}^{\tau}_{X/k}$ is an open group subscheme, and forming it commutes with extending k. Moreover, if X is projective, then $\mathbf{Pic}^{\tau}_{X/k}$ is closed and of finite type.*

PROOF. By Proposition 9.4.17, $\mathbf{Pic}_{X/S}$ is locally of finite type. So owing to Lemma 9.6.9, we need only prove $\mathbf{Pic}^{\tau}_{X/k}$ is quasi-compact when X is projective. Since forming $\mathbf{Pic}^{\tau}_{X/k}$ commutes with extending k, we may also assume k is algebraically closed.

At the very end of the proof of Theorem 9.6.3, we proved that the invertible sheaves $\mathcal{L}$ numerically equivalent to $\mathcal{O}_X$ form a bounded family. In other words, there is a k-scheme T of finite type and an invertible sheaf $\mathcal{M}$ on X_T such that the $\mathcal{L}$ appear among the fibers $\mathcal{M}_t$.

Then $\mathcal{M}$ defines a map $\theta\colon T \to \mathbf{Pic}_{X/k}$. Owing to Theorem 9.6.3 and to Exercise 9.6.11, we have $\theta(T) \supset \mathbf{Pic}^{\tau}_{X/k}$. Since T is Noetherian, so is $\theta(T)$; whence, so is any subspace of $\theta(T)$. Thus $\mathbf{Pic}^{\tau}_{X/k}$ is quasi-compact, as needed. □

EXERCISE 9.6.13. Assume S is the spectrum of a field k. Assume X is projective and geometrically integral. Show $\mathbf{Pic}^{\tau}_{X/k}$ is quasi-projective. If also X is geometrically normal, show $\mathbf{Pic}^{\tau}_{X/k}$ is projective.

REMARK 9.6.14. In Proposition 9.6.12, if X is projective, then $\mathbf{Pic}_{X/k}$ does exist and represent $\mathrm{Pic}_{(X/k)\,(\mathrm{fppf})}$ according to Corollary 9.4.18.3. In fact, this corollary asserts $\mathbf{Pic}_{X/k}$ exists and represents $\mathrm{Pic}_{(X/k)\,(\mathrm{fppf})}$ whenever X is complete; X need not be projective. Furthermore, although we used projective methods to prove Proposition 9.6.12, we can infer it whenever X is complete, as follows.

Assume X is complete. By Chow's lemma, there is a projective variety X' and a surjective map $\gamma\colon X' \to X$. By Theorem 9.4.18.5, the induced map

$$\gamma^*\colon \mathbf{Pic}_{X/k} \to \mathbf{Pic}_{X'/k}$$

is of finite type. Set $H := (\gamma^*)^{-1}\,\mathbf{Pic}^{\tau}_{X'/k}$. Then H is of finite type since $\mathbf{Pic}^{\tau}_{X'/k}$ is by Proposition 9.6.12. Now, plainly $\gamma^*\,\mathbf{Pic}^{0}_{X/k} \subset \mathbf{Pic}^{0}_{X'/k}$. So since γ^* is a homomorphism, $\gamma^*\,\mathbf{Pic}^{\tau}_{X/k} \subset \mathbf{Pic}^{\tau}_{X'/k}$; whence, $\mathbf{Pic}^{\tau}_{X/k} \subset H$ (in fact, the two are equal by Exercise 9.6.10). Since $\mathbf{Pic}^{\tau}_{X/k}$ is open, it is therefore a subscheme of finite type.

Similarly, in Theorem 9.6.3, Conditions (a)–(e) continue to make sense and to remain equivalent whenever X is complete. Indeed, our proofs of the implications

$$\text{(c)} \Longrightarrow \text{(a)} \Longrightarrow \text{(d)} \Longrightarrow \text{(e)} \Longleftrightarrow \text{(b) and (c)} \Longrightarrow \text{(f)} \Longrightarrow \text{(b)}$$

work without change. However, we used projective methods to prove (b) $\Longrightarrow$ (c). Nevertheless, we can infer this implication whenever X is complete, as follows.

Let $\mathcal{L}$ be numerically equivalent to $\mathcal{O}_X$. Let $\gamma\colon X' \to X$ be as above. Then $\gamma^*\mathcal{L}$ is numerically equivalent to $\mathcal{O}_{X'}$. Indeed, Let $Y' \subset X'$ be a closed curve. Then $\int c_1\gamma^*\mathcal{L}\cdot[Y']$ is a multiple of $\int c_1\mathcal{L}\cdot[\gamma Y']$ by the Projection Formula; hence, the former number vanishes as the latter does.

Let $\lambda \in \mathbf{Pic}_{X/k}$ represent $\mathcal{L}$. Then $\gamma^*\lambda \in \mathbf{Pic}_{X'/k}$ represents $\gamma^*\mathcal{L}$. Now, $\gamma^*\mathcal{L}$ is numerically equivalent to $\mathcal{O}_{X'}$. Hence $\gamma^*\lambda \in \mathbf{Pic}^{\tau}_{X'/k}$ owing to Theorem 9.6.3 and to Exercise 9.6.11. So $\lambda \in H := (\gamma^*)^{-1}\,\mathbf{Pic}^{\tau}_{X'/k}$.

The inclusion $H \hookrightarrow \mathbf{Pic}_{X/k}$ is defined by an invertible sheaf $\mathcal{M}$ on X_T for some fppf covering $T \to H$. (Although k is algebraically closed, possibly $\mathcal{O}_S \xrightarrow{\sim} f_*\mathcal{O}_X$ does not hold universally, so we cannot simply take $T := H$.) Replace T by an open subscheme so that $T \to H$ is of finite type and surjective. Since H is of finite type, so is T.

Let $t \in T$ be a k-point that maps to $\lambda \in H$. Then $\mathcal{M}_t \simeq \mathcal{L}$. Now, for every $p \in \mathbb{Z}$, plainly $\mathcal{L}^{\otimes p}$ is numerically equivalent to $\mathcal{O}_X$. So similarly $\mathcal{L}^{\otimes p} \simeq \mathcal{M}_{t_p}$ for some k-point $t_p \in T$. Thus (c) holds.

EXERCISE 9.6.15. Assume $\mathbf{Pic}_{X/S}$ exists and represents $\mathrm{Pic}_{(X/S)\,(\mathrm{fppf})}$. Let Λ be an arbitrary subset of $\mathbf{Pic}_{X/S}$, and L the corresponding family of classes of invertible sheaves on the fibers of X/S in the sense of Exercise 9.4.5. Show Λ is quasi-compact (with the induced topology) if and only if L is *bounded* in the following sense: there exist an S-scheme T of finite type and an invertible sheaf $\mathcal{M}$ on X_T such that every class in L is represented by a fiber $\mathcal{M}_t$ for some $t \in T$.

THEOREM 9.6.16. Assume $f\colon X \to S$ is projective Zariski locally over S, and flat with integral geometric fibers. Then $\mathbf{Pic}^{\tau}_{X/S}$ is an open and closed group subscheme of finite type, and forming it commutes with changing S. If also X/S is projective, and S is Noetherian, then $\mathbf{Pic}^{\tau}_{X/S}$ is quasi-projective.

PROOF. The second assertion follows from the first and Exercise 9.4.11. The first assertion is local on S; so, to prove it, we may assume X/S is projective.

Theorem 9.4.8 asserts $\mathbf{Pic}_{X/S}$ exists and is locally of finite type. So the $\mathbf{Pic}^{\tau}_{X_s/k_s}$ are subgroups of the $\mathbf{Pic}_{X_s/k_s}$, and forming $\mathbf{Pic}^{\tau}_{X_s/k_s}$ commutes with extending k_s by Exercise 9.4.4 and Lemma 9.6.9. Plainly, $\mathbf{Pic}^{\tau}_{X/S} = \bigcup_{s\in S}\mathbf{Pic}^{\tau}_{X_s/k_s}$ as sets. Hence, forming $\mathbf{Pic}^{\tau}_{X/S}$ commutes with changing S; moreover, in order to infer $\mathbf{Pic}^{\tau}_{X/S}$ is a group subscheme, we need only prove it is open.

Plainly, a subset A of a topological space B is open or closed if (and only if), for every member B_i of an open covering of B, the intersection $A \cap B_i$ is so in B_i. Hence, in order to infer $\mathbf{Pic}^{\tau}_{X/S}$ is an open and closed group subscheme, we need only prove that, for any affine open subscheme U of $\mathbf{Pic}_{X/k}$, the intersection $U \cap \mathbf{Pic}^{\tau}_{X/S}$ is open and closed in U.

Theorem 9.4.8 also asserts $\mathbf{Pic}_{X/k}$ represents $\mathrm{Pic}_{(X/S)\,(\text{ét})}$. Thus the inclusion $U \hookrightarrow \mathbf{Pic}_{X/S}$ is defined by an invertible sheaf $\mathcal{M}$ on X_T for some étale covering $T \to U$. Replace T by an open subscheme so that $T \to U$ is of finite type and surjective. Let T^{τ} be the set of $t \in T$ where M_t is τ-equivalent to $\mathcal{O}_{X_t}$. Then T^{τ} is the preimage of $\mathbf{Pic}^{\tau}_{X/S}$ in T owing to Exercise 9.6.11. So we have to prove T^{τ} is open and closed.

Since U is affine, it is quasi-compact, so of finite type. Hence T is of finite type. So it has only finitely many connected components. But, for every pair p, n, the function $t \mapsto \chi(\mathcal{M}_t^{\otimes p}(n))$ is constant on each connected component of T. Therefore, Theorem 9.6.3 implies T^{τ} is open and closed, as desired.

It remains to prove $\mathbf{Pic}^{\tau}_{X/S}$ is of finite type. Let L be the corresponding family of classes of invertible sheaves on the fibers of X/S. By Exercise 9.6.15, we need only prove L is bounded. At the very end of the proof of Theorem 9.6.3, we proved essentially this statement when S is the spectrum of an algebraically closed field, and X is projective, but not necessarily integral. We can argue similarly here, but must make two important modifications.

First, if a class in L is represented by an invertible sheaf $\mathcal{L}$ on a fiber X_k where k is a field containing the field k_s of a point $s \in S$, then $\chi(\mathcal{L}(n)) = \chi(\mathcal{O}_{X_s}(n))$ for all n, owing to Lemma 9.6.6. But $\chi(\mathcal{O}_{X_s}(n))$ can vary with s. Nevertheless, it must remain the same on each connected component of S. And the matter in question is local on S. So we may and must assume S is connected.

Second, at the end of the proof of Lemma 9.6.6, when we modified Mumford's work, we used the bound $B_{\mathcal{F}}$ with $\mathcal{F} := \mathcal{O}_X$ of Lemma 9.6.5; in fact, in the induction step, we implicitly used the corresponding bounds for various subschemes of X. Unfortunately, it is not clear, in general, how these bounds vary with X. But in place of Lemma 9.6.6, we can use Exercise 9.6.7, which provides a uniform m such that $\mathcal{L}$ is m-regular for every $\mathcal{L}$ representing a class in L. □

COROLLARY 9.6.17. Assume S is Noetherian. Assume $f: X \to S$ is projective Zariski locally over S, and is flat with geometrically integral fibers. For each $s \in S$, let k'_s be the algebraic closure of the residue field k_s. Then the torsion group

$$\mathbf{Pic}^{\tau}_{X_{k'_s}/k'_s}(k'_s) \;/\; \mathbf{Pic}^{0}_{X_{k'_s}/k'_s}(k'_s) \tag{9.6.17.1}$$

is finite, and its order is bounded.

PROOF. Since $\mathbf{Pic}^{0}_{X_{k'_s}/k'_s}$ is open in $\mathbf{Pic}^{\tau}_{X_{k'_s}/k'_s}$ by Proposition 9.5.3, the order of their quotient is equal to the number of connected components of $\mathbf{Pic}^{\tau}_{X_{k'_s}/k'_s}$. This number is finite because $\mathbf{Pic}^{\tau}_{X_{k'_s}/k'_s}$ is of finite type by Proposition 9.6.12.

Moreover, $\mathbf{Pic}^{\tau}_{X_{k'_s}/k'_s}$ is equal to $\mathbf{Pic}^{\tau}_{X_s/k_s} \otimes k'_s$ again by Proposition 9.6.12, so equal to $\mathbf{Pic}^{\tau}_{X/S} \otimes k'_s$ essentially by Definition 9.6.8. But, since $\mathbf{Pic}^{\tau}_{X/S}$ is of finite type by Theorem 9.6.16, the number of connected components of $\mathbf{Pic}^{\tau}_{X/S} \otimes k'_s$ is constant for s in a nonempty open subset of S by [**EGAIV3**, 9.7.9]. Hence the number is bounded by Noetherian induction. □

EXERCISE 9.6.18. Assume X/S is projective and smooth, its geometric fibers are irreducible, and S is Noetherian. Show that $\mathbf{Pic}^{\tau}_{X/S}$ is projective.

REMARK 9.6.19. Assume X/S is proper. If $\mathbf{Pic}_{X/S}$ exists and it represents $\mathrm{Pic}_{(X/k)\,(\mathrm{fppf})}$, then $\mathbf{Pic}^{\tau}_{X/S}$ is an open group subscheme of finite type. This fact can be derived from Theorem 9.6.16 through a series of reduction steps; see [**Kle71**, Thm. 4.7, p. 647].

Assume S is Noetherian in addition. Then, whether or not $\mathbf{Pic}_{X/k}$ exists, the torsion group (9.6.17.1) is finite and its order is bounded. This fact follows from the preceding one via the proof of Corollary 9.6.17, since there is an nonempty open subscheme V of S_{red} such that $\mathbf{Pic}_{X_V/V}$ exists and represents $\mathrm{Pic}_{(X_V/V)\,(\mathrm{fppf})}$ by Grothendieck's Theorem 9.4.18.2.

Furthermore, the rank of the corresponding "Néron–Severi" group

$$\mathbf{Pic}_{X_{k'_s}/k'_s}(k'_s) \;/\; \mathbf{Pic}^{\tau}_{X_{k'_s}/k'_s}(k'_s)$$

is finite and bounded. This fact is far deeper; see [**Kle71**, Thm. 5.1, p. 650, and Rem. 5.3, p. 652], and see [**Zar71**, pp. 121–124]. Moreover, the rank is arithmetic in nature: its value need not be a constructible function of $s \in S$; a standard example is discussed in [**BLR90**, p. 235].

THEOREM 9.6.20. Assume X/S is projective and flat with integral geometric fibers. Given a polynomial $\varphi \in \mathbb{Q}[n]$, let $\mathbf{Pic}^{\varphi}_{X/S} \subset \mathbf{Pic}_{X/S}$ be the set of points representing invertible sheaves $\mathcal{L}$ such that $\chi(\mathcal{L}(n)) = \varphi(n)$ for all n. Then the $\mathbf{Pic}^{\varphi}_{X/S}$ are open and closed subschemes of finite type; they are disjoint and cover; and forming them commutes with changing S. If also S is Noetherian, then $\mathbf{Pic}^{\varphi}_{X/S}$ is quasi-projective.

PROOF. Plainly, the $\mathbf{Pic}^{\varphi}_{X/S}$ are disjoint and cover, and forming them commutes with changing S. The rest of the proof is similar to the proof of Theorem 9.6.16. In fact, the present case is simpler because the sheaves in question have the same Hilbert polynomials by hypothesis; there is no need to appeal to Theorem 9.6.3 nor to Lemma 9.6.6. □

EXERCISE 9.6.21. Assume X/S is locally projective over S and flat, and its geometric fibers are integral curves. Given an integer m, let $\mathbf{Pic}^{m}_{X/S} \subset \mathbf{Pic}_{X/S}$ be the set of points representing invertible sheaves $\mathcal{L}$ of degree m.

Show the $\mathbf{Pic}^{m}_{X/S}$ are open and closed subschemes of finite type; show they are disjoint and cover; and show that forming them commutes with changing S.

Show there is no abuse of notation: the fiber of $\mathbf{Pic}^{0}_{X/S}$ over $s \in S$ is the connected component of $0 \in \mathbf{Pic}_{X_s/k_s}$. Show there is no torsion: $\mathbf{Pic}^{0}_{X/S} = \mathbf{Pic}^{\tau}_{X/S}$. Show each $\mathbf{Pic}^{m}_{X/S}$ is an fppf-torsor under $\mathbf{Pic}^{0}_{X/S}$; that is, the latter acts naturally on the former, and the two become isomorphic after base change by an fppf-covering.

Show the $\mathbf{Pic}^{m}_{X/S}$ are quasi-projective if X/S is projective and S is Noetherian.

REMARK 9.6.22. There is another important case where $\mathbf{Pic}^{0}_{X/S} = \mathbf{Pic}^{\tau}_{X/S}$, namely, when X is an Abelian S-scheme. Indeed, the equation holds if it does on each geometric fiber of X/S; so we may assume that S is the spectrum of an algebraically closed field. In this case, a modern proof was given by Mumford [**Mum70**, Cor.2, p. 178].

EXAMPLE 9.6.23. Theorem 9.6.20 can fail if a geometric fiber of X/S is reducible. For example, let S be the spectrum of a field k, and let X be the union of two disjoint lines. For each pair $a, b \in \mathbb{Z}$, let $\mathcal{L}_{a,b}$ be the invertible sheaf that restricts to $\mathcal{O}(a)$ on the first line and to $\mathcal{O}(b)$ on the second.

By Riemann's Theorem, $\chi(\mathcal{L}_{a,b}(n)) = (a+b) + 2n + 1$. And it is easy to see that $\mathrm{Pic}_{X/k}$ is the disjoint union of copies of $\mathrm{Spec}(k)$ indexed by $\mathbb{Z} \times \mathbb{Z}$; compare with Exercise 9.4.15. Moreover, $\mathcal{L}_{a,b}$ is represented by the point with index (a, b). Hence, each set $\mathbf{Pic}^{\varphi}_{X/k}$ is infinite, and so not of finite type.

REMARK 9.6.24. Theorem 9.6.20 can be modified as follows. Assume S is Noetherian. Assume X/S is projective and flat with integral geometric fibers of dimension r. Then a subset $\Lambda \subset \mathbf{Pic}_{X/S}$ is quasi-compact if, in the Hilbert polynomials $\sum_{i=0}^{r} a_i\binom{n+i}{i}$ of the corresponding invertible sheaves, a_{r-1} remains bounded from above and below, and a_{r-2} remains bounded from below alone. Moreover, Λ is quasi-compact if, instead, $\int \ell h^{r-1}$ remains bounded from above and below, and $\int \ell^2 h^{r-2}$ remains bounded from below alone, where ℓ and h are as usual.

These facts can be derived from Theorem 9.6.20 by reducing to the case where the fibers are normal and by showing, in this case using simple elementary means, that the given bounds imply bounds on all the a_i; see [**Kle71**, Thm. 3.13, p. 641].

The first fact is essentially equivalent, given Theorems 9.6.20 and 9.4.18.5, to the following fact. Assume S is Noetherian, and X/S is projective and flat with fibers whose irreducible components have dimension at least 3. Let Y be a relative effective divisor whose associated sheaf is $\mathcal{O}_X(1)$. Then the induced map of Picard functors is representable by maps of finite type. See [**Kle71**, Thm. 3.8, p. 636].

The latter fact was first proved directly using the "equivalence criterion" mentioned in Remark 9.5.8 by Grothendieck [**FGA**, p. C-10].

COROLLARY 9.6.25. *Assume X/S is projective and flat with integral geometric fibers. Then the connected components of $\mathbf{Pic}_{X/S}$ are open and closed subschemes of finite type.*

PROOF. By construction, $\mathbf{Pic}_{X/S}$ is locally Noetherian. Hence, its connected components are open and closed; see the proof of Lemma 9.5.1. Now, a connected component is always contained in any open and closed set it meets. Hence the connected components of $\mathbf{Pic}_{X/S}$ are of finite type owing to Theorem 9.6.20. □

REMARK 9.6.26. In Corollary 9.6.25, X/S must be projective, not simply proper, nor even projective Zariski locally over S. Indeed, Grothendieck [**FGA**, Rem. 3.3, p. 232-07] gave an example where $\mathbf{Pic}_{X/S}$ has a connected component that is not of finite type: here S is a curve with two components that meet in two points, such as the union of a smooth conic and a line in the plane over an algebraically closed field, and X is projective over a neighborhood of each component of S.

COROLLARY 9.6.27. *Assume X/S is projective and flat with integral geometric fibers. For $n \neq 0$, the nth power map $\varphi_n\colon \mathbf{Pic}_{X/S} \to \mathbf{Pic}_{X/S}$ is of finite type.*

PROOF. Owing to Corollary 9.6.25, we need only prove that, given any connected component U of $\mathbf{Pic}_{X/S}$, the preimage $\varphi_n^{-1}U$ is of finite type too. Since $\mathbf{Pic}_{X/k}$ represents $\mathrm{Pic}_{(X/S)\,(\text{ét})}$ by Theorem 9.4.8, the inclusion $U \hookrightarrow \mathbf{Pic}_{X/S}$ is defined by an invertible sheaf $\mathcal{M}$ on X_T for some étale covering $T \to U$.

Fix $t \in T$, and set $\psi(p,q) := \chi(\mathcal{M}_t^p(q))$. Fix p, q, and form the set T' of points t' of T such that $\chi(\mathcal{M}_{t'}^p(q)) = \psi(p,q)$. By [**EGAIII2**, 7.9.4], the set T' is open, and so is its complement. Hence their images are open in U, and plainly these images are disjoint. But U is connected. Hence $T' = T$.

Let $\lambda \in \varphi_n^{-1}U$. Represent λ by an invertible sheaf $\mathcal{L}$. Set $\theta(m,q) := \chi(\mathcal{L}^m(q))$. Say $\theta(m,q) = \sum a_i(m)\binom{q+i}{i}$ where $a_i(m)$ is a polynomial. Now, $\varphi_n(\lambda) \in U$. So $\theta(mn,q) = \psi(m,q)$. So $a_i(mn)$ is independent of the choice of λ for all m. Hence $a_i(m)$ is too. Set $\varphi(n) := \theta(1,q)$. Then $\varphi_n^{-1}U \subset \mathbf{Pic}_{X/S}^{\varphi}$. Hence Theorem 9.6.20 implies $\varphi_n^{-1}U$ is of finite type. □

REMARK 9.6.28. Corollary 9.6.27 holds in greater generality. Assume X/S is proper, and assume $\mathbf{Pic}_{X/S}$ exists and represents $\mathrm{Pic}_{(X/k)\,(\text{fppf})}$. Then, for $n \neq 0$, the nth power map $\varphi_n\colon \mathbf{Pic}_{X/S} \to \mathbf{Pic}_{X/S}$ is of finite type. This fact can be derived from the corollary through a series of reduction steps similar to those used to generalize Theorem 9.6.16; see [**Kle71**, Thm. 3.6, p. 635].

EXERCISE 9.6.29. Assume S is Noetherian, and X/S is projective and flat. Assume $\mathbf{Pic}_{X/S}$ exists and represents $\mathrm{Pic}_{(X/S)\,(\mathrm{fppf})}$. Let Λ be an arbitrary subset of $\mathbf{Pic}_{X/S}$, and Π the corresponding set of Hilbert polynomials. Show Π is finite if Λ is quasi-compact. Show Π has only one element if Λ is connected.

Appendix A. Answers to all the exercises

The exercises are not meant to be tricky, but are designed to help you check, solidify, and expand your understanding of the ideas and methods. So to promote your own mathematical health, try seriously to do each exercise before you read its answer here. Note that the answer key is the same number as the exercise key.

ANSWER 9.2.3. Since A is local, $\mathrm{Pic}(T)$ is trivial; so Definitions 9.2.1 and 9.2.2 yield the first isomorphism, $\mathrm{Pic}_X(A) \xrightarrow{\sim} \mathrm{Pic}_{X/S}(A)$.

Let $\mathcal{L}$ be an invertible sheaf on X_A. Suppose the isomorphism class of $\mathcal{L}$ maps to 0 in $\mathrm{Pic}_{(X/S)\,(\mathrm{zar})}(A)$. Then there is a Zariski covering $T' \to T$ and an isomorphism $v' \colon \mathcal{L}_{T'} \xrightarrow{\sim} \mathcal{O}_{X_{T'}}$. Now, T' is a disjoint union of Zariski open subschemes of T. One of them contains the closed point, so is equal to T. Restricting v' yields an isomorphism $\mathcal{L} \xrightarrow{\sim} \mathcal{O}_{X_A}$. Thus $\mathrm{Pic}_X(A) \to \mathrm{Pic}_{(X/S)\,(\mathrm{zar})}(A)$ is injective.

Given $\lambda \in \mathrm{Pic}_{(X/S)\,(\mathrm{zar})}(A)$, represent λ by an invertible sheaf $\mathcal{L}'$ on $X_{T'}$ for a suitable Zariski covering $T' \to T$. Again, T' contains a copy of T as an open and closed subscheme. Restricting $\mathcal{L}'$ yields an invertible sheaf $\mathcal{L}$ on X_A. Since $\mathcal{L}'$ represent λ, there is an isomorphism v'' between the two pullbacks of $\mathcal{L}'$ to $T' \times_T T'$. Restricting v'' to $T' \times_T T$, or T', yields an isomorphism between $\mathcal{L}'$ and $\mathcal{L}_{T'}$. Hence $\mathcal{L}$ too represents λ. Thus $\mathrm{Pic}_X(A) \to \mathrm{Pic}_{(X/S)\,(\mathrm{zar})}(A)$ is surjective too.

Assume A is Artin local with algebraically closed residue field. Then every étale A-algebra B of finite type is a direct product of Artin local algebras, each isomorphic to A, owing to [**EGAIV4**, 17.6.2 and 17.6.3]. So if $T' \to T$ is any étale covering, then T' is a disjoint union of open subschemes, each a copy of T. Hence, reasoning as above, we conclude $\mathrm{Pic}_X(A) \xrightarrow{\sim} \mathrm{Pic}_{(X/S)\,(\text{ét})}(A)$.

Assume $A = k$ where k is an algebraically closed field. Then any fppf covering $T' \to T$ has a section; indeed, at any closed point of T', the local ring is essentially of finite type over k, and so the residue field is equal to k by the Hilbert Nullstellensatz. This point is not necessarily isolated in T'; nevertheless, reasoning essentially as above, we conclude $\mathrm{Pic}_X(k) \xrightarrow{\sim} \mathrm{Pic}_{(X/S)\,(\mathrm{fppf})}(k)$. $\square$

ANSWER 9.2.4. The extension $\mathbb{C}/\mathbb{R}$ is étale. So if the two pullbacks of $\varphi^*\mathcal{O}(1)$ to $X_{\mathbb{C}\otimes_{\mathbb{R}}\mathbb{C}}$ are isomorphic, then $\varphi^*\mathcal{O}(1)$ defines an element λ in $\mathrm{Pic}_{(X/\mathbb{R})\,(\text{ét})}(\mathbb{R})$.

Take an indeterminate z, and identify $\mathbb{C}$ with $\mathbb{R}[z]/(z^2+1)$. Then, by extension of scalars, $\mathbb{C} \otimes_{\mathbb{R}} \mathbb{C}$ becomes identified with $\mathbb{C}[z]/(z-1)(z+1)$. So, by the Chinese Remainder Theorem, $\mathbb{C}\otimes_{\mathbb{R}}\mathbb{C}$ is isomorphic to the product $\mathbb{C}\times\mathbb{C}$. (Correspondingly, $c\otimes 1$ and $1\otimes c$ are identified with (c,c) and $(c,\bar{c})$ where $\bar{c}$ is the conjugate of c, but this fact is not needed here.)

Thus $X_{\mathbb{C}\otimes_{\mathbb{R}}\mathbb{C}}$ is isomorphic, over $\mathbb{R}$, to the disjoint union of two copies of $X_{\mathbb{C}}$. Now, for any field k, an invertible sheaf $\mathcal{L}$ on $\mathbf{P}^1_k$ is determined, up to isomorphism, by a single integer its Euler characteristic $\chi(\mathcal{L})$ by [**Har77**, Cor. 6.17, p. 145]. Hence, the two pullbacks of $\varphi^*\mathcal{O}(1)$ to $X_{\mathbb{C}\otimes_{\mathbb{R}}\mathbb{C}}$ are isomorphic. Thus $\varphi^*\mathcal{O}(1)$ defines a λ in $\mathrm{Pic}_{(X/\mathbb{R})\,(\text{ét})}(\mathbb{R})$.

(Similarly, the isomorphism class of $\varphi^*\mathcal{O}(1)$ on $X_{\mathbb{C}}$ is independent of the choice of the isomorphism $\varphi\colon X_{\mathbb{C}} \xrightarrow{\sim} \mathbf{P}^1_{\mathbb{C}}$. So λ is independent too. But this fact too is not needed here.)

Finally, we must show λ is not in the image of $\mathrm{Pic}_{(X/\mathbb{R})\,(\mathrm{zar})}(\mathbb{R})$. By way of contradiction, suppose λ is. Then λ arises from an invertible sheaf $\mathcal{L}$ on X. A priori, the pullback $\mathcal{L}|X_{\mathbb{C}}$ need not be isomorphic to $\varphi^*\mathcal{O}(1)$. Rather, these two invertible sheaves need only become isomorphic after they are pulled back to X_A where A is some étale $\mathbb{C}$-algebra.

However, cohomology commutes with flat base change. So

$$\dim_{\mathbb{R}}\mathrm{H}^0(\mathcal{L}) = \mathrm{rank}_A\,\mathrm{H}^0(\mathcal{L}|X_A) = \dim_{\mathbb{C}}\mathrm{H}^0(\varphi^*\mathcal{O}(1)) = 2.$$

Hence, $\mathcal{L}$ has a nonzero section. It defines an exact sequence

$$0 \to \mathcal{O}_X \to \mathcal{L} \to \mathcal{O}_D \to 0.$$

Similarly $\mathrm{H}^1(\mathcal{O}_X) = 0$. Hence $\dim_{\mathbb{R}}\mathrm{H}^0(\mathcal{O}_D) = 1$. Therefore, D is an $\mathbb{R}$-point of X. But X has no $\mathbb{R}$-point. Thus λ is not in the image of $\mathrm{Pic}_{(X/\mathbb{R})\,(\mathrm{zar})}(\mathbb{R})$. □

ANSWER 9.2.6. First of all, we have $\mathrm{Pic}_{(X/S)\,(\mathrm{fppf})}(k) = \mathrm{Pic}_{(X_k/k)\,(\mathrm{fppf})}(k)$ essentially by definition, because a map $T' \to T$ of k-schemes is an fppf-covering if and only if it is an fppf-covering when viewed as a map of S-schemes. And a similar analysis applies to the other three functors. Now, $f_k\colon X_k \to k$ has a section; indeed, f_k is of finite type and k is algebraically closed, and so any closed point of X has residue field k by the Hilbert Nullstellensatz. Hence, by Part 2 of Theorem 9.2.5, the k-points of all four functors are the same. Finally, $\mathrm{Pic}_{X/S}(k) = \mathrm{Pic}(X_k)$ because $\mathrm{Pic}(T)$ is trivial whenever T has only one closed point.

Whether or not $\mathcal{O}_S \xrightarrow{\sim} f_*\mathcal{O}_X$ holds universally, all four functors have the same geometric points by Exercise 9.2.3; in fact, given an algebraically closed field k, the k-points of these functors are just the elements of $\mathrm{Pic}(X_k)$. □

ANSWER 9.3.2. By definition, a section in $\mathrm{H}^0(X,\mathcal{L})_{\mathrm{reg}}$ corresponds to an injection $\mathcal{L}^{-1} \hookrightarrow \mathcal{O}_X$. Its image is an ideal $\mathcal{I}$ such that $\mathcal{L}^{-1} \xrightarrow{\sim} \mathcal{I}$. So $\mathcal{I}$ is the ideal of an effective divisor D. Then $\mathcal{O}_X(-D) = \mathcal{I}$. So $\mathcal{L}^{-1} \xrightarrow{\sim} \mathcal{O}_X(-D)$. Taking inverses yields $\mathcal{O}_X(D) \simeq \mathcal{L}$. So $D \in |\mathcal{L}|$. Thus we have a map $\mathrm{H}^0(X,\mathcal{L})_{\mathrm{reg}} \to |\mathcal{L}|$.

If the section is multiplied by a unit in $\mathrm{H}^0(X,\mathcal{O}_X^*)$, then the injection $\mathcal{L}^{-1} \hookrightarrow \mathcal{O}_X$ is multiplied by the same unit, so has the same image $\mathcal{I}$; so then D is unaltered. Conversely, if D arises from a second section, corresponding to a second isomorphism $\mathcal{L}^{-1} \xrightarrow{\sim} \mathcal{I}$, then these two isomorphism differ by an automorphism of $\mathcal{L}^{-1}$, which is given by multiplication by a unit in $\mathrm{H}^0(X,\mathcal{O}_X^*)$; so then the two sections differ by multiplication by this unit. Thus $\mathrm{H}^0(X,\mathcal{L})_{\mathrm{reg}}/\mathrm{H}^0(X,\mathcal{O}_X^*) \hookrightarrow |\mathcal{L}|$.

Finally, given $D \in |\mathcal{L}|$, by definition there exist an isomorphism $\mathcal{O}_X(D) \simeq \mathcal{L}$. Since $O_X(-D)$ is the ideal $\mathcal{I}$ of D, the inclusion $\mathcal{I} \hookrightarrow \mathcal{O}_X$ yields an injection $\mathcal{L}^{-1} \hookrightarrow \mathcal{O}_X$. The latter corresponds to a section in $\mathrm{H}^0(X,\mathcal{L})_{\mathrm{reg}}$, which yields D via the procedure of the first paragraph. Thus $\mathrm{H}^0(X,\mathcal{L})_{\mathrm{reg}}/\mathrm{H}^0(X,\mathcal{O}_X^*) \xrightarrow{\sim} |\mathcal{L}|$. □

ANSWER 9.3.5. Let $x \in D+E$. If $x \notin D\cap E$, then $D+E$ is a relative effective divisor at x, as $D+E$ is equal to D or to E on a neighborhood of x. So suppose $x \in D\cap E$. Then Lemma 9.3.4 says X is S-flat at x, and each of D and E is cut out at x by one element that is regular on the fiber X_s through x. Form the product of the two elements. Plainly, it cuts out $D+E$ at x, and it too is regular on X_s. Hence $D+E$ is a relative effective divisor at x by Lemma 9.3.4 again. □

ANSWER 9.3.8. Consider a relative effective divisor D on X_T/T. Each fiber D_t is of dimension 0. So its Hilbert polynomial $\chi(\mathcal{O}_{D_t}(n))$ is constant. Its value is $\dim \mathrm{H}^0(\mathcal{O}_{D_t})$, which is just the degree of D_t.

The assertions are local on S; so we may assume X/S is projective. Then $\mathrm{Div}_{X/S}$ is representable by an open subscheme $\mathbf{Div}_{X/S} \subset \mathbf{Hilb}_{X/S}$ by Theorem 9.3.7. And $\mathbf{Hilb}_{X/S}$ is the disjoint union of open and closed subschemes of finite type $\mathbf{Hilb}^{\varphi}_{X/S}$ that parameterize the subschemes with Hilbert polynomial φ. Set

$$\mathbf{Div}^m_{X/S} := \mathbf{Div}_{X/S} \cap \mathbf{Hilb}^m_{X/S}\,.$$

Then the $\mathbf{Div}^m_{X/S}$ have all the desired properties.

In general, whenever X/S is separated, X represents $\mathrm{Hilb}^1_{X/S}$, and the diagonal subscheme $\Delta \subset X \times X$ is the universal subscheme. Indeed, the projection $\Delta \to X$ is an isomorphism, so $\Delta \in \mathrm{Hilb}^1_{X/S}(X)$. Now, given any S-map $g : T \to X$, note $(1 \times g)^{-1}\Delta = \Gamma_g$ where $\Gamma_g \subset X \times T$ is the graph subscheme of g, because the T'-points of both $(1\times g)^{-1}\Delta$ and Γ_g are just the pairs (gp, p) where $p\colon T' \to T$. So $\Gamma_g \in \mathrm{Hilb}^1_{X/S}(T)$.

Conversely, let $\Gamma \in \mathrm{Hilb}^1_{X/S}(T)$. So Γ is a closed subscheme of $X \times T$. The projection $\pi\colon \Gamma \to T$ is proper, and its fibers are finite; hence, it is finite by Chevalley's Theorem [**EGAIII1**, 4.4.2]. So $\Gamma = \mathrm{Spec}(\pi_*\mathcal{O}_\Gamma)$. Moreover, $\pi_*\mathcal{O}_\Gamma$ is locally free, being flat and finitely generated over $\mathcal{O}_T$. And forming $\pi_*\mathcal{O}_\Gamma$ commutes with passing to the fibers, so its rank is 1. Hence $\mathcal{O}_T \xrightarrow{\sim} \pi_*\mathcal{O}_\Gamma$. Therefore, π is an isomorphism. Hence Γ is the graph of a map $g\colon T \to X$. So, $(1\times g)^{-1}\Delta = \Gamma$ by the above; also, g is the only map with this property, since a map is determined by its graph. Thus X represents $\mathrm{Hilb}^1_{X/S}$, and $\Delta \subset X \times X$ is the universal subscheme.

In the case at hand, $\mathrm{Div}^1_{X/S}$ is therefore representable by an open subscheme $U \subset X$ by Theorem 9.3.7. In fact, its proof shows U is formed by the points $x \in X$ where the fiber Δ_x is a divisor on X_x. Now, Δ_x is a k_x-rational point for any $x \in X$; so Δ_x is a divisor if and only if X_x is regular at Δ_x. Since X/S is flat, X_x is regular at Δ_x if and only if $x \in X_0$. Thus $X_0 = \mathbf{Div}^1_{X/S}$.

Finally, set $T := X_0^m$ and let $\Gamma_i \subset X \times T$ be the graph subscheme of the ith projection. By the above analysis, $\Gamma_i \in \mathrm{Div}^1_{X/S}(T)$. Set $\Gamma := \sum \Gamma_i$. Then $\Gamma \in \mathrm{Div}^m_{X/S}(T)$ owing to Exercise 9.3.5 and to the additivity of degree. Plainly Γ represents the desired T-point of $\mathbf{Div}^m_{X/S}$. □

ANSWER 9.3.11. Let $s \in S$. Let K be the algebraically closure of k_s, and set $A := \mathrm{H}^0(X_K, \mathcal{O}_{X_K})$. Since f is proper, A is finite dimensional as a K-vector space; so A is an Artin ring. Since X_K is connected, A is not a product of two nonzero rings by [**EGAIII2**, 7.8.6.1]; so A is an Artin local ring. Since X_K is reduced, A is reduced; so A is a field, which is a finite extension of K. Since K is algebraically closed, therefore $A = K$. Since cohomology commutes with flat base change, consequently $k_s \xrightarrow{\sim} \mathrm{H}^0(X, \mathcal{O}_{X_s})$.

The isomorphism $k_s \xrightarrow{\sim} \mathrm{H}^0(X, \mathcal{O}_{X_s})$ factors through $f_*(\mathcal{O}_X) \otimes k_s$:

$$k_s \to f_*(\mathcal{O}_X) \otimes k_s \to \mathrm{H}^0(X_s, \mathcal{O}_{X_s}).$$

So the second map is a surjection. Hence this map is an isomorphism by the implication (iv)$\Rightarrow$(iii) of Subsection 9.3.10 with $\mathcal{F} := \mathcal{O}_X$ and $\mathcal{N} := k_s$. Therefore, the first map is an isomorphism too.

It follows that $\mathcal{O}_S \to f_*\mathcal{O}_X$ is surjective at s. Indeed, denote its cokernel by $\mathcal{G}$. Since tensor product is right exact and since $k_s \to f_*(\mathcal{O}_X) \otimes k_s$ is an isomorphism, $\mathcal{G} \otimes k_s = 0$. So by Nakayama's lemma, the stalk $\mathcal{G}_s$ vanishes, as claimed

Let $\mathcal{Q}$ be the $\mathcal{O}_S$-module associated to $\mathcal{F} := \mathcal{O}_X$ as in Subsection 9.3.10. Then $\mathcal{Q}$ is free at s by the implication (iv)$\Rightarrow$(i) of Subsection 9.3.10. And rank $\mathcal{Q}_s = 1$ owing to the isomorphism in (9.3.10.1) with $\mathcal{N} := k_s$. But, with $\mathcal{N} := \mathcal{O}_S$, the isomorphism becomes $\mathcal{H}om(\mathcal{Q}, \mathcal{O}_X) \xrightarrow{\sim} f_*\mathcal{O}_X$. Hence $f_*\mathcal{O}_X$ too is free of rank 1 at s. Therefore, the surjection $\mathcal{O}_S \to f_*\mathcal{O}_X$ is an isomorphism at s. Since s is arbitrary, $\mathcal{O}_S \xrightarrow{\sim} f_*\mathcal{O}_X$ everywhere.

Finally, let T be an arbitrary S-scheme. Then $f_T\colon X_T \to T$ too is proper and flat, and its geometric fibers are reduced and connected. Hence, by what we just proved, $\mathcal{O}_T \xrightarrow{\sim} f_T * \mathcal{O}_{X_T}$. □

ANSWER 9.3.14. By Theorem 9.3.13, L represents $\mathrm{LinSys}_{\mathcal{L}/X/S}$. So by Yoneda's Lemma [**EGAG**, (0,1.1.4), p. 20], there exists a $W \in \mathrm{LinSys}_{\mathcal{L}/X/S}(L)$ with the required universal property. And W corresponds to the identity map $p\colon L \to L$. The proof of Theorem 9.3.13 now shows $\mathcal{O}_{X_L}(W) = (\mathcal{L}|X_L) \otimes f_L^*\mathcal{O}_L(1)$. □

ANSWER 9.4.2. The structure sheaf $\mathcal{O}_X$ defines a section $\sigma\colon S \to \mathbf{Pic}_{X/S}$. Its image is a subscheme, which is closed if $\mathbf{Pic}_{X/S}$ is separated, by [**EGAG**, Cors. (5.1.4), p. 275, and (5.2.4), p. 278]. Let $N \subset T$ be the pullback of this subscheme under the map $\lambda\colon T \to \mathbf{Pic}_{X/S}$ defined by $\mathcal{L}$. Then the third property holds.

Both $\mathcal{L}_N$ and $\mathcal{O}_X$ define the same map $N \to \mathbf{Pic}_{X/S}$. So, since $\mathcal{O}_S \xrightarrow{\sim} f_*\mathcal{O}_X$ holds universally, the Comparison Theorem, Theorem, Theorem 9.2.5, implies that there exists an invertible sheaf $\mathcal{N}$ on N such that the first property holds.

Consider the second property. Then $\mathcal{L}_{T'} \simeq f_{T'}^*\mathcal{N}'$. So $\lambda t\colon T' \to \mathbf{Pic}_{X/S}$ is also defined by $\mathcal{O}_{X_{T'}}$; hence, λt factors through $\sigma\colon S \to \mathbf{Pic}_{X/S}$. Therefore, $t\colon T' \to T$ factors through N. So, since the first property holds, $\mathcal{L}_{T'} \simeq f_{T'}^* t^*\mathcal{N}$. Hence $\mathcal{N}' \simeq t^*\mathcal{N}$ by Lemma (9.2.7). Thus the second property holds.

Finally, suppose the pair $(N_1, \mathcal{N}_1)$ also possesses the first property. Taking t to be the inclusion of N_1 into T, we conclude that $N_1 \subset N$ and $\mathcal{N}_1 \simeq \mathcal{N}|N$. Suppose $(N_1, \mathcal{N}_1)$ possess the second property too. Then, similarly, $N \subset N_1$. Thus $N = N_1$ and $\mathcal{N}_1 \simeq \mathcal{N}$, as desired. □

ANSWER 9.4.3. By Yoneda's Lemma [**EGAG**, (0,1.1.4), p. 20], a universal sheaf $\mathcal{P}$ exists if and only if $\mathbf{Pic}_{X/S}$ represents $\mathrm{Pic}_{X/S}$. Set $P := \mathbf{Pic}_{X/S}$.

Assume $\mathcal{P}$ exists. Then, for any invertible sheaf $\mathcal{N}$ on P, plainly $\mathcal{P} \otimes f_P^*\mathcal{N}$ is also a universal sheaf. Moreover, if $\mathcal{P}'$ is also a universal sheaf, then $\mathcal{P}' \simeq \mathcal{P} \otimes f_P^*\mathcal{N}$ for some invertible sheaf $\mathcal{N}$ on P by the definition with $h := 1_P$.

Assume $\mathcal{O}_S \xrightarrow{\sim} f_*\mathcal{O}_X$ holds universally. If $\mathcal{P} \otimes f_P^*\mathcal{N} \simeq \mathcal{P} \otimes f_P^*\mathcal{N}'$ for some invertible sheaves $\mathcal{N}$ and $\mathcal{N}'$ on P, then $\mathcal{N} \simeq \mathcal{N}'$ by Lemma 9.2.7.

By Part 2 of Theorem 9.2.5, if also f has a section, then $\mathbf{Pic}_{X/S}$ does represent $\mathrm{Pic}_{X/S}$; so then $\mathcal{P}$ exists. Furthermore, the curve $X/\mathbb{R}$ of Exercise 9.2.4 provides an example where no $\mathcal{P}$ exists, because $\mathrm{Pic}_{(X/\mathbb{R})\,(\text{ét})}$ is representable by Theorem 9.4.8, but $\mathrm{Pic}_{X/\mathbb{R}}$ is not since the two functors differ. □

ANSWER 9.4.4. Say $\mathbf{Pic}_{X/S}$ represents $\mathrm{Pic}_{(X/S)\,(\text{ét})}$. For any S'-scheme T,

$$\mathrm{Pic}_{(X_{S'}/S')\,(\text{ét})}(T) = \mathrm{Pic}_{(X/S)\,(\text{ét})}(T),$$

which holds essentially by definition, since a map of S'-schemes is an étale-covering if and only if it is an étale-covering when viewed as a map of S-schemes. However,

$$(\mathbf{Pic}_{X/S} \times_S S')(T) = \mathbf{Pic}_{X/S}(T)$$

because the structure map $T \to S'$ is fixed. Since the right-hand sides of the two displayed equations are equal, so are their left-hand sides. Thus $\mathbf{Pic}_{X/S} \times_S S'$ represents $\mathrm{Pic}_{(X_{S'}/S')\,(\text{ét})}$. Of course, a similar analysis applies when $\mathbf{Pic}_{X/S}$ represents one of the other relative Picard functors.

An example is provided by the curve $X \subset \mathbf{P}^2_{\mathbb{R}}$ of Exercise 9.2.4. Indeed, since the functors $\mathrm{Pic}_{X/\mathbb{R}}$ and $\mathrm{Pic}_{(X/\mathbb{R})\,(\text{ét})}$ differ, $\mathrm{Pic}_{X/\mathbb{R}}$ is not representable. But $\mathrm{Pic}_{(X/\mathbb{R})\,(\text{ét})}$ is representable by the Main Theorem, 9.4.8. Finally, since $X_{\mathbb{C}}$ has a $\mathbb{C}$-point, all its relative Picard functors are equal by the Comparison Theorem, 9.2.5. □

ANSWER 9.4.5. An $\mathcal{L}$ on an X_k defines a map $\mathrm{Spec}(k) \to \mathbf{Pic}_{X/S}$; assign its image to $\mathcal{L}$. Then, given any field k'' containing k, the pullback $\mathcal{L}|X_{k''}$ is assigned the same scheme point of $\mathbf{Pic}_{X/S}$.

Consider an $\mathcal{L}'$ on an $X_{k'}$. If $\mathcal{L}$ and $\mathcal{L}'$ represent the same class, then there is a k'' containing both k and k' such that $\mathcal{L}|X_{k''} \simeq \mathcal{L}'|X_{k''}$; hence, then both $\mathcal{L}$ and $\mathcal{L}'$ are assigned the same scheme point of $\mathbf{Pic}_{X/S}$. Conversely, if $\mathcal{L}$ and $\mathcal{L}'$ are assigned the same point, take k'' to be any algebraically closed field containing both k and k'. Then $\mathcal{L}|X_{k''}$ and $\mathcal{L}'|X_{k''}$ define the same map $\mathrm{Spec}(k'') \to \mathbf{Pic}_{X/S}$. Hence $\mathcal{L}|X_{k''} \simeq \mathcal{L}'|X_{k''}$ by Exercise 9.2.3 or 9.2.6.

Finally, given any scheme point of $\mathbf{Pic}_{X/S}$, let k be the algebraic closure of its residue field. Then $\mathrm{Spec}(k) \to \mathbf{Pic}_{X/S}$ is defined by an $\mathcal{L}$ on X_k by Exercise 9.2.3 or 9.2.6. So the given point is assigned to $\mathcal{L}$. Thus the classes of invertible sheaves on the fibers of X/S correspond bijectively to the scheme points of $\mathbf{Pic}_{X/S}$. □

ANSWER 9.4.7. An S-map $h\colon T \to \mathbf{Div}_{X/S}$ corresponds to a relative effective divisor D on X_T. So the composition $\mathbf{A}_{X/S}h\colon T \to P$ corresponds to the invertible sheaf $\mathcal{O}_{X_T}(D)$. Hence $\mathcal{O}_{X_T}(D) \simeq (1 \times \mathbf{A}_{X/S}h)^*\mathcal{P} \otimes f_P^*\mathcal{N}$ for some invertible sheaf $\mathcal{N}$ on T. Therefore, if T is viewed as a P-scheme via $\mathbf{A}_{X/S}h$, then D defines a T-point η of $\mathrm{LinSys}_{\mathcal{P}/X\times P/P}$. Plainly, the assignment $h \mapsto \eta$ is functorial in T. Thus if $\mathbf{Div}_{X/S}$ is viewed as a P-scheme via $\mathbf{A}_{X/S}$, then there is a natural map Λ from its functor of points to $\mathrm{LinSys}_{\mathcal{P}/X\times P/P}$.

Furthermore, Λ is an isomorphism. Indeed, let T be a P-scheme. A T-point η of $\mathrm{LinSys}_{\mathcal{P}/X\times P/P}$ is given by a relative effective divisor D on X_T such that $\mathcal{O}_{X_T}(D) \simeq \mathcal{P}_T \otimes f_T^*\mathcal{N}$ for some invertible sheaf $\mathcal{N}$ on T. Then $\mathcal{O}_{X_T}(D)$ and $\mathcal{P}_T$ define the same S-map $T \to P$. But $\mathcal{P}_T$ defines the structure map. And $\mathcal{O}_{X_T}(D)$ defines the composition $\mathbf{A}_{X/S}h$ where $h\colon T \to \mathbf{Div}_{X/S}$ is the map defined by D. Thus $\eta = \Lambda(h)$, and h is determined by η; hence, Λ is an isomorphism.

In other words, $\mathbf{Div}_{X/S}$ represents $\mathrm{LinSys}_{\mathcal{P}/X\times P/P}$. But $\mathbf{P}(\mathcal{Q})$ too represents $\mathrm{LinSys}_{\mathcal{P}/X\times P/P}$ by Theorem 9.3.13. Therefore, $\mathbf{P}(\mathcal{Q}) = \mathbf{Div}_{X/S}$ as P-schemes. □

ANSWER 9.4.10. First, suppose $F \to G$ is a surjection. Given a map of étale sheaves $\varphi\colon F \to H$ such that the two maps $F \times_G F \to H$ are equal, we must show there is one and only one map $G \to H$ such that $F \to G \to H$ is equal to φ.

Let $\eta \in G(T)$. By hypothesis, there exist an étale covering $T' \to T$ and an element $\zeta' \in F(T')$ such that ζ' and η have the same image in $G(T')$. Set $T'' := T' \times_T T'$. Then the two images of ζ' in $F(T'')$ define an element ζ'' of $(F \times_G F)(T'')$. Since the two maps $F \times_G F \to H$ are equal, the two images of ζ'' in $H(T'')$ are equal. But these two images are equal to those of $\varphi(\zeta') \in H(T')$. Since H is a sheaf, therefore $\varphi(\zeta')$ is the image of a unique element $\theta \in H(T)$.

Note $\theta \in H(T)$ is independent of the choice of T' and $\zeta' \in F(T')$. Indeed, let $\zeta_1' \in F(T_1')$ be a second choice. Arguing as above, we find $\varphi(\zeta_1') \in H(T_1')$ and $\varphi(\zeta') \in H(T')$ have the same image in $H(T_1' \times_T T')$. So ζ_1' also leads to θ.

Define a map $G(T) \to H(T)$ by $\eta \mapsto \theta$. Plainly this map behaves functorially in T. Thus there is a map of sheaves $G \to H$. Plainly, $F \to G \to H$ is equal to $\varphi\colon F \to H$. Finally, $G \to H$ is the only such map, since the image of η in $H(T)$ is determined by the image of η in $G(T')$, and the latter must map to $\varphi(\zeta') \in H(T')$. Thus G is the coequalizer of $F \times_G F \rightrightarrows F$.

Conversely, suppose G is the coequalizer of $F \times_G F \rightrightarrows F$. Form the étale subsheaf $H \subset G$ associated to the presheaf whose T-points are the images in $G(T)$ of the elements of $F(T)$. Then the map $F \to G$ factors through H. So the two maps $F \times_G F \to H$ are equal. Since G is the coequalizer, there is a map $G \to H$ so that $F \to G \to H$ is equal to $F \to H$. Hence $F \to G \to H \hookrightarrow G$ is equal to $F \to G$. So $G \to H \hookrightarrow G$ is equal to 1_G by uniqueness. Therefore, $H = G$. Thus $F \to G$ is a surjection. □

ANSWER 9.4.11. Theorem 9.4.8 implies each connected component Z' of Z lies in an increasing union of open quasi-projective subschemes of $\mathbf{Pic}_{X/S}$. So Z' lies in one of them since Z' is quasi-compact. So Z' is quasi-projective. But Z has only finitely many components Z'. Therefore, Z is quasi-projective. □

ANSWER 9.4.12. Set $P := \mathbf{Pic}_{X/S}$, which exists by Theorem 9.4.8. If $\mathcal{P}$ exists, then $\mathbf{A}_{X/S}$ is, by Exercise 9.4.7, the structure map of the bundle $\mathbf{P}(\mathcal{Q})$ where $\mathcal{Q}$ denotes the coherent sheaf on $\mathbf{Pic}_{X/S}$ associated to $\mathcal{P}$ as in Subsection 9.3.10. In particular, $\mathbf{A}_{X/S}$ is projective Zariski locally over S.

In general, forming P commutes with extending S by Exercise 9.4.4. Similarly, forming $\mathbf{A}_{X/S}$ does too. But a map is proper if it is after an fppf base extension by [**EGAIV2**, 2.7.1(vii)].

However, $f\colon X \to S$ is fppf. Moreover, $f_X\colon X \times X \to X$ has a section, namely, the diagonal. So use f as a base extension. Then, by Exercises 9.3.11 and 9.4.3, a universal sheaf $\mathcal{P}$ exists. Therefore, $\mathbf{A}_{X/S}$ is proper by the first case. □

ANSWER 9.4.13. Let's use the ideas and notation of Answer 9.4.12. Now, X_0 represents $\mathrm{Div}^1_{X/S}$ by Exercise 9.3.8. Hence the Abel map $\mathbf{A}_{X/S}$ induces a natural map $A\colon X_0 \to P$, and forming A commutes with extending S. But a map is a closed embedding if it is after an fppf base extension by [**EGAIV2**, 2.7.1(xii)]. So we may assume $\mathbf{P}(\mathcal{Q}) = \mathbf{Div}_{X/S}$.

The function $\lambda \mapsto \deg \mathcal{P}_\lambda$ is locally constant. Let $W \subset P$ be the open and closed subset where the function's value is 1. Plainly $\mathbf{P}(\mathcal{Q}_W) = \mathbf{Div}^1_{X/S}$ owing to the above. Therefore, $X_0 = \mathbf{P}(\mathcal{Q}_W)$, and $A\colon X_0 \to P$ is equal to the structure map of $\mathbf{P}(\mathcal{Q}_W)$. So it remains to show that this structure map is a closed embedding.

Fix $\lambda \in W$. Then $\dim_{k_\lambda}(\mathcal{Q} \otimes k_\lambda) = \dim_{k_\lambda} \mathrm{H}^0(X_\lambda, \mathcal{P}_\lambda)$. Suppose P_λ has two independent global sections. Each defines an effective divisor of degree 1, which is a k_λ-rational point x_i. Since neither section is a multiple of the other, the x_i are distinct. Hence the sections generate P_λ. So they define a map $h\colon X_\lambda \to \mathbf{P}^1_{k_\lambda}$ by [**EGAII**, 4.2.3] or [**Har77**, Thm. II, 7.1, p. 150]. Then h is birational since each x_i is the scheme-theoretic inverse image of a k_λ-rational point of $\mathbf{P}^1_{k_\lambda}$. Hence h is an isomorphism. But, by hypothesis, X_λ is of arithmetic genus at least 1. So there is a contradiction. Therefore, $\dim_{k_\lambda}(\mathcal{Q} \otimes k_\lambda) \le 1$.

By Nakayama's lemma, $\mathcal{Q}$ can be generated by a single element on a neighborhood $V \subset W$ of λ. So there is a surjection $\mathcal{O}_V \twoheadrightarrow \mathcal{Q}_V$. It defines a closed embedding $\mathbf{P}(\mathcal{Q}_V) \hookrightarrow \mathbf{P}(\mathcal{O}_V)$. But the structure map $\mathbf{P}(\mathcal{O}_V) \to V$ is an isomorphism. Hence $\mathbf{P}(\mathcal{Q}_V) \to V$ is a closed embedding. But $\lambda \in W$ is arbitrary. So $\mathbf{P}(\mathcal{Q}_W) \to W$ is indeed a closed embedding. □

ANSWER 9.4.15. Representing $\mathrm{Pic}_{X/S}$ is similar to representing $\mathrm{Pic}_{X'/S'}$ in Example 9.4.14, but simpler. Indeed, On $X \times_S \mathbb{Z}_S$, form an invertible sheaf $\mathcal{P}$ by placing $\mathcal{O}_X(n)$ on the nth copy of X. Then it suffices to show this: given any S-scheme T and any invertible sheaf $\mathcal{L}$ on X_T, there exist a unique S-map $q\colon T \to \mathbb{Z}_S$ and some invertible sheaf $\mathcal{N}$ on T such that $(1 \times q)^*\mathcal{P} = \mathcal{L} \otimes f_T^*\mathcal{N}$.

Plainly, we may assume T is connected. Then the function $s \mapsto \chi(X_t, \mathcal{L}_t)$ is constant on T by [**EGAIII2**, 7.9.11]. Now, X_t is a projective space of dimension at least 1 over the residue field k_t; so $\mathcal{L}_t \simeq \mathcal{O}_{X_t}(n)$ for some n by [**Har77**, Prp. 6.4, p. 132, and Cors. 6.16 and 6.17, p. 145]. Hence n is independent of t.

Set $\mathcal{M} := \mathcal{L}^{-1}(n)$. Then $\mathcal{M}_t \simeq \mathcal{O}_{X_t}$ for all $t \in T$. Hence $\mathrm{H}^1(X_t, \mathcal{M}_t) = 0$ and $\mathrm{H}^0(X_t, \mathcal{M}_t) = k_t$ by Serre's explicit computation [**EGAIII1**, 2.1.12]. Hence $f_{T*}\mathcal{M}$ is invertible, and forming it commutes with changing the base T, owing to the theory in Subsection 9.3.10.

Set $\mathcal{N} := f_{T*}\mathcal{M}$. Consider the natural map $u\colon f_T^*\mathcal{N} \to \mathcal{M}$. Forming u commutes with changing T, since forming $\mathcal{N}$ does. But u is an isomorphism on the fiber over each $t \in T$. So $u \otimes k_t$ is an isomorphism. Hence u is surjective by Nakayama's lemma. But both source and target of u are invertible; so u is an isomorphism. Hence $\mathcal{L} \otimes f_T^*\mathcal{N} = \mathcal{O}_{X_T}(n)$.

Let $q\colon T \to \mathbb{Z}_S$ be the composition of the structure map $T \to S$ and the nth inclusion $S \hookrightarrow \mathbb{Z}_S$. Plainly $(1 \times q)^*\mathcal{P} = \mathcal{O}_{X_T}(n)$, and q is the only such S-map. Thus $\mathbb{Z}_S$ represents $\mathrm{Pic}_{X/S}$, and $\mathcal{P}$ is a universal sheaf. □

ANSWER 9.4.16. First of all, $\mathbf{Pic}_{X/\mathbb{R}}$ exists by Theorem 9.4.8. Now, $X_\mathbb{C} \simeq \mathbf{P}^1_\mathbb{C}$. Hence $\mathbf{Pic}_{X/\mathbb{R}} \times_\mathbb{R} \mathbb{C} \simeq \mathbb{Z}_\mathbb{C}$ by Exercises 9.4.4 and 9.4.15. The induced automorphism of $\mathbb{Z}_{\mathbb{C}\otimes_\mathbb{R}\mathbb{C}}$ is the identity; indeed, a point of this scheme corresponds to an invertible sheaf on $\mathbf{P}^1_\mathbb{C}$, and every such sheaf is isomorphic to its pullback under any $\mathbb{R}$-automorphism of $\mathbf{P}^1_\mathbb{C}$. Hence, by descent theory, $\mathbf{Pic}_{X/\mathbb{R}} = \mathbb{Z}_\mathbb{R}$.

The above reasoning leads to a second proof that $\mathrm{Pic}_{(X/\mathbb{R})\,(\text{ét})}$ is representable. Indeed, set $P := \mathrm{Pic}_{(X/\mathbb{R})\,(\text{ét})}$. By the above reasoning, the pair

$$(P \otimes_\mathbb{R} \mathbb{C}) \otimes_\mathbb{C} (\mathbb{C} \otimes_\mathbb{R} \mathbb{C}) \rightrightarrows P \otimes_\mathbb{R} \mathbb{C}$$

is representable by the pair $\mathbb{Z}_{\mathbb{C}\otimes_{\mathbb{R}}\mathbb{C}} \rightrightarrows \mathbb{Z}_{\mathbb{C}}$, whose coequalizer is $\mathbb{Z}_{\mathbb{R}}$. On the other hand, in the category of étale sheaves, the coequalizer is P owing to Exercise 9.4.10.

Notice in passing that $\mathbf{Pic}_{X/\mathbb{R}} = \mathbf{Pic}_{\mathbf{P}^1_{\mathbb{R}}/\mathbb{R}}$. But $\mathrm{Pic}_{X/\mathbb{R}}$ is not representable due to Exercise 9.2.4, where as $\mathrm{Pic}_{\mathbf{P}^1_{\mathbb{R}}/\mathbb{R}}$ is representable owing to Exercise 9.4.15. □

ANSWER 9.5.7. Exercise 9.4.11 implies Z is quasi-projective. Hence Z is projective if Z is proper. By [**EGAIV2**, 2.7.1], an S-scheme is proper if it is so after an fppf base change, such as $f\colon X \to S$. But $f_X : X \times X \to X$ has a section, namely, the diagonal. Thus we may assume f has a section.

Using the Valuative Criterion for Properness [**Har77**, Thm. 4.7, p. 101], we need only check this statement: given an S-scheme T of the form $T = \mathrm{Spec}(A)$ where A is a valuation ring, say with fraction field K, every S-map $u\colon \mathrm{Spec}(K) \to Z$ extends to an S-map $T \to Z$. We do not need to check the extension is unique if it exists; indeed, this uniqueness holds by the Valuative Criterion for Separatedness [**Har77**, Thm. 4.3, p. 97] since Z is quasi-projective, so separated.

Since f has a section, u arises from an invertible sheaf $\mathcal{L}$ on X_K by Theorem 9.2.5. We have to extend $\mathcal{L}$ over X_T. Indeed, this extension defines a map $t\colon T \to \mathbf{Pic}_{X/S}$ extending u, and t factors through Z because Z is closed and T is integral.

Plainly it suffices to extend $\mathcal{L}(n)$ for any $n \gg 0$. So replacing $\mathcal{L}$ if need be, we may assume $\mathcal{L}$ has a nonzero section. It is regular since X_K is integral. So X_K has a divisor D such that $\mathcal{O}(D) = \mathcal{L}$.

Let $D' \subset X_T$ be the closure of D. Now, X/S is smooth and T is regular, so X_T is regular by [**EGAIV2**, 6.5.2], so factorial by [**EGAIV2**, 21.11.1]. Hence D' is a divisor. And $\mathcal{O}(D')$ extends $\mathcal{L}$. □

ANSWER 9.5.16. Owing to Serre's Theorem [**Har77**, Thm. 5.2, p. 228], we have $\mathrm{H}^i(\Omega^2_X(n)) = 0$ for $i > 0$ and $n \gg 0$. So $\varphi(n) = \chi(\Omega^2_X(n))$. Hence

$$q = \mathrm{H}^1(\Omega^2_X) - \mathrm{H}^2(\Omega^2_X) + 1.$$

Serre duality [**Har77**, Cor. 7.13, p. 247] yields $\dim \mathrm{H}^i(\Omega^i_X) = \dim \mathrm{H}^{2-i}(\mathcal{O}_X)$ for all i. And $\dim \mathrm{H}^0(\mathcal{O}_X) = 1$ since X is projective and geometrically integral. So

$$q = \dim \mathrm{H}^1(\mathcal{O}_X).$$

Hence Corollary 9.5.14 yields dim $\mathbf{Pic}_{X/S} \leq q$, with equality in characteristic 0. □

ANSWER 9.5.17. Set $P := \mathbf{Pic}_{X/S}$, which exists by Theorem 9.4.8. By Exercises 9.3.11 and 9.4.3, there exists a universal sheaf $\mathcal{P}$ on $X \times P$

Suppose $q = 0$. Then P is smooth of dimension 0 everywhere by Corollary 9.5.13. Let D be a relative effective divisor on X_T/T where T is a connected S-scheme. Then $\mathcal{O}_{X_T}(D)$ defines a map $\tau : T \to P$, and

$$\mathcal{O}_{X_T}(D) \simeq (1 \times \tau)^*\mathcal{P} \otimes f_T^*\mathcal{N}$$

for some invertible sheaf $\mathcal{N}$ on T. Now, T is connected and P is discrete and reduced; so τ is constant. Set $\lambda := \tau T$, and view $\mathcal{P}_\lambda$ as an invertible sheaf $\mathcal{L}$ on X. Then $\mathcal{L}_T = (1 \times \tau)^*\mathcal{P}$. So $\mathcal{O}_{X_T}(D) \simeq \mathcal{L}_T \otimes f_T^*\mathcal{N}$, as required.

Consider the converse. Again by Exercise 9.4.3, there is a coherent sheaf $\mathcal{Q}$ on P such that $\mathbf{P}(\mathcal{Q}) = \mathbf{Div}_{X/S}$. Furthermore, $\mathcal{Q}$ is nonzero and locally free at any closed point λ representing an invertible sheaf $\mathcal{L}$ on X such that $\mathrm{H}^1(\mathcal{L}) = 0$ by Subsection 9.3.10; for example, take $\mathcal{L} := \mathcal{O}_X(n)$ for $n \gg 0$.

Let $U \subset P$ be a connected open neighborhood of λ on which Q is free. Let $T \subset \mathbf{P}(\mathcal{Q})$ be the preimage of U, and let D be the universal relative effective divisor on X_T/T. Then the natural map $A\colon T \to U$ is smooth with irreducible fibers. So T is connected. Moreover, A is the map defined by $\mathcal{O}_{X_T}(D)$.

Suppose $\mathcal{O}_{X_T}(D) \simeq \mathcal{M}_T \otimes f_T^*\mathcal{N}$ for some invertible sheaves $\mathcal{M}$ on X and $\mathcal{N}$ on T. Then $A\colon T \to U$ is also defined by $\mathcal{M}_T$. Say $\mu \in P$ represents $\mathcal{M}$. Then A factors through the inclusion of the closed point μ. Hence $\mu = \lambda$; moreover, since A is smooth and surjective, its image, the open set U, is just the reduced closed point λ. Now, there is an automorphism of P that carries 0 to λ, namely, "multiplication" by λ. So P is smooth of dimension 0 at 0. Therefore, $q = 0$ by Corollary 9.5.13.

In characteristic 0, a priori P is smooth by Corollary 9.5.14. Now, $A\colon T \to U$ is smooth. Hence, T is smooth too. But the preceding argument shows that, if the condition holds for this T, then $q = 0$, as required. □

ANSWER 9.5.23. By hypothesis, $\dim X_s = 1$ for $s \in S$; so $\mathrm{H}^2(\mathcal{O}_{X_s}) = 0$. Hence the $\mathbf{Pic}^0_{X_s/k_s}$ are smooth by Proposition 9.5.19, so of dimension p_a by Proposition 9.5.13. Hence, by Proposition 9.5.20, the $\mathbf{Pic}^0_{X_s/k_s}$ form a family of finite type, whose total space is the open subscheme $\mathbf{Pic}^0_{X/S}$ of $\mathbf{Pic}_{X/S}$. And $\mathbf{Pic}_{X/S}$ is smooth over S again by Proposition 9.5.19.

Hence $\mathbf{Pic}^0_{X/S}$ is quasi-projective by Exercise 9.4.11.

If X/S is smooth, then $\mathbf{Pic}^0_{X/S}$ is projective over S by Exercise 9.5.7. Alternatively, use Theorem 9.5.4 and Proposition 9.5.20 again to conclude $\mathbf{Pic}^0_{X/S}$ is proper, so projective since it is quasi-projective.

Conversely, assume $\mathbf{Pic}^0_{X/S}$ is proper, and let us prove X/S is smooth, Since X/S is flat, we need only prove each X_s is smooth. So we may replace S by the spectrum of the algebraic closure of k_s. If $p_a = 0$, then X is smooth, indeed $X = \mathbf{P}^1$, by [**Har77**, Ex. 1.8(b), p. 298].

Suppose $p_a > 0$. Let X_0 be the open subscheme where X is smooth. Then there is a closed embedding $A\colon X_0 \hookrightarrow \mathbf{Pic}_{X/S}$ by Exercise 9.4.13. Its image consists of points λ representing invertible sheaves of degree 1. Fix a rational point λ, and define an automorphism β of $\mathbf{Pic}_{X/S}$ by $\beta(\kappa) := \kappa\lambda^{-1}$. Then βA is a closed embedding of X_0 in $\mathbf{Pic}^0_{X/S}$.

By assumption, $\mathbf{Pic}^0_{X/S}$ is proper. So X_0 is proper. Hence $X_0 \hookrightarrow X$ is proper since X is separated. Hence X_0 is closed in X. But X_0 is dense in X since X is integral and the ground field is algebraically closed. Hence $X_0 = X$; in other words, X is smooth. □

ANSWER 9.6.4. As before, by Lemma 9.6.6, there is an m such that every $\mathcal{N}(m)$ is generated by its global sections. So there is a section that does not vanish at any given associated point of X; since these points are finite in number, if σ is a general linear combination of the corresponding sections, then σ vanishes at no associated point. So σ is regular, whence defines an effective divisor D such that $\mathcal{O}_X(-D) = \mathcal{N}^{-1}(-m)$.

Plainly $\mathcal{N}^{-1}$ is numerically equivalent to $\mathcal{O}_X$ too. So $\chi(\mathcal{N}^{-1}(n)) = \chi(\mathcal{O}_X(n))$ by Lemma 9.6.6. Hence the sequence $0 \to \mathcal{O}_X(-D) \to \mathcal{O}_X \to \mathcal{O}_D \to 0$ yields

$$\chi(\mathcal{O}_D(n)) = \psi(n) \text{ where } \psi(n) := \chi(\mathcal{O}_X(n)) - \chi(\mathcal{O}_X(n-m)).$$

Let $T \subset \mathbf{Div}_{X/k}$ be the open and closed subscheme parameterizing the effective divisors with Hilbert polynomial $\psi(n)$. Then T is a k-scheme of finite type. Let $\mathcal{M}'$ be the invertible sheaf on X_T associated to the universal divisor; set $\mathcal{M} := \mathcal{M}'(-n)$. Then there exists a rational point $t \in T$ such that $\mathcal{N} = \mathcal{M}_t$. Thus the $\mathcal{N}$ numerically equivalent to $\mathcal{O}_X$ form a bounded family. □

ANSWER 9.6.7. Suppose $a < a_r$. Suppose $\mathcal{L}(-1)$ has a nonzero section. It defines an effective divisor D, possibly 0. Hence

$$0 \le \int h^{r-1}[D] = \int h^{r-1}\ell - \int h^r = a - a_r < 0,$$

which is absurd. Thus $\mathrm{H}^0(\mathcal{L}(-1)) = 0$.

Let H be a hyperplane section of X. Then there is an exact sequence

$$0 \to \mathcal{L}(n-1) \to \mathcal{L}(n) \to \mathcal{L}_H(n) \to 0.$$

It yields the following bound:

$$\dim \mathrm{H}^0(\mathcal{L}(n)) - \dim \mathrm{H}^0(\mathcal{L}(n-1)) \le \dim \mathrm{H}^0(\mathcal{L}_H(n)). \tag{A.9.6.7.1}$$

Since $\binom{n+i}{i} - \binom{n-1+i}{i} = \binom{n+i-1}{i-1}$, the sequence also yields the following formula:

$$\chi(\mathcal{L}_H(n)) = \textstyle\sum_{0 \le i \le r-1} a_{i+1}\binom{n+i}{i}.$$

Suppose $r = 1$. Then $\dim \mathrm{H}^0(\mathcal{L}_H(n)) = \chi(\mathcal{L}_H(n)) = a_1$. Therefore, owing to Equation (A.9.6.7.1), induction on n yields $\dim \mathrm{H}^0(\mathcal{L}(n)) \le a_1(n+1)$, as desired.

Furthermore, $\mathcal{L}_H$ is 0-regular. Set $m := \dim \mathrm{H}^1(\mathcal{L}(-1))$. Then $\mathcal{L}$ is m-regular by Mumford's conclusion at the bottom of [**Mum66**, p. 102]. But

$$m = \dim \mathrm{H}^0(\mathcal{L}(-1)) - \chi(\mathcal{L}(-1)) = 0 - a_1(-1+1) - a_0 = -a_0.$$

Thus we may take $\Phi_1(u_0) := -u_0$ where u_0 is an indeterminate.

Suppose $r \ge 2$. Then we may take H irreducible by Bertini's Theorem [**Sei50**, Thm. 12, p. 374] or [**Jou79**, Cor. 6.7, p. 80]. Set $h_1 := c_1 O_H(1)$ and $\ell_1 := c_1\mathcal{L}_H$. Then $\int \ell_1 h_1^{r-2} = \int \ell h^{r-2}[H] = a < a_r$. So by induction on r, we may assume

$$\dim \mathrm{H}^0(\mathcal{L}_H(n)) \le a_r\tbinom{n+r-1}{r-1}.$$

Therefore, owing to Equation (A.9.6.7.1), induction on n yields the desired bound.

Furthermore, we may assume $\mathcal{L}_H$ is m_1-regular where $m_1 := \Phi_{r-1}(a_1 \dots, a_{r-1})$. Set $m := m_1 + \dim \mathrm{H}^1(\mathcal{L}(m_1 - 1))$. By Mumford's same work, $\mathcal{L}$ is m-regular. But

$$\begin{aligned} m &= m_1 + \dim \mathrm{H}^0(\mathcal{L}(m_1-1)) - \chi(\mathcal{L}(m_1-1)) \\ &\le m_1 + a_r\tbinom{m_1-1+r}{r} - \textstyle\sum_{0\le i\le r} a_i\tbinom{m_1-1+i}{i}. \end{aligned}$$

The latter expression is a polynomial in $a_0, \dots, a_{r-1}$ and m_1. So it is a polynomial Φ_r in $a_0, \dots, a_{r-1}$ alone, as desired.

In general, consider $\mathcal{N} := \mathcal{L}(-a)$. Then

$$\chi(\mathcal{N}(n)) = \textstyle\sum_{0\le i\le r} b_i\binom{n+i}{i} \text{ where } b_i := \sum_{j=0}^{r-i} a_{i+j}(-1)^j\binom{a-i-j}{j}.$$

Set $\nu := c_1\mathcal{N}$ and $b := \int \nu h^{r-1}$. Then

$$b = \int \ell h^{r-1} - a\int h^r = a - aa_r \le 0 < a_r.$$

Hence $\mathcal{N}$ is m-regular where $m := \Phi_r(b_0, \ldots, b_{r-1})$. But the b_i are polynomials in $a_0, \ldots, a_r$ and a. Hence there is a polynomial Ψ_r depending only on r such that $m := \Psi_r(a_0, \ldots, a_r; a)$, as desired. □

ANSWER 9.6.10. Let k' be the algebraic closure of k. If $H \otimes k' \subset G^\tau \otimes k'$, then $H \subset G^\tau$. But $G^\tau \otimes k' = (G \otimes k')^\tau$ by Lemma 9.6.10. Thus we may assume $k = k'$.

Then $H \subset \bigcup_{h \in H(k)} hG^0$. But G^0 is open, so hG^0 is too. And H is quasi-compact. So H lies in finitely many hG^0. So $G^0(k)$ has finite index in $H(k)G^0(k)$, say n. Then $h^n \in G^0(k)$ for every $h \in H(k)$. So $\varphi_n(H) \subset G^0$. Thus $H \subset G^\tau$. □

ANSWER 9.6.11. Given n, plainly $\mathcal{L}^{\otimes n}$ corresponds to $\varphi_n \lambda$. And $\mathcal{L}^{\otimes n}$ is algebraically equivalent to $\mathcal{O}_X$ if and only if $\varphi_n \lambda \in \mathbf{Pic}^0_{X/k}$ by Proposition 9.5.10. So $\mathcal{L}$ is τ-equivalent to $\mathcal{O}_X$ if and only if $\lambda \in \mathbf{Pic}^\tau_{X/k}$ by Definitions 9.6.1 and 9.6.8. □

ANSWER 9.6.13. Theorem 9.4.8 says $\mathbf{Pic}_{X/k}$ exists and represents $\mathrm{Pic}_{(X/k)\,(\text{ét})}$. So $\mathbf{Pic}^\tau_{X/k}$ is of finite type by Proposition 9.6.12. Hence $\mathbf{Pic}^\tau_{X/k}$ is quasi-projective by Exercise 9.4.11.

Suppose X is also geometrically normal. Since $\mathbf{Pic}^\tau_{X/k}$ is quasi-projective, to prove it is projective, it suffices to prove it is complete. By Proposition 9.6.12, forming $\mathbf{Pic}^0_{X/k}$ commutes with extending k. And by [**EGAIV2**, 2.7.1(vii)], a k-scheme is complete if (and only if) it is after extending k. So assume k is algebraically closed.

As λ ranges over the k-points of $\mathbf{Pic}^\tau_{X/k}$, the cosets $\lambda\,\mathbf{Pic}^0_{X/k}$ cover $\mathbf{Pic}^\tau_{X/k}$. So finitely many cosets cover, since $\mathbf{Pic}^0_{X/k}$ is an open by Proposition 9.5.3 and since $\mathbf{Pic}^\tau_{X/k}$ is quasi-compact, Now, $\mathbf{Pic}^0_{X/k}$ is projective by Theorem 9.5.4, so complete, And $\mathbf{Pic}^\tau_{X/k}$ is closed, again by Proposition 9.6.12. Hence $\mathbf{Pic}^\tau_{X/k}$ is complete. □

ANSWER 9.6.15. First, suppose that L is bounded. Then $\mathcal{M}$ defines a map $\theta\colon T \to \mathbf{Pic}_{X/S}$, and $\theta(T) \supset \Lambda$. Since T is Noetherian, plainly so is $\theta(T)$; whence, plainly so is any subspace of $\theta(T)$. Thus Λ is quasi-compact.

Conversely, suppose Λ is quasi-compact. Since $\mathbf{Pic}_{X/S}$ is locally of finite type by Proposition 9.4.17, there is an open subscheme of finite type containing any given point of Λ. So finitely many of the subschemes cover Λ. Denote their union by U.

The inclusion $U \hookrightarrow \mathbf{Pic}_{X/S}$ is defined by an invertible sheaf $\mathcal{M}$ on X_T for some fppf covering $T \to U$. Replace T be an open subscheme so that $T \to U$ is of finite type and surjective. Since U is of finite type, so is T. Given $\lambda \in \Lambda$, let $t \in T$ map to λ. Then λ corresponds to the class of $\mathcal{M}_t$. □

ANSWER 9.6.18. Theorem 9.6.16 asserts $\mathbf{Pic}^\tau_{X/S}$ is of finite type. So it is projective by Exercise 9.5.7. □

ANSWER 9.6.21. Plainly, replacing S by an open subset, we may assume X/S is projective and S is connected. Given $s \in S$, set $\psi(n) := \chi(\mathcal{O}_{X_s}(n))$. Then $\psi(n)$ is independent of s. Given m, set $\varphi(n) := m + \psi(n)$.

Let $\lambda \in \mathbf{Pic}_{X/S}$. Then $\lambda \in \mathbf{Pic}^m_{X/S}$ if and only if λ represents an invertible sheaf $\mathcal{L}$ of degree m. And $\lambda \in \mathbf{Pic}^\varphi_{X/S}$ if and only if $\chi(\mathcal{L}(n)) = \varphi(n)$. But,

$$\chi(\mathcal{L}(n)) = \deg(\mathcal{L}(n)) + \psi(0) = \deg(\mathcal{L}) + \psi(n)$$

by Riemann's Theorem and the additivity of $\deg(\bullet)$. Hence $\mathbf{Pic}^m_{X/S} = \mathbf{Pic}^\varphi_{X/S}$. So Theorem 9.6.20 yields all the assertions, except for the two middle about $\mathbf{Pic}^0_{X/S}$.

To show $\mathbf{Pic}^0_{X/S} = \mathbf{Pic}^\tau_{X/S}$, similarly we need only show $\deg \mathcal{L} = 0$ if and only if $\mathcal{L}$ is τ-equivalent to $\mathcal{O}_X$, for, by Exercise 9.6.11, the latter holds if and only if $\lambda \in \mathbf{Pic}^\tau_{X/S}$. Plainly, we may assume $\mathcal{L}$ lives on a geometric fiber of X/S. Then the two conditions on $\mathcal{L}$ are equivalent by Theorem 9.6.3.

Since deg is additive, multiplication carries $\mathbf{Pic}^0_{X/S} \times \mathbf{Pic}^m_{X/S}$ set-theoretically into $\mathbf{Pic}^m_{X/S}$. So $\mathbf{Pic}^0_{X/S}$ acts on $\mathbf{Pic}^m_{X/S}$ since these two sets are open in $\mathbf{Pic}_{X/S}$.

Since X/S is flat with integral geometric fibers, its smooth locus X_0 provides an fppf covering of S. Temporarily, make the base change $X_0 \to S$. After it, the new map $X_0 \to S$ has a section. Its image is a relative effective divisor D, and tensoring with $\mathcal{O}_X(mD)$ defines the desired isomorphism from $\mathbf{Pic}^0_{X/S}$ to $\mathbf{Pic}^m_{X/S}$.

Finally, to show there is no abuse of notation, we must show the fiber $(\mathbf{Pic}^0_{X/S})_s$ is connected. To do so, we instead make the base change to the spectrum of an algebraically closed field $k \supset k_s$. Then X_0 has a k-rational point D, and again tensoring with $\mathcal{O}_X(mD)$ defines an isomorphism from $\mathbf{Pic}^0_{X/k}$ to $\mathbf{Pic}^m_{X/k}$. So it suffices to show $\mathbf{Pic}^m_{X/k}$ is connected for some $m \geq 1$.

Let $\beta\colon X_0^m \to \mathbf{Div}^m_{X/k} \to \mathbf{Pic}^m_{X/S}$ be composition of the map α of Exercise 9.3.8 and the Abel map. Since X is integral, so is the m-fold product X_0^m. Hence it suffices to show β is surjective for some $m \geq 1$.

By Exercise 9.6.7, there is an $m_0 \geq 1$ such that every invertible sheaf on X of degree 0 is m_0-regular. Set $m := \deg(\mathcal{O}_X(m_0))$. Then every invertible sheaf $\mathcal{L}$ on X of degree m is 0-regular, so generated by its global sections.

In particular, for each singular point of X, there is a global section that does not vanish at it. So, since k is infinite, a general linear combination of these sections vanishes at no singular point. This combination defines an effective divisor E such that $\mathcal{O}_X(E) = \mathcal{L}$. It follows that β is surjective, as desired. □

ANSWER 9.6.29. Suppose Λ is quasi-compact. Then, owing to Exercise 9.6.15, there exist an S-scheme T of finite type and an invertible sheaf $\mathcal{M}$ on X_T such that every polynomial $\varphi \in \Pi$ is of the form $\varphi(n) = \chi(\mathcal{M}_t(n))$ for some $t \in T$. Hence, by [**EGAIII2**, 7.9.4], the number of φ is at most the number of connected components of T. Thus Π is finite.

Suppose Λ is connected. Then its closure is too. So we may assume Λ is closed. Give Λ its reduced subscheme structure. Then the inclusion $\Lambda \hookrightarrow \mathbf{Pic}_{X/S}$ is defined by an invertible sheaf $\mathcal{M}$ on X_T for some fppf covering $T \to \Lambda$. Fix $t \in T$ and set $\varphi(n) := \chi(\mathcal{M}_t(n))$. Fix n, and form the set T' of points t' of T such that $\chi(\mathcal{M}_{t'}(n)) = \varphi(n)$. By [**EGAIII2**, 7.9.4], the set T' is open, and so is its complement. Hence their images are open in Λ, and plainly these images are disjoint. But Λ is connected. Hence $T' = T$. Thus $\Pi = \{\varphi\}$. □

Appendix B. Basic intersection theory

This appendix contains an elementary treatment of basic intersection theory, which is more than sufficient for many purposes, including the needs of Section 9.6. The approach was originated in 1959–60 by Snapper. His results were generalized and his proofs were simplified immediately afterward by Cartier [**Car60**]. Their work was developed further in fits and starts by the author.

The Index Theorem was proved by Hodge in 1937. Immediately afterward, B. Segre [**Seg37**, § 1] gave an algebraic proof for surfaces, and this proof was rediscovered by Grothendieck in 1958. Their work was generalized a tad in [**Kle71**, p. 662], and a variation appears below in Theorem B.27. From the index theorem, Segre [**Seg37**, §6] derived a connectedness statement like Corollary B.29 for surfaces, and the proof below is basically his.

DEFINITION B.1. Let $\mathbf{F}(X/S)$ or $\mathbf{F}$ denote the Abelian category of coherent sheaves $\mathcal{F}$ on X whose support $\operatorname{Supp}\mathcal{F}$ is proper over an Artin subscheme of S, that is, a 0-dimensional Noetherian closed subscheme. For each $r \geq 0$, let $\mathbf{F}_r$ denote the full subcategory of those $\mathcal{F}$ such that $\dim \operatorname{Supp}\mathcal{F} \leq r$.

Let $\mathbf{K}(X/S)$ or $\mathbf{K}$ denote the "Grothendieck group" of $\mathbf{F}$, namely, the free Abelian group on the $\mathcal{F}$, modulo short exact sequences. Abusing notation, let $\mathcal{F}$ also denote its class. And if $\mathcal{F} = \mathcal{O}_Y$ where $Y \subset X$ is a subscheme, then let $[Y]$ also denote the class. Let $\mathbf{K}_r$ denote the subgroup generated by $\mathbf{F}_r$.

Let $\chi\colon \mathbf{K} \to \mathbb{Z}$ denote the homomorphism induced by the Euler characteristic, which is just the alternating sum of the lengths of the cohomology groups.

Given $\mathcal{L} \in \operatorname{Pic}(X)$, let $c_1(\mathcal{L})$ denote the endomorphism of $\mathbf{K}$ defined by the following formula:

$$c_1(\mathcal{L})\mathcal{F} := \mathcal{F} - \mathcal{L}^{-1} \otimes \mathcal{F}.$$

Note that $c_1(\mathcal{L})$ is well defined since tensoring with $\mathcal{L}^{-1}$ preserves exact sequences.

LEMMA B.2. *Let $\mathcal{L} \in \operatorname{Pic}(X)$. Let $Y \subset X$ be a closed subscheme with $\mathcal{O}_Y \in \mathbf{F}$. Let $D \subset Y$ be an effective divisor such that $\mathcal{O}_Y(D) \simeq \mathcal{L}_Y$. Then*

$$c_1(\mathcal{L}) \cdot [Y] = [D].$$

PROOF. The left side is defined since $\mathcal{O}_Y \in \mathbf{F}$. The equation results from the sequence $0 \to \mathcal{O}_Y(-D) \to \mathcal{O}_Y \to \mathcal{O}_D \to 0$ since $\mathcal{O}_Y(-D) \simeq \mathcal{L}^{-1} \otimes \mathcal{O}_Y$. □

LEMMA B.3. *Let $\mathcal{L}$, $\mathcal{M} \in \operatorname{Pic}(X)$. Then the following relations hold:*

$$c_1(\mathcal{L})c_1(\mathcal{M}) = c_1(\mathcal{L}) + c_1(\mathcal{M}) - c_1(\mathcal{L} \otimes \mathcal{M});$$
$$c_1(\mathcal{L})c_1(\mathcal{L}^{-1}) = c_1(\mathcal{L}) + c_1(\mathcal{L}^{-1});$$
$$c_1(\mathcal{O}_X) = 0.$$

Furthermore, $c_1(\mathcal{L})$ and $c_1(\mathcal{M})$ commute.

PROOF. Let $\mathcal{F} \in \mathbf{K}$. By definition, $c_1(\mathcal{O}_X)\mathcal{F} = 0$; thus the third relation holds. Plainly, each side of the first relation carries $\mathcal{F}$ into

$$\mathcal{F} - \mathcal{L}^{-1} \otimes \mathcal{F} - \mathcal{M}^{-1} \otimes \mathcal{F} + \mathcal{L}^{-1} \otimes \mathcal{M}^{-1} \otimes \mathcal{F}.$$

Thus the first relation holds. It and the third relation imply the second. Furthermore, the first relation implies that $c_1(\mathcal{L})$ and $c_1(\mathcal{M})$ commute. □

LEMMA B.4. Given $\mathcal{F} \in \mathbf{F}_r$, let $Y_1, \ldots, Y_s$ be the r-dimensional irreducible components of $\operatorname{Supp}\mathcal{F}$ equipped with their induced reduced structure, and let l_i be the length of the stalk of $\mathcal{F}$ at the generic point of Y_i. Then, in $\mathbf{K}_r$,

$$\mathcal{F} \equiv \sum l_i \cdot [Y_i] \bmod \mathbf{K}_{r-1}.$$

PROOF. The assertion holds if it does after we replace S by a neighborhood of the image of $\operatorname{Supp}\mathcal{F}$. So we may assume S is Noetherian.

Let $\mathbf{F}' \subset \mathbf{F}_r$ denote the family of $\mathcal{F}$ for which the assertion holds. Since length($\bullet$) is an additive function, $\mathbf{F}'$ is "exact" in the following sense: for any short exact sequence $0 \to \mathcal{F}' \to \mathcal{F} \to \mathcal{F}'' \to 0$ such that two of the $\mathcal{F}$s belong to $\mathbf{F}'$, then the third does too. Trivially, $\mathcal{O}_Y \in \mathbf{F}'$ for any closed integral subscheme $Y \subset X$ such that $\mathcal{O}_Y \in \mathbf{F}_r$. Hence $\mathbf{F}' = \mathbf{F}_r$ by the "Lemma of Dévissage," [**EGAIII1**, Thm. 3.1.2]. □

LEMMA B.5. Let $\mathcal{L} \in \operatorname{Pic}(X)$. Then $c_1(\mathcal{L})\mathbf{K}_r \subset \mathbf{K}_{r-1}$ for all r.

PROOF. Let $\mathcal{F} \in \mathbf{F}_r$. Then $\mathcal{F}$ and $\mathcal{L}^{-1} \otimes \mathcal{F}$ are isomorphic at the generic point of each component of $\operatorname{Supp}\mathcal{F}$. So Lemma B.4 implies $c_1(\mathcal{L})\mathcal{F} \in \mathbf{K}_{r-1}$. □

LEMMA B.6. Let $\mathcal{L} \in \operatorname{Pic}(X)$, let $\mathcal{F} \in \mathbf{K}_r$, and let $m \in \mathbb{Z}$. Then

$$\mathcal{L}^{\otimes m} \otimes \mathcal{F} = \textstyle\sum_{i=0}^{r} \binom{m+i-1}{i} c_1(\mathcal{L})^i \mathcal{F}.$$

PROOF. Let x be an indeterminate, and consider the formal identity

$$(1-x)^n = \textstyle\sum_{i \geq 0} (-1)^i \binom{n}{i} x^i.$$

Replace x by $1 - y^{-1}$, set $n := -m$, and use the familiar identity

$$(-1)^i \tbinom{n}{i} = \tbinom{m+i-1}{i},$$

to obtain the formal identity

$$y^m = \textstyle\sum \binom{m+i-1}{i} (1 - y^{-1})^i.$$

It yields the assertion, because $c_1(L)^i \mathcal{F} = 0$ for $i > r$ owing to Lemma B.5. □

THEOREM B.7 (Snapper). Let $\mathcal{L}_1, \ldots, \mathcal{L}_n \in \operatorname{Pic}(X)$, let $m_1, \ldots, m_n \in \mathbb{Z}$, and let $\mathcal{F} \in \mathbf{K}_r$. Then the Euler characteristic $\chi\big(\mathcal{L}_1^{\otimes m_1} \otimes \cdots \otimes \mathcal{L}_n^{\otimes m_n} \otimes \mathcal{F}\big)$ is given by a polynomial in the m_i of degree at most r. In fact,

$$\chi\big(\mathcal{L}_1^{\otimes m_1} \otimes \cdots \otimes \mathcal{L}_n^{\otimes m_n} \otimes \mathcal{F}\big) = \textstyle\sum a(i_1, \ldots, i_n) \binom{m_1+i_1-1}{i_1} \cdots \binom{m_n+i_n-1}{i_n}$$

where $i_j \geq 0$ and $\sum i_j \leq r$ and where $a(i_1, \ldots, i_n) := \chi\big(c_1(\mathcal{L}_1)^{i_1} \cdots c_1(\mathcal{L}_n)^{i_n} \mathcal{F}\big)$.

PROOF. The theorem follows from Lemmas B.6 and B.5. □

DEFINITION B.8. Let $\mathcal{L}_1, \ldots, \mathcal{L}_r \in \operatorname{Pic}(X)$, repetitions allowed. Let $\mathcal{F} \in \mathbf{K}_r$. Define the *intersection number* or *intersection symbol* by the formula

$$\textstyle\int c_1(\mathcal{L}_1) \cdots c_1(\mathcal{L}_r)\mathcal{F} := \chi\big(c_1(\mathcal{L}_1) \cdots c_1(\mathcal{L}_r)\mathcal{F}\big) \in \mathbb{Z}.$$

If $\mathcal{F} = \mathcal{O}_X$, then also write $\int c_1(\mathcal{L}_1) \cdots c_1(\mathcal{L}_r)$ for the number. If $\mathcal{L}_j = \mathcal{O}_X(D_j)$ for a divisor D_j, then also write $(D_1 \cdots D_r \cdot \mathcal{F})$, or just $(D_1 \cdots D_r)$ if $\mathcal{F} = \mathcal{O}_X$.

THEOREM B.9. Let $\mathcal{L}_1, \ldots, \mathcal{L}_r \in \operatorname{Pic}(X)$ and $\mathcal{F} \in \mathbf{K}_r$.

(1) If $\mathcal{F} \in \mathbf{F}_{r-1}$, then $\int c_1(\mathcal{L}_1) \cdots c_1(\mathcal{L}_r)\mathcal{F} = 0$.

(2) (symmetry and additivity) The symbol $\int c_1(\mathcal{L}_1) \cdots c_1(\mathcal{L}_r)\mathcal{F}$ is symmetric in the $\mathcal{L}_j$. Furthermore, it is a homomorphism separately in each $\mathcal{L}_j$ and in $\mathcal{F}$.

(3) Set $\mathcal{E} := \mathcal{L}_1^{-1} \oplus \cdots \oplus \mathcal{L}_r^{-1}$. Then

$$\textstyle\int c_1(\mathcal{L}_1) \cdots c_1(\mathcal{L}_r)\mathcal{F} = \sum_{i=0}^{r} (-1)^i \chi\big((\bigwedge^i \mathcal{E}) \otimes \mathcal{F}\big).$$

PROOF. Part (1) results from Lemma B.5. So the symbol is a homomorphism in each $\mathcal{L}_j$ owing to the relations asserted in Lemma B.3. Furthermore, the symbol is symmetric owing to the commutativity asserted in Lemma B.3. Part (3) results from the definitions. □

COROLLARY B.10. Let $\mathcal{L}_1$, $\mathcal{L}_2 \in \operatorname{Pic}(X)$ and $\mathcal{F} \in \mathbf{K}_2$. Then

$$\textstyle\int c_1(\mathcal{L}_1)c_1(\mathcal{L}_2)\mathcal{F} = \chi(\mathcal{F}) - \chi(\mathcal{L}_1^{-1} \otimes \mathcal{F}) - \chi(\mathcal{L}_2^{-1} \otimes \mathcal{F}) + \chi(\mathcal{L}_1^{-1} \otimes \mathcal{L}_2^{-1} \otimes \mathcal{F}).$$

PROOF. The assertion is a special case of Part (3) of Proposition B.9. □

COROLLARY B.11. Let $D_1, \ldots, D_r$ be effective divisors on X, and $\mathcal{F} \in \mathbf{F}_r$. Set $Z := D_1 \cap \cdots \cap D_r$. Suppose $\boldsymbol{Z} \cap \operatorname{Supp} F$ is finite, and at each of its points, F is Cohen–Macaulay. Then

$$(D_1 \cdots D_r \cdot \mathcal{F}) = \operatorname{length} \mathrm{H}^0(\mathcal{F}_Z) \text{ where } \mathcal{F}_Z := \mathcal{F} \otimes \mathcal{O}_Z.$$

PROOF. For each j, set $\mathcal{L}_j := \mathcal{O}_X(D_j)$ and let $\sigma_j \in \mathrm{H}^0(\mathcal{L}_j)$ be the section defining D_j. Set $\mathcal{E} := \mathcal{L}_1^{-1} \oplus \cdots \oplus \mathcal{L}_r^{-1}$. Form the corresponding Koszul complex $(\bigwedge^\bullet \mathcal{E}) \otimes \mathcal{F}$ and its cohomology sheaves $\mathcal{H}^i\big((\bigwedge^\bullet \mathcal{E}) \otimes \mathcal{F}\big)$. Then

$$\textstyle\int c_1(\mathcal{L}_1) \cdots c_1(\mathcal{L}_r)\mathcal{F} = \sum_{i=0}^{r} (-1)^i \chi\big(\mathcal{H}^i\big((\bigwedge^\bullet \mathcal{E}) \otimes \mathcal{F}\big)\big).$$

owing to Part (3) of Proposition B.9 and to the additivity of χ. Furthermore, essentially by definition, $\mathcal{H}^0\big((\bigwedge^\bullet \mathcal{E}) \otimes \mathcal{F}\big) = \mathcal{F}_Z$. And by standard local algebra, the higher $\mathcal{H}^i$ vanish. Thus the assertion holds. □

LEMMA B.12. Let $\mathcal{L}_1, \ldots, \mathcal{L}_r \in \operatorname{Pic}(X)$ and $\mathcal{F} \in \mathbf{F}_r$. Let $Y_1, \ldots, Y_s$ be the r-dimensional irreducible components of $\operatorname{Supp} \mathcal{F}$ given their induced reduced structure, and let l_i be the length of the stalk of $\mathcal{F}$ at the generic point of Y_i. Then

$$\textstyle\int c_1(\mathcal{L}_1) \cdots c_1(\mathcal{L}_r)\mathcal{F} = \sum_i l_i \int c_1(\mathcal{L}_1) \cdots c_1(\mathcal{L}_r)[Y_i].$$

PROOF. Apply Lemma B.4 and Parts (1) and (2) of Proposition B.9. □

LEMMA B.13. Let $\mathcal{L}_1, \ldots, \mathcal{L}_r \in \operatorname{Pic}(X)$ and $Y \subset X$ a closed subscheme with $\mathcal{O}_Y \in \mathbf{F}$. Let $D \subset Y$ be an effective divisor such that $\mathcal{O}_Y(D) \simeq \mathcal{L}_r|Y$. Then

$$\textstyle\int c_1(\mathcal{L}_1) \cdots c_1(\mathcal{L}_r)[Y] = \int c_1(\mathcal{L}_1) \cdots c_1(\mathcal{L}_{r-1})[D].$$

PROOF. Apply Lemma B.2. □

PROPOSITION B.14. *Let $\mathcal{L}_1, \ldots, \mathcal{L}_r \in \operatorname{Pic}(X)$ and $\mathcal{F} \in \mathbf{F}_r$. If all the $\mathcal{L}_j$ are relatively ample and if $\mathcal{F} \notin \mathbf{K}_{r-1}$, then*

$$\textstyle\int c_1(\mathcal{L}_1) \cdots c_1(\mathcal{L}_r)\mathcal{F} > 0.$$

PROOF. Proceed by induction on r. If $r = 0$, then $\int \mathcal{F} = \dim \mathrm{H}^0(\mathcal{F})$ essentially by definition, and $H^0(\mathcal{F}) \neq 0$ since $\mathcal{F} \notin \mathbf{K}_{r-1}$ by hypothesis.

Suppose $r \geq 1$. Owing to Proposition B.12, we may assume $\mathcal{F} = \mathcal{O}_Y$ where Y is integral. Owing to Part (2) of Theorem B.9, we may replace $\mathcal{L}_r$ by a multiple, and so assume it is very ample. Then, for the corresponding embedding of Y, a hyperplane section D is a nonempty effective divisor such that $\mathcal{O}_Y(D) \simeq \mathcal{L}_r|Y$. Hence the assertion results from Proposition B.13 and the induction hypothesis. □

LEMMA B.15. Let $g\colon X' \to X$ be an S-map. Let $\mathcal{L}_1, \ldots, \mathcal{L}_r \in \mathrm{Pic}(X)$ and let $\mathcal{F} \in \mathbf{F}_r(X'/S)$. Then

$$\int c_1(g^*\mathcal{L}_1) \cdots c_1(g^*\mathcal{L}_r)\mathcal{F} = \int c_1(\mathcal{L}_1) \cdots c_1(\mathcal{L}_r) g_*\mathcal{F}.$$

PROOF. Let $\mathcal{G} \in \mathbf{F}_r(X'/S)$. Then, by hypothesis, $\mathrm{Supp}\,\mathcal{G}$ is proper over an Artin subscheme of S, and $\dim \mathrm{Supp}\,\mathcal{G} \leq r$; furthermore, X/S is separated. Hence, the restriction $g|\,\mathrm{Supp}\,\mathcal{G}$ is proper; so $g(\mathrm{Supp}\,\mathcal{G})$ is closed. And by the dimension theory of schemes of finite type over Artin schemes, $\dim g(\mathrm{Supp}\,\mathcal{G}) \leq r$. Therefore, $\mathrm{R}^i g_*\mathcal{G} \in \mathbf{F}_r(X/S)$ for all i.

Define a map $\mathbf{F}_r(X'/S) \to \mathbf{K}_r(X/S)$ by $\mathcal{G} \mapsto \sum_{i=0}^r (-1)^i \mathrm{R}^i g_*\mathcal{G}$. It induces a homomorphism $\mathrm{R}g_*\colon \mathbf{K}_r(X'/S) \to \mathbf{K}_r(X/S)$. And $\chi(Rg_*(\mathcal{G})) = \chi(\mathcal{G})$ owing to the Leray Spectral Sequence [**EGAIII1**, **0**-12.2.4] and to the additivity of χ [**EGAIII1**, **0**-11.10.3]. Furthermore, $\mathcal{L} \otimes \mathrm{R}^i g_*(\mathcal{G}) \xrightarrow{\sim} \mathrm{R}^i g_*(g^*\mathcal{L} \otimes \mathcal{G})$ for any $\mathcal{L} \in \mathrm{Pic}(X)$ by [**EGAIII1**, **0**-12.2.3.1]. Hence $c_1(\mathcal{L})Rg_*(\mathcal{G}) = Rg_*(c_1(g^*\mathcal{L})\mathcal{G})$. Therefore,

$$\int c_1(g^*\mathcal{L}_1) \cdots c_1(g^*\mathcal{L}_r)\mathcal{F} = \int c_1(\mathcal{L}_1) \cdots c_1(\mathcal{L}_r)\mathrm{R}g_*\mathcal{F}.$$

Finally, $\mathrm{R}g_*\mathcal{F} \equiv \mathrm{g}_*\mathcal{F} \bmod \mathbf{K}_{r-1}(X/S)$, because $\mathrm{R}^i g_*\mathcal{F} \in \mathbf{F}_{r-1}$ for $i \geq 1$ since, if $W \subset X'$ is the locus where $\mathrm{Supp}\,\mathcal{F} \to X$ has fibers of dimension at least 1, then $\dim g(W) \leq r-1$. So Part (1) of Theorem B.9 yields the asserted formula. □

PROPOSITION B.16 (Projection Formula). *Let $g\colon X' \to X$ be an S-map. Let $\mathcal{L}_1, \ldots, \mathcal{L}_r \in \mathrm{Pic}(X)$. Let $Y' \subset X'$ be an integral subscheme with $\mathcal{O}_{Y'} \in \mathbf{F}_r(X'/S)$. Set $Y := gY' \subset X$, give Y its induced reduced structure, and let $\deg(Y'/Y)$ be the degree of the function field extension if finite and be 0 if not. Then*

$$\int c_1(g^*\mathcal{L}_1) \cdots c_1(g^*\mathcal{L}_r)[Y'] = \deg(Y'/Y) \int c_1(\mathcal{L}_1) \cdots c_1(\mathcal{L}_r)[Y].$$

PROOF. Lemma B.4 yields $g_*\mathcal{O}_{Y'} \equiv \deg(Y'/Y)[Y] \bmod \mathbf{K}_{r-1}(X/S)$. So the assertion results from Lemma B.15 and from Part (1) of Theorem B.9. □

PROPOSITION B.17. *Assume S is the spectrum of a field, and let T be the spectrum of an extension field. Let $\mathcal{L}_1, \ldots, \mathcal{L}_r \in \mathrm{Pic}(X)$ and $\mathcal{F} \in \mathbf{F}_r(X/S)$. Then*

$$\int c_1(\mathcal{L}_{1,T}) \cdots c_1(\mathcal{L}_{r,T})\mathcal{F}_T = \int c_1(\mathcal{L}_1) \cdots c_1(\mathcal{L}_r)\mathcal{F}.$$

PROOF. The base change $T \to S$ preserves short exact sequences. So it induces a homomorphism $\kappa\colon \mathbf{K}_r(X/S) \to \mathbf{K}_r(X_T/T)$. Plainly κ preserves the Euler characteristic. The assertion now follows. □

PROPOSITION B.18. *Let* $\mathcal{L}_1, \ldots, \mathcal{L}_r \in \operatorname{Pic}(X)$. *Let* $\mathcal{F}$ *be a flat coherent sheaf on* X. *Assume* $\operatorname{Supp} \mathcal{F}$ *is proper and of relative dimension* r. *Then the function*

$$y \mapsto \textstyle\int c_1(\mathcal{L}_1) \cdots c_1(\mathcal{L}_r) \mathcal{F}_y$$

is locally constant.

PROOF. The assertion results from Definition B.8 and [**Mum70**, Cor., top p. 50]. □

DEFINITION B.19. Let $\mathcal{L}, \mathcal{N} \in \operatorname{Pic}(X)$. Call them *numerically equivalent* if $\int c_1(\mathcal{L})[Y] = \int c_1(\mathcal{N})[Y]$ for all closed integral curves $Y \subset X$ with $\mathcal{O}_Y \in \mathbf{F}_1$.

PROPOSITION B.20. *Let* $\mathcal{L}_1, \ldots, \mathcal{L}_r; \mathcal{N}_1, \ldots, \mathcal{N}_r \in \operatorname{Pic}(X)$ *and* $\mathcal{F} \in \mathbf{K}_r$. *If* $\mathcal{L}_j$ *and* $\mathcal{N}_j$ *are numerically equivalent for each* j, *then*

$$\textstyle\int c_1(\mathcal{L}_1) \cdots c_1(\mathcal{L}_r) \mathcal{F} = \int c_1(\mathcal{N}_1) \cdots c_1(\mathcal{N}_r) \mathcal{F}.$$

PROOF. If $r = 1$, then the assertion results from Lemma B.12. Suppose $r \geq 2$. Then $c_1(\mathcal{L}_2) \cdots c_1(\mathcal{L}_r) \mathcal{F} \in \mathbf{K}_1$ by Lemma B.5. Hence

$$\textstyle\int c_1(\mathcal{L}_1) c_1(\mathcal{L}_2) \cdots c_1(\mathcal{L}_r) \mathcal{F} = \int c_1(\mathcal{N}_1) c_1(\mathcal{L}_2) \cdots c_1(\mathcal{L}_r) \mathcal{F}.$$

Similarly, $c_1(\mathcal{N}_1) c_1(\mathcal{L}_3) \cdots c_1(\mathcal{L}_r) \mathcal{F} \in \mathbf{K}_1$, and so

$$\textstyle\int c_1(\mathcal{N}_1) c_1(\mathcal{L}_2) c_1(\mathcal{L}_3) \cdots c_1(\mathcal{L}_r) \mathcal{F} = \int c_1(\mathcal{N}_1) c_1(\mathcal{N}_2) c_1(\mathcal{L}_3) \cdots c_1(\mathcal{L}_r) \mathcal{F}.$$

Continuing in this fashion yields the assertion. □

PROPOSITION B.21. *Let* $g \colon X' \to X$ *be an* S-*map. Let* $\mathcal{L}, \mathcal{N} \in \operatorname{Pic}(X)$.

(1) *If* $\mathcal{L}$ *and* $\mathcal{N}$ *are numerically equivalent, then so are* $g^*\mathcal{L}$ *and* $g^*\mathcal{N}$.

(2) *Conversely, when* g *is proper and surjective, if* $g^*\mathcal{L}$ *and* $g^*\mathcal{N}$ *are numerically equivalent, then so are* $\mathcal{L}$ *and* $\mathcal{N}$.

PROOF. Let $Y' \subset X'$ be a closed integral curve with $\mathcal{O}_{Y'} \in \mathbf{F}_1(X'/S)$. Set $Y := g(Y')$ and give Y its induced reduced structure. Then Proposition B.16 yields

$$\textstyle\int c_1(g^*\mathcal{L})[Y'] = \deg(Y'/Y) \int c_1(\mathcal{L})[Y] \text{ and}$$
$$\textstyle\int c_1(g^*\mathcal{N})[Y'] = \deg(Y'/Y) \int c_1(\mathcal{N})[Y].$$

Part (1) follows.

Conversely, suppose g is proper and surjective. Let $Y \subset X$ be a closed integral curve with $\mathcal{O}_Y \in \mathbf{F}_1(X/S)$. Then Y is a complete curve in the fiber X_s over a closed point $s \in S$. Hence, since g is proper, there exists a complete curve Y' in X'_s mapping onto Y. Indeed, let $y \in Y$ be the generic point, and $y' \in g^{-1}Y$ a closed point; let Y' be the closure of y' given Y' its induced reduced structure. Plainly $\mathcal{O}_{Y'} \in \mathbf{F}_1(X'/S)$ and $\deg(Y'/Y) \neq 0$. The two equations displayed above now yield Part (2). □

DEFINITION B.22. Assume S is Artin, and X a proper curve. Let $\mathcal{L} \in \operatorname{Pic}(X)$. Define its *degree* $\deg(\mathcal{L})$ by the formula

$$\deg(\mathcal{L}) := \textstyle\int c_1(\mathcal{L}).$$

Let D be a divisor on X. Define its *degree* $\deg(D)$ by $\deg(D) := \deg(\mathcal{O}_X(D))$.

PROPOSITION B.23. *Assume S is Artin, and X a proper curve.*
(1) *The map* $\deg\colon \operatorname{Pic}(X) \to \mathbb{Z}$ *is a homomorphism.*
(2) *Let $D \subset X$ be an effective divisor. Then*

$$\deg(D) = \dim \mathrm{H}^0(\mathcal{O}_D).$$

(3) (Riemann's Theorem) *Let $\mathcal{L} \in \operatorname{Pic}(X)$. Then*

$$\chi(\mathcal{L}) = \deg(\mathcal{L}) + \chi(\mathcal{O}_X).$$

(4) *Suppose X is integral, and let $g\colon X' \to X$ be the normalization map. Then*

$$\deg(\mathcal{L}) = \deg(g^*\mathcal{L}).$$

PROOF. Part (1) results from Theorem B.9 (2). And Part (2) results from Lemma B.13. As to Part (3), note $\deg(\mathcal{L}^{-1}) = -\deg(\mathcal{L})$ by Part (1). And the definitions yield $\deg(\mathcal{L}^{-1}) = \chi(\mathcal{O}_X) - \chi(\mathcal{L})$. Thus Part (3) holds. Finally, Part (3) results from the definition and the Projection Formula. □

DEFINITION B.24. Assume S is Artin, and X a proper surface. Given a divisor D on X, set

$$p_a(D) := 1 - \chi\big(c_1(\mathcal{O}_X(D))\,\mathcal{O}_X\big).$$

PROPOSITION B.25. *Assume S is Artin, and X a proper surface. Let D and E be divisors on X. Then*

$$p_a(D+E) = p_a(D) + p_a(E) + (D \cdot E) - 1.$$

Furthermore, if D is effective, then

$$p_a(D) = 1 - \chi(\mathcal{O}_D);$$

in other words, $p_a(D)$ is equal to the arithmetic genus *of D.*

PROOF. The assertions result from Lemmas B.3 and B.2. □

PROPOSITION B.26 (Riemann–Roch for surfaces). *Assume S is the spectrum of a field, and X is a reduced, projective, equidimensional, Cohen–Macaulay surface. Let ω be a dualizing sheaf, and set $\mathcal{K} := \omega - \mathcal{O}_X$. Let D be a divisor on X. Then $\mathcal{K} \in \mathbf{K}_1$; furthermore,*

$$p_a(D) = \frac{(D^2) + (D\cdot\mathcal{K})}{2} + 1 \quad \text{and} \quad \chi(\mathcal{O}_X(D)) = \frac{(D^2) - (D\cdot\mathcal{K})}{2} + \chi(\mathcal{O}_X).$$

If X/S is Gorenstein, that is, $\omega = \mathcal{O}_X(K)$ for some "canonical" divisor K, then

$$p_a(D) = \frac{(D\cdot(D+K))}{2} + 1 \quad \text{and} \quad \chi(\mathcal{O}_X(D)) = \frac{(D\cdot(D-K))}{2} + \chi(\mathcal{O}_X).$$

PROOF. Since X is reduced, ω is isomorphic to $\mathcal{O}_X$ on a dense open subset of X by [**AK70**, (2.8), p. 8]. Hence $\mathcal{K} \in \mathbf{K}_1$.

Set $\mathcal{L} := \mathcal{O}_X(D)$. Then $(D^2) := \int c_1(\mathcal{L})^2 = -\int c_1(\mathcal{L})c_1(\mathcal{L}^{-1})$ by Parts (1) and (2) of Theorem B.9. Now, the definitions yield

$$\begin{aligned} c_1(\mathcal{L})(-c_1(\mathcal{L}^{-1})\mathcal{O}_X + \mathcal{K}) &= c_1(\mathcal{L})(\mathcal{L} - 2\mathcal{O}_X + \omega) \\ &= \mathcal{L} + \omega - 3\mathcal{O}_X + 2\mathcal{L}^{-1} - \mathcal{L}^{-1}\otimes\omega \quad \text{and} \end{aligned}$$

$$c_1(\mathcal{L})(-c_1(\mathcal{L}^{-1})\mathcal{O}_X - \mathcal{K}) = c_1(\mathcal{L})(\mathcal{L} - \omega) = \mathcal{L} - \omega - \mathcal{O}_X + \mathcal{L}^{-1}\otimes\omega.$$

But, $\mathrm{H}^i(\mathcal{L})$ is dual to $\mathrm{H}^{2-i}(\mathcal{L}^{-1}\otimes\omega)$ by duality theory; see [**Har77**, Cor. 7.7, p. 244], where k needn't be taken algebraically closed. So $\chi(\mathcal{L}) = \chi(\mathcal{L}^{-1}\otimes\omega)$. Similarly, $\chi(\mathcal{O}_X) = \chi(\omega)$. Therefore,

$$(D^2) + (D\cdot\mathcal{K}) = 2(\chi(\mathcal{L}^{-1}) - \chi(\mathcal{O}_X)) \quad\text{and}\quad (D^2) - (D\cdot\mathcal{K}) = 2(\chi(\mathcal{L}) - \chi(\mathcal{O}_X)).$$

Now, $-c_1(\mathcal{O}_X(D))\,\mathcal{O}_X = \mathcal{L}^{-1} - \mathcal{O}_X$. The first assertion follows.

Suppose ω is invertible. Then $-c_1(\omega^{-1})\mathcal{O}_X = \mathcal{K}$ owing to the definitions. And $-\int c_1(\mathcal{L})c_1(\omega^{-1}) = \int c_1(\mathcal{L})c_1(\omega)$ by Part (2) of Theorem B.9. Therefore, $(D\cdot\mathcal{K}) = (D\cdot K)$. Hence Part (2) of Theorem B.9 yields the second assertion. □

THEOREM B.27 (Hodge Index). *Assume S is the spectrum of a field, and X is a geometrically irreducible complete surface. Assume there is an $\mathcal{H}\in\operatorname{Pic}(X)$ such that $\int c_1(\mathcal{H})^2 > 0$. Let $\mathcal{L}\in\operatorname{Pic}(X)$. Assume $\int c_1(\mathcal{L})c_1(\mathcal{H}) = 0$ and $\int c_1(\mathcal{L})^2 \ge 0$. Then $\mathcal{L}$ is numerically equivalent to $\mathcal{O}_X$.*

PROOF. We may extend the ground field to its algebraic closure owing to Proposition B.17. Furthermore, we may replace X by its reduction; indeed, the hypotheses are preserved due to Lemma B.12, and the conclusion is preserved due to Definition B.19.

By Chow's Lemma, there is a surjective map $g\colon X'\to X$ where X' is an integral projective surface. Furthermore, we may replace X' by its normalization. Now, we may replace X by X' and $\mathcal{H}$ and $\mathcal{L}$ by $g^*\mathcal{H}$ and $g^*\mathcal{L}$. Indeed, the hypotheses are preserved due to the Projection Formula, Proposition B.16. And the conclusion is preserved due to Part (2) of Proposition B.21.

By way of contradiction, assume that there exists a closed integral subscheme $Y\subset X$ such that $\int c_1(\mathcal{L})\mathcal{O}_Y \ne 0$. Let $g\colon X'\to X$ be the blowing-up along Y, and $E := g^{-1}Y\subset X$ the exceptional divisor. Let $E_1,\dots,E_s$ be the irreducible components of E, and give them their induced reduced structure.

Since X is normal, it has only finitely may singular points. Off them, Y is a divisor, and g is an isomorphism. Hence one of the E_i, say E_1 maps onto Y, and the remaining E_i map onto points. Therefore, $\int c_1(g^*\mathcal{L})[E_1] = \int c_1(g\mathcal{L})[Y]$ and $\int c_1(g^*\mathcal{L})[E_i] = 0$ for $i\ge 2$ by the Projection Formula. Hence Lemma B.12 yields $\int c_1(g^*\mathcal{L})[E] = \int c_1(\mathcal{L})[Y]$. The latter is nonzero by the new assumption, and the former is equal to $\int c_1(g^*\mathcal{L})c_1(\mathcal{O}_{X'}(E))$ by Lemma B.13.

Set $\mathcal{M} := \mathcal{O}_{X'}(E)$. Then $\int c_1(g^*\mathcal{L})c_1(\mathcal{M}) \ne 0$. Moreover, by the Projection Formula, $\int c_1(g^*\mathcal{H}) > 0$ and $\int c_1(g^*\mathcal{L})c_1(g^*\mathcal{H}) = 0$ and $\int c_1(g^*\mathcal{L})^2 \ge 0$. Let's prove this situation is absurd. First, replace X by X' and $\mathcal{H}$ and $\mathcal{L}$ by $g^*\mathcal{H}$ and $g^*\mathcal{L}$.

Let $\mathcal{G}$ be an ample invertible sheaf on X. Set $\mathcal{H}_1 := \mathcal{G}^{\otimes m}\otimes\mathcal{M}$. Then

$$\int c_1(\mathcal{L})c_1(\mathcal{H}_1) = m\int c_1(\mathcal{L})c_1(\mathcal{G}) + \int c_1(\mathcal{L})c_1(\mathcal{M})$$

by additivity (see Part (2) of Theorem B.9). Now, $\int c_1(\mathcal{L})c_1(\mathcal{M}) \ne 0$. Hence there is an $m>0$ so that $\int c_1(\mathcal{L})c_1(\mathcal{H}_1) \ne 0$ and so that $\mathcal{H}_1$ is ample.

Set $\mathcal{L}_1 := \mathcal{L}^{\otimes p}\otimes\mathcal{H}^{\otimes q}$. Since $\int c_1(\mathcal{L})c_1(\mathcal{H}) = 0$, additivity yields

$$\int c_1(\mathcal{L}_1)^2 = p^2\int c_1(\mathcal{L})^2 + q^2\int c_1(\mathcal{H})^2, \text{ and}$$
$$\int c_1(\mathcal{L}_1)c_1(\mathcal{H}_1) = p\int c_1(\mathcal{L})c_1(\mathcal{H}_1) + q\int c_1(\mathcal{H})c_1(\mathcal{H}_1).$$

Since $\int c_1(\mathcal{L})c_1(\mathcal{H}_1) \ne 0$, there are p, q with $q\ne 0$ so that $\int c_1(\mathcal{L}_1)c_1(\mathcal{H}_1) = 0$. Then $\int c_1(\mathcal{L}_1)^2 > 0$ since $\int c_1(\mathcal{L})^2 \ge 0$ and $\int c_1(\mathcal{H})^2 > 0$. Replace $\mathcal{L}$ by $\mathcal{L}_1$ and $\mathcal{H}$ by $\mathcal{H}_1$. Then $\mathcal{H}$ is ample, $\int c_1(\mathcal{L})c_1(\mathcal{H}) = 0$ and $\int c_1(\mathcal{L})^2 > 0$.

Set $\mathcal{N} := \mathcal{L}^{\otimes n} \otimes \mathcal{H}^{-1}$ and $\mathcal{H}_1 := \mathcal{L} \otimes \mathcal{H}^a$. Take $a > 0$ so that $\mathcal{H}_1$ is ample. By additivity,

$$\textstyle\int c_1(\mathcal{N})c_1(\mathcal{H}_1) = n\int c_1(\mathcal{L})^2 - a\int c_1(\mathcal{H})^2.$$

Take $n > 0$ so that $\int c_1(\mathcal{N})c_1(\mathcal{H}_1) > 0$. Then additivity and Proposition B.14 yield

$$\textstyle\int c_1(\mathcal{N})c_1(\mathcal{H}) = -\displaystyle\int c_1(\mathcal{H})^2 < 0,$$
$$\int c_1(\mathcal{N})^2 = n^2\int c_1(\mathcal{L})^2 + \int c_1(\mathcal{H})^2 > 0.$$

But this situation stands in contradiction to the next lemma. □

LEMMA B.28. Assume S is the spectrum of a field, and X is an integral surface. Let $\mathcal{N} \in \operatorname{Pic}(X)$, and assume $\int c_1(\mathcal{N})^2 > 0$. Then these conditions are equivalent:

(i) For every ample sheaf $\mathcal{H}$, we have $\int c_1(\mathcal{N})c_1(\mathcal{H}) > 0$.
(i′) For some ample sheaf $\mathcal{H}$, we have $\int c_1(\mathcal{N})c_1(\mathcal{H}) > 0$.
(ii) For some $n > 0$, we have $\mathrm{H}^0(\mathcal{N}^{\otimes n}) \neq 0$.

PROOF. Suppose (ii) holds. Then there exists an effective divisor D such that $\mathcal{N}^{\otimes n} \simeq \mathcal{O}_X(D)$. And $D \neq 0$ since $\int c_1(\mathcal{N})^2 > 0$. Hence (i) results as follows:

$$\textstyle\int c_1(\mathcal{N})c_1(\mathcal{H}) = \int c_1(\mathcal{H})c_1(\mathcal{N}) = \int c_1(\mathcal{H})[D] > 0$$

by symmetry, by Lemma B.13, and by Proposition B.14.

Trivially, (i) implies (i′). Finally, assume (i′), and let's prove (ii). Let ω be a dualizing sheaf for X; then ω is torsion free of rank 1, and $\mathrm{H}^2(\mathcal{L})$ is dual to $\operatorname{Hom}(\mathcal{L}, \omega)$ for any coherent sheaf $\mathcal{F}$ on X; see [**FGA**, p. 149-17], [**AK70**, (1.3), p. 5, and (2.8), p. 8], and [**Har77**, Prp, 7.2, p. 241]. Set $\mathcal{K} := \omega - \mathcal{O}_X \in \mathbf{K}_1$.

Suppose $\mathcal{L}$ is invertible and $\mathrm{H}^2(\mathcal{L})$ is nonzero. Then there is a nonzero map $\mathcal{L} \to \omega$, and it is injective since X is integral. Let $\mathcal{F}$ be its cokernel. Then

$$\mathcal{K} = \mathcal{F} - c_1(\mathcal{L}^{-1})\mathcal{O}_X \text{ in } \mathbf{K}_1.$$

Hence Proposition B.14, symmetry, and additivity yield

$$\textstyle\int c_1(\mathcal{H})\mathcal{K} \geq \int c_1(\mathcal{L})c_1(\mathcal{H}).$$

Take $\mathcal{L} := \mathcal{N}^{\otimes n}$. Then $\int c_1(\mathcal{H})c_1(\mathcal{L}) = n\int c_1(\mathcal{H})c_1(\mathcal{N})$ by additivity. But $\int c_1(\mathcal{N})c_1(\mathcal{H}) > 0$ by hypothesis. Hence $\mathrm{H}^2(\mathcal{N}^{\otimes n})$ vanishes for $n \gg 0$. Now,

$$\textstyle\chi(\mathcal{N}^{\otimes n}) = \int c_1(\mathcal{N})^2\binom{n+1}{2} + a_1 n + a_0$$

for some a_1, a_0 by Snapper's Theorem, Theorem B.7. But $\int c_1(\mathcal{N})^2 > 0$ by hypothesis. Therefore, (ii) holds. □

COROLLARY B.29. Assume S is the spectrum of a field, and X a geometrically irreducible projective r-fold with $r \geq 2$. Let D, E be effective divisors, with E possibly trivial. Assume D is ample. Then $D + E$ is connected.

PROOF. Plainly we may assume the ground field is algebraically closed and X is reduced. Fix $n > 0$ so that $nD + E$ is ample; plainly we may replace D and E by $nD + E$ and 0. Proceeding by way of contradiction, assume D is the disjoint union of two closed subschemes D_1 and D_2. Plainly D_1 and D_2 are divisors; so $D = D_1 + D_2$.

Proceed by induction on r. Suppose $r = 2$. Then, since D_1 and D_2 are disjoint, $(D_1 \cdot D_2) = 0$ by Lemma B.13. Now, D is ample. Therefore, Proposition B.14 yields

$$(D_1^2) = (D.D_1) > 0 \text{ and } (D_2^2) = (D.D_2) > 0.$$

These conclusions contradict Theorem B.27 with $\mathcal{H} := \mathcal{O}_X(D_1)$ and $\mathcal{L} := \mathcal{O}_X(D_2)$.

Finally, suppose $r \geq 3$. Let H be a general hyperplane section of X. Then H is integral by Bertini's Theorem [**Sei50**, Thm. 12, p. 374]. And H is not a component of D. Set $D' := D \cap H$ and $D_i' := D_i \cap H$. Plainly D_1' and D_2' are disjoint, and $D' = D_1' + D_2'$; also, D' is ample. So induction yields the desired contradiction. □

Bibliography

[AK70] Allen Altman and Steven Kleiman, *Introduction to Grothendieck duality theory*, Lecture Notes in Mathematics, Vol. 146, Springer-Verlag, Berlin, 1970.

[AK74] ———, *Algebraic systems of linearly equivalent divisor-like subschemes*, Compositio Math. **29** (1974), 113–139.

[AK79] ———, *Compactifying the Picard scheme. II*, Amer. J. Math. **101** (1979), no. 1, 10–41.

[AK80] ———, *Compactifying the Picard scheme*, Adv. in Math. **35** (1980), no. 1, 50–112.

[AM69] M. F. Atiyah and I. G. Macdonald, *Introduction to commutative algebra*, Addison-Wesley Publishing Co., Reading, Mass.-London-Don Mills, Ont., 1969.

[Art69a] Michael Artin, *Algebraization of formal moduli. I*, Global Analysis (Papers in Honor of K. Kodaira), Univ. Tokyo Press, Tokyo, 1969, pp. 21–71.

[Art69b] ———, *The implicit function theorem in algebraic geometry*, Algebraic Geometry (Internat. Colloq., Tata Inst. Fund. Res., Bombay, 1968), Oxford Univ. Press, London, 1969, pp. 13–34.

[Art71] ———, *Algebraic spaces*, Yale University Press, New Haven, Conn., 1971, A James K. Whittemore Lecture in Mathematics given at Yale University, 1969, Yale Mathematical Monographs, 3.

[Art73] ———, *Théorèmes de représentabilité pour les espaces algébriques*, Les Presses de l'Université de Montréal, Montreal, Quebec, 1973, En collaboration avec Alexandru Lascu et Jean-François Boutot, Séminaire de Mathématiques Supérieures, No. 44 (Été, 1970).

[Art74a] ———, *Lectures on deformations of singularities*, Tata Institute of Fundamental Research, 1974.

[Art74b] ———, *Versal deformations and algebraic stacks*, Invent. Math. **27** (1974), 165–189.

[Ati57] M. F. Atiyah, *Complex analytic connections in fibre bundles*, Trans. Amer. Math. Soc. **85** (1957), 181–207.

[Bat99] Victor V. Batyrev, *Birational Calabi-Yau n-folds have equal Betti numbers*, New trends in algebraic geometry (Warwick, 1996), London Math. Soc. Lecture Note Ser., vol. 264, Cambridge Univ. Press, Cambridge, 1999, pp. 1–11.

[BB73] A. Białynicki-Birula, *Some theorems on actions of algebraic groups*, Ann. of Math. (2) **98** (1973), 480–497.

[BBD82] A. A. Beĭlinson, J. Bernstein, and P. Deligne, *Faisceaux pervers*, Analysis and topology on singular spaces, I (Luminy, 1981), Astérisque, vol. 100, Soc. Math. France, Paris, 1982, pp. 5–171.

[BCP04] Aldo Brigaglia, Ciro Ciliberto, and Claudio Pedrini, *The Italian school of algebraic geometry and Abel's legacy*, The legacy of Niels Henrik Abel, Springer, Berlin, 2004, pp. 295–347.

[Bea83] Arnaud Beauville, *Variétés Kähleriennes dont la première classe de Chern est nulle*, J. Differential Geom. **18** (1983), no. 4, 755–782 (1984).

[BLR90] Siegfried Bosch, Werner Lütkebohmert, and Michel Raynaud, *Néron models*, Ergebnisse der Mathematik und ihrer Grenzgebiete (3), vol. 21, Springer-Verlag, Berlin, 1990.

[BLR00] Jean-Benoît Bost, François Loeser, and Michel Raynaud (eds.), *Courbes semi-stables et groupe fondamental en géométrie algébrique*, Progress in Mathematics, vol. 187, Basel, Birkhäuser Verlag, 2000.

[Bor69] Armand Borel, *Linear algebraic groups*, Notes taken by Hyman Bass, W. A. Benjamin, Inc., New York-Amsterdam, 1969.

[Bor94a] Francis Borceux, *Handbook of categorical algebra. 1*, Encyclopedia of Mathematics and its Applications, vol. 50, Cambridge University Press, Cambridge, 1994, Basic category theory.

[Bor94b] ———, *Handbook of categorical algebra. 2*, Encyclopedia of Mathematics and its Applications, vol. 51, Cambridge University Press, Cambridge, 1994, Categories and structures.

[Bor94c] ———, *Handbook of categorical algebra. 3*, Encyclopedia of Mathematics and its Applications, vol. 52, Cambridge University Press, Cambridge, 1994, Categories of sheaves.

[Bou61] Nicolas Bourbaki, *Éléments de mathématique. Fascicule XXVIII. Algèbre commutative. Chapitre 3: Graduations, filtra- tions et topologies. Chapitre 4: Idéaux premiers associés et décomposition primaire*, Actualités Scientifiques et Industrielles, No. 1293, Hermann, Paris, 1961.

[Bou64] ———, *Éléments de mathématique. Fasc. XXX. Algèbre commutative. Chapitre 5: Entiers. Chapitre 6: Valuations*, Actualités Scientifiques et Industrielles, No. 1308, Hermann, Paris, 1964.

[Bou78] Jean-François Boutot, *Schéma de Picard local*, Lecture Notes in Mathematics, vol. 632, Springer, Berlin, 1978.

[BR70] Jean Bénabou and Jacques Roubaud, *Monades et descente*, C. R. Acad. Sci. Paris Sér. A-B **270** (1970), A96–A98.

[Bri77] Joël Briançon, *Description de* $H\mathrm{ilb}^n C\{x,y\}$, Invent. Math. **41** (1977), no. 1, 45–89.

[Car60] Pierre Cartier, *Sur un théorème de Snapper*, Bull. Soc. Math. France **88** (1960), 333–343.

[Che96] Jan Cheah, *On the cohomology of Hilbert schemes of points*, J. Algebraic Geom. **5** (1996), no. 3, 479–511.

[Cho57] Wei-Liang Chow, *On the projective embedding of homogeneous varieties*, Algebraic geometry and topology. A symposium in honor of S. Lefschetz, Princeton University Press, Princeton, N. J., 1957, pp. 122–128.

[Con02] Brian Conrad, *A modern proof of Chevalley's theorem on algebraic groups*, J. Ramanujan Math. Soc. **17** (2002), no. 1, 1–18.

[CS01] Pierre Colmez and Jean-Pierre Serre (eds.), *Correspondance Grothendieck-Serre*, Documents Mathématiques (Paris), Société Mathématique de France, Paris, 2001.

[Dan70] V. I. Danilov, *Samuel's conjecture*, Mat. Sb. (N.S.) **81 (123)** (1970), 132–144.

[dCM00] Mark Andrea A. de Cataldo and Luca Migliorini, *The Douady space of a complex surface*, Adv. Math. **151** (2000), no. 2, 283–312.

[dCM02] ———, *The Chow groups and the motive of the Hilbert scheme of points on a surface*, J. Algebra **251** (2002), no. 2, 824–848.

[Del68] Pierre Deligne, *Théorème de Lefschetz et critères de dégénérescence de suites spectrales*, Inst. Hautes Études Sci. Publ. Math. (1968), no. 35, 259–278.

[Del73] ———, *Cohomologie à supports propres*, Théorie des topos et cohomologie étale des schémas. Tome 3, Springer-Verlag, Berlin, 1973, Séminaire de Géométrie Algébrique du Bois-Marie 1963–1964 (SGA 4), Dirigé par M. Artin, A. Grothendieck et J. L. Verdier. Avec la collaboration de P. Deligne et B. Saint-Donat, Lecture Notes in Mathematics, Vol. 305.

[Del81] ———, *Relèvement des surfaces $K3$ en caractéristique nulle*, Algebraic surfaces (Orsay, 1976–78), Lecture Notes in Math., vol. 868, Springer, Berlin, 1981, prepared for publication by Luc Illusie, pp. 58–79.

[DG70] Michel Demazure and Pierre Gabriel, *Groupes algébriques. Tome I: Géométrie algébrique, généralités, groupes commutatifs*, Masson & Cie, Éditeur, Paris, 1970, Avec un appendice *Corps de classes local* par Michiel Hazewinkel.

[DHVW85] L. Dixon, J. A. Harvey, C. Vafa, and E. Witten, *Strings on orbifolds*, Nuclear Phys. B **261** (1985), no. 4, 678–686.

[DHVW86] L. Dixon, J. Harvey, C. Vafa, and E. Witten, *Strings on orbifolds. II*, Nuclear Phys. B **274** (1986), no. 2, 285–314.

[DI87] Pierre Deligne and Luc Illusie, *Relèvements modulo p^2 et décomposition du complexe de de Rham*, Invent. Math. **89** (1987), no. 2, 247–270.

[DIX68] *Dix exposés sur la cohomologie des schémas*, Advanced Studies in Pure Mathematics, Vol. 3, North-Holland Publishing Co., Amsterdam, 1968.

[DL99] Jan Denef and François Loeser, *Germs of arcs on singular algebraic varieties and motivic integration*, Invent. Math. **135** (1999), no. 1, 201–232.

[DM69] Pierre Deligne and David Mumford, *The irreducibility of the space of curves of given genus*, Inst. Hautes Études Sci. Publ. Math. **36** (1969), 75–109.

[EG00] Geir Ellingsrud and Lothar Göttsche, *Hilbert schemes of points and Heisenberg algebras*, School on Algebraic Geometry (Trieste, 1999), ICTP Lect. Notes, vol. 1, Abdus Salam Int. Cent. Theoret. Phys., Trieste, 2000, pp. 59–100.

[EGAI] Alexander Grothendieck, *Éléments de géométrie algébrique. I. Le langage des schémas*, Inst. Hautes Études Sci. Publ. Math. (1960), no. 4, 228.

[EGAII] ———, *Éléments de géométrie algébrique. II. Étude globale élémentaire de quelques classes de morphismes*, Inst. Hautes Études Sci. Publ. Math. (1961), no. 8, 222.

[EGAIII1] ———, *Éléments de géométrie algébrique. III. Étude cohomologique des faisceaux cohérents. I*, Inst. Hautes Études Sci. Publ. Math. (1961), no. 11, 167.

[EGAIII2] ———, *Éléments de géométrie algébrique. III. Étude cohomologique des faisceaux cohérents. II*, Inst. Hautes Études Sci. Publ. Math. (1963), no. 17, 91.

[EGAIV1] ———, *Éléments de géométrie algébrique. IV. Étude locale des schémas et des morphismes de schémas. I*, Inst. Hautes Études Sci. Publ. Math. (1964), no. 20, 259.

[EGAIV2] ———, *Éléments de géométrie algébrique. IV. Étude locale des schémas et des morphismes de schémas. II*, Inst. Hautes Études Sci. Publ. Math. (1965), no. 24, 231.

[EGAIV3] ———, *Éléments de géométrie algébrique. IV. Étude locale des schémas et des morphismes de schémas. III*, Inst. Hautes Études Sci. Publ. Math. (1966), no. 28, 255.

[EGAIV4] ———, *Éléments de géométrie algébrique. IV. Étude locale des schémas et des morphismes de schémas IV*, Inst. Hautes Études Sci. Publ. Math. (1967), no. 32, 361.

[EGAG] ———, *Eléments de géométrie algébrique. I*, Grundlehren Math. Wiss., vol. 166, Springer-Verlag, 1971.

[EGK02] Eduardo Esteves, Mathieu Gagné, and Steven Kleiman, *Autoduality of the compactified Jacobian*, J. London Math. Soc. (2) **65** (2002), no. 3, 591–610.

[Eke86] Torsten Ekedahl, *Diagonal complexes and F-gauge structures*, Travaux en Cours, Hermann, Paris, 1986.

[Ell] Geir Ellingsrud, *Irreducibility of the punctual Hilbert scheme of a surface*, unpublished.

[ES87] Geir Ellingsrud and Stein Arild Strømme, *On the homology of the Hilbert scheme of points in the plane*, Invent. Math. **87** (1987), no. 2, 343–352.

[ES98] ______, *An intersection number for the punctual Hilbert scheme of a surface*, Trans. Amer. Math. Soc. **350** (1998), no. 6, 2547–2552.

[Fal03] Gerd Faltings, *Finiteness of coherent cohomology for proper fppf stacks*, J. Algebraic Geom. **12** (2003), no. 2, 357–366.

[FGA] Alexander Grothendieck, *Fondements de la géométrie algébrique. Extraits du Séminaire Bourbaki*, 1957–1962, Secrétariat mathématique, Paris, 1962.

[FK88] Eberhard Freitag and Reinhardt Kiehl, *Étale cohomology and the Weil conjecture*, Ergebnisse der Mathematik und ihrer Grenzgebiete (3), vol. 13, Springer-Verlag, Berlin, 1988.

[FK95] Kazuhiro Fujiwara and Kazuya Kato, *Logarithmic étale topology theory*, (incomplete) preprint, 1995.

[FM98] Barbara Fantechi and Marco Manetti, *Obstruction calculus for functors of Artin rings. I*, J. Algebra **202** (1998), no. 2, 541–576.

[Fog68] John Fogarty, *Algebraic families on an algebraic surface*, Amer. J. Math **90** (1968), 511–521.

[Fos81] Robert Fossum, *Formes différentielles non fermées*, Séminaire sur les Pinceaux de Courbes de Genre au Moins Deux (Lucien Szpiro, ed.), Astérisque, vol. 86, Société Mathématique de France, Paris, 1981, pp. 90–96.

[Gir64] Jean Giraud, *Méthode de la descente*, Bull. Soc. Math. France Mém. **2** (1964), viii+150.

[Gir71] ______, *Cohomologie non abélienne*, Springer-Verlag, Berlin, 1971, Die Grundlehren der mathematischen Wissenschaften, Band 179.

[God58] Roger Godement, *Topologie algébrique et théorie des faisceaux*, Actualit'es Sci. Ind. No. 1252. Publ. Math. Univ. Strasbourg. No. 13, Hermann, Paris, 1958.

[Göt90] Lothar Göttsche, *The Betti numbers of the Hilbert scheme of points on a smooth projective surface*, Math. Ann. **286** (1990), no. 1-3, 193–207.

[Göt01] ______, *On the motive of the Hilbert scheme of points on a surface*, Math. Res. Lett. **8** (2001), no. 5-6, 613–627.

[Göt02] ______, *Hilbert schemes of points on surfaces*, Proceedings of the International Congress of Mathematicians, Vol. II (Beijing, 2002) (Beijing), Higher Ed. Press, 2002, pp. 483–494.

[Gra66] John W. Gray, *Fibred and cofibred categories*, Proc. Conf. Categorical Algebra (La Jolla, Calif., 1965), Springer, New York, 1966, pp. 21–83.

[Gro95a] Alexander Grothendieck, *Géométrie formelle et géométrie algébrique*, Séminaire Bourbaki, Vol. 5, Soc. Math. France, Paris, 1995, pp. Exp. No. 182, 193–220, errata p. 390.

[Gro95b] ______, *Technique de descente et théorèmes d'existence en géométrie algébrique. I. Généralités. Descente par morphismes fidèlement plats.*, Séminaire Bourbaki, Exp. No. 190, vol. 5, Soc. Math. France, Paris, 1995, pp. 299–327.

[Gro95c] ______, *Techniques de construction et théorèmes d'existence en géométrie algébrique. IV. Les schémas de Hilbert*, Séminaire Bourbaki, Vol. 6, Soc. Math. France, Paris, 1995, pp. Exp. No. 221, 249–276.

[Gro96] I. Grojnowski, *Instantons and affine algebras. I. The Hilbert scheme and vertex operators*, Math. Res. Lett. **3** (1996), no. 2, 275–291.

[GS93] Lothar Göttsche and Wolfgang Soergel, *Perverse sheaves and the cohomology of Hilbert schemes of smooth algebraic surfaces*, Math. Ann. **296** (1993), no. 2, 235–245.

[Har66] Robin Hartshorne, *Residues and duality*, Lecture notes of a seminar on the work of A. Grothendieck, given at Harvard 1963/64. With an appendix by P. Deligne. Lecture Notes in Mathematics, No. 20, Springer-Verlag, Berlin, 1966.

[Har70] ———, *Ample subvarieties of algebraic varieties*, Notes written in collaboration with C. Musili. Lecture Notes in Mathematics, Vol. 156, Springer-Verlag, Berlin, 1970.

[Har77] ———, *Algebraic geometry*, Graduate Texts in Mathematics, no. 52, Springer-Verlag, New York, 1977.

[Har92] Joe Harris, *Algebraic geometry. a first course*, Graduate Texts in Mathematics, vol. 133, Springer-Verlag, New York, 1992.

[Hir62] Heisuke Hironaka, *An example of a non-Kählerian complex-analytic deformation of Kählerian complex structures*, Ann. of Math. (2) **75** (1962), 190–208.

[Hir99] Masayuki Hirokado, *A non-liftable Calabi-Yau threefold in characteristic* 3, Tohoku Math. J. (2) **51** (1999), no. 4, 479–487.

[HS] André Hirschowitz and Carlos Simpson, *Descente pour les n-champs*, arXiv: math. AG/9807049.

[Iar72] Anthony Iarrobino, *Reducibility of the families of* 0*-dimensional schemes on a variety*, Invent. Math. **15** (1972), 72–77.

[Iar77] ———, *Punctual Hilbert schemes*, Mem. Amer. Math. Soc. **10** (1977), no. 188, viii+112.

[Iar85] ———, *Compressed algebras and components of the punctual Hilbert scheme*, Algebraic geometry, Sitges (Barcelona), 1983, Lecture Notes in Math., vol. 1124, Springer, Berlin, 1985, pp. 146–165.

[Iar87] ———, *Hilbert scheme of points: overview of last ten years*, Algebraic geometry, Bowdoin, 1985 (Brunswick, Maine, 1985), Proc. Sympos. Pure Math., vol. 46, Amer. Math. Soc., Providence, RI, 1987, pp. 297–320.

[Igu55] Jun-ichi Igusa, *On some problems in abstract algebraic geometry*, Proc. Nat. Acad. Sci. U. S. A. **41** (1955), 964–967.

[Ill71] Luc Illusie, *Complexe cotangent et déformations. I*, Springer-Verlag, Berlin, 1971, Lecture Notes in Mathematics, Vol. 239.

[Ill72a] ———, *Complexe cotangent et déformations. II*, Springer-Verlag, Berlin, 1972, Lecture Notes in Mathematics, Vol. 283.

[Ill72b] ———, *Cotangent complex and deformations of torsors and group schemes*, Toposes, algebraic geometry and logic (Conf., Dalhousie Univ., Halifax, N.S., 1971), Springer, Berlin, 1972, pp. 159–189. Lecture Notes in Math., Vol. 274.

[Ill85] ———, *Déformations de groupes de Barsotti-Tate (d'après A. Grothendieck)*, Astérisque (1985), no. 127, 151–198, Seminar on arithmetic bundles: the Mordell conjecture (Paris, 1983/84).

[Ill02] ———, *An overview of the work of K. Fujiwara, K. Kato, and C. Nakayama on logarithmic étale cohomology*, Astérisque (2002), no. 279, 271–322, Cohomologies p-adiques et applications arithmétiques, II.

[Ive70] Birger Iversen, *Linear determinants with applications to the Picard scheme of a family of algebraic curves*, Springer-Verlag, Berlin, 1970, Lecture Notes in Mathematics, Vol. 174.

[Jou79] Jean-Pierre Jouanolou, *Théorèmes de Bertini et applications*, Université Louis Pasteur Département de Mathématique Institut de Recherche Mathématique Avancée, Strasbourg, 1979.

[JT84] André Joyal and Myles Tierney, *An extension of the Galois theory of Grothendieck*, Mem. Amer. Math. Soc. **51** (1984), no. 309, vii+71.

[Kis00] Mark Kisin, *Prime to p fundamental groups, and tame Galois actions*, Ann. Inst. Fourier (Grenoble) **50** (2000), no. 4, 1099–1126.

[Kle71] Steven L. Kleiman, *Les théorèmes de finitude pour le foncteur de Picard*, Théorie des intersections et théorème de Riemann-Roch, Springer-Verlag, 1971, Séminaire de Géométrie Algébrique du Bois-Marie 1966–1967 (SGA 6), Dirigé par P. Berthelot, A. Grothendieck et L. Illusie. Avec la collaboration de D. Ferrand, J. P. Jouanolou, O. Jussila, S. Kleiman, M. Raynaud et J-P. Serre, Lecture Notes in Mathematics, Vol. 225.

[Kle73] ———, *Completeness of the characteristic system*, Advances in Math. **11** (1973), 304–310.

[Kle04] ———, *What is Abel's theorem anyway?*, The legacy of Niels Henrik Abel, Springer, Berlin, 2004, pp. 395–440.

[Knu71] Donald Knutson, *Algebraic spaces*, Springer-Verlag, Berlin, 1971, Lecture Notes in Mathematics, Vol. 203.

[KO74] Max-Albert Knus and Manuel Ojanguren, *Théorie de la descente et algèbres d'Azumaya*, vol. 389, Springer-Verlag, Berlin, 1974, Lecture Notes in Mathematics.

[Kod05] Kunihiko Kodaira, *Complex manifolds and deformation of complex structures*, english ed., Classics in Mathematics, Springer-Verlag, Berlin, 2005, Translated from the 1981 Japanese original by Kazuo Akao.

[KR04a] Maxim Kontsevich and Alexander Rosenberg, *Noncommutative spaces and flat descent*, Max Planck Institute of Mathematics preprint, 2004.

[KR04b] ———, *Noncommutative stacks*, Max Planck Institute of Mathematics preprint, 2004.

[KS04] Kazuya Kato and Takeshi Saito, *On the conductor formula of Bloch*, Publ. Math. Inst. Hautes Études Sci. (2004), no. 100, 5–151.

[Lan79] William E. Lang, *Quasi-elliptic surfaces in characteristic three*, Ann. Sci. École Norm. Sup. (4) **12** (1979), no. 4, 473–500.

[Leh99] Manfred Lehn, *Chern classes of tautological sheaves on Hilbert schemes of points on surfaces*, Invent. Math. **136** (1999), no. 1, 157–207.

[Leh04] ———, *Symplectic moduli spaces*, Intersection Theory and Moduli, ICTP Lect. Notes, Abdus Salam Int. Cent. Theoret. Phys. (Trieste), 2004, pp. 139–184.

[Lip74] Joseph Lipman, *Picard schemes of formal schemes; application to rings with discrete divisor class group*, Classification of algebraic varieties and compact complex manifolds, Springer, Berlin, 1974, pp. 94–132. Lecture Notes in Math., Vol. 412.

[Lip75] ———, *Unique factorization in complete local rings*, Algebraic geometry (Proc. Sympos. Pure Math., Vol. 29, Humboldt State Univ., Arcata, Calif., 1974), Amer. Math. Soc., Providence, R.I., 1975, pp. 531–546.

[Lip76] ———, *The Picard group of a scheme over an Artin ring*, Inst. Hautes Études Sci. Publ. Math. (1976), no. 46, 15–86.

[LLR04] Qing Liu, Dino Lorenzini, and Michel Raynaud, *Néron models, Lie algebras, and reduction of curves of genus one*, Invent. Math. **157** (2004), no. 3, 455–518.

[LMB00] Gérard Laumon and Laurent Moret-Bailly, *Champs algébriques*, Ergebnisse der Mathematik und ihrer Grenzgebiete. 3. Folge., vol. 39, Springer-Verlag, Berlin, 2000.

[LN80] William E. Lang and Niels O. Nygaard, *A short proof of the Rudakov-Šafarevič theorem*, Math. Ann. **251** (1980), no. 2, 171–173.

[LQW02] Wei-ping Li, Zhenbo Qin, and Weiqiang Wang, *Vertex algebras and the cohomology ring structure of Hilbert schemes of points on surfaces*, Math. Ann. **324** (2002), no. 1, 105–133.

[LS67] S. Lichtenbaum and M. Schlessinger, *The cotangent complex of a morphism*, Trans. Amer. Math. Soc. **128** (1967), 41–70.

[LS01] Manfred Lehn and Christoph Sorger, *Symmetric groups and the cup product on the cohomology of Hilbert schemes*, Duke Math. J. **110** (2001), no. 2, 345–357.

[Mac62] I. G. Macdonald, *The Poincaré polynomial of a symmetric product*, Proc. Cambridge Philos. Soc. **58** (1962), 563–568.

[Mat89] Hideyuki Matsumura, *Commutative ring theory*, second ed., Cambridge Studies in Advanced Mathematics, vol. 8, Cambridge University Press, Cambridge, 1989, Translated from the Japanese by M. Reid.

[MFK94] David Mumford, John Fogarty, and Frances C. Kirwan, *Geometric invariant theory*, third ed., Ergebnisse der Mathematik und ihrer Grenzgebiete (2), vol. 34, Springer-Verlag, Berlin, 1994.

[Mil80] James S. Milne, *Étale cohomology*, Princeton Mathematical Series, vol. 33, Princeton University Press, Princeton, N.J., 1980.

[ML98] Saunders Mac Lane, *Categories for the working mathematician*, second ed., Graduate Texts in Mathematics, vol. 5, Springer-Verlag, New York, 1998.

[Moe02] Ieke Moerdijk, *Introduction to the language of stacks and gerbes*, arXiv: math.AT/ 0212266, 2002.

[Mum] David Mumford, *Letter to Serre*, May 1961.

[Mum61a] ———, *Pathologies of modular algebraic surfaces*, Amer. J. Math. **83** (1961), 339–342.

[Mum61b] ———, *The topology of normal singularities of an algebraic surface and a criterion for simplicity*, Inst. Hautes Études Sci. Publ. Math. (1961), no. 9, 5–22.

[Mum65] ———, *Geometric invariant theory*, Ergebnisse der Mathematik und ihrer Grenzgebiete, Neue Folge, Band 34, Springer-Verlag, Berlin, 1965.

[Mum66] ———, *Lectures on curves on an algebraic surface*, With a section by G. M. Bergman. Annals of Mathematics Studies, No. 59, Princeton University Press, Princeton, N.J., 1966.

[Mum67] ———, *Pathologies. III*, Amer. J. Math. **89** (1967), 94–104.

[Mum69] ———, *Bi-extensions of formal groups*, Algebraic Geometry (Internat. Colloq., Tata Inst. Fund. Res., Bombay, 1968), Oxford Univ. Press, London, 1969, pp. 307–322.

[Mum70] ———, *Abelian varieties*, Tata Institute of Fundamental Research Studies in Mathematics, No. 5, Published for the Tata Institute of Fundamental Research, Bombay; Oxford University Press, London, 1970.

[Mur64] J. P. Murre, *On contravariant functors from the category of pre-schemes over a field into the category of abelian groups (with an application to the Picard functor)*, Inst. Hautes Études Sci. Publ. Math. (1964), no. 23, 5–43.

[Mur95] ———, *Representation of unramified functors. Applications (according to unpublished results of A. Grothendieck)*, Séminaire Bourbaki, Vol. 9, Soc. Math. France, Paris, 1995, pp. Exp. No. 294, 243–261.

[Nag56] Masayoshi Nagata, *A general theory of algebraic geometry over Dedekind domains. I. The notion of models*, Amer. J. Math. **78** (1956), 78–116.

[Nak97] Hiraku Nakajima, *Heisenberg algebra and Hilbert schemes of points on projective surfaces*, Ann. of Math. (2) **145** (1997), no. 2, 379–388.

[Nak99] ———, *Lectures on Hilbert schemes of points on surfaces*, University Lecture Series, vol. 18, American Mathematical Society, Providence, RI, 1999.

[Nyg79] Niels O. Nygaard, *A p-adic proof of the nonexistence of vector fields on K3 surfaces*, Ann. of Math. (2) **110** (1979), no. 3, 515–528.

[OB72] Arthur Ogus and George Bergman, *Nakayama's lemma for half-exact functors*, Proc. Amer. Math. Soc. **31** (1972), 67–74.

[Ols] Martin Olsson, *On proper coverings of artin stacks*, to appear in Adv. Math.

[Oor62] Frans Oort, *Sur le schéma de Picard*, Bull. Soc. Math. France **90** (1962), 1–14.

[Oor71] ———, *Finite group schemes, local moduli for abelian varieties, and lifting problems*, in "Proc. 5th Nordic Summer School Oslo, 1970," Wolters-Noordhoff, 1972, pp. 223–254 · = · Compositio Math. **23** (1971), 265–296.

[OV00] Fabrice Orgogozo and Isabelle Vidal, *Le théorème de spécialisation du groupe fondamental*, Courbes semi-stables et groupe fondamental en géométrie algébrique (Luminy, 1998), Progr. Math., vol. 187, Birkhäuser, Basel, 2000, pp. 169–184.

[Ray66] Michel Raynaud, *Faisceaux amples sur les schémas en groupes et les espaces homogènes*, C. R. Acad. Sci. Paris Sér. A-B **262** (1966), A1313–A1315.

[Ray70] ———, *Faisceaux amples sur les schémas en groupes et les espaces homogènes*, Lecture Notes in Mathematics, no. 119, Springer-Verlag, 1970.

[Ray71] ———, *Un théorème de représentabilité pour le foncteur de Picard*, Théorie des intersections et théorème de Riemann-Roch, Springer-Verlag, 1971, Exposé XII, written up by S. Kleiman. In: Séminaire de Géométrie Algébrique du Bois-Marie 1966–1967 (SGA 6), Dirigé par P. Berthelot, A. Grothendieck et L. Illusie. Avec la collaboration de D. Ferrand, J. P. Jouanolou, O. Jussila, S. Kleiman, M. Raynaud et J-P. Serre, Lecture Notes in Mathematics, Vol. 225.

[Ray94] ———, *Revêtements de la droite affine en caractéristique $p > 0$ et conjecture d'Abhyankar*, Invent. Math. **116** (1994), no. 1-3, 425–462.

[RŠ76] A. N. Rudakov and I. R. Šafarevič, *Inseparable morphisms of algebraic surfaces*, Izv. Akad. Nauk SSSR Ser. Mat. **40** (1976), no. 6, 1269–1307, 1439.

[Sch68] Michael Schlessinger, *Functors of Artin rings*, Trans. Amer. Math. Soc. **130** (1968), 208–222.

[Seg37] Beniamino Segre, *Intorno ad un teorema di hodge sulla teoria della base per le curve di una superficie algebrica*, Ann. di Mat. pura e applicata **16** (1937), 157–163.

[Sei50] A. Seidenberg, *The hyperplane sections of normal varieties*, Trans. Amer. Math. Soc. **69** (1950), 357–386.

[Ser55] Jean-Pierre Serre, *Faisceaux algébriques cohérents*, Ann. of Math. (2) **61** (1955), 197–278 (n. 29 in [**Ser**]).

[Ser56] ———, *Géométrie algébrique et géométrie analytique*, Ann. Inst. Fourier, Grenoble **6** (1955–1956), 1–42 (n. 32 in [**Ser**]).

[Ser58] ———, *Sur la topologie des variétés algébriques en caractéristique p*, Symposium internacional de topología algebraica, Universidad Nacional Autónoma de México and UNESCO, Mexico City, 1958, pp. 24–53 (n. 38 in [**Ser**]).

[Ser59] ———, *Groupes algébriques et corps de classes*, Publications de l'institut de mathématique de l'université de Nancago, VII. Hermann, Paris, 1959.

[Ser61] ———, *Exemples de variétés projectives en caractéristique p non relevables en caractéristique zéro*, Proc. Nat. Acad. Sci. U.S.A. **47** (1961), 108–109 (n. 50 in [**Ser**]).

[Ser] ———, *Oeuvres (Collected Papers)*, I, II, III, Springer-Verlag (1986); IV, Springer-Verlag (2000).

[SGA1] Alexander Grothendieck, *Revêtements étales et groupe fondamental (SGA 1)*, Lecture Notes in Math., 224, Springer-Verlag, Berlin, 1964, Séminaire de géométrie algébrique du Bois Marie 1960–61, Directed by A. Grothendieck, With two papers by M. Raynaud.

[SGA2] ———, *Cohomologie locale des faisceaux cohérents et théorèmes de Lefschetz locaux et globaux (SGA 2)*, North-Holland Publishing Co., Amsterdam, 1968, Augmenté d'un exposé par Michèle Raynaud, Séminaire de Géométrie Algébrique du Bois-Marie, 1962, Advanced Studies in Pure Mathematics, Vol. 2.

[SGA4] Michael Artin, Alexander Grothendieck, and Jean-Louis Verdier, *Théorie des topos et cohomologie étale des schémas, 1, 2, 3*, Springer-Verlag, Berlin, 1972–73, Séminaire de Géométrie Algébrique du Bois-Marie 1963–1964 (SGA 4), Dirigé par M. Artin, A. Grothendieck et J. L. Verdier. Avec la collaboration de N. Bourbaki, P. Deligne et B. Saint-Donat, Lecture Notes in Mathematics, Voll. 270, 305, 569.

[SGA4-1/2] Pierre Deligne, *Cohomologie étale*, Springer-Verlag, Berlin, 1977, Séminaire de Géométrie Algébrique du Bois-Marie SGA $4\frac{1}{2}$, Avec la collaboration de J. F. Boutot, A. Grothendieck, L. Illusie et J. L. Verdier, Lecture Notes in Mathematics, Vol. 569.

[SGA5] *Cohomologie l-adique et fonctions L*, Springer-Verlag, Berlin, 1977, Séminaire de Géometrie Algébrique du Bois-Marie 1965–1966 (SGA 5), Edité par Luc Illusie, Lecture Notes in Mathematics, Vol. 589.

[SGA6] *Théorie des intersections et théorème de Riemann-Roch*, Springer-Verlag, Berlin, 1971, Séminaire de Géométrie Algébrique du Bois-Marie 1966–1967 (SGA 6), Dirigé par P. Berthelot, A. Grothendieck et L. Illusie. Avec la collaboration de D. Ferrand, J. P. Jouanolou, O. Jussila, S. Kleiman, M. Raynaud et J-P. Serre, Lecture Notes in Mathematics, Vol. 225.

[SGA7] *Groupes de monodromie en géométrie algébrique. I*, Springer-Verlag, Berlin, 1972, Séminaire de Géométrie Algébrique du Bois-Marie 1967–1969 (SGA 7 I), Dirigé par A. Grothendieck. Avec la collaboration de M. Raynaud et D. S. Rim, Lecture Notes in Mathematics, Vol. 288.

[Str96] Stein Arild Strømme, *Elementary introduction to representable functors and Hilbert schemes*, Parameter spaces (Warsaw, 1994), Banach Center Publ., vol. 36, Polish Acad. Sci., Warsaw, 1996, pp. 179–198.

[Str03] Ross Street, *Categorical and combinatorial aspects of descent theory*, arXiv: math. CT/0303175, 2003.

[Tam04] Akio Tamagawa, *Finiteness of isomorphism classes of curves in positive characteristic with prescribed fundamental groups*, J. Algebraic Geom. **13** (2004), no. 4, 675–724.

[Tho87] Robert W. Thomason, *Algebraic K-theory of group scheme actions*, Algebraic topology and algebraic K-theory (Princeton, N.J., 1983), Ann. of Math. Stud., vol. 113, Princeton Univ. Press, Princeton, NJ, 1987, pp. 539–563.

[Vid01] Isabelle Vidal, *Contributions à la cohomologie étale des schémas et des log schémas*, Ph.D. thesis, Paris XI Orsay, 2001.

[Vis97] Angelo Vistoli, *The deformation theory of local complete intersections*, arXiv: alg-geom/9703008.

[VW94] Cumrun Vafa and Edward Witten, *A strong coupling test of S-duality*, Nuclear Phys. B **431** (1994), no. 1-2, 3–77.

[Wat79] William C. Waterhouse, *Introduction to affine group schemes*, Graduate Texts in Mathematics, vol. 66, Springer-Verlag, New York, 1979.

[Wei54] André Weil, *Sur les critères d'équivalence en géométrie algébrique*, Math. Ann. **128** (1954), 95–127.

[Wei79] André Weil, *Commentaire*, Collected papers, vol. Vol. I, pp. 518–574; Vol. II, pp. 526–553; and Vol. III, pp. 443–465., Springer-Verlag, 1979.

[Zar58] Oscar Zariski, *Introduction to the problem of minimal models in the theory of algebraic surfaces*, Publications of the Mathematical Society of Japan, no. 4, The Mathematical Society of Japan, Tokyo, 1958.

[Zar71] ———, *Algebraic surfaces*, supplemented ed., Springer-Verlag, New York, 1971, With appendices by S. S. Abhyankar, J. Lipman, and D. Mumford, Ergebnisse der Mathematik und ihrer Grenzgebiete, Band 61.

[ZS60] Oscar Zariski and Pierre Samuel, *Commutative algebra. Vol. II*, The University Series in Higher Mathematics, D. Van Nostrand Co., Inc., Princeton, N. J.-Toronto-London-New York, 1960.

Index

Titles in This Series

123 **Barbara Fantechi, Lothar Göttsche, Luc Illusie, Steven L. Kleiman, Nitin Nitsure, and Angelo Vistoli,** Fundamental algebraic geometry: Grothendieck's FGA explained, 2005

122 **Antonio Giambruno and Mikhail Zaicev, Editors,** Polynomial identities and asymptotic methods, 2005

121 **Anton Zettl,** Sturm-Liouville theory, 2005

120 **Barry Simon,** Trace ideals and their applications, 2005

119 **Tian Ma and Shouhong Wang,** Geometric theory of incompressible flows with applications to fluid dynamics, 2005

118 **Alexandru Buium,** Arithmetic differential equations, 2005

117 **Volodymyr Nekrashevych,** Self-similar groups, 2005

116 **Alexander Koldobsky,** Fourier analysis in convex geometry, 2005

115 **Carlos Julio Moreno,** Advanced analytic number theory: L-functions, 2005

114 **Gregory F. Lawler,** Conformally invariant processes in the plane, 2005

113 **William G. Dwyer, Philip S. Hirschhorn, Daniel M. Kan, and Jeffrey H. Smith,** Homotopy limit functors on model categories and homotopical categories, 2004

112 **Michael Aschbacher and Stephen D. Smith,** The classification of quasithin groups II. Main theorems: The classification of simple QTKE-groups, 2004

111 **Michael Aschbacher and Stephen D. Smith,** The classification of quasithin groups I. Structure of strongly quasithin K-groups, 2004

110 **Bennett Chow and Dan Knopf,** The Ricci flow: An introduction, 2004

109 **Goro Shimura,** Arithmetic and analytic theories of quadratic forms and Clifford groups, 2004

108 **Michael Farber,** Topology of closed one-forms, 2004

107 **Jens Carsten Jantzen,** Representations of algebraic groups, 2003

106 **Hiroyuki Yoshida,** Absolute CM-periods, 2003

105 **Charalambos D. Aliprantis and Owen Burkinshaw,** Locally solid Riesz spaces with applications to economics, second edition, 2003

104 **Graham Everest, Alf van der Poorten, Igor Shparlinski, and Thomas Ward,** Recurrence sequences, 2003

103 **Octav Cornea, Gregory Lupton, John Oprea, and Daniel Tanré,** Lusternik-Schnirelmann category, 2003

102 **Linda Rass and John Radcliffe,** Spatial deterministic epidemics, 2003

101 **Eli Glasner,** Ergodic theory via joinings, 2003

100 **Peter Duren and Alexander Schuster,** Bergman spaces, 2004

99 **Philip S. Hirschhorn,** Model categories and their localizations, 2003

98 **Victor Guillemin, Viktor Ginzburg, and Yael Karshon,** Moment maps, cobordisms, and Hamiltonian group actions, 2002

97 **V. A. Vassiliev,** Applied Picard-Lefschetz theory, 2002

96 **Martin Markl, Steve Shnider, and Jim Stasheff,** Operads in algebra, topology and physics, 2002

95 **Seiichi Kamada,** Braid and knot theory in dimension four, 2002

94 **Mara D. Neusel and Larry Smith,** Invariant theory of finite groups, 2002

93 **Nikolai K. Nikolski,** Operators, functions, and systems: An easy reading. Volume 2: Model operators and systems, 2002

92 **Nikolai K. Nikolski,** Operators, functions, and systems: An easy reading. Volume 1: Hardy, Hankel, and Toeplitz, 2002

91 **Richard Montgomery,** A tour of subriemannian geometries, their geodesics and applications, 2002

90 **Christian Gérard and Izabella Łaba,** Multiparticle quantum scattering in constant magnetic fields, 2002

89 **Michel Ledoux,** The concentration of measure phenomenon, 2001

88 **Edward Frenkel and David Ben-Zvi,** Vertex algebras and algebraic curves, second edition, 2004

87 **Bruno Poizat,** Stable groups, 2001

86 **Stanley N. Burris,** Number theoretic density and logical limit laws, 2001

85 **V. A. Kozlov, V. G. Maz′ya, and J. Rossmann,** Spectral problems associated with corner singularities of solutions to elliptic equations, 2001

84 **László Fuchs and Luigi Salce,** Modules over non-Noetherian domains, 2001

83 **Sigurdur Helgason,** Groups and geometric analysis: Integral geometry, invariant differential operators, and spherical functions, 2000

82 **Goro Shimura,** Arithmeticity in the theory of automorphic forms, 2000

81 **Michael E. Taylor,** Tools for PDE: Pseudodifferential operators, paradifferential operators, and layer potentials, 2000

80 **Lindsay N. Childs,** Taming wild extensions: Hopf algebras and local Galois module theory, 2000

79 **Joseph A. Cima and William T. Ross,** The backward shift on the Hardy space, 2000

78 **Boris A. Kupershmidt,** KP or mKP: Noncommutative mathematics of Lagrangian, Hamiltonian, and integrable systems, 2000

77 **Fumio Hiai and Dénes Petz,** The semicircle law, free random variables and entropy, 2000

76 **Frederick P. Gardiner and Nikola Lakic,** Quasiconformal Teichmüller theory, 2000

75 **Greg Hjorth,** Classification and orbit equivalence relations, 2000

74 **Daniel W. Stroock,** An introduction to the analysis of paths on a Riemannian manifold, 2000

73 **John Locker,** Spectral theory of non-self-adjoint two-point differential operators, 2000

72 **Gerald Teschl,** Jacobi operators and completely integrable nonlinear lattices, 1999

71 **Lajos Pukánszky,** Characters of connected Lie groups, 1999

70 **Carmen Chicone and Yuri Latushkin,** Evolution semigroups in dynamical systems and differential equations, 1999

69 **C. T. C. Wall (A. A. Ranicki, Editor),** Surgery on compact manifolds, second edition, 1999

68 **David A. Cox and Sheldon Katz,** Mirror symmetry and algebraic geometry, 1999

67 **A. Borel and N. Wallach,** Continuous cohomology, discrete subgroups, and representations of reductive groups, second edition, 2000

66 **Yu. Ilyashenko and Weigu Li,** Nonlocal bifurcations, 1999

65 **Carl Faith,** Rings and things and a fine array of twentieth century associative algebra, 1999

64 **Rene A. Carmona and Boris Rozovskii, Editors,** Stochastic partial differential equations: Six perspectives, 1999

63 **Mark Hovey,** Model categories, 1999

62 **Vladimir I. Bogachev,** Gaussian measures, 1998

For a complete list of titles in this series, visit the AMS Bookstore at **www.ams.org/bookstore/**.